Medieval History and

General Editors
JOHN BLAIR   HELENA

The Open Fields of England

The Open Fields of England describes the open-field system of agriculture that operated in Medieval England before the establishment of present-day farms surrounded by hedges or walls. The volume encompasses a wide range of primary data not previously assembled, to which are added the results of new research based upon a fifty-year study of open-field remains and their related documents. The whole of England is examined, describing eight different kinds of field-system that have been identified, and relating them to their associated land-use and settlement. Details of field structure are explained, such as the demesne, the lord's land, and the tenants' holdings, as well as tenurial arrangements and farming methods.

Previous explanations of open-field origins and possible antecedents to medieval fields are discussed. Various types of archaeological and historical evidence relating to Saxon-period settlements and fields are presented, followed by the development of a new theory to explain the lay-out and planned nature of many field systems found in the central belt of England.

Of particular interest is the Gazetteer, which is organized by historic counties. Each county has a summary of its fields, including tabulated data and sources for future research, touching on the demesne, yardland size, work-service, assarts, and physical remains of ridge and furrow. The Gazetteer acts as a national hand-list of field systems, opening the subject up to further research and is essential to scholars of medieval agriculture.

MEDIEVAL HISTORY AND ARCHAEOLOGY

General Editors
John Blair   Helena Hamerow

The volumes in this series bring together archaeological, historical, and visual methods to offer new approaches to aspects of medieval society, economy, and material culture. The series seeks to present and interpret archaeological evidence in ways readily accessible to historians, while providing a historical perspective and context for the material culture of the period.

RECENTLY PUBLISHED IN THIS SERIES

PERCEPTIONS OF THE PREHISTORIC IN
ANGLO-SAXON ENGLAND
Religion, Ritual, and Rulership in the Landscape
Sarah Semple

TREES AND TIMBER IN THE ANGLO-SAXON WORLD
Edited by Michael D. J. Bintley and Michael G. Shapland

VIKING IDENTITIES
Scandinavian Jewellery in England
Jane F. Kershaw

LITURGY, ARCHITECTURE, AND SACRED PLACES
IN ANGLO-SAXON ENGLAND
Helen Gittos

RURAL SETTLEMENTS AND SOCIETY IN ANGLO-SAXON ENGLAND
Helena Hamerow

PARKS IN MEDIEVAL ENGLAND
S. A. Mileson

ANGLO-SAXON DEVIANT BURIAL CUSTOMS
Andrew Reynolds

BEYOND THE MEDIEVAL VILLAGE
The Diversification of Landscape Character in Southern Britain
Stephen Rippon

WATERWAYS AND CANAL-BUILDING IN MEDIEVAL ENGLAND
Edited by John Blair

FOOD IN MEDIEVAL ENGLAND
Diet and Nutrition
Edited by C. M. Woolgar, D. Serjeantson, and T. Waldron

THE OPEN FIELDS
OF ENGLAND

DAVID HALL

OXFORD

UNIVERSITY PRESS

OXFORD
UNIVERSITY PRESS

Great Clarendon Street, Oxford, OX2 6DP,
United Kingdom

Oxford University Press is a department of the University of Oxford.
It furthers the University's objective of excellence in research, scholarship,
and education by publishing worldwide. Oxford is a registered trade mark of
Oxford University Press in the UK and in certain other countries

First published 2014
First published in paperback 2020

Impression: 1

Published in the United States of America by Oxford University Press
198 Madison Avenue, New York, NY 10016, United States of America

British Library Cataloguing in Publication Data
Data available

Library of Congress Cataloging in Publication Data
Data available

ISBN 978–0–19–870295–5 (Hbk.)
ISBN 978–0–19–885548–4 (Pbk.)

Printed and bound by
CPI Group (UK) Ltd, Croydon, CR0 4YY

Contents

List of Plates

List of Figures

List of Tables

List of Abbreviations

B	W. de Gray Birch (ed.), *Cartularium Saxonicum*, 3 vols. (London, 1885–93); cited by charter number
CBA	Council for British Archaeology
CRO	County Record Office
CUCAP	Cambridge University Collection of Aerial Photographs
ER	East Riding (Yorkshire)
HER	Historic Environment Record
HLC	Historic Landscape Characterisation, see Rippon 2004 for discussion
K	J. M. Kemble (ed.), *Cartularium Saxonicum*, 6 vols. (1839–48); cited by charter number
Magd.	Estate archives of Magdalen College, Oxford
NR	North Riding (Yorkshire)
RCHM(E)	Royal Commission on Historical Monuments (England)
S	P. H. Sawyer, *Anglo-Saxon Charters: An Annotated List and Bibliography* (London, 1968); cited from the revised online edition at <http://www.esawyer.org.uk> by catalogue number
TNA	The National Archives, Kew
VCH	Victoria County History
Worc. Cath. Mun.	Worcester Cathedral Muniments
WR	West Riding (Yorkshire)

Preface and Acknowledgements

This book has drawn upon and owes great debt to the work of very many people. To those currently active in fieldwork or unravelling the complexities of field systems which are described in documentary sources, and to early authors, long gone, who have left their thoughts and evidences in national and county publications. Some of the sources are an author's single lifetime publication in a county journal; others are from well-known works written by nationally known authorities. The amount of study published by some workers is astonishing, such as the contribution of the Reverend Herbert Edward Salter (1863–1951) to the history of Oxford and the county, George Herbert Fowler (1861–1940) to Bedfordshire, and William Farrer (1861–1924) to Lancashire and Yorkshire. And, of course, the book owes much to the pioneering studies of H. L. Gray, *English Field Systems* (1915) and to *Field Systems in the British Isles* edited by Alan Baker and Robin Butlin (1973), as well as to the work of Maurice Beresford, who inspired my early interest in fieldwork with his *History on the Ground* (1957).

Gray's 1915 work covered the whole of England and provided detailed appendices that are still invaluable sources of primary data. Much more information has become available since 1915, and Gray's largely 'ethnic' interpretation of regional differences in field systems has been superseded—the variations may be explained in other ways. The 1973 regional essays presented by Baker and Butlin leave gaps in the west and south, and new approaches and interpretations have emerged. It therefore seems appropriate to present a review of the field systems of the whole of England.

The account given here differs from Gray and from Baker and Butlin, in that, as well as written historical information, use has been made of physical evidence. This comprises plans of field systems produced by archaeological survey as well as other archaeological evidence. The book is organized into two sections. Firstly, a series of chapters examines the wide variety of field system types and shows how they operated, how they differed regionally, and finally discusses what their origins might be. The second part consists of a gazetteer that summarizes the open fields of each county and gives the detail and references to material used in the main text. The counties are the 'historic' ones as tidied up by the Victorians in the 1880s; most counties do not differ greatly from the medieval arrangements. It is necessary to work with the old counties because both the original records and the literature are organized through them. Particular types of field system are more inclined to lie in regions, but these are readily apparent as each county is studied—for instance the fields of the Chilterns clearly differ from the northern parts of Oxford-shire, Buckinghamshire, and Bedfordshire. Fine detail of variations within a county can come only from local studies.

The text will often refer and relate to Northamptonshire, not because it is necessarily a model county but because both fieldwork and historical analyses have been undertaken county wide, township by township, and any particular feature or statement can be put into a county context. Since there is no reason to suppose that Northamptonshire is very different from other East Midland counties, comparisons and contrasts with counties farther away are likely to be valid comparisons with the whole of the East Midlands.

As a summary interpretation of the open fields of the whole of England, this book can only be based on limited amounts of primary written data. It has included, where possible, evidence from photographs and maps. The subject is very amenable to a geographic approach—much more so than has been done hitherto. For instance, the three-field system types that occur in Dorset, the East Midlands, and Eastern Yorkshire all lie in Gray's central area, where two- and three-field systems predominate, but a glance at Plates 3 and 4, and Figures 1.2 and 2.1, make it obvious that there are great differences between them.

Data collection has fallen into two main classes, the physical record and his-torical evidence. Most rural county sites and monuments records (historic en-vironment records) have been visited to ascertain what information is readily available, what research is currently being undertaken in a county, and where there are sites worthy of a visit. National coverage of aerial photography rele-vant to open-fields was studied at Cambridge in the University Collection of Aerial Photographs and at the English Heritage collection held in the National Monuments Record, Swindon, which includes copies of RAF vertical photog-raphy of the 1940s and later.

A full understanding of field systems can only come from detailed studies of individual examples by minute examination of original documents, and many good descriptions can be found in the national and county literature. A view supporting such an approach to historical research was given by the formidable Reverend E. H. Bates Harbin in 1912 when reviewing an excel-lent book on the history of Exmoor published by E. T. MacDermot in 1911, *The History of the Forest of Exmoor*. Bates Harbin (sometime priest of Puck-ington, squire of Newton Surmaville, editor of five Somerset Record Society volumes, and local editor of the *Victoria County History*) pointed out that there were already several books about Exmoor and wondered if another was really necessary, but conceded that the book did have value because it was 'compiled from original documents, and therefore avoiding the twin sins of padding and futile speculation'. A similar warning had come from Dorset in 1903. Discussing 'the problem' of lynchets, H. Colley March (1903, 67) wrote 'theorists have dogmatised as usual, and have complacently generalised on insufficient data'.

The use of original documents has been my approach in analysing the field systems of Northamptonshire (Hall 1995). However, for a national synthesis, the amount of detail provided by studying original sources had to be limited, with reliance made on the availability of suitable printed material. County record society publications offer primary data that allow independent inter-pretations to be made. Published articles and monographs have also been used as sources of information, especially those that supply original data. Informa-tion has been pursued back to data that are being quoted for the first time from original manuscripts, but not often to the manuscripts themselves. Sometimes conclusions different from those provided by the authors and editors are sug-gested. I have tried to give acknowledgement to the authors of original discov-eries or interpretations—the literature is full of statements by authors that do not properly acknowledge sources of information and present it as their own.

Each county had as much as possible of its literature studied—journals, record society volumes, published cartularies and surveys, village history monographs, etc. This was a variable task—some counties have limited information, others have overwhelming quantities. At that stage it was clear where there were gaps in the published record that might be improved by a visit to the relevant record office. For instance, many counties have no published works describing the im-portant information provided by glebe terriers. Selected tasks were undertaken where there were serious gaps as it was impossible to work through every county in detail. No doubt much improvement could be made by, say, a two-month visit to each county record office. But with the Record Offices of thirty-nine historic counties, the British Library, Cambridge and Oxford university libraries and their College estate records, and many other important private and public arch-ives, the task is enormous. It has been estimated that there are 40,000 medieval charters at Oxford, yet they receive little study. We have to keep in mind W. G. Hoskins' (1958–9) exhortation, when reviewing Beresford's *History on the Ground* that 'the minute study of local topography can become a bottomless morass...[in which]...one is overwhelmed in loving but futile detail'. It would be easy to fall into the temptation of undertaking too much and not have time to present a complete national view, leaving some counties unstudied.

One difficulty for open-field studies is that agricultural information does not get published or is not fully catalogued. Old indexes tend to concentrate on famous individuals or events. In the Compton Muniments at Castle Ashby, Northamptontshire, a bundle of eighteenth-century material relating to one of its parishes had written on it 'farming papers of no interest whatever'. The latter contains an enclosure act and a 'quality' survey of 1780 listing all the furlongs, as well as other material relevant to this study. It is often not possible to overcome such problems without recourse to the original manuscript collec-tions. There has been lack of interest in open-field matters. For Ballingham, Herefordshire, when discussing a collection of 137 charters dated 1237–71, the editor reported that they 'are not very interesting because they are mainly concerned with transfers of land and property' (Martin 1952–4, 70). For Wortley, Gloucestershire, it was noted that 'the land is mostly strips in the common fields, which it would not be of interest to locate even if that were possible' (Lindley 1952, 93).

The subject of open fields overlaps with many other closely related themes: estates, settlement types, the farming economy and crop yields, work-service, tenure, tithes and rates, and enclosure. These have been touched upon where relevant, but not exhaustively—each item is a specialist topic and has its own publications. The following chapters deal with various aspects of open fields, among them field structure and field numbers; the type of field system and how it operated; the physical disposition of the demesne, meadows, pastures, and other resources within a township; the amount of the area under cultivation; the disposition of holdings (yardlands and oxgangs); and tenurial arrangements. It concludes with thoughts about their likely origin as suggested by the surviving evidence. Much more work could be done. There is no substitute for a detailed study of all the townships in a county, both on the ground and through documentary sources. When that is complete for the whole country, we can dispense with the dangers of Bates Harbin's 'futile speculation'.

It is hoped that this volume will be of assistance to those undertaking research into field systems and encourage interest among all who have not previously been aware of scope of the subject and its relevance to settlement and economic history. I hope to hear from readers about detailed historical records and research programmes for a particular area so that county gazetteers may be updated. I can be contacted through the Northamptonshire Record Society, Wootton Hall Park, Northampton NN4 8BQ (e-mail: enquiries@northamptonshirerecordsociety.org.uk).

I am most grateful to all the county archaeologists and their staff who have kindly received me and sometimes undertaken field trips and provided detailed print-outs and distribution maps. I have often been introduced to information that could not possibly have been discovered through the published literature. Specific acknowledgements are given in each county section of the Gazetteer; they refer to people met at various dates, some of whom have since moved into retirement or beyond. I also thank David Pelteret and two anonymous readers who made suggestions that have helped to improve the text.

I am especially grateful for access to Cambridge University Library, to the Cambridge University Collection of Aerial Photographs at the Department of Geography, and to the Haddon Library at the Department of Archaeology, University of Cambridge. My thanks are also due to Dr Robin Darwall-Smith, archivist, and the President and Fellows of Magdalen College, Oxford, for access to the Magdalen Estate Archives, to John Hardaker for preparing the illustrations, and to Tracey Partida for the plans produced by means of a Geographical Information System using the Northamptonshire Historic Database.

Introduction: medieval fields
and the landscape

This book gives an account of the open fields of England. They formed the system of agriculture that operated before the establishment of present-day farms surrounded by hedged or walled fields and held in single ownership. Formerly, two main types of field system could be found, which will be fully described in chapters 2 and 3.

In much of central England the earlier agricultural system consisted of extensive arable 'open' fields divided into narrow strips without hedges or fences. A given holding had its land scattered in many strips, and was farmed from a homestead located centrally in a village. Cropping was regulated by a manorial lord or the village community, the whole area commonly being grouped into two or three parts called 'fields', which formed the basis of a simple crop rotation. This type of arrangement is now called a 'regular field system'.

The east and west of the country had less extensive areas of strip fields which were associated with small and dispersed settlements. Strips were grouped into several or many small blocks, either called 'fields' or identified by topographical names. There was little or no communal regulation and cropping was varied. There were some enclosures amongst the arable and often much associated pasture. In hilly regions the area of arable could be quite small, being confined to valleys. Such field systems are known as 'irregular systems'.

The work of previous authors on the subject of open fields will be discussed in some detail later, but the main sources are introduced here. Seebohm (1883, 1–7) and Vinogradoff (1892, 235–8, 244) were among the first to write about open fields. Gray's detailed study of the whole of England (1915) classified fields into several main types. The two- and three-field arrangement, found by Gray to occur primarily in the Midlands, became known as the 'Midland system'. Bishop proposed an assarting mechanism to explain the formation of intermixed fields in 1935–6. The Orwins (1938, 37–40) approached open fields through the practicalities of ploughing techniques. In 1964 Thirsk defined four essential elements that were characteristic of mature fully developed common-field systems. Baker and Butlin edited a series of essays in 1973 which showed the variety and complexity of British field systems, and a volume of papers specifically considering possible origins of open fields was edited by Rowley in 1981.

More recently, Lewis, Mitchell-Fox, and Dyer's *Village, Hamlet and Field: Changing Medieval Settlements in Central England* (1997) examined four East Midland counties as a regional case study. Roberts and Wrathmell's *Region and Place: A Study of English Rural Settlement* (2002) set the context for the whole country, viewed in the first instance from early nineteenth-century settlement patterns. Williamson, in his *Shaping Medieval Landscapes* (2003), studied a large area of East Anglia and the East Midlands and presents a view that soils based on Midland clays encouraged the creation of extensive open fields, with a concomitant settlement nucleation at the end of the first millennium AD. Rippon's account of settlement and fields in the South-West, *Beyond the Medieval Village* (2008), gives important recent information for a region outside of the Midlands, which hitherto have tended to dominate research themes. He showed that south-western England developed efficient fields, mostly within a dispersed settlement regime, and should not be regarded as a 'backward' region that failed to adopt a Midland system.

Although less obvious now than say three centuries ago, England consists of two main types of settlement and countryside—not a north and south divide, but more a central region contrasting with those of the east and west, here designated the Western, Central, and Eastern Regions (Figure 0.1). The regions have been defined by the occurrence of regular two- or three-field systems according to the evidence presented in the Gazetteer. The Central Region is similar to, but not the same as, the extent of the two- and three-field region shown in Gray's 1915 *frontispiece* and the 'Central Province' of Roberts and Wrathmell (2000). It is the intention in these pages to try to ascertain why there was this great difference between regions and examine in some detail how the regions varied in terms of field systems.

In the thirteenth century, the landscape of much of the lowland Central Region was open, apart from an occasional wood or clusters of trees in paddocks adjacent to a village or hamlet; exceptions were large areas of royal forests. Although there was more woodland than in the High Middle Ages, much of the countryside was arable, particularly in the East Midlands, and sometimes the only grasslands were permanent meadows and pastures lying next to rivers and brooks. Arable land was subdivided into many small parcels that were long in proportion to their width. Only rarely did a hedge, fence, or ditch demarcate these arable strips, furrows left by the plough being deemed a sufficient boundary.

Arable parcels of the Central Region did not differ much from those in the remainder of England. Everywhere, they were most commonly one or two roods in size (0.1 ha), as attested by charters and other documents from the twelfth century onwards. Each parcel was commonly called a 'land' by the

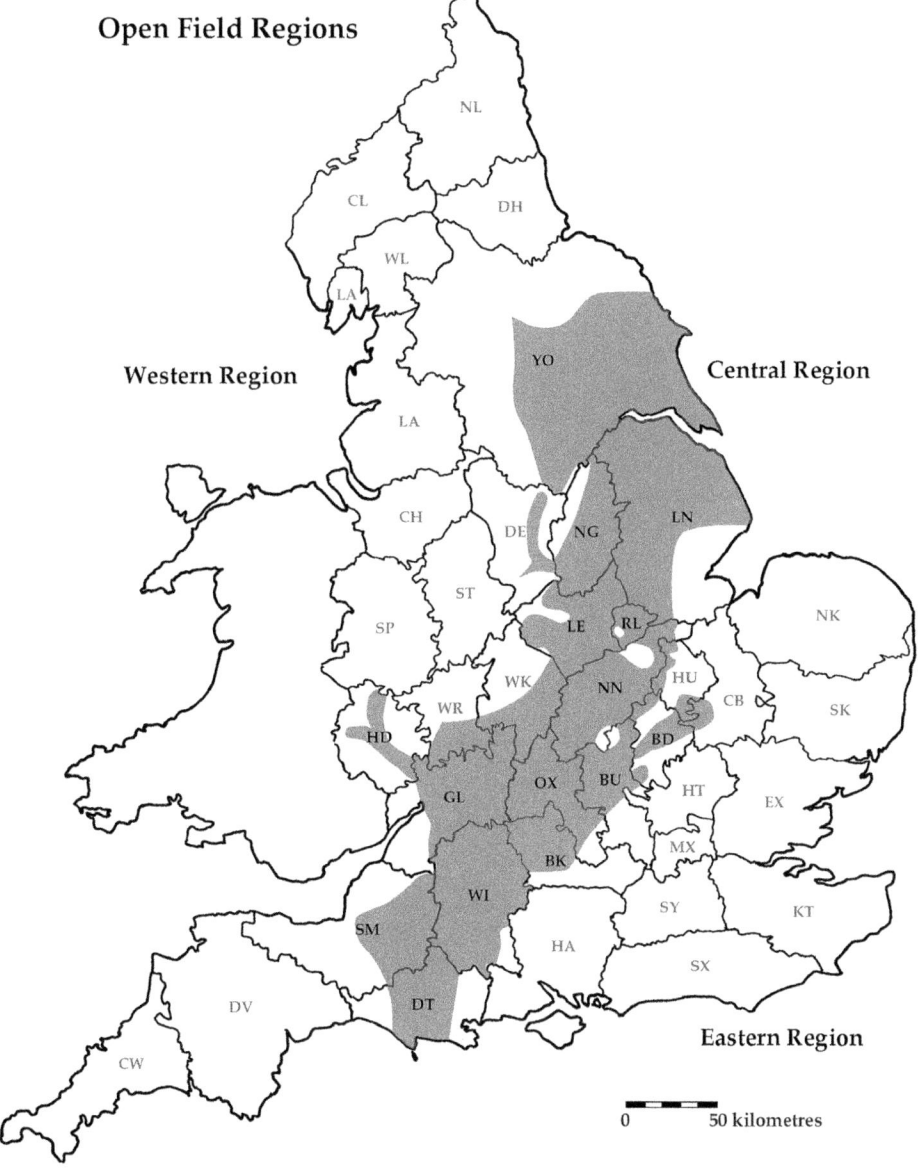

Open Field Regions

Western Region

Central Region

Eastern Region

0   50 kilometres

Fig. 0.1. The three open-field regions of England, based on evidence given in the Gazetteer.

people who worked them (a 'selio' in Medieval Latin, with many other regional English names; see the Glossary). Its size was long in proportion to its width, being 5¼ by 220 yards (4.8 by 200 m) in the case of rood lands. Collectively they lay together in blocks with strips running parallel to each other called 'furlongs' (or 'flats', 'bydales', etc.). An individual holding, called a 'yardland' ('virgate') in the south and an 'oxgang' ('bovate') in the north, was not normally a block of land but consisted of a number of strips dispersed through part, or all, of the system. More will be explained about yardlands later. Parts of the Isle of Axholme, Lincolnshire, are still cultivated in narrow lands now ploughed flat, as on the south side of Belton (see satellite images). Figure 0.2 shows curved strips at Epworth in *c*.1935, with rows of shocks of grain. Braunston, North-amptonshire, affords good examples of open-field ridged lands fossilized in their eighteenth-century state when they were abandoned for farms based on newly hedged closes (Plate 1). In the Central Region, the overall management of the fields diverges from the other two regions, as does the physical distribution of the holdings.

Central Region holdings were cultivated and regulated in various complex ways that involved some form of communal cooperation. It was the regulation of farming arrangements that was communal, not possession of the land—the crops produced were the private property of the farmer or tenant. Seebohm, in 1883, was the first to describe the details of what came to be called the 'Midland system'. The wide, open landscape with little high ground, associated with this type of cultivation was not to every traveller's taste. In 1744 Philip Yorke stayed at Easton Neston House, west of Northampton, and 'travelled through an unpleasant country [mostly open field] to Leicester'

Fig. 0.2. Curved strip fields at Epworth, Isle of Axholme, Lincolnshire, in *c*.1935 (Lipson 1937, frontispiece).

(Godber 1968, 126). William Cobbett journeyed along the Old North Road from Royston to Huntingdon in 1822 and saw 'those very ugly things, common fields, which have all the nakedness, without any of the smoothness, of Downs' (Cobbett 1822, 73).

Strips of arable were often ridged up to varying degrees in most parts of the Central Region to create a naturally draining seedbed. These survive in some pasture fields or urban parks as 'ridge and furrow', as at Braunston and Wollaston, Northamptonshire (Plates 1 and 2). Ridge and furrow, although much destroyed by ploughing for cereal production since 1940, has a wider distribution than the Central Region, once being common in the lower-lying land of south-western and western counties, from eastern Somerset to Herefordshire, and in the north from Staffordshire to Northumberland. In most of East Anglia, south-eastern England, and Devon and Cornwall, broad ridges of 'Midland type' (five to ten yards wide with slightly curved ends) are not found. There was probably some slight ridging in part of the Western Region (a Devon example occurs at Thorverton near Exeter: see Plate 16), but almost all ridges have been obliterated by centuries of severalty farming within enclosures. Cultivation ridges of early modern date occur in many areas. Straight narrow furrows of mainly nineteenth century date are easily identified, but curved severalty ridges ploughed in the seventeenth and eighteenth centuries are much more difficult to distinguish from ancient tenurial ridges (see Hall 2013, 49–50).

Dispersed strip-field systems in most areas were called 'fields', 'open fields', 'common fields', or 'town fields'. The term 'subdivided' has been used in recent literature to move away from such complications as enclosed fields that had common rights and open unhedged fields that did not have common rights. There is, however, still potential confusion that the term 'subdivided fields' might refer to small hedged fields made within larger closes during the eighteenth and nineteenth centuries. This book will use the term 'open fields', often synonymous with 'open and commonable fields', a description used in the East Midlands during the eighteenth century. The areas where open fields were not commonable, such as the Wash Silt Fens and Kent, will be made clear in Chapter 3.

Counties away from the Central Region differ primarily in that they had land, often high ground in the west, that was 'waste': heaths, downs, hills, fells, moors, and mosses. The east was low-lying for the most part, excepting the chalklands, but also had areas of poor quality ground such as the Weald of Kent and the Bagshot Beds of the Thames Basin. These 'wastes' were used as commons to graze animals and for sources of fuel. On low-lying ground there were different types of waste—large areas of fen, marsh, and lowland moor—that were used for the same purposes. Common rights on these wastelands were sometimes shared by adjacent and nearby townships. The area of a township under the plough was generally far less than the 90 per cent sometimes found in parts of the Central Region. Normally, there was no township communal cropping arrangement and holdings were not necessarily dispersed throughout. Because of a less rigid agricultural system and the presence of small settlements and individual farms, these areas were enclosed at a much earlier date than the Central Region—often as early as the fourteenth and fifteenth centuries. The presence of extensive old enclosures made a marked visual contrast with the open spaces of the Central Region in the seventeenth and eighteenth centuries. It is still apparent today in counties such as Essex or Devon, where the crooked roads and species-rich hedges differ from the straight lines of predominantly hawthorn hedges characteristic of the eighteenth- and nineteenth-century enclosed landscapes found in parts of the Central Region.

This book is arranged in two parts. First, the land-use found in contrasting examples of townships in different geographical locations across the country is described (Chapter 1). After examination of the main types of regional and sub-regional field systems (Chapter 2), their detailed structure in terms of manorial demesne, tenurial arrangements, and management are examined (Chapters 3 and 4). Early evidence for settlements and fields is then collected together using archaeological and historical sources (Chapter 5), after which Chapter 6 discusses processes that may account for differing field-system types and their origins, drawing upon the data presented. The second part of the book is a Gazetteer providing a summary account of field systems organized by the historic counties. This gives much of the detailed evidence used in the main text and serves as a handlist for further local studies.

1

Townships and land-use

This chapter begins by discussing the township and its relationship with the parish. Examples of contrasting field types within townships will draw attention to the great variation in the proportions of pasture and arable land that once existed. Townships containing a single settlement in the East Midlands, Dorset, and Derbyshire will then be compared with areas of complex dispersed settlement and open field in Essex and Herefordshire. The importance of meadow and its management in areas with extensive arable will be outlined, showing how the lack of adequate pastures in the Central Region was partly alleviated by the conversion of arable to leys (temporary grassland). Outside of the Central Region, in contrast, pasture was available in the form of commons, greens, and droves, and in upland regions there were ample and extensive rough pastures.

A. TOWNSHIP AND SETTLEMENT

The term 'township' has already been used without any explanation or definition. It was a basic land-unit characterized by one complete, self-contained field system. A township was normally a single block of land, but sometimes there were additional detached pieces of meadow, woodland, or pasture that supplemented its resources. In the Central Region, townships were often of moderate size (*c*.800–3,000 acres) and were frequently coextensive with a single parish containing one settlement and one church. In the Eastern and Western Regions, especially in northern parts, parishes could be very large and often contained many townships, as well as severalty farm holdings (i.e. holdings in private ownership). In Salford parish, Lancashire, there were twenty-nine townships in 1212 (Farrer 1902, *frontispiece*). A survey of the Dean and Chapter of Worcester's lands, made in 1649, found eight townships at Warton in the same county (Cave and Wilson 1924, 234).

Outside of the Central Region, where there is dispersed settlement, several small estates (possibly 'townships') can sometimes be identified. An example is the parish of Chittlehampton, Devon. Analysis of the settlements and enclosed field names recorded on the tithe map demonstrate the existence of discrete areas and farms held in severalty, even though open-field data for them is lacking. Chittlehampton (SS 82 63), lying in the north of the county, was studied by Andrews (1962). Using place-name evidence and Andrews' version of the 1842 tithe data, it is possible to reconstruct the complicated structure, assigning territories for these settlements. One of these territories, Bradbury, in 1842 had small strip-like enclosed fields lying around it that were in intermixed ownership. In addition to the settlement of Chittlehampton itself within the parish of that name, there are thirty further small settlements and farms bearing the names of medieval owners (Gover et al. 1931–2, ii: 337–40); the information is summarized with dates in Figure 1.1. Chittlehampton vill has very little land assigned to it and may have developed from an early Christian site dedicated to St Urith rather than originally being an agricultural settlement. The existence of the Anglo-Saxon 'cote' name-forms and the three small estates recorded in 1086 show that the fragmented settlement pattern is old and not the result of farm dispersal after enclosure. Although the analysis is based on late evidence from enclosed field systems, it shows there had been many small 'townships' each containing a farm.

No information survives about the founding of the severalty estates at Chittlehampton, but in other places there are records of farms that had been taken out of the waste at an early date. The farm of Creacombe (SX 59 49) in Holbeton, south of Dartmoor, had its block boundaries described in 1185 (Weaver 1909, 158). In Dorset, severalty farms and small estates were recorded in the late twelfth and thirteenth centuries at Broadwindsor and Thorcombe (Hobbs 1998, 31–48, 78, and 91). The Surrey Downs likewise have identifiable small compact farms (Gardiner 2011, 103). Some of these severalty holdings may be eleventh- or twelfth-century assarts or equally they may be the land units belonging to small settlements of earlier date.

Elsewhere, in regions with dispersed settlement, land near to any particular place was associated with it, although no 'township' can be established. Examples are Hanbury, Worcestershire, in the Western Region, studied by Christopher Dyer (1991). Extreme dispersion occurs in Cornwall, where Beresford investigated medieval settlement in three Duchy of Cornwall manors comprising five parishes in different parts of the county (Helston in Trigg, Tremanton, and Tybeste). In the 1356 there were 203 messuages dispersed in 57 separate places. Only eleven places consisted of a single farm, sixteen had two messuages, and the larger hamlets (excluding boroughs) contained groups of messuages. The whole estate was mapped with a schedule in 1819 and areas can be assigned to each place that is likely to correspond closely to pre-enclosure 'townships' (Beresford 1964). These examples from the Western Region demonstrate that large parishes often contain many small townships, each with its

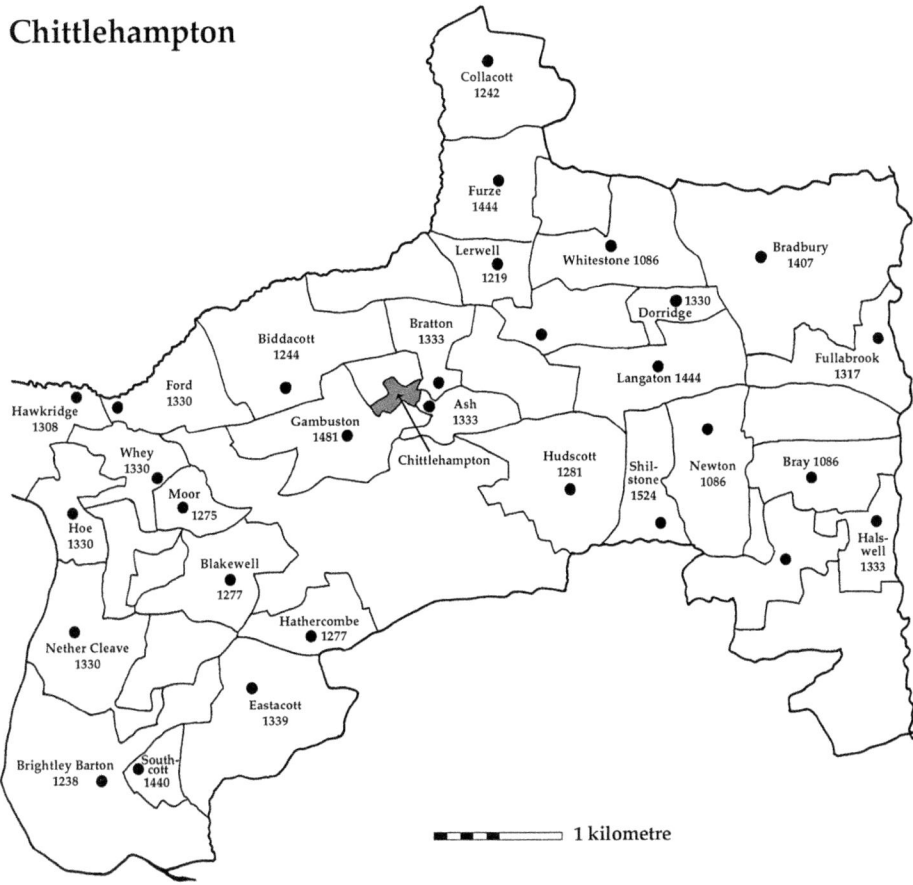

Chittlehampton

Fig. 1.1. Chittlehampton, Devon (SX 65 25), showing settlements and their territories (after Andrews 1962 from the 1842 Tithe Map; settlement dates from Gover et al. 1931–2, ii: 337–40).

own fields. They are often complicated by the presence of medieval assarts and older severalty holdings.

The widespread practice of modern intensive agriculture means that it is not always easy to appreciate the regional variation that existed before substantial farming improvements were made in the early modern period. It is important to understand how great the differences were, and the often limiting effect that topography, soil quality, and climate had on the potential and management of the local farming economy. A few examples are given next, drawing as far as possible on contemporary maps or surveys, and choosing places uncluttered by the masking effects of early modern changes. These differences of land-use go a long way to explain the variation in regional field systems and settlement types.

B. CONTRASTING SETTLEMENT AND FIELD TYPES

Villages, the dominant settlement form of the Central Region, lie in a long central swathe running the length of the country approximately from Alnwick, Northumberland, to Dorchester, Dorset. They were associated with large areas of arable, and about half of them remained open and relatively hedgeless until the eighteenth century. Both the Western and Eastern Regions had more dispersed settlement. The Western Region includes most of the high ground of England and the Eastern has much 'ancient countryside' resulting from early enclosure.

The central area has been defined and mapped from the evidence of early nineteenth-century settlement patterns by Brian Roberts and Stuart Wrathmell (2000), who called it the Central Province. The approximate extent of the region has been apparent for some time, measured with greater or lesser accuracy by several inter-related parameters, which have been illustrated by Roberts and Wrathmell (2000, 28–31) using a variety of sources. Among them are the extent of Parliamentary Enclosures (Slater 1907, 73), the distribution of deserted medieval settlements (Beresford and Hurst 1971, 67), and the area where woodland was absent or rare in the late Saxon period, *c*.730–1086. There is some overlapping of types—many woodland areas within the Central Region had dispersed settlement and some areas outside of it had large villages associated with a considerable amount of arable.

A mixed farm needs a suitable combination of fertile arable land for cropping and pasture for animals to graze. Animals need hay for winter fodder. This was obtained from meadows that flooded annually, which thereby drew from arable land nutrients that encouraged abundant spring growth. If there was little or no meadow, then fodder had to be produced less efficiently from a larger area of pasture. Farms also needed access to woods for timber and fuel, and the requirement for all these resources played a part in marking out the territory belonging to a settlement and its fields. The results of such decisions are evident in many of today's township boundaries, where extended linear shapes lying parallel to each other cut across meadows, over belts of fertile arable, and up to higher ground that had been pasture, wood, or heath. Examples in Lincolnshire and Berkshire were given by the Orwins, with narrow parishes from four to nine miles in length, ranging from low ground and fen to heath or chalk downland. The boundary of Hardwell (SU 28 87) in Compton Beacham, Berkshire, described in a charter of 903, was illustrated to demonstrate that linear townships were laid out as early as the late Saxon period to utilize a range of soil types (Orwin and Orwin 1938, 25–9; S 369). Similar narrow, long townships on the Down edge occur farther east, from Wantage to East Hendred, where seven are together only five miles wide and each about the same in length (Dils 1998, p.viii, from the 1887 Ordnance Survey). The tithe boundaries of central Wiltshire show narrow townships (Kain and Oliver, 1995, 549) and other Berkshire and Wiltshire examples have been discussed by Hooke (1988). Some townships in Sussex are similarly linear, taking in resources from the Weald, for example West Taring in Milepost, 400 yards by 9 miles (Chapman and Gardiner 2005, 42).

Woodlands as well as meadows were desirable resources and large woods were often shared by adjacent townships, with some pieces belonging to far distant townships. Details have been provided for Northamptonshire (Hall 1995, 99–106). Detached pieces of the Weald belonging to parishes far away are notably characteristic of Sussex (Gardiner 1984; see the Gazetteer). In Staffordshire, parishes with detached parts providing wood pastures in the Forest of Needwood were mapped on First Edition 1:10,560 scale Ordnance Survey maps (Hooke 1998, 218). Although common pasture and pannage rights in woodland formed a resource that benefited an open-field economy, the theme of woodland will not be pursued here, however, since it is not directly relevant to fields and field-structure in the same way that meadow and pasture are. Meadows lying detached from the main township will be discussed shortly (see p. 18).

Land-use in townships with a single settlement

The defining characteristics of the Central Region in terms of intensive arable usage and a simple settlement pattern have long been known, but they have not often been stressed adequately with detailed maps using original source material. It is, therefore, worth discussing the examples of two townships of simple structure in the Central Region that display these characteristics, after which more complex arrangements elsewhere will be examined.

Rothersthorpe and Twywell, lying in the East Midlands part of the Central Region, illustrate the intensive use of arable in an area of gently undulating countryside with variable permeable and semi-permeable soils. Many open-field townships in this region were characterized by a very high proportion of arable (up to 90 per cent) and a severe shortage of permanent meadow and pasture. In some places this system of extensive arable lasted for a very long time, with large areas of intermixed open-field arable inextricably locked into a system of communal operation that was rigidly controlled by legal and social regulations and by common rights.

Rothersthorpe, Northamptonshire (SP 71 56), is a parish and township of 1,232 acres. It remained open-field until 1809 and the enclosure map (Northamptonshire CRO, Map 2872) shows a single nucleated settlement hemmed in by a small area of old enclosure surrounded by a wide expanse of open-field land (Plate 3). The vill and its closes occupied 6 per cent of the total township. There is no open-field map, but a plan was easily constructed from the remains surviving on the ground in 1977, much of which can be seen on aerial photographs of 1947 (RAF CPE UK/1926). By relating enclosed field names to terriers it was possible to work out the open-field structure. In 1803 there were three great fields occupying 86 per cent of the open-field area, leaving only 8 per cent as meadow and pasture alongside brooks. The fields were run on a simple three-course rotation; in 1344 there were three fields. It is clear that grassland was in short supply. In the Middle Ages there was a small extra grazing resource in a piece of detached woodland belonging to the manor lying near Roade, three miles distant (still called Thorpewood Farm), as well as a piece of meadow, now incorporated into the neighbouring parish of Wootton (references in Hall 1995, 338–9). By 1803 these no longer contributed to Rothersthorpe's economy, but the provision of grazing was not quite as limited as the plan might indicate, because there had been a widespread introduction of grass into open-field arable land in the form of leys and cow pasture from the fifteenth century onwards (see p. 22).

Another similar example is the parish and township of Twywell, Northamptonshire (SP 95 78), 998 acres, not enclosed until 1765 (Northamptonshire CRO, BSL 29). It has excellent records including an open-field map and an accompanying field book (a survey of a whole township strip by strip) made in 1736 (Northamptonshire CRO, Maps 1409 and 4323, and survey Buccleuch 24/5 in Northamptonshire CRO, Box X8670). There was a single settlement that had a core of old enclosure and a few small closes in the open fields which amounted to 7.6 per cent of the township total. The remainder was open-field strips apart from fifty-two acres (5 per cent) of meadow, leaving 87 per cent as strips, mostly arable with a few grass strips called leys. Twywell's furlongs were grouped into three great fields, and the arable belonging to the church, the glebe, was dispersed evenly throughout (Plate 4). Additional resources were two closes at Kirtley, approximately thirty-five acres, which had been seigneurial woodland, lying detached half a mile from the main township on the edge of Rockingham Forest.

The next three examples are single-settlement townships with a high percentage of pasture: Chilfrome, Dorset, lying on chalk at the southern end of the Central Region; Iford, Sussex, situated in the Eastern Region, but a chalk downland township similar to Chilfrome; and Sheldon, Derbyshire, located on limestone in the Peak District, part of the Western Region. The small parish of Chilfrome extends to 940 acres (SY 55 99). It was mapped in *c*.1823 when it was still open (Dorset History Centre, D11/1); 822 acres were enclosed in 1823–4 (Tate and Turner 1978, 103). To the north, downland pasture comprised 51 per cent of the township and there were three great fields in the centre: West, Middle, and North Fields; at its south lay the village with an envelope of old enclosure adjacent to the River Frome (Figure 1.2). Chilfrome is a good example of open-field land-use on the chalklands of the southern counties, with a large area of common on the downs. The open-field arable occupied 29.2 per cent, the settlement 2.8 per cent, meadow 2.9, old enclosure 14.6, and downland 50.5 per cent. The old enclosure probably had been arable, making the medieval arable 43.8 per cent of land-use.

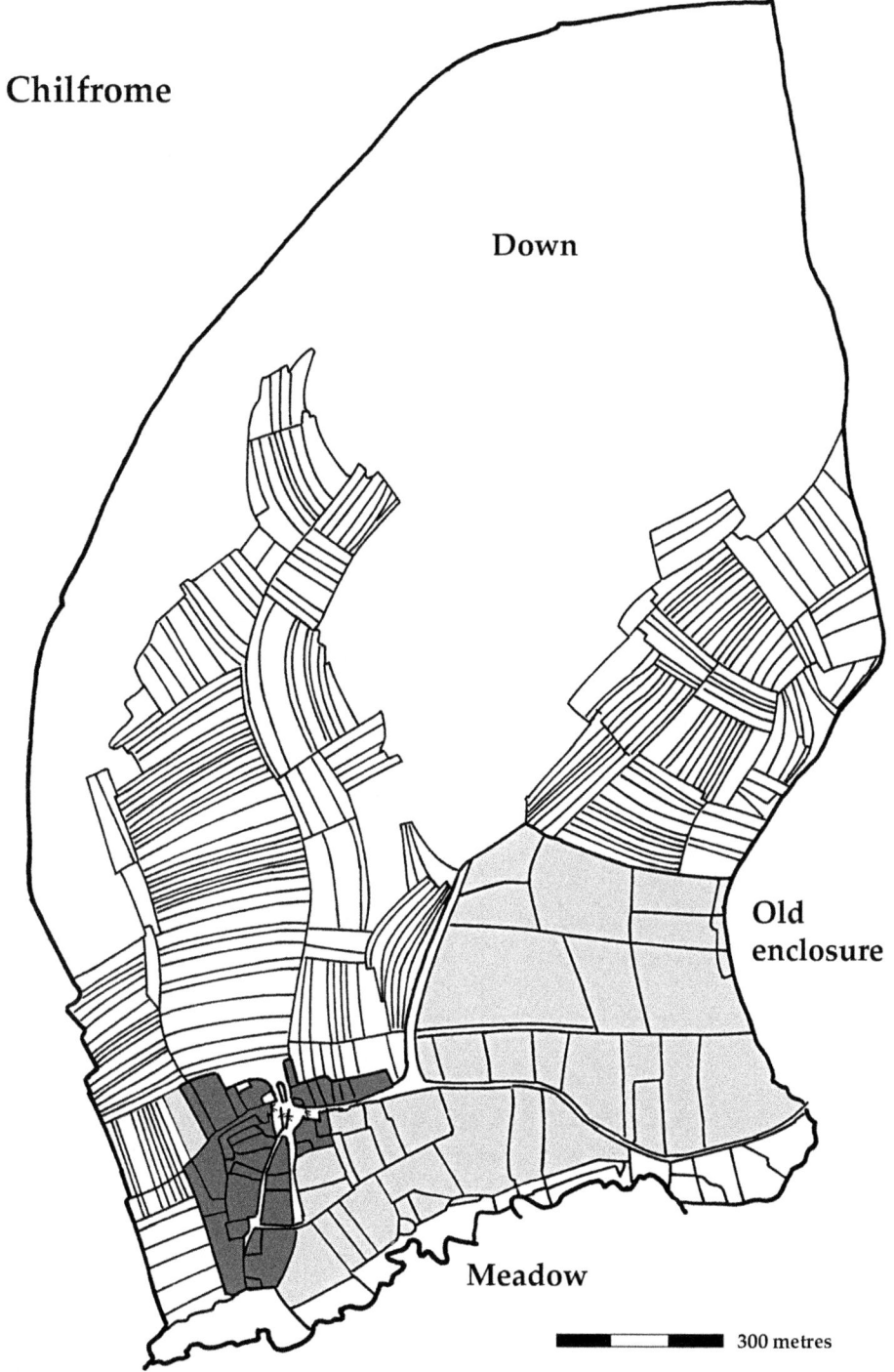

Chilfrome

Down

Old
enclosure

Meadow

300 metres

Fig. 1.2. Chilfrome open fields, Dorset (SY 55 99), in *c*.1823 (after Dorset History Centre, map D11/1).

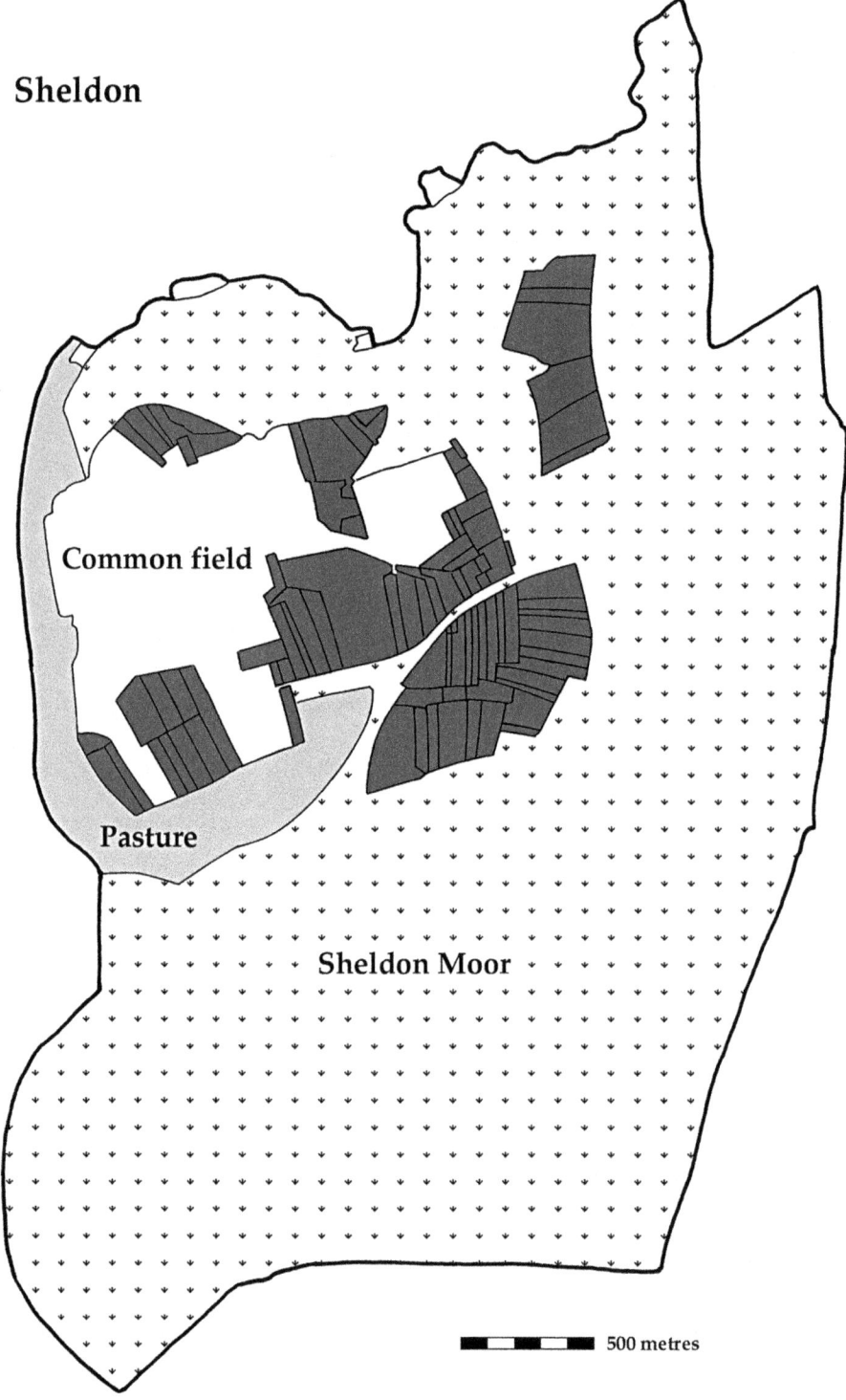

Sheldon

Common field

Pasture

Sheldon Moor

500 metres

Fig. 1.3. Sheldon, Derbyshire (SK 18 68), in 1617. The shaded areas are the vill and old enclosures (after a negative of the original Chatsworth map held at the University of Nottingham, William Senior Volume, fol. 51r).

The second similar chalk downland example is Iford, south of Lewes, Sussex. In 1842 it was largely open and the land-use of its 2,173 acres was open-field 12 per cent, old enclosure 13, meadow 24, with the remaining 51 per cent being open downland pasture of about 1,100 acres (East Sussex CRO, Tithe Map TD/E 61).

The third example, Sheldon township, 2,068 acres lying in Ashford parish in the White Peak of Derbyshire, had much pasture (Figure 1.3). It was mapped and surveyed by William Senior in 1617 for the Earl of Devonshire (Fowkes and Potter 1988, 123–6). The area of settlement closes and arable amounted to only 601 acres (29 per cent). Each holding consisted of about half open field and half enclosures lying in intermixed ownership; the open field was only 9 per cent of the total and probably operated as a one-field system with convertible husbandry (see Wigtwizzle, Yorkshire (WR), p. 88). The plan shows the core of houses and crofts with a few separate blocks of narrow closes that look like enclosed and divided furlongs, one on the west one called *Stadfold*. Closes to the north-east on the Moor are likely to be intakes or outfield; they also had intermixed ownership.

Land-use and complex dispersed settlement

The previous five examples illustrated arrangements in small- or medium-sized townships containing a single settlement with differing percentages of arable. Quite different are large and complex parishes containing several settlements and farmsteads, which are characteristic of many in the Eastern and Western Regions. Chittlehampton has already been described in terms of its settlement and component 'townships', but the details of land-use in its open-field state are not known. There are, however, examples where some details of open fields can be determined.

Writtle, Essex, lying west of Chelmsford (TL 67 06) on the lowlands of the Eastern Region, with mixed soils, had many settlements and enclosures. The parish extended to 8,672 acres, which included detached parts near Chignall Smealy (once probably woodland (Reaney 1935, 247)). In Writtle parish was the chapelry of Roxwell lying to the north-west (TL 64 08), comprising another 4,755 acres. Apart from the main two vills there were six settlements associated with greens, all of medieval date. There were also thirteen isolated medieval manors and severalty farms. The fields that operated within this parish were as complex as the settlements. They had all been enclosed before the late sixteenth century, but in the fourteenth century they were still open (Newton 1970, 23–37). The main Writtle manor had a large arable block of demesne, 1,300 acres, lying adjacent to the vill near to the River Can and Roxwell Brook. There were small areas of meadow and pasture (thirty-two and fifty-six acres). Two emparked demesne woods lay to the south: High Wood, lying between them, and Great Edney Wood were commons. Greens, such as Little Edney and Cooksmill, were commons for the individual hamlets.

The Writtle block demesne was divided into seven named component fields and cultivated on a three-course season or tilth in 1328. The fields making up each tilth were not adjacent but were composed of several blocks. In contrast to the demesne, the tenants' lands in the Middle Ages, according to charter and court roll evidence, were made up of small enclosures. There is no overall view until surveys made in 1564 and 1595, when for example a 270-acre holding lay in twenty-seven crofts; another 26-acre holding was made up of eight crofts. In the south, a few medieval charters refer to hedges around strips lying in crofts. The pattern here was probably one of assarting and formation of closes at an early date. The other thirteen manors in the parish also had block demesnes lying in one or a few places and, where located, the freeholds belonging to them lay together in blocks. The tenants' lands were not precisely identified and were probably partly dispersed holdings lying near to each manor or vill. Newton's map V shows the extraordinary complexity of the fourteen manors with greens and woods in the 13,473 acres of the two adjacent townships. It is possible that some of these manors were separate townships.

Further examples of complex multiple settlements that had substantial amounts of open field surviving to the early modern period are Eversholt, Bedfordshire, and Marden, Herefordshire, which have detailed maps or surveys. Eversholt lies in a region of complex fields and settlement in the west of the Eastern Region. In its 2,119 acres were seventeen named ends and farmsteads when mapped in 1764 and surveyed in a field book. It then had appreciable enclosure and eleven named open fields, of which eight were stated to be 'common'. Some of the proprietors' lands tended to be concentrated at one of the many 'ends' (Fowler 1936, 43–6). This suggests that there were several townships or that some holdings were never dispersed, as evidence of 1200 indicates, when it was recorded that half a yardland (an open-field holding) lay in a croft (Fowler 1926, 130).

Herefordshire, in the Western Region, had many examples of small settlements with multiple open fields and severalty farms (see the Gazetteer). A good example is the manor of Marden, 3,671 acres lying north of Hereford on undulating ground, which had detailed maps and surveys made in *c*.1720 (Sheppard 1979). Eight settlements have medieval origins: Sutton, Marden, Fromanton, Amberley, Venn, Vald, Vern, and Wistaston, and these were specified as being separate townships in 1608 (Gray 1915, 95 n. 4). There were twenty named fields in 1720. Eight other hamlets and farmsteads at the north are not medieval, but were formed after partial enclosure. The main vill at the south was Sutton with three named large fields and a core of old enclosure with curved hedges. The largest holding had a block of demesne in the north of Upper Field, plus at least one strip in every furlong. Smaller holdings had twenty to thirty acres with a toft, one or two closes, and several scattered strips in each arable field and some meadow strips. Very small holdings had a toft and a few strips lying in one part of the open fields only. At the north, only Marden was substantial. Small open fields extended from 10 to 170 acres, some owned by one person. The common waste was fully open, and woods were held in severalty. Closes with curved hedges were mainly used for pasture. Most farmsteads had large blocks of land near to them and a few scattered parcels of arable. Smallholdings had parcels within 800 yards of a dwelling; thus a series of arable fields was being farmed from different hamlets lying around it. The plan of Marden manor is not exactly comparable to the simple townships first described, but it does show a very complex dispersed settlement pattern lying in several townships with multiple fields.

The contrast between these field systems thus is considerable. Rothersthorpe and Twywell are small single-townships extensively cultivated and were en-closed late; Sheldon and Chilfrome had single settlements but with a high proportion of pasture. Writtle, Eversholt, and Marden had many settlements, woods, and closes present before the sixteenth century, soon becoming fully enclosed in the case of Writtle. Freehold and demesne lands at Writtle be-longing to the smaller manors seem to be fairly compact in area and possibly make up townships for many of the green settlements, but this is not explicitly stated. Eversholt and Marden are similar in principle to Writtle. How other regions and counties compare with each type can be seen in the evidence pro-vided by the Gazetteer.

C. LAND-USE AND MANAGEMENT OF MEADOWS AND PASTURE WITHIN TOWNSHIPS

The previous plans of townships and manors illustrate the great variety of land-use viewed on a large scale. It is obvious that those places with a very high percentage of arable would often have difficulties with supply of sufficient hay and provision of grazing. It is appropriate, therefore, to look in more detail at how these difficulties were addressed within the main types already described. The focus will be mainly on the Central Region.

Meadow

Meadow was one of the most important resources, as evidenced by its high value in medieval and later valuations. The largest and most extensive meadows lay in the valleys of major rivers, but relatively small belts of col-luvial soil lying adjacent to brooks in the many townships located distant from rivers were used for hay and were also called 'meadows'. As already mentioned, they gained fertility from nutrients washed from neighbouring arable which encouraged the growth of luxuriant grass suitable for preparing winter fodder, after which they became available for grazing when the hay had been removed.

Townships of sufficient standing, but without adjacent meadowland, had detached meadows belonging to them from an early date. A charter of Waltham, Berkshire, of AD 940 describes meadow at Cookham on the Thames, five miles away (S 1477; Gelling 1976, 636–8). Cheselbourne, Dorset (SX 76 99), lying on chalk upland, has a charter of 869 referring to six strips of meadow that lay by the River Frome, identified to be in Puddletown parish six miles distant (Grundy 1934, 127–8; S 342). In the thirteenth century, the abbot of Thorney, Cambridgeshire, had twenty acres of meadow at Elton, Huntingdonshire, upstream in the Nene Valley, by exchange with the abbot of Ramsey, for which Elton received pasturing rights at Finsett Marsh in the fen. Abbots Ripton, cut off from river meadows, had the detached *Dorkemede* in Houghton near Huntingdon in 1252 (Hart and Lyons 1884, i: 257, 321). Croughton, Northamptonshire, had thirty-five acres of meadow in a block lying in the meadows of Aynho next to the River Cherwell, three miles away, as mapped on Aynho enclosure award of 1793 (Northamptonshire CRO, Map 2861). Croughton meadows, lying next to the Cherwell, were recorded in *c.*1260 (Magd. Aynho Charter 66), when the rents received were used to maintain an altar lamp in Aynho Hospital.

Most meadows were divided into strips and had intermixed ownership exactly like the arable land. Since they were not ploughed, there are no earthwork remains on the ground other than drainage channels, both natural and artificial, which divided separate meadows. There were various complicated ways of establishing where the strips lay at hay time so that each owner could mow his swathe. In Northamptonshire, often subdivisions, called 'hides', were marked with stones or a hole. At Wollaston, in 1788, meadowmen put boughs into the holes so that they could be found at hay time when the grass was long. The hides were then divided by treading a path between each strip for the farmers to mow their own piece (Hall 1977, 144). Stones were used for markers at Aynho (Hall 1995, 76–8, 184). Meadow divisions were also called 'hides' in Bedfordshire at Aspley Guise in 1745 (Fowler 1936, 29).

Some places had one or two meadows that were marked out in strips and let by lot, hence called 'lot meadows'. Each proprietor drew a marked piece of stick to identify his strip for the year. Meadows at Pavenham and Wymington, Bedfordshire, were assigned in this way (Fowler 1936, 29). At Warkworth, Northamptonshire, the mowing was accompanied by a supply of ale followed by a feast of cheese and more ale in the evening (Bridges 1791, i, 219). The lot meadows of Paulerspury, in the same county, were divided into eight parts, in 1727, using markers inscribed with a 'crown, rele, headlays cross, crowfoot, ladel, pitt, garrett, and hayhook' (Northamptonshire CRO, G1631 p. 92; see Hall 2013, 32, for more examples). At Haddenham, Buckinghamshire, pieces of dockweed were marked with signs such as crows' feet and hog troughs, and placed in a hat ready for drawing lots (Gomme 1890, 270).

Yarnton and Begbroke, Oxfordshire, had similar arrangements; meadow strips (marked on the tithe map of 1840) were still open in 1910 and assigned annually. This was done using thirteen balls marked with names of village families who had been resident in the thirteenth century, so establishing the medieval origin of the procedure. The smallest unit was a quarter of a 'ball', meaning a quarter of a strip split lengthways. Sometimes the quarters were known as 'yards' (Gretton 1912). The strips were marked by stones or posts at each end and after they had been allotted were marked out by treading from one end to the other (Gretton 1910). The division process was described by depositions made in 1936, and photographs of the balls and of the farmers 'drawing' balls in 1947 are given by Gill Hey (2004, 219–23; VCH, *Oxfordshire*, xii (1990): 478–82).

The thirteen balls comprised nine for Yarnton and four for Begbroke—the meadows of both villages were intermixed. It is likely that they correspond to the 10 hides at Yarnton and 4¼ hides at Begbrooke recorded in 1086 (Williams and Martin 2003, 426, 428, 443), the slight shortfall possibly being explained by some demesne meadow not accounted as lot meadow. One ball represented the meadow belonging to four yardlands in later centuries, with the obvious similarity to hides and virgates. Aston and Cote, a township in Bampton, Oxfordshire, had sixty-four yardlands in 1657 and each had lot-meadow. The meadow was divided into sixty-four parts and assigned by lottery using wooden markers that were drawn from a hat. Money raised by selling the hay crop of a few hams (plots of pasture land, also known as 'nooks') was used to buy ale for 'merry-meeting' (Fisher 1911, ii: 344–5). At Wootton Underwood, Buckinghamshire, a field book of 1657 provides details of marking out the lot meadow making use of meerstones 'of old time' (Eland 1946, 145). The meadows of Brailes (1661) and Binton (1714), Warwickshire, were changed by lot annually (Barratt 1955–71, i, p. lvii).

Meadows (called 'brookland') at Kingston by Lewes, Sussex, were divided by treading marker lines in the grass before mowing. There was a piece of meadow called a 'drinker' that was held in turn by groups of four yardlanders who paid for a drinking party after the division was complete (c.1580). A ten-year cycle for holding the drinker ensured that all the forty yardlanders paid in turn. A similar 'drinker' meadow was used at nearby Southease until it was enclosed in 1842. The meadows there were divided into hides and annually allotted by marking sixteen sticks with notches and other characters such as 'doter, dung hook, cross, drinker'. The sticks were drawn at random and placed in the hides in sequence. The drinker hide was let to the highest bidder and half the rent was spent to provide the tenants with a dinner and the other half used to buy drink for the labourers (Figg 1851, 305–8). Some of the Southease hides belonged to Heighton and Telscombe, which lie on the downs and had no meadow of their own. This arrangement was also recorded in 1623 (Figg 1850, 250), again demonstrating the use of detached resources. Treading was used to mark out lands in the sheep pastures of Norfolk in 1516, when twenty acres were set out at Syderstone, and lands were marked out at Tittleshall (Moreton and Rutledge 1997, 113, 119). Meadows ('ings') at Walkeringham, Nottinghamshire, were set out with a 'wand' of two yards and a foot in 1777 because there were no boundary marks between them (Nichols 1987, vii: 144).

A third type of meadow tenure was that called 'parting meadows', where a few meadow strips were shared between two holdings. In any one year one holder cropped half of the strips and in the next year he cropped the other half, so alternating annually with his partner. An early example is recorded at Pilsgate in Barnack, Northamptonshire, in the late thirteenth century (Briston and Halliday 2008, item 144). The practice operated in other Northamptonshire villages in later centuries; Whiston glebe terrier of 1633 refers to a partible rood of meadow (Northamptonshire CRO, X603). At Aynho in 1744 it was clearly spelt out that the 'doles...[of]...meadow ground...[belonging to two holders] change round and alter their possession...alternately every year' (Northamptonshire CRO, C(A) 572). Church Langton, Leicestershire, had meadow pieces shifting every two years (Hill 1867, 134–5).

Meadows in southern England were managed by a network of channels and were called 'water meadows'. In the Frome Valley, Dorset, feed channels were made along the brow of slight ridges during the seventeenth and eighteen centuries (Whitehead 1968). Similar systems were used in the Stirchel Valley (Ross 1995). Damerham glebelands, Wiltshire, had six acres of water meadow in 1680 (Hobbs 2003, 130). At Wylye, Wiltshire, in 1632, tenants agreed to dig trenches, make water meadows, and maintain timber works and flood hatches (Kerridge 1953, 138–40). A plan of the water meadows of Norton Bavant is illustrated by Attwood (1963, pl. 1). The importance of the meadows was to take advantage of water that had drained through chalk downland and was warmer than freshly fallen rainwater in March and April, so promoting early growth of grass. Sheep and cattle could feed early and be in good time for Christmas markets. The technique has a long history. At Otterbourne, Winchester, there was a dispute in 1260 about a stream that had been dammed up for a fishpond by the bishop's men. It was resolved by putting a perforated board one foot wide in the pond so that meadows could be irrigated if there was a summer drought (Magd. charters Hants, Skyres 55).

Pasture in townships with little or no meadow land

In townships lying away from major rivers that did not have the benefit of detached resources, meadows were often limited to long narrow belts alongside a brook. Braybrooke, Northamptonshire, had only slade pasture of this type. It was called 'meadow' in 1767, when it amounted to a stated 370 acres, which accounts for the whole of the area of the slades (Northamptonshire CRO, ZA

716). At Kilsby in the same county, narrow pieces of colluviated land lying in slades were called 'meadows' in 1348, for example *Smythmedewe, Ryelandes Medew*, and *Southbrekmedewe* (Oxford, Queen's Coll., MS 366, fol. 58). Earlier still, the slades of Astcote in Pattishall had doles of meadow named *Le Holmes, Sheteredonemede, Toggelake*, and *Le More* in the thirteenth century (Ross, 1964, ii: no. 660).

These belts were difficult to manage when nearby arable was fallow and animals were likely to damage the hay crop. At Littlethorpe, Leicestershire, in 1397, pasture was held in severalty when crops were sown on the adjacent arable, but when it was fallow, the pasture was in common (Farnham 1930–3, ii: 102). An earlier example can be found at Toft, Lincolnshire, where in 1258 the manor had ten acres of demesne meadow on the west side of the town and eight acres on the east side, each mown alternately in the year when the field was sown, being common pasture in the other year (*Calendar of Inquisitions Miscellaneous*, i: 1219–1307 (1916), no. 242). Similarly at Geddington, Northamptonshire, in 1461, small pieces of demesne meadow were not let out to tenants for hay when the adjacent field was fallow, but left as an addition to the fallow grazing (Northamptonshire CRO, X351A, Buccleuch10/30, rental 1 Edw. 4).

A solution was to fence the meadows off and this was done at Syresham, Northamptonshire. An estate map of 1764 marks a hedge along the edge of the narrow belt of meadows by the southern brook (Magd. MP/1/54). It was enclosed and held in severalty alongside the open fields—an arrangement of long standing. Yardlands of arable and yardlands of meadow were always treated separately at Syresham, as in a survey of Brackley Hospital estate in 1424 (Magd. 36/1) and a copyhold transaction of 1375 (Magd. 35/54). The earliest reference at Syresham is in *c*.1210 when a yardland of meadow was sold for five shillings (Magd. Syresham Charter 39). In 1258 the abbots of Leicester and Biddlesden (Buckinghamshire) agreed with the freeholders of Syresham that meadows should be divided and each proprietor should have his own piece (Magd. Syresham Charter 25).

Pasture and leys in the Central Region

The general solution to the problem of pasture shortage in the open-field townships of the Central Region was by provision of rough grazing on stubble that was left as fallow, a practice that was often unsatisfactory. It could be assisted by the temporary reversion of arable to pasture. In many townships outside of the Central Region this was a standard practice known as 'convertible husbandry' and will be fully described in Chapter 3, E. Within the Central Region this variable land-use was not practised, but there were several procedures used to convert permanently small, and sometimes large, amounts of arable into pasture. These are illustrated by a few examples.

One way to increase grassland was to form cow pastures by the permanent conversion of arable strips to grass, called 'leys'. At Oxhill, Warwickshire, in 1595 it was noted 'there is a parcel of leys used in comen for a beste pasture where fewe or none of the parishe knoweth there owne leyes' (Barratt 1955–71, ii: 27). In Northamptonshire there are many examples recorded over a wide date-range (Hall 1995, 21–2). After the Black Death of the mid fourteenth century there was arable land to spare, and there were already many neglected and abandoned arable lands immediately before the pestilence, as shown by the returns of the taxation of 1340. Much land had been thrown out of cultivation; in 21 Cambridgeshire parishes the total left as 'leys and uncultivated' was 4,530 acres. At Gamlingay, 430 acres were left 'by the poverty of the tenants and there had once been 200 sheep from "foreign parts" folded' (Royal Commission 1807, 201–15). At Bedford, the northern high ground of Brickhill near Clapham, called *Wilshamdowne* in 1507, had a hundred acres of leys lying in several furlongs (Henman 1947).

East Midland cow pastures were usually leys converted from arable, but not always specified to be so. The creation of cow pastures was frequent during the sixteenth to eighteenth centuries and the results of agreements can be seen on estate maps or is identifiable as 'cow pasture' in parliamentary enclosure awards. Details of the 1712 cow pasture agreement for Ashby St Ledgers, Northamptonshire, have been published (Hall 1995, 21). The practice was widespread in that county. Morton (1712, 14) noted the existence of these pastures, in the outskirts of the fields, most formerly ploughed. Leys located next to meadows at Weston and Weedon, Northamptonshire, were marked as 'lottable' in 1593 (All Souls College, Oxford, Hovenden Portfolio 1, Map 9). This was presumably so that a crop of hay could be taken before the pasture became common.

Smaller parcels of pasture were introduced into the open-fields by several methods from the fourteenth century (Hall 1995, 22–3). Parts of furlongs, and sometimes whole furlongs, were left to grass over permanently as leys. The word used for leys in regions where convertible husbandry was or had been practised, was *friscus* in medieval Latin, from classical Latin *frio*, 'crumble', but the word is often misleadingly translated in the literature. In the Middle Ages, many places outside of the Central Region had arable left for several years to grass over, which was broken or crumbled up when it was due for cultivation again. Eventually the term was applied to describe lands permanently left to grass and lost its connection with arable, coming to mean 'pasture' that had been created from arable strips. An inquisition at Aynho, Northamptonshire, in 1367, stated that the demesne had 300 acres 'frisce et non culture', which shows that leys were not part of a convertible husbandry system (*Calendar of Inquisitions Post Mortem 39–41 Edward 3*, 1938, 135; extent in Northamptonshire CRO, C(A) 1205). An earlier example is found at Keasbeck in Harwood Dale, Yorkshire (NR), in 1231, where closes, meadow, and *frussum* were let out for hay, suggesting that the word *frussum* already meant pasture rather than arable (Atkinson 1878–81, 218).

Leys appear frequently in records of the fourteenth century. For instance, the 1340 returns of the 'ninth' tax for central Bedfordshire state for Segenhoe there were 200 acres of ley and land was for the most part sandy and would not grow corn. At Houghton Conquest 100 acres of ley belonged to several men (Royal Commission 1807, 14–15). Cambridgeshire had similar areas of leys that have already been mentioned. In Northamptonshire, leys occur at similar dates: Broughton in 1316 and Draughton in 1375. In 1477, Preston Capes had 20 per cent of its arable strips put to ley (Hall 1995, 22–3). Tenants at Bromfield, Shropshire, were ordered to cultivate leys in 1383; ley was recorded at Ercall Magna in 1479 and the manor of Oak had permanently enclosed ley in 1407 (Kettle 1979, 84). Grass headlands are referred to in the late fourteenth century at Long Itchington and Lower Shuckburgh, Warwickshire, and leys were mentioned at Alveston and Bishops Tachbrook in the same county during the fifteenth century (Dyer 1981, 11–12).

A 373-acre estate in Wardley, Rutland, had 263 acres of arable, 79 of leys, and 31 of meadow, hence leys amounted to 23 per cent of what had once been arable (Northamptonshire CRO, Bru. H.xi.30, dated 1600). Glaston, in the same county, had 42 per cent leys amongst its arable in 1635 (Glebe at Northamptonshire CRO). Deddington, Duns Tew, and Steeple Aston, Oxfordshire, had converted some clayland arable to leys in the sixteenth to eighteenth centuries. At Duns Tew it amounted to 25 per cent of the former arable (VCH, *Oxfordshire*, xi (1983): 31, 100, 218). The extent of leys for Deddington and its members in 1728 was Deddington 17–25 per cent (with the demesne 38 per cent), Clifton 27–33 per cent, and Hempton 22–27 per cent (Northamptonshire CRO, C(A) 3309, pp. 1–79). Land-use for Ashby St Ledgers in 1715 (Hall 1995, fig. 2) shows the dates when leys were first recorded from the fifteenth to eighteenth centuries, where they contributed significant addition of pasture to the limited amount of meadow.

Leys differed from cow pasture in that they were 'known land,' being part of a yardland and recorded in terriers. They were presumably cropped by individual owners for hay before being thrown open for common grazing. There is no evidence that leys in the Midlands were part of a convertible husbandry regime, that is, ploughed for a few years and then left to grass over again for several years, as in the brecklands of Norfolk, Nottinghamshire, Staffordshire, and elsewhere. Analyses of Northamptonshire glebe terriers for many townships show that the number and positions of leys remained constant. Leys in Cambridgeshire were also left as permanent pasture and no evidence was found that they were used in a convertible husbandry system (Postgate 1973, 305; Spufford 2000, 75). Yelling thought leys were permanent pasture in eastern Worcestershire (quoted by Roberts 1973, 203).

Leys occur in areas where the ground was less suitable for arable cultivation and especially in townships where there was little meadow. In Northamptonshire, the proportion of leys to arable was high in the west of the county, frequently over 40 per cent, where the soil was predominantly clay-based and there were limited amounts of meadow alongside brooks. On the other hand, townships along the Nene Valley, with abundant meadow, had leys among the arable ranging from only 5 to 10 per cent (Hall 1995, 25). A similar county distribution was found for Cambridgeshire. Grass in the form of leys was introduced amongst the intensively cultivated arable clays in the west. The glebe of Toft and Comberton in 1638 contained 2–3 per cent pasture (Spufford 2000, 75). The remainder of the county had pastures near the fens or on the southern ridge. Chippenham and Snailwell leys were located on the edge of heath, showing there had been an abandonment of arable on the poorest ground. In Leicestershire, grass was similarly introduced into fields: Denton had a grass headland in 1402, and leys were recorded at Allington in 1468 and at Great Bowden in 1507 (Thompson 1933, 212, 107, 185). Arnesby, in *c.*1550, had 42 per cent leys and pasture on a sixty-acre holding, dispersed in three fields (Farnham 30–33, i: 41–2). An agreement was made at Wigston Magna in 1679 to increase the amount of greensward (Hoskins 1937, 179). Ley ground was in the range 11–21 per cent of the arable for holdings at Lutterworth in 1607 (Gray 1915, 445). Hoskins discussed leys in occurring in open fields of Leicestershire (1950, 138–54)

Balks, green furrows, and rick-places

Smaller amounts of pasture were introduced into open field arable in the form of balks, green furrows, grass ends, and rick places, terms which in this section will be explained in detail with illustrative examples drawn from the written sources (Hall 1995, 23–38).

Narrow strips of grass, sometimes used as access routes, were called 'balks'; they are recorded from the thirteenth century and later. An early reference to a balk (1254) occurs at Sharnbrook, Bedfordshire, where the 'rowbalk' divided the Templars' land from that of the lord (Fowler 1919, 157). At Cuxham, Oxfordshire, Paul Harvey found there was a balk between the lands of a manorial tenant and the demesne in 1303 (Harvey 1965, 23). No early agreement or order for the formation of balks has been discovered, but they were often used to mark off significant pieces of land, especially groups of several lands in a furlong that formed part of a demesne. For example, a balk (*divisa*) was used to describe a boundary marking seigneurial land at Croglin, Cumberland, in *c.*1180 (Prescott 1897, 251), and the demesnes of Ashby St Ledgers and Brockhall, Northamptonshire, were marked in this way in 1418. From the fifteenth century balks are frequently mentioned as reference points in terriers, and were usually said to be 'common'. They are marked on most open-field maps from the 1580s onwards (Plate 5; Hall 1995, 23–6). At Little Oakley, Northamptonshire, the purpose of balks is spelled out, confirming the above observations. A series of field regulations for 1554 describes the normal range of open-field items, and orders the tenants to make balks to mark off the demesne and the parsonage lands (Northamptonshire CRO, Buccleuch 16/98 in Box X379).

Balks were formed by ploughing a narrow strip away from the ridge of a particular land, leaving it to grass over. It is surprising to see them considered as a significant source of pasture since they occurred only infrequently and were very narrow. Nevertheless, manorial court orders from several counties refer to cattle and horses being tied on balks or mention that they were mowed for hay. Seaton Delaval, Northumberland, had orders for 1559–79 stating that horses or mares were to be tethered only on the tenants' riggs; and that grass growing upon a balk was to be cut for hay; Hartley beasts were not to be pastured on balks in the sixteenth century (Craster 1909, 194–6). On the other hand, at Cambridge horses were pastured on balks in 1583. In Stratton, Dorset, which remained open until 1900, horses were still tethered for grazing on the 'lanchets' (Kerr 1968).

Grass balks were probably of considerable antiquity outside the Central Region, where strips were ploughed flat and the balks were used for demarcation. In Cumberland, balks were called 'ranes' or 'reanes', such as at Holme Cultram (Graham 1908, 342–3; 1913, 1–31) and Warwick-on-Eden (Beecham 1956, 41); likewise in various parts of Lancashire and Northumberland (at Hartley). Some riggs in Northumberland were separated by balks left as grass by the seventeenth century at Denwick and Acklington (Butlin 1973, 131–2). In 1379 the division (balk) between the fields of Upper and Lower Wardon was called a 'reyne' (Raine 1864, 26). In East Anglia, Hapton Hole, Norfolk, had three strips called 'londs' separated by grass balks about a yard wide, still surviving in 1955 (Beecham 1956, 43). Four flat-ploughed lands still used at Westhorpe, Suffolk, are separated by a grass balk (illustrated by Martin and Satchell 2008, 29). These are rare survivals. Balks were used at Martham, Norfolk, in the late Middle Ages and in 1506 all tenants were ordered to leave one furrow width unploughed to act as boundaries between each tenant and fee (Campbell 1981–3, 16). A balk 'in East Anglia' in the process of being destroyed is illustrated by Venn (1923, pl. xi). At Clothall, Hertfordshire, flat-ploughed lands were separated by upstanding grass balks (illustrated in Cunningham 1910a, frontispiece). At Westcote in Gloucestershire there were flat ploughed strips with grass balks illustrated in a photograph published by Finberg (1975, 40). In the south-west of the country in Devon, balks between flat-ploughed strips at Braunton and Brixham were called 'landsherds'. At Portland, Dorset, strips (called 'lawns') were separated by grass balks called 'lawnsherds'. Balks at Forrabury, Cornwall, are shown in Plate 15. It can be seen that balks of varying type occur in all parts of the country outside of the Central Region.

In the Central Region by the seventeenth and eighteenth centuries the amount of grass in some townships with extensive open arable fields was further increased by introducing balks in all the furrows between the lands—leaving unploughed a strip roughly proportional to the width of the land (one or two feet). These are often referred to as 'greensward' or 'green furrows' and would not have been common but held with a particular land. When the adjacent lands were under crop the green furrows were used first for a crop of hay, and afterwards for grazing, the animals being tethered, as with balks in the other regions. When the field was fallow, the good grass base of a balk or green furrow would improve the overall quality of the grazing. Manorial court orders in Northamptonshire record the creation of greensward in every furrow at Badby 1592, Byfield 1660, East Haddon north fields 1664, and Great Billing 1684 (Hall 1995, 26). At Kirtlington, Oxfordshire, the manor court for 1590 ordered balks to be made between strips (Griffiths 1980, 260), which probably means greensward. The creation of greensward was an open-field improvement made as late as 1730 at Twywell, Northamptonshire. Green furrows 1½ feet wide throughout one great field and 1 foot in the other two fields were made, along with 5 yards of greensward left against every headland and furlong boundary (Northamptonshire CRO, Buccleuch 24/6 in Box X8670).

A photograph of Crimscote, Warwickshire, illustrating green furrows was published by Venn in 1933 (a similar one was reproduced by the Orwins in 1938, pl. 11). Green furrows have been observed by fieldwork at North Littleton, Worcestershire (SP 083 478) (2004), and Shenton, Leicestershire (SK 395 001) (2005). Similar introduction of grass occurred in Buckinghamshire; in 1722, at Sherington, the commons were improved by the addition of grass sward by several methods (Chibnall 1965, 286).

The creation and use of green furrows was widespread and a method of making productive the otherwise rather barren subsoil of the furrow; excess water draining from the ridge helped the grass growth. At Orston, Nottinghamshire, a court of 1753 ordered that grass in green furrows should be mowed before the first of August. This would serve the purpose of removing hay and opening up spaces between the ridges to access light and air to ripen and dry the corn crops. Furrows still had to be drained even though covered with grass; in the same vill, in 1641, a 'grip' was ploughed on the sun side of every green furrow (Barnes 1997, 128).

There have been rather pointless controversies and confusion between balks and green furrows in the past. Seebohm (1883, 2–3) referring to the strip fields of Hitchin, 1816, said that throughout the country strips were separated from each other by green balks of unploughed turf. This, however, was not the original arrangement in most parts of the country, being a late modification introduced mainly during the seventeenth and eighteenth centuries in some townships within the Central Region. Many places never had them according to the surviving ground evidence. In the case of flat-ploughed lands in the Western Region, grass balks were probably always used for demarcation between every strip. Beecham reviewed aspects of balks in 1956, and his county appendix provides many useful examples. Some are balks and others are greensward, but called balks; the following counties are represented: Bedfordshire, Buckinghamshire, Dorset, Gloucestershire, Hampshire, Nottinghamshire, Rutland, and Wiltshire (Beecham 1956).

Another method of introducing grass that first appeared in the sixteenth century was by forshortening lengths of arable lands against a headland or joint (the two types of furlong boundary; see Plate 2 and Aston et al 1989, 184–5 for excellent aerial photographs by David Wilson). Thus instead of ploughing the full length of a land, several yards were left at the end, which would rapidly grass over. Many examples can be found in Northamptonshire court orders such as at Hargrave 1580, Badby 1592, and Maidford 1696. At Maidford eight yards were to be left at each headland and twelve yards at every joint. After a few years a series of 'heads,' or soil heaps, was formed at the new turning places (illustrated in Hall 2001, 30, 49). The cases of Sherington 1722 and Twywell in 1730 have been noted. Grassing down was made by common agreement and enforced by the manorial or village court. In cases where all the headlands and joints were treated in this manner the net result was that a belt of grass surrounded every furlong. The draft enclosure map of Little Addington, Northamptonshire, 1830, shows grass pasture on all the ridgeways between furlongs (illustrated in Hall 1995, pl. 4).

The early occurrence of grass ends, called 'heads' *(chevesca)*, was noted at Somersby, Lincolnshire, *c.*1200 (Major, 1950, 138).They were referred to at Willoughby, Warwickshire, in *c.*1243 (Magd. Charter Willoughby 75), and fully detailed at Chacombe, Northamptonshire, in a terrier of monastic possessions made in *c.*1350 (TNA, E315/358). Grass heads were sometimes made on furlongs lying next to meadows, as at Biddenham, Bedfordshire, in 1506 (Crook 1947), which, in effect, increased the size of the pasture. Biddenham furlongs and meadows are mapped on an open-field plan of 1794 (Bedfordshire CRO, ME 51).

Small grass areas called 'rick-places' occurred in some open fields. They were used as platforms for stacks or ricks, and were formed at the end of a group of lands by flattening the ridges. Several are marked on a map of Brixworth, Northamptonshire, dated 1688, and until 1978 one survived in earthwork form (illustrated in Hall 1982, 40). The original course of the furrows was just discernible within the square. At the edge away from the furlong boundary the forshortened lands had developed new heads next to the rick-place. Rick-places are referred to in other Northamptonshire townships (Hall 1995, 26–7): Clipston (1767), Crick (1632), Walgrave (1776), Ashton by Oundle (1678), and Wollaston (1633). The only surviving example currently known is at Toft, Cambridgeshire (TL 3610 5583), discovered by Susan Oosthuizen. A Brixworth by-law of 1577 states that 'none having a rick in the field shall put any sheep or hog fed at the said rick out into the field' (Northamptonshire CRO, Th 802). A court roll of 1578 from Luddington, Northamptonshire, explains rick-place usage, which was to feed hogs or sheep from the stacks of peas or beans (when they were on the fallow field: Northamptonshire CRO, Buccleuch 25/51 no. 262 in X889). Rick places in Scaldwell, Northamptonshire, in 1577 were to be opened up for common usage as soon as possible after corn or 'wood' (probably hurdles) were removed (Northamptonshire CRO, IL 123 m.20). The Ashton by Oundle rick-place had a fence and hedge around it (Northamptonshire CRO, Smith of Oundle 515/10).

In the Middle Ages most of the townships in the Eastern and Western Regions had ample pasture, so that the complicated creation of scraps of pasture within the open-field strips was unnecessary and did not commonly occur. Those that used one-field or irregular systems often had enclosed arable closes as part of an otherwise open holding. Examples are found at Winnersh and Sonning, Berkshire, in *c.*1600; at Egham, Surrey, in 1336, where 1½ acres of arable were enclosed at *Elderudene* (Torns 1954); and at Gillingham, Dorset, 1608, as well as at places in Wiltshire such as Bishopstone. These closes were probably used for winter corn and held in severalty.

D. GREENS AND DROVES IN THE EASTERN AND WESTERN REGIONS

Outside of the Central Region many places had no access to riverine meadow or suitable peripheral colluvium belts and very different systems of land-use operated. They fall into two main groups, one based on large greens, the other type having extensive pastures on 'waste' grounds.

The claylands of East Anglia, although having some areas of ill-drained moors, were generally devoid of good quality meadowland. Hay sometimes had to come from pastures left unploughed and not used as part of the regular arable land. If most of the land in a township was of similar quality, it did not much matter where the land allotted as pasture was to be located. It was therefore often placed next to settlements for convenience. Arable land, especially if not very good quality, does not produce the same quantity of fodder as rich alluviated meadows and therefore commons needed to be fairly large or numerous. Land in many parts of the country that is, or was once, relatively marginal such as clayland or heath, was characterized by the presence of large, often oval-shaped or triangular commons linked by lanes or wide droves. The commons attracted dispersed settlement around their edges and in many cases were probably intended to have settlement disposed in such a manner from the time of the first colonization.

Large commons (often called 'greens') can be seen in Norfolk, many with dispersed medieval settlements lying around them. These are mapped on

Faden's map made in 1797 and often, with linking droves, run for several miles (1797 map reproduced by Norfolk Record Society (1975); see the complex around Attleborough in Macnair and Williamson 2010, 106). Chris Barringer (in Wade-Martins 1994, 80–1) plotted the data from Faden's map. In 1796 there were 200,000 acres of unimproved common, marsh, warrens, or sheep-walks in Norfolk. Peter Wade-Martins (1980b, 19–20) published the plans of the large greens of East Bilney and Brisley. The precise locations of the commons vary. Some were in low wet places alongside minor streams, and therefore are not in principle much different from the colluviated slades of the Midlands and elsewhere. Others were 'high commons' on areas of poorly drained land or infertile sand. A third type of common was located on land of quality indistinguishable from the neighbouring arable. They were characteristically curvilinear or convex in shape narrowing into droves (Williamson 1993, 167–9; Williamson 2003, 91–104 and 160–79 for a wider East Anglian discussion).

Wade-Martins' (1980b) archaeological survey work shows how complicated the history of green settlement can be. It was found that Middle Saxon Period settlements (as revealed by pottery evidence) lay around what are now isolated Norfolk churches, but had been drawn away by the twelfth century to adjacent greens. The attraction was presumably the convenience of the pasture lying next to a farm and the opportunity to erect squatter dwellings. The commons to a certain extent characterized the later settlement pattern since a township may have had several commons, each of which attracted medieval settlement around it, even if it had not been the initial intention to locate dwellings around every green.

Suffolk greens and commons, often interconnecting, similarly existed on clay-lands, and have been discussed by Peter Warner (1987, 1–17). Many greens are recorded on Hodkinson's Map of 1783 (facsimile reproduction by Suffolk Record Society, XV, 1972). They are associated with much greenside settlement of probable medieval origin. The interlinked greens and broad lanes are well illustrated by plans of Fressingfield, and for the Wrentham and Reydon areas with marshes, heaths, greens, and commons (Warner 1987, figs. 4 and 5). Few of the green settlements were named in the Domesday Survey of 1086 but were probably assessed with nearby major centres sited along valley sides. Some churches are located, however, in or near to the greens where the secondary settlements became the most important, as at Linstead Parva, Stoven, and Westhall (ibid. fig. 6). The clayland soils were of mixed permeability and were attractive to farmers. There were also ill-drained moors, some shared with neighbouring parishes, such as Southerton Moor, which was over 200 acres (ibid. fig. 2). Hinton may be a primary settlement based on a demesne with triangular greens and tenements outside them (ibid. fig. 9). Some of the greens were over 1,000 acres and occur predominantly on the clayland plateau of the centre and west. From pottery evidence most of the settlement around them dates from the twelfth century and later. The stinting of each common belonged to the bordering settlement (Dymond and Martin 1999, 36–7).

Mark Bailey (2007, 70–1) discussed the increased amount of settlement around greens during the thirteenth century at Capel St Andrew, Woolpit, Hopton, Ickworth, and elsewhere in Suffolk. Greens, commons, and tyes (the name for small greens in Suffolk and Essex) have been classified by Edward Martin (in Martin and Satchell 2008, 15–17; Martin, in Christie and Stamper 2011, 238). Archaeological evidence at South Elmham and Bardwell shows that settlement around the greens developed in the twelfth century and that therefore the greens must have been established by that date. Oosthuizen (1993) has studied Cambridgeshire greens at Knapwell, Whaddon, and Wilburton. From the disposition of eleventhth-century churches inserted within the greens it seems possible that these greens have a pre-Conquest origin.

The extents of greens, together with droves, commons, and meadows, for various East Anglian places have been reconstructed from historical sources, accompanied with large scale plans, by Martin and Satchell. In all cases the greens and droves are an important and integral part of the field systems. Large greens occurred at Scole, Norfolk, and in Suffolk at St Michael, South Elmham, Worlingworth, and Walsham le Willows (Martin and Satchell 2008, 105, 96, 175). The overall distribution of East Anglian greens shows they are more numerous in Suffolk and Essex (Martin and Satchell 2008, 202). At Writtle in Essex there are many small settlements associated with their own greens, and where commons were very large, there was intercommoning by many adjacent villages, as occurred at Tiptree Heath near Witham (Britnell 1983, 53). John Hunter (1999, 99–104) has discussed green settlement; many are marked on Chapman and Andre's map of 1777.

Greens were not confined to East Anglia. A particularly fine example of large triangular and linear greens connected by sinuous droves into which they funnel, occurs at Semley near Shaftesbury, Wiltshire (ST 88 26; 2,945 acres). The network covers the entire parish and has seven major greens (Figure 1.4), which was mapped on the Tithe Map of *c*.1839 and by Denman in 1967 (Denman et al. 1967, map 19); it survives almost intact. In 1922 the 300 acres of the common were used to feed 245 cattle. Settlement is very dispersed, the church being located in the centre of Church Green, with more settlement at the large Gutch Common on high ground to the south. Semley was given to Wilton Abbey in 955 (S 582) as part of an estate of small dwellings. In the Middle Ages, settlements were located in seven different parts of the parish. In 1599 the lands were held by three yardlanders and fourteen half-yardlanders, but it was mainly enclosed by this date. The stint of sheep was forty per yardland (Freeman 1987, 70–5). Semley lies relatively low on Kimmeridge Clay except at the south, where there is Gault and Greensand. In spite of being assessed in yardlands, the field pattern in the areas between the greens is very irregular and looks like the result of assarting. Not all the greens occupy low wet ground; at

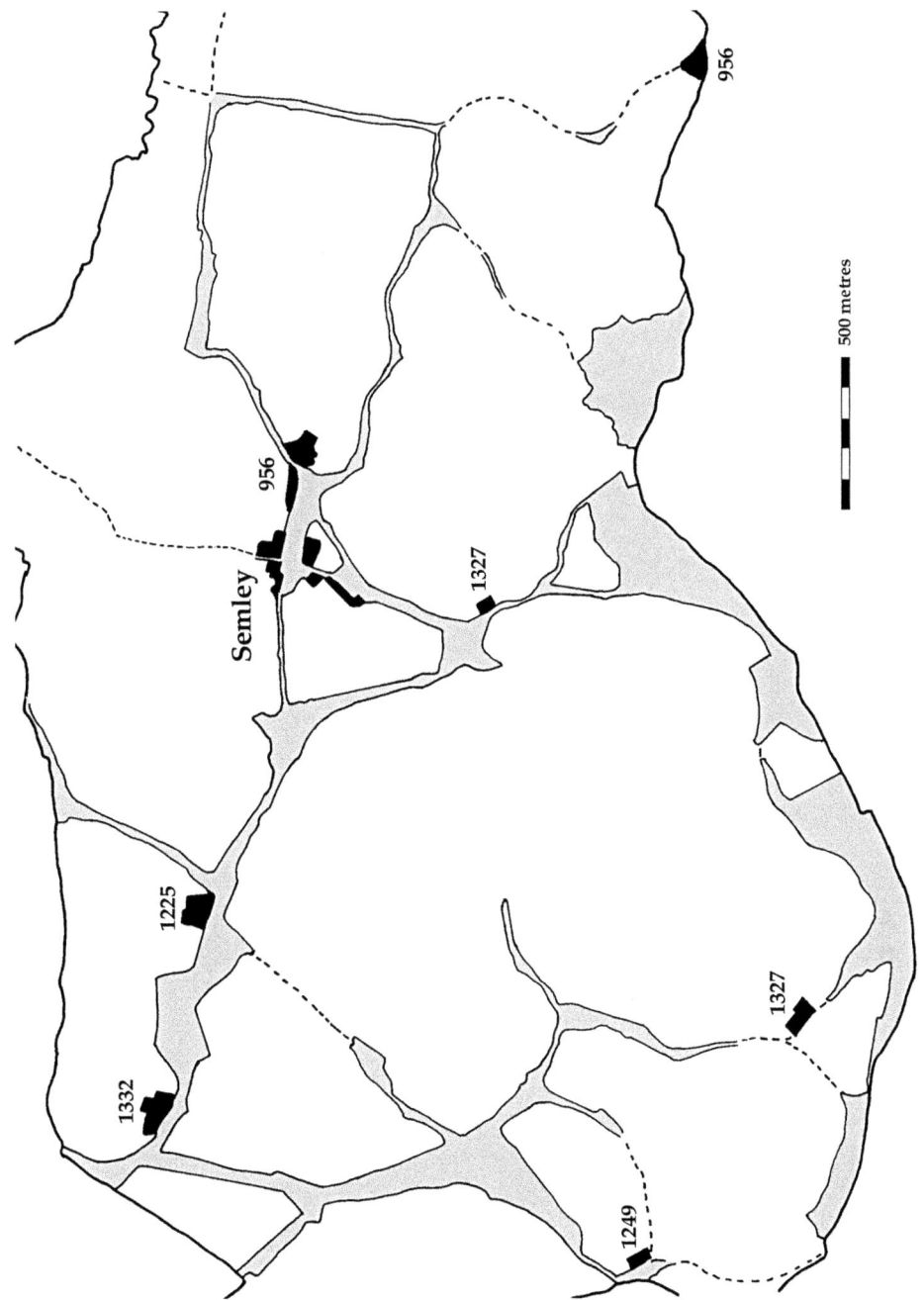

Fig. 1.4. Semley, Wiltshire (ST 88 26), showing linear commons (after Denman et al. 1967, map 19; settlement dates from Gover et al. 1939, 209–10).

the north-west, a very wide linear green survives (2009) and can be seen to traverse undulating ground. The overall layout is very similar to some of the East Anglian greens.

Middlesex had dispersed settlement with a multiplicity of roads and greens. For instance at Hendon a map of 1754 shows many lanes at the north making a cell-like pattern with dispersed settlement lying along them around Mill Hill, and at the south were Church End, Brent Street, and Golders Green (Tyack 1976, 4). A map of Kingsbury, 1597, shows a similar arrangement, with multiple fields (Tyack 1976, 52). Edmonton in *c*.1600 had woods at the west, settlements with several greens, and, again, multiple fields (Tyack 1976, 132). Tottenham in *c*.1619 had multiple settlements with some woodland and greens (Tyack 1976,310). A large common lay at the north of Finchley, in 1814, with settlement at the edge (T. F. T. Bolton 1980, 40).

Victor Skipp's studies of the Arden in the north-west of Warwickshire near Solihull showed there had been complex irregular fields associated with commons and greens and very dispersed settlement (mapped in Skipp 1970, facing p. 26; 1981, 168, 172). A 1562 terrier of Greete in Yardley described 470 acres of land held in severalty of which 85 per cent was pasture (Skipp 1970, 87). A similar pattern of irregular fields, commons wastes, and woods occurred at Tanworth (Roberts 1968, 111; Roberts and Wrathmell 2002, 168). Yates's map of Staffordshire, 1775, shows dispersed settlement, greens, and heaths in the south of the county around Walsall and Wolverhampton, for instance, Garrets Green, Lyndon Green, Wells Green, and Marston Green (Pawson 1979).

It would seem, therefore, that in areas of semi-marginal land, especially those with a clay soil base, and where there was no land particularly suitable for meadow, greens were set out to provide hay and pasture. This appears to be the case with some East Anglian greens that include low wet areas. Alternatively, other greens may have been areas left that had not been assarted into arable cultivation, as suggested for Semley. Inevitably they would attract settlement, partly because the edges were a convenient place to be, where hay could be gathered and grazing was available after hay time, and also because there would be little resistance to minor encroachments such as the establishment of a house and small paddock.

It can be seen from the above evidence that large greens forming part of the open-field structure in terms of grazing resources (as opposed to small greens in settlements) are mainly a feature of the Eastern and Western Regions. Some parts of the Central Region have similar areas with greens and irregular fields, as in the northern, boulder-clay covered parts of Buckinghamshire and Bedfordshire. They too have the characteristic dispersed settlement, as can be seen at Hanslope, Buckinghamshire, where can be found Tathall End 1227, Stocking Green 1254, Upper and Lower Balney 1281, Hungate End 1616, and Bullington End 1663 (Mawer and Stenton 1925, 6–7). A detailed example of the type of settlement in this sort of terrain has been published for Buckinghamshire at Shenley Brook

End (Paul Everson, in Ivens et al. 1995, 79), where paddocks and buildings spread 1,100 metres along various hollow ways, interspersed with ridge and furrow. The Thurleigh, Bedfordshire, enclosure map of 1805 shows a parish more than half enclosed (Bedfordshire Record Office, MA 47/2; Roberts and Wrathmell 2002, 88) with eleven named fields lying in eight separate blocks with many dispersed settlements and farmsteads. The occurrence of the name 'green' has been plotted nationally (Roberts and Wrathmell 2000, 38). Not all of them will represent the type of green discussed above. Nevertheless, it can be seen that they occur predominantly in the Eastern and Western Regions, with very few in the Central Region. The area with no, or very few, 'green' names corresponds closely with the distribution of extensive two- and three-field systems.

Wide droves were not always pasture resources, but sometimes served as access routes for large numbers of animals to get to large areas of grazing. This can be seen on the edge of the North York Moors at Goathland, where settlements are scattered along wide droves that lead to moorland pastures. A rental of 1599 names thirty-two separate farmsteads, which are mostly identifiable on a modern map (Spratt and Harrison 1989, 102). A similar pattern occurs in the contrasting level terrain of the Marshland in the Norfolk Fens, where nine major settlements have droves up to three miles in length leading to fenland summer pastures. Minor settlements dispersed along the droves were discovered by Robert Silvester (1988, fig. 124).

Extensive pastures

Hilly regions provided pastures, sometimes in very large amounts. Who had rights on the pastures was a matter of custom, and was often in dispute. The spacious commons of the Cumbrian Fells were used for cattle fattening from the seventeenth century and were over-stocked by graziers from outside. The lords and smaller landholders wished to prevent such activities by enclosure, since there was no control over the outsiders. Both parties gained at the expense of the outsiders, since they acquired severalty pastures, and tenure was changed from copyhold to freehold (Searle 1993).

At Wensleydale, Yorkshire, small areas of open fields survived in areas recorded in a detailed survey of 1605, amounting to 7.5 per cent of the good-quality land lying in the dale bottom around West Witton and Thoralby. On the dale sides were large stinted pastures, and the high ground above them had unstinted commons and wastes (Fieldhouse 1980, 171). Dartmoor in Devon had large areas of common around the edge that belonged to each adjacent village. The main core of the moor was commonable to the whole county (mapped by Denman et al. 1967). Sherwood Forest in Nottinghamshire was open to the nearby villages for making brecks (Fowkes 1977, 56–7). On the hundredal scale, a pasture in Colneis Hundred, Suffolk, was common to all men of the Hundred in 1086 (Williams and Martin 2003, 1226).

The chalk downs of southern England were used for sheepfolding by the lord and by the village of each township (described in Chapter 2, pp. 53–7). In these cases, folding was intimately linked with manuring common arable fields. Overall, meadows, as opposed to commons, were more abundant in the Central Region than in the East and West, partly because there were many large rivers with meadows in their flood plains.

'Commons'

The discussion began with meadows and has moved to pasture and commons. It is important to be clear about what is meant by 'common' in both original documents and in the published literature of the last century or so. There has been confusion between 'common right', often called 'commons' in primary sources, and physical areas called 'commons', on which there were access and grazing rights for specified people at specified times. The notion that all places had one or more physical commons to which almost everybody had the right to enter is quite untrue.

Common right was the right, agreed by the custom of a given township, for freeholders and tenants to put their animals into physical commons (if there were any) and in particular, to use meadows and stubble after hay and corn had been removed. The times that such lands were open was regulated. All major tenants with yardlands and oxgangs had such rights and the numbers of animals allowed ('stints') were specified. Smallholders with only a few acres of land had rights for smaller numbers of animals, as did the 'cottagers' or holders of cottage commons. Cottage commons belonged to a limited and fixed, unchanging number of dwellings, or dwelling sites, that had the right to graze a few animals, even if the holders of the right had no land. The rights did not extend to additional houses built over the years in medieval plots or on the waste—hence the frequent reference to the rights of 'ancient cottages' in the eighteenth century. These cottages and their commons tended to trade as real estate and were largely subsumed into larger holdings by purchase during later centuries. At Finedon, Northamptonshire (Northamptonshire CRO, ZA 3127 [1806]), there were 43 cottage commons; 10 people with no land had 9 commons, and the lords had acquired 8 cottage commons with their 524 acres of open field and 100 acres of old enclosure. The remainder had been acquired by yardlanders.

In contrast to many parts of the Eastern and Western Regions, the champion parts of the Central Region had very few physical commons, greens, or pastures that could be mapped as discrete areas. In Northamptonshire 'commons' other than those made out of land that had formerly been arable were rare, being mainly confined to greens adjacent to or lying within the royal forests (Hall 2013, 33 and Atlas). Many places had acute shortages of pasture and in the early modern period various methods of increasing it were undertaken, as has been described in this chapter.

2

Field-system types: extensive fields
of the Central Region

This chapter describes the characteristics of representative field systems found in the Central Region at the height of their development in or around the thirteenth century. The extensive fields of the region did not have exactly the same form throughout, but fall into three distinctive types, largely depending on their location.

First, new examples of the familiar Midland field systems are described. Evidence for changes from two to three fields shows that this transformation occurred over a long period and was not the result of fourteenth-century population pressure as suggested by Gray. There was further development of regular four-field and five-field systems in townships having less fertile soil.

A second type of field arrangement, known in the Yorkshire Wolds and Holderness, had extensive fields consisting of a few blocks of parallel strips occupying much of a township. Further examples from other parts of Yorkshire have been newly identified by archaeological survey and from documentary evidence.

A third field-system type was practised on the chalk downland of Southern England from the thirteenth century and probably earlier. The proportion of arable land in a given township was lower than in the other sub-regions, but its cultivation was highly regulated, using a system of sheepfolding, with daily movements of animals to downland pastures and their return in the evening.

A. THE MIDLANDS

The extensive fields of the undulating landscape of the Midland part of the Central Region are the well-known two- and three-field systems, otherwise known as the 'Midland type', after the work of Gray (1915). The fields were highly organized and are well documented because of their late survival. The main characteristics in terms of field numbers and overall operation will be explained below. Within Gray's central belt of the 'Midland type' system, there are significant variations from the norm in Yorkshire and in the southern chalklands. These will be explained separately. One of the distinguishing characteristics of the chalklands was the use of intensive sheepfolding, and this topic will be discussed again, when it will be contrasted with the form of sheepfolding used in East Anglia.

The internal structure of the fields describing the demesne, yardlands, tenurial arrangements within the furlongs, and the regulations by which the manorial court, or the village community, controlled the management of the fields, will be discussed after field system types found in all parts of the country have been examined.

Two and three regular fields

The two- and three-field system of the extensive arable or Midland type has received much study since Seebohm's 1883 publication, which used the example of Hitchin, Hertfordshire, and Gray's classic 1915 work, which studied the whole of England. Under this arrangement, townships with extensive arable had their furlongs grouped into two or three large, approximately equal-sized, blocks called 'fields' that related to cropping schemes. One of the fields was left bare fallow every two or three years in turn. There were common grazing rights on the fallow after harvest and on the meadows after hay was taken. Communal regulations controlled details of agricultural practices, animal stinting rights, and related matters.

The system has been described many times. Twywell, Northamptonshire, is an example much better than most of those cited in the literature, being a remarkable survival of a system mapped in 1736 that likely looked little different 500 years earlier. In 1735 it was still completely open, with no partial enclosure potentially causing changes to the field structure. The township had the same boundaries as 1735 as when it was described in 1013 (S 863; Hart 1966, 193–8). A single holding, called the 'yardland', consisted of many strips distributed fairly uniformly over the furlongs of the township, as shown for the glebe in 1735, which consisted of 61 strips containing 66 lands with an average size 0.77 rood, 12.75 acres in all (Plate 4). The other holdings at Twywell were similarly dispersed over all the furlongs. Land was rested to regain fertility by having a bare fallow every three years. The fallow was also used for rough grazing, since there was so little meadow and pasture. It was not possible for one farmer to have a land cropped and the adjacent farmer to have land fallow with animals ranging over it, for then the crop would be eaten. The potential problem was solved by having a communal arrangement whereby a group of furlongs, called a 'field', was left fallow at the same time. The herd of the whole village ranged over the stubble and fallow to graze, making it relatively easy for herdsmen to prevent animals wandering onto crops in the sown areas.

Elsewhere in the Central Region, cropping was done on a cycle of two or three years, depending on whether there had been a division into two or three great fields. Medieval evidence that fields were cultivated in this way is clear from many sources, especially manorial extents, surveys and valuations that make statements such as half or a third of the demesne was fallow ('waste') and lay in common with the tenants, when it was worth nothing (in terms of cropping). Twywell had three fields during the seventeenth and eighteenth centuries. Manorial court orders to control the management of the fields are given in Chapter 4 (pp. 128–9). A reconstructed medieval three-field system has also been posited for Brixworth, Northamptonshire, as it was in 1422, based on an outline map of 1688 filled out by fieldwork and using the evidence provided by very detailed terriers of ten yardlands made in 1422 (Hall 1995, fig. 1; reproduced in Williamson 2003, 64).

Midland-type fields are easily identified from contemporary maps, field books, surveys, or terriers that show evidence of two or three large compact fields. The characteristic feature is the more-or-less equal partition of the strips between the fields. However, field numbers and the number of cropping 'seasons' or 'tilths' sometimes need careful interpretation in their relation to geographic 'great fields'. In the Midlands, a terrier referring to pieces in furlongs lying in two or three fields is likely to mean that there was a two- or three-year rotation of crops with a bare fallow, especially if the pieces are equally distributed. However, for valid conclusions to be drawn, the information should relate to more than just a few acres—some grants only concern a few pieces that might lie in only two of three regular fields, or be part of an irregular system.

Provided a terrier is given in full or there is a clear statement of the type ('so many acres lie in one field and the same number of acres lie in the other'), the interpretation is likely to be sound. If, however, the estate is a demesne, then caution is required. The demesne may be in block form and have its own fields that have no relation to the fields of the tenants and freeholders, as at Mears Ashby, Northamptonshire. In the seventeenth century there were multiple grouped fields called the Town Fields, and a separate and independently cultivated demesne, called the Hall Field (Hall 1995, 177–80). There are many examples of demesnes outside the Central Region, like Writtle, Essex, that were cultivated in three seasons, but did not lie in a three-field system.

Gray (1915, 422–509) in his Appendices gave numerous examples of Midland two- and three-field systems, mostly of early date. As has been explained, some of the sources are descriptions of demesnes and the interpretation may not always be clear. Likewise the number of fields operating in townships that had been more than half enclosed may not bear much relation to an earlier, more open state. Other satisfactory examples are provided in the Gazetteer, which provides the evidence for the extent of the two- and three-field system, as mapped in Figure 0.1.

The contrast between the complicated fields in the Western Region and the regular fields of the Central Region is brought out by a detailed survey of the bishop of Worcester's estate in 1299 which describes the demesne lands. The Worcestershire estates are mostly blocks and pieces having a lengthy description; they may represent dispersed demesnes which it was felt necessary to locate accurately. For example, Hartlebury had a demesne of 246½ acres lying in 26 named pieces ranging in size from 2 to 29½ acres. Equally, the described pieces may have been grouped in a block. However, when the survey reaches outlying manors in the Midlands, simplicity is immediately apparent. So for Hampton Lucy with Hatton, near Stratford on Avon, Warwickshire, the demesne was 288½ acres and 'divided into two fields of which one is called the *overefelde* and the other is called the *netherefelde*, and each is worth in alternate years 6d [per acre]. The pasture is separate until mowed when it falls in common' (Hollings, 1934–50, 189, 263). Whether the demesne of Hampton was dispersed or not (it was probably dispersed), this description was felt to be adequate. It was self-evident where the demesne lay: either in a block next to the manor-house, cultivated in its own two fields, or uniformly dispersed at regular intervals throughout the two fields of the tenants. The other Warwickshire and Gloucestershire manors have similar brief descriptions.

Similarly, Senior's North Derbyshire surveys of the Earl of Devonshire's estates have limited references to small areas that may be great fields (details in the Gazetteer), or to furlongs forming part of one-field systems, but when he describes the Lincolnshire estates, great fields predominate and the lands are identified by their numbered position in each furlong.

Early examples of two-field arrangements can be found in several counties. Two twelfth-century examples in Lincolnshire are described below. Edworth, Bedfordshire, had two fields in *c*.1185 (Lees 1935, 223) and Bloxworth, Dorset, had a half-yardland of sixteen acres divided exactly between the East and West Fields in *c*.1190 (Historical Manuscripts Commission 1911, 31). Gloucestershire had many early examples of two-field systems, such as Rodmarton, which in about 1200 had twenty-four parcels in each of the North and South Fields (Hilton 1966, 116, from Ross 1964, 344–5). Kempsey, Worcestershire, was two-field in 1240 (Hilton 1966, 122).

The following is a summary working from north to south of what few complete county studies have been made, with other comments drawn from the Gazetteer data.

Yorkshire: Beresford studied the glebe terriers of the whole county. In the West Riding there were 186 parishes or chapelries, in the East Riding 121, and in the North Riding 97. The results apply principally to seventeenth-century terriers (Beresford 1948–51, 348–9). The East Riding had most surviving open field, with parishes ranging from 57 to 74 per cent open. As would be expected, there were few open fields in the uplands of the north and west; Cleveland had only 12 per cent of its parishes in the open state. In terms of named fields satisfactory information was supplied by 157 places, of which there were 20 (13 per cent) with one field, 46 (29 per cent) with two fields, 66 (42 per cent) three, 19 (11 per cent) four, and 3 (2 per cent) with five fields and more. This gives an overall view sufficient to say that two and three fields predominated. Without more analysis it cannot be determined whether places with four named fields had two fields grouped together to make a three-course tilth or whether there was a Cotswold type four-field system (see p. 42). This is an overall view of Yorkshire and does not distinguish the long strip fields of the East Riding from Midland types in the Vale of York. In the West Riding, away from the Vale of York, there were very few regular field systems and much pasture. A sub-regional view is blurred because all three Ridings share the central lowland belt, where regular field systems would be expected.

Lincolnshire: Stenton (1922, xx, 47) published twelfth-century charters of Lindsey and Kesteven that described two fields, at Oxcombe and elsewhere. Another twelfth-century example is Snelland, where lands were distributed as thirty-one selions in the South Side and twenty-five in the North (Stenton 1920, 157–9). Stenton thought there were still many two-field townships in the early seventeenth century. No examples of three-fields were known before 1300 (Stenton 1920, xxix–xxxii). For Hallam's analysis see the Gazetteer. The Lindsey district of Lincolnshire in the late eighteenth and early nineteenth centuries had seventy-one villages with two fields; twenty-four with three; eleven with four; and twelve with either one or five or more fields (Thirsk 1973, 257). Study of seventeen villages around Corringham in the north-west of the county, showed there were nine with three fields, six with two, and two had four fields (Beckwith 1967, 111). Harold Fox (1981, 102–4) prepared lists of townships with early evidence (1150–1250) for regular fields of Midland type in the counties of Gloucester, Cambridge, and Lincoln.

Leicestershire: Beresford (1948) studied the glebe terriers of Leicestershire. Of the 250 parishes that were still open in the seventeenth century, where the evidence is clear 14 (9 per cent) had two fields, 118 (79 per cent) three, 12 (8 per cent) four fields, and 4 and 2 parishes had five and six fields listed respectively.

Buckinghamshire: In a similar study of Buckinghamshire glebe terriers, Beresford found that open-field parishes had twenty townships (19 per cent) with two fields, sixty (59 per cent) with three fields, nineteen (18 per cent) with four, and five (4 per cent) with five fields. Only one had six fields (Beresford 1947–52 and 1953–60).

Rutland: Glebe terriers (Northamptonshire CRO, see Gazetteer) of the seventeenth to eighteenth centuries yield the following information for fifty townships. Only two were two-field (Barrow and Lyndon), forty-six three-field, Empingham had four fields, and Whissendine had five. Barrow had changed to three fields by 1651. Most of the glebe strips in the three-field systems were very evenly distributed, except for Barrowden, Market Overton, Ryhall, and Tinwell.

Northamptonshire: A study of 270 townships showed that two or three fields were the predominant type. In the fifteenth century of the forty-five townships with open-field information, twelve were two-field, thirty-one three-field, and two had complex fields. By the eighteenth century, 161 open-field townships comprised 11 two-field, 120 three-field, and 30 with more than three (Hall 1995, 51–5).

Bedfordshire: The published information referring to the field systems of Bedfordshire, together with those of *Buckinghamshire*, *Leicestershire*, and *Northamptonshire* before

1500, has been mapped (Carenza Lewis et al. 1997, 173). Most of them are two- or three-field, with a clustering of two-field arrangements in western Northamptonshire and central Bedfordshire and Buckinghamshire.

Oxfordshire: This county was investigated in detail by Gray (1915, 486–93). He recorded fifty-two places that had two-field systems, many of early date, and twenty-six with three fields. Analysis of the information in the *Oxfordshire* Victoria County History volumes available in 1972, found 148 places that afforded suitable information dated before 1300, of which 118 were two-field and only 28 had three fields (Roden 1973, 348). Early examples of two-field systems are Sibford, in *c.*1190–1220 (Leys 1938–41, 257–63), Brize Norton in 1187, with a yardland described in a terrier (Salter 1947, 68), and Great Tew (with land lying in 'one field and another') in 1249 (*Calendar of Inquisitions Miscellaneous*, i: *1219–1307*, no. 86 (1916). Two-field systems surviving into the eighteenth and nineteenth centuries were common in Oxfordshire, including Kencot (1767), Hook Norton and Southrop (1774), Arncott (1816), and Taynton (1822) (Gray 1915, 123).

Changes from two to three fields

There were changes in field systems over the centuries. In the eastern part of the Midland region there were commonly two fields in the thirteenth century with some three-field arrangements, slowly changing to mainly three-field by the eighteenth century.

A few examples are known of early agreements or other evidence showing there had been changes from two to three fields. These did not involve any physical changes on the ground, merely changes in the way furlongs were grouped into blocks to form the great fields. Gray (1915, 80) quoted two examples of changes from two to three fields between 1240 and 1350, at South Stoke, Oxfordshire (1240–1) and Puddletown, Dorset (1291–2) These examples were part of his assertion that there had been a widespread change from two to three fields before 1350 in response to population pressure. Harold Fox (1986, 529–44) discussed another case at Podimore, Somerset, in 1333, showing that there were in all only ten clear cases then known of conversion of two to three fields before 1350.

It is doubtful that there was ever a widespread change caused by population pressure at this date. Walter of Henley, in the mid thirteenth century, made no comment on the number of great fields, even though he was writing a manual on how to run profitable agriculture. Theoretically a change from two to three fields would increase the yield by 33 per cent. But this is only true if the soils are sufficiently fertile to maintain an intensified use. Since so many townships retained two fields in the Cotswolds and elsewhere, it must be assumed that farmers thought the soils were not suitable for a three-field arrangement. Many of the Cotswold townships have high limestone ground that drains easily and contains sand.

An agreement for establishing a three-field system survives for Dunton and Mursley, Buckinghamshire. In 1345, three manorial lords, the villeins, and the freeholders agreed that two-thirds of their land should be cultivated every year, leaving the rest fallow. Previously the vills had intercommoned. The document expressly states that the arable and meadows were held in severalty until crops and hay had been gathered, at which point all would fall in common (Gurney 1946). Newton Blossomville was the only village in the county to retain two open fields until enclosure in 1810; it also had forty-one acres continuously cultivated called 'every year's land' (Chibnall 1979, 52). Changes from two to three fields were not always smooth. Steeple Claydon remained two-field until 1794. It had been changed to three fields in *c*.1780, but in *c*.1790 one of the farmers objected and turned his cattle into the new crops (James and Malcolm 1794, 30).

A fifteenth-century court order for Marton, in Dishforth, Yorkshire (NR) recorded that the representatives of the lord, the abbot of Fountains, freeholders, and tenants agreed to cast the field into three parts. There is no record of what the arrangement had been before (Homans 1941, 56–7). Great Corringham, Lincolnshire, had two fields in *c*.1200 that had been changed to four fields before 1601, when they were changed again to create a three-field system (Beckwith 1967).

A detailed study of Northamptonshire shows changes that occurred in an 80 per cent sample of the county townships (Hall 1995, 53–4). No three-field systems were identified during the twelfth century and there are only five cases (12 per cent) of three fields known out of a total of forty-three townships that have thirteenth-century information. The number of three-course tilths continued to increase steadily, being the predominant form in the eighteenth century, when it reached 74 per cent of the total (118 out of 159). The proportion was probably even greater, because 15 per cent (twenty townships) had multiple fields of unknown cropping regimes of which many may have been three-course. There was no evidence of any widespread change from two-field to three-field between 1250 and 1350, as proposed by Gray (1915, 81–2). The more probable explanation is that the farmers found the soil was sufficiently fertile to withstand the heavier cropping regime and created three fields when it was convenient to do so. The changes in the Newnham field system were definitely made as a result of experimentation in successive years (see the discussion under *Multiple Fields*, p. 44).

Gray (1915, *frontispiece*) produced a map showing the extent of regular two- and three-field systems. It now needs some modification. There are, as would be expected, examples of irregular fields in the Central Region and some regular ones outside of it. It is doubtful if much of Herefordshire should be included, since there is only one satisfactory example given by Gray (1915, 37, 447), at Hennor in 1608, and only four in all. The Sussex South Downs region, although often having organized fields, differs significantly from the East Midlands. Nevertheless, Gray's map is a good interpretation of the evidence then available. Roberts and Wrathmell (2002, 124) show that Gray's region is similar to their Central Province.

The boundaries of the three national regions resulting from the present study are given in Figure 0.1. It is based on data provided in the Gazetteer county tables supplemented by selected evidence taken from Gray. Included are the Yorkshire Wolds and the southern chalklands of Wiltshire, even though these systems have differences from fields of the Midlands. Many counties hitherto included in the region have areas of irregular fields, sometimes with dispersed settlement, such as the northern claylands of Buckinghamshire and Bedfordshire, the Charnwood Forest of Leicestershire, the Sherwood Forest region of Nottinghamshire, etc. The corrected 'classic' region with examples as claimed by Seebohm for Hitchin and discussed by Gray for Oxfordshire is fairly limited in extent. It runs from Yorkshire through Lincolnshire and the East and South Midlands, taking in south Warwickshire and much of Gloucestershire, Wiltshire, northern Berkshire, eastern Somerset, and central Dorset. The Central Region differs from the Central Province of Roberts and Wrathmell (which is not a plot of open fields) in that the far north-east is omitted and most of Wiltshire and Dorset are included. Wiltshire has been placed in the Central Region because it had a predominance of regular fields with very few irregular fields recorded, as discussed in the Gazetteer. The evidence, derived from the published glebe terriers and two Pembroke surveys, is fairly detailed for this county.

Four regular fields

Four regular fields were used to grow three crops successively, leaving a fallow in the fourth year. They occur frequently from the sixteenth century onwards and are mostly found in the Cotswolds and adjacent areas. This was a region where two-field arrangements lasted for a long time and the soil was not suitable to produce two consecutive crops as in the three-field systems of the East Midlands. Four fields developed from the practice of 'hitching' or sowing pulses, turnips, or grass on some of the fallow of a two-field system, followed by the creation of four-field areas, often known as 'quarters'. The crop was generally used for animal feed, but it also had the advantage of resting the land from grain production and increasing the nitrogen content.

The practice of hitching may have had early origins in Oxfordshire: at Upper Heyford in 1200 and at Waterperry in 1263 the lords put temporary fences around part of open-field demesne, which they did not fallow for lack of pasture (Salter 1934, 216 and 383). Gray (1915, 493) records eight examples of four-field systems in the county. In Tackley, from the late sixteenth century, half the fallow area was used for peas, beans, and vetches. In 1745 there were four fields, two in the north and two in the south. Nethercott had four fields by 1798 and Whitehills four fields in 1605 (VCH, *Oxfordshire* xi (1983): 201). Deddington and its townships of Clifton and Hempton were each divided into nearly equal quarters by 1728 (Northamptonshire CRO, C(A) 3309, pp. 24–58).

Lincolnshire also provides some good examples of four-field systems. Billing-hay has 'inhoc' evidence for 1262. ('Inhoc' was land temporarily enclosed from the fallow and put under cultivation.) Its two-field system had land sown every other year with commoning on the fallow. One furlong, by consent of the neighbours, was sown when the field lay fallow, as it had been four times in the previous twenty years (Hallam 1965, 172). An example of four-field system with fallowing of one quarter each year is found at Crowle in the Isle of Axholme in 1381 (Thirsk 1973, 258).

There are various early examples of cropping parts of what should have been fallow. A small area of fallow had been sown in the thirteenth century at Broughton, Huntingdonshire (Homans 1941, 57–8). Hitching (not so called) was undertaken on the fallow fields of Bletchley and Newton Longville, Buck-inghamshire, in 1311. The townships had intercommoned on their fallows but agreed to allow each other to sow several furlongs on their own fallow. After the crops were carried, intercommoning could be resumed as anciently (Salter 1921, 18–20). In Bedfordshire, the nuns of Harrold were allowed in 1304 to have crops every year on land that normally would be fallow but could be used for hitching (*hyech*: Fowler 1935, 30). The demesne of the Gloucester Abbey manor of Littleton (possibly Gloucestershire), in 1266, was run as a three-tilth system according to the amount of fallow recorded. Every other year sixty acres (of the fallow) were used as an 'inhok' (Hart 1863–7, iii: 35–6).

Broughton Hackett, Worcestershire, intercommoned with Upton Snodsbury, in 1287. When Broughton common was due to be open, it was found that a furlong in the fallow had been sown and the owners tried to impound the feeding cattle, causing an affray in which one of them was killed (*Calendar of Inquisitions Miscellaneous*, i (1916): 613). Peachley in Hallow had a field called *Inhiech* in 1312 (Worc. Cath. Mun., B648). Elmley and its hamlets, in 1376, controlled inhoking by a payment to the manorial court at Ashton under Hill (for twenty acres), Netherton (three acres), Cropthorne (twelve), and Little Comberton (three). In 1449 the court ordered that the inhokes should be sur-veyed (Field 2004, 29, 122). The conversion of two fields into four fields is re-corded in a 1585 glebe terrier of Naunton Beachamp, Worcestershire. It consisted of twenty-five acres of arable plus forty-six lands lying in the North Field and forty-one in the South Field, 'which said two fields are divided into 4 parts according to the course of husbandry for three crops of corn and the 4th to lie fallow' (Roberts 1973, 200). Southern Worcestershire has several ex-amples of four-field systems. Glebe terriers of 1584–5 provide eight examples—Flyford Flavell had 'one yardland…40 acres…lying in the four fields'. Eight other parishes were four-field in the early seventeenth century. Most of this part of the county appears to have converted to four fields by the middle of the six-teenth century (Yelling 1968, 31–3). Roberts (1973, 195–205, with county plans) showed that there were many four-field systems in east Worcestershire and south Warwickshire in the early modern period and many examples of four-fields are given in glebe terriers (Barratt 1955–71, i: 49, 52; ii: 89–90). Gray (1915, 88, 516) noted four-field arrangements in Gloucestershire at Welford and Marston Sicca in *c.*1552. Eighteenth-century examples are Aston Blank (Dyer 1987, 173) and Cold Aston (VCH, *Gloucestershire*, xi (2001): 15).

More complicated examples of four named fields are known. Some were grouped and run on a three-course tilth. Others may represent two townships, each with two independent two-field systems. This can be confirmed by study of the furlong names to see if there are two different sets, or by plotting them on a map if sufficient topographical detail is known. Four fields could also merely be locational names for irregular areas that may sometimes have been grouped together to make approximately equal two- or three-season arrangements. Only written evidence can decide this, with either an explicit statement of 'seasons' or where there are sequential annual details of crops and fallow that identify furlongs. The four fields of Crick, Northamptonshire, are different again; they were run as two overlapping three-field systems and are fully explained on p. 188; Aynho (Hall 2006) was similar.

Multiple fields in the Central Region

Multiple fields are a complex class. In those regions where centrally controlled arrangements normally operated, most multiple systems occur in the early-modern period. For Northamptonshire, twenty-six multiple field systems (including some with four named fields) were run as three-course tilths in the eighteenth century, the names being of locational significance only. At Finedon, the eleven fields of 1739 seem to be the result of dealing with disparate soil types. Although cultivated in three tilths, the fields did not lie in three contiguous blocks, the tilths consisting of fields scattered in various parts of the township (Hall 1995, fig. 7). Wootton Underwood, Buckinghamshire, had five fields in 1649, but they were run on a three-year tilth. Roden (1973, 352) gives a land-use diagram showing the fields.

Among regularly organized multiple field systems, a fairly common type had five fields, of which Newnham, Northamptonshire, in 1625 forms a well-recorded example (Hall 1995, 59–61). It had been two-field until a three-field arrangement was created to last for three years, which was followed by a four-field system that lasted for four years. Each time the system came back to what it had been before 1618. The village farmers were then able to compare the results of successive two-, three-, and four-course tilths between the years 1615 and 1625; they chose a three-course system for the best land and left the lighter land as a two-field course. Such arrangements were double field systems, a two-field and three-field running side by side. Newnham fields developed into this system because of the need to deal with differing soil types, as with Finedon. A rather barren 'red land' (sandy ironstone) was used to grow rye and required fallowing every other year, but clay and limestone soils were able to support a three-year rotation, with less fallowing. Several other Northamptonshire town-ships with five fields had two named 'rye fields'. Spratton affords a good example: the enclosure award of 1766 relates the new allotments to the great fields and shows there were five in all, the South Rye Field and the New Rye Field being nearly the same size as the other three and occupying all the ironstone soils (Plate 6, from Heaton 2009, 8–9). A similar arrangement probably operated at Kirton, Lincolnshire, where, in 1776, the low fields were fallowed every third year and the high fields with light and shallow soil were fallowed every second year (Thirsk 1973, 257).

There are various possible reasons for the existence of late multiple fields, apart from the deliberate improvements of the Finedon and Newnham type. Multiple fields may represent two townships that had been combined. Assarting may be an explanation for multiple fields in townships lying within or close to woodland. In the Rockingham Forest region of Northamptonshire, multiple fields of unknown cultivation regime are found at Apethorpe, Islip, and Lowick. Three-course tilths with multiple fields occurred at the woodland townships of Aldwincle, Brigstock, Bulwick, and Glapthorn (Hall 1995; 2009, 33, 36).

At the extremities of the Central Region there were field systems that had arrangements with some similarities to those of the Midlands, but existed in different local environments. The differences arise from the physical layout, or from the way the fields were managed in relation to the pasture resources. These will be discussed next.

B. YORKSHIRE PLANNED FIELDS

Some Yorkshire field systems exhibit a very remarkable and simple physical form, consisting of no more than a few blocks of parallel strips that fill up much of a township, sometimes running for up to one mile in length from the vill to the township boundary. They occur primarily in Eastern York-shire on the high ground of the Wolds, the lower land of Holderness, and near Pickering, Doncaster, and Towton. Except for Holderness, they occur on topography consisting of relatively large areas of near-planar ground with well-drained soils based on chalk and limestone.

Harris (1959) discussed field systems in the East Riding but he did not comment on these simple plans or long strips. His reproduction of the Burton Agnes 1840 tithe map showed three large furlongs, with aligned strips, being one mile in length (Harris 1961, 57). He found medieval evidence for the existence of long strips running from the vill tofts to the township boundary (Harris 1959, 31). June Sheppard (1973, 150) published part of the plan of Great Kelk, in Holder-ness, East Riding, dated 1789, with strips 1,200 yards long (see also Hall 2012a, 282). Examples from Holderness and the Wolds have been discussed by Mary

Harvey. Much of Holderness remained open until the eighteenth century and contemporary maps were characterized by often having only two fields, some with strips of great length, up to a mile. There were few subdivisions in the fields, that is, very few furlongs (called 'flats'). Plans are given of Skeffling (1721) and Preston (1750). Terriers show there were similar long lands in the Middle Ages; in the thirteenth century Long Riston and East Halsham had strips extending from the village closes to the neighbouring townships (Harvey 1981). Very long lands are similarly found at Preston, Kilnsea, Long Riston, and elsewhere (Harvey 1980, 4–5, with plans).

Wilhelm Maztat (1988, 133–4) further considered the Yorkshire long strips, and reproduced a plan of South Cave on the Wolds made in 1759. He also reproduced the remarkable 1842 plan of Great Kelk, Holderness (the tithe version), showing a regular tenurial cycle through the whole township. Almost all the strips lay on an east–west alignment either side of the central linear vill, stretching for more than half a mile on each side. Long lands occurred over most of the Wolds. Butterwick and Wetwang had strips more than a mile long. A plan of Butterwick in Foxholes was reconstructed from a field book of 1563, suggesting there were some very long strips. There is medieval evidence for long lands in many places, for example Ganton, Potter Brompton, Willerby, and Staxton (Harvey 1982).

Several Wolds field systems have been reconstructed from fieldwork (by D. Hall and P. Martin during 1976–82). There were very few 'furlong' boundaries; the strips stretch from dale to dale over plateaux of high, fairly level ground. The 'mapping' consisted of identifying what few boundaries existed and recording surviving ridge and furrow in spinneys and farm paddocks that still lay in pasture. The former presence of strips was confirmed on land that was then arable by reference to ridge and furrow visible on air photographs of the 1940s, when some very long ridges still survived. Ground observation of bare weathered ploughland in January or February revealed long parallel belts of flinty soil where the modern plough had cut into previously undisturbed subsoil under the ridges (the 1970s agricultural system used bare winter fallow with little autumn planting). Plans have also been published as part of the Wharram Percy Project (Hall 2012, 285–7, 293) and the Burdale plan was available earlier (Hall 1982, fig. 32).

Butterwick, in Foxholes, reconstructed by Harvey from a 1563 field book, was also mapped by fieldwork in 1983 and produced much the same plan (Figure 2.1). The fieldwork evidence adds the topographical detail missing from the historical reconstruction. The major blocks of strips (called 'flatts') were divided by valleys, probably dry, called 'dales', and by large gullies. These would have provided pasture on the slopes and hay from small linear meadows. There was a division of long strips into two shorter lengths of 400 yards each ('furlongs') west of the vill, proved by the linear bank of a furlong boundary. North of the vill there also seems to be more complicated flatts than the field book suggested. The long alignment of the strips of Kirkdales (2,500 yards) at the north-east is broken by several small dales, so that the overall length was reduced. The dry dales may, however, have been over-ploughed. The long north–south strips suggested by the field book at the south-west were not visible on the ground. Nevertheless, the normal length of strips in Butterwick was 600–800 yards. The 1563 field book describes Kirk-dales as a single block and is evidence that the landscape had been planned and laid out on a large scale, ignoring the details of the topography. The well-drained chalk soils would allow this, whereas the clays of the Vale of York

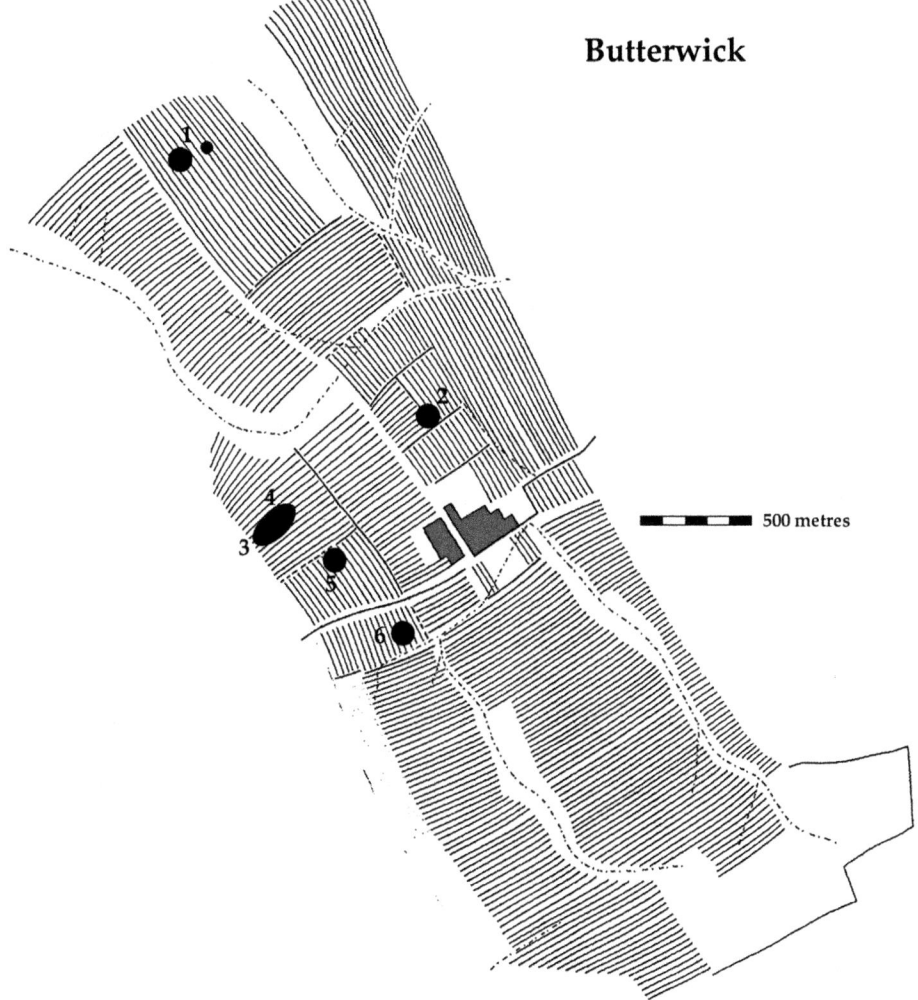

Fig. 2.1. Butterwick, Yorkshire (ER) (SE 99 71). The dark spots locate settlements with Iron Age and Romano-British sherd scatters. Site 6 at the south has Roman- and Saxon-period sherds (field survey by D. Hall and P. Martin 1981).

would not. The serious absence of pasture and meadow for hay was probably ameliorated to a certain extent by practising convertible husbandry or out-field, as was done at nearby Helperthorpe and Kilham in the eighteenth century (Harris 1959, 6–7, and Chapter 3, p. 94). Roman and prehistoric settlement was evidenced over many parts of the township from cropmarks and from finds of flints and pottery. The cropmarks had no relation to the long strip fields; see p. 157 for a fuller discussion (commented upon by Roberts and Wrathmell 2002, 93–5).

Settrington, lying on the northern edge of the Wolds, has a plan intermediate between the Wolds townships and a Midland-type chequerboard. There are some long strips and some smaller furlongs. Field patterns in the Vale of York have furlongs of the usual size, with strips about 220 yards in length. Fangfoss was studied as an example (SE 76 53). There are many furlongs with lands 200–300 yards long, the pattern interrupted in places by small areas of meadow ground, similar to Ludgershall, Buckinghamshire (SP 66 19), lying on similar low, poorly drained ground (Settrington plan in Hall 2012a, 287, and Fangfoss (Hall and Martin, unpublished). Planned fields in Yorkshire are not limited to the Wolds and Holderness. The Ordnance Survey map of enclosed field systems west of Pickering show patterns of walls and hedges lying in narrow fields that make up very long curved lines (up to 2,200 yards), suggesting that they reflect similar simple strip fields (e.g. Middleton, Aislaby, and Wrelton on Ordnance Survey Sheet SE 78, scale 1:25,000 (illustrated in Hall 1982, 51, and easily viewed on satellite images)). At Barton le Street, west of Malton, Yorkshire (NR) (SE 72 83), hedged boundaries reflect curved strips 1,100 yards long (Figure 2.2).

There are other areas of the county with planned systems that have been subdivided into smaller furlong units but where the overall alignment and original large-scale planning is unmistakably evident. Sutton township in Campsall parish, west of Doncaster, had a very late enclosure (1858) and draft enclosure maps of 1854 show the open fields (Northamptonshire CRO, Maps 2595–6). The eastern part of the township was enclosed, but at the north-west were strip fields lying in four blocks ('furlongs') with north–south strips all exactly aligned. On the east and west extremities were furlongs orientated east–west (Figure 2.3). The glebe had some alignment through the furlongs and lay in regular cycles. The lands in the four main fields were 352–440 yards in length and the overall length of all four lying in alignment is 1,400 yards. A site visit to Sutton in 2002 identified linear earthwork boundaries along the lines of the furlong edges as marked on the map. The soil is limestone loam. On the east the ground falls to peaty ground where there were meadows called 'ings'.

Towton (SE 48 39), famous for a very bloody battle in 1461 during the Wars of the Roses, has a remarkable hedge pattern now almost completely destroyed by modern agriculture, but preserved on the Ordnance Survey First Edition 1:10,560 scale map (see Figure 2.4). The hedgelines suggest a simple plan and

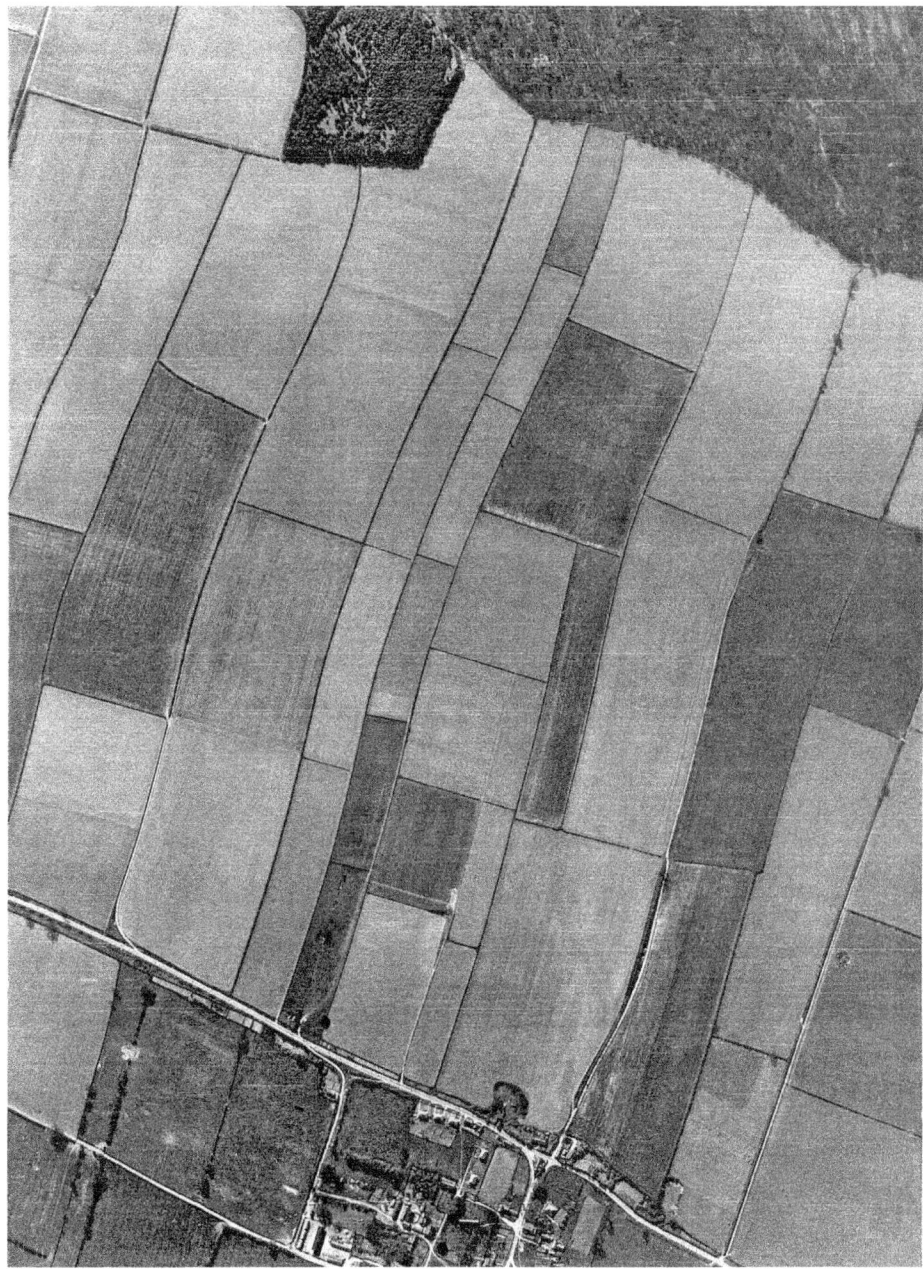

Fig. 2.2. Barton le Street, Yorkshire (NR) (SE 72 83), 15 April 1947. The curved hedges are 1,100 yards in length (RAF 106G/UK/1417 frame 4389; English Heritage (RAF photography)).

Fig. 2.3. Sutton in Campsall, Yorkshire (WR) (SE 56 12), in 1854. The black line represents the glebe strips (after Northamptonshire CRO, Map 2596).

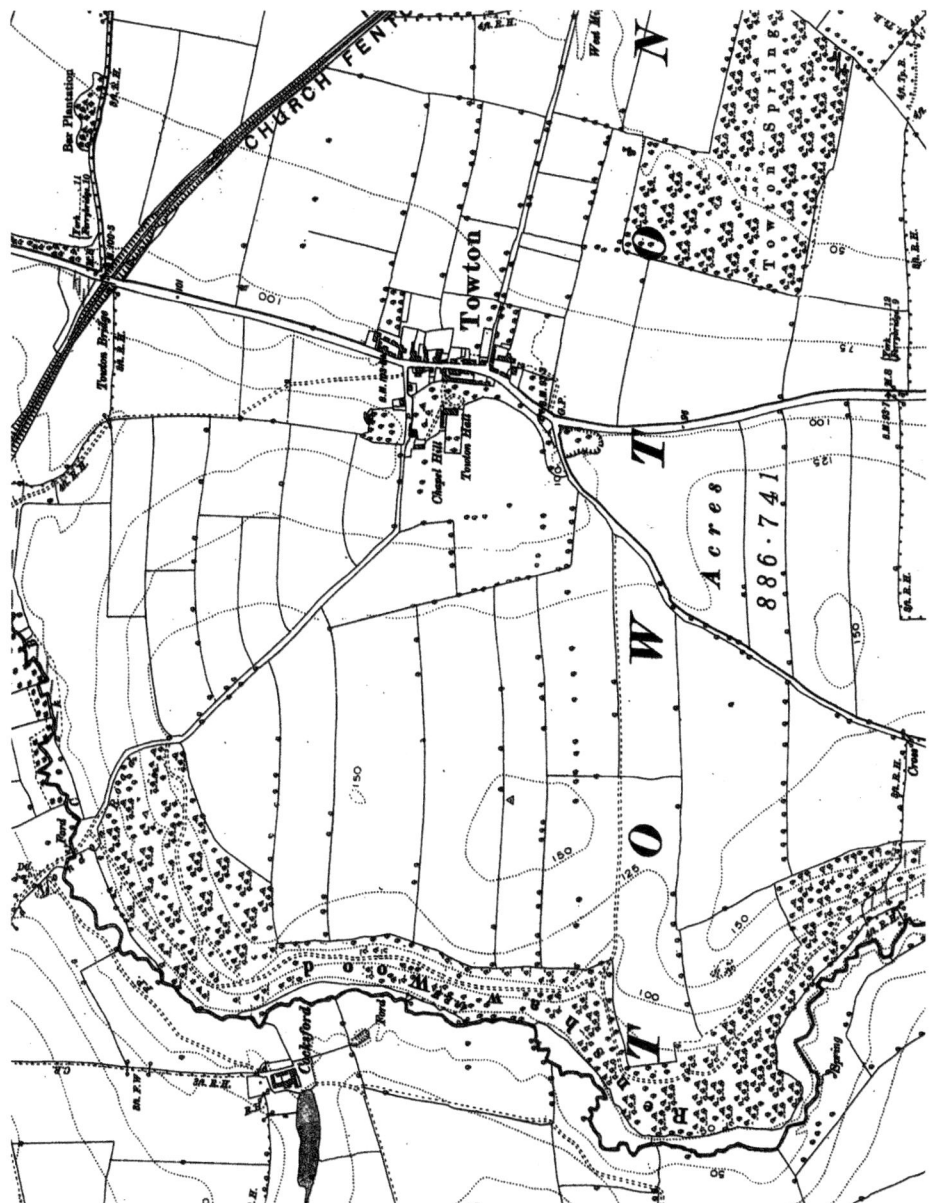

Fig. 2.4. Towton, Yorkshire (WR), in 1894, showing curved hedge lines (First Edition Ordnance Survey 1:10560 scale).

this was proved by fieldwork that showed there were very few furlong cross-boundaries. In 2002 only small amounts of ridge and furrow survived, but RAF vertical aerial photographs of the 1940s show much more, and additional ridges were discovered in spinneys to the east during a fieldwork survey (Figure 2.5). Saxton and Barkston, lying adjacent to the south, have a similar

Towton

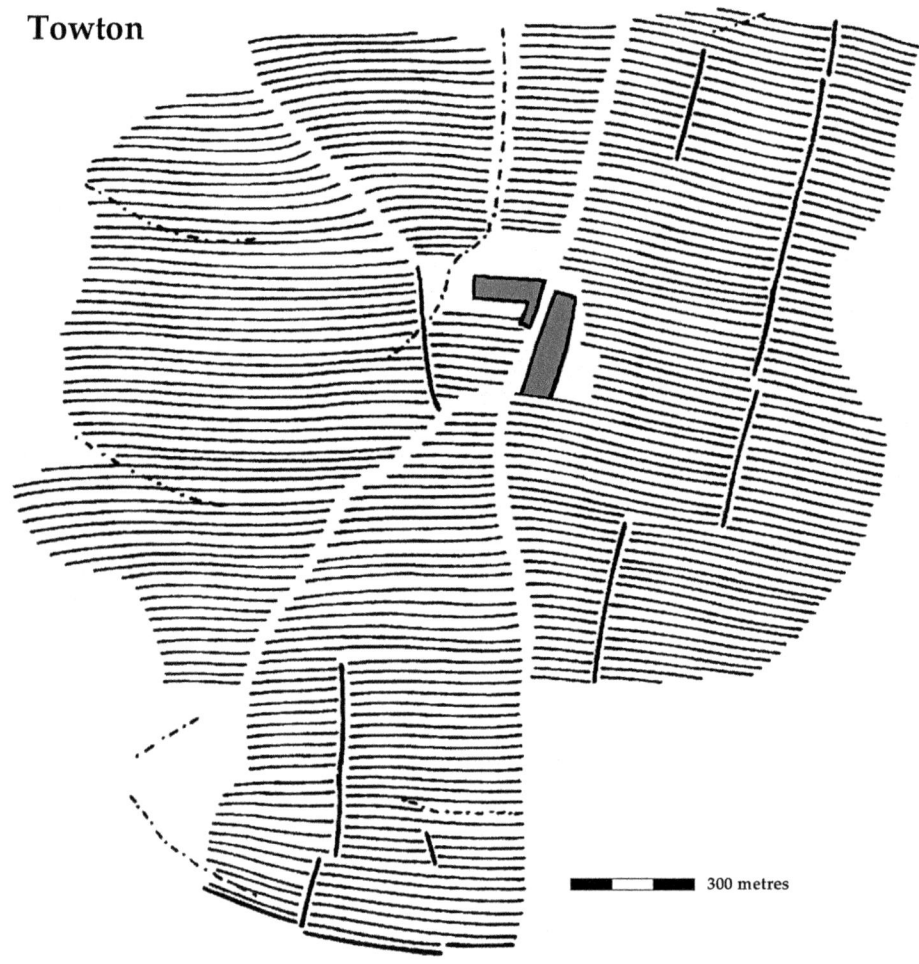

300 metres

Fig. 2.5. The furlong pattern of Towton, Yorkshire (WR) (field survey by D. Hall 2002).

layout. The overall plan of the open fields was simple, with very long align-ments, giving a pattern intermediate between the Wolds type and Sutton. Except for the smaller scale, these fields resemble Great Kelk.

Agricultural practices in the east of the county were similar to the East Midlands, with regulated two- and three-named field blocks that were the crop-rotation units, and like the Midlands, many survived until the era of parliamentary enclosure. Some had a bare fallow with communal field regu-lations for stinting and stubble grazing, etc. (Sheppard 1973, 152, 177; see Gazetteer, under Yorkshire). It was the physical layout of the fields, rather than their operation, that differed from the Midlands, although the use of convertible husbandry to help solve the shortage of pasture does indicate another significant difference.

The large-scale, planned nature of these Yorkshire field systems is indisputable from the evidence of simple patterns from five different parts of the county. The long lines of the furrows were most probably set out in a single operation. One of the chief items of interest is their dating. From charter evidence they were in existence by the thirteenth century—but how much before? Their obviously planned form, seemingly different from the rest of the country, suggests a date when lordship was dominant, or when there was great political change. The 'harrying of the north' by the Normans in 1069–70 is an obvious possible time for such a change. Large areas of the countryside were laid waste and Anglo-Saxon estates were granted to just a few new owners. But nowhere else are the Normans shown to be agricultural innovators—they accepted what they found. Sheppard (1974) thought that the regular planned villages of central Yorkshire were laid out in the later eleventh century, based on the correlation of house-row length in many vills with their fiscal assessment, but settlement replanning does not necessarily mean that the fields were altered at the same time. Harvey (1982, 39) suggested that the tenurial structure closely associated with planned fields may indicate their origins if it could be dated—perhaps to the Danish settlement. Danish settlement of the early tenth century was another time of change, but Lincolnshire and Norfolk had a similar upheaval and there is no evidence of a great landscape replanning associated with it. Long strips up to 880 yards in length can be found outside of Yorkshire, in Northumberland and Westmorland. Roberts (1996) has discussed them and given plans of Cumwhitton, Great Asby, Waitby, and elsewhere.

C. CENTRAL REGION FIELD SYSTEMS INCORPORATING SHEEPFOLDING

Sheepfolding was an important method of managing the fields of much of Southern England. Many of the field systems of the chalkland counties fall into Gray's Midland type, with regulated two- and three-field systems that lasted into the eighteenth century. They were superficially similar to the Midlands but had more pasture resources and hence a somewhat different economy. The description that follows draws upon evidence from Hampshire and Sussex, where downland sheepfolding was practised, even though few of their field systems were simple two- or three-field, and the counties have not been assigned to the Central Region. Sheepfolding will also be discussed in other areas, with the exception of East Anglia.

The higher parts of the chalk downlands were for the most part not used as arable in the Middle Ages, and instead provided extensive grazing for sheep. The Glastonbury Abbey survey of 1189 records that the Wiltshire manors of Winterbourne Monkton and Idmiston could have 400 and 1,000 sheep respectively (Jackson 1882, 124, 126). Sussex fourteenth-century taxation returns show that the chalkland region was the most prosperous part of the county, with many sheep. Wool merchants were located at towns lying along the downland edge (Pelham 1931, 156–67; Pelham 1934, 129–35). Sheep were grazed on the Dorset downs in large numbers (Bettey 1982, 1–5). Seven Dorset monasteries had 24,941 sheep belonging to them, recorded in the *Valor Ecclesiasticus* of 1535. The extensive downland pastures typical of the townships are illustrated by the plan of Chilfrome (Figure 1.2).

In addition to the intrinsic value of the sheep themselves, their manure was an important component of arable farming. It was managed by a complex 'mobile' method. Manors had a common fold made of hurdles that was set up on the arable fallow in the evening. It could contain 1,000 sheep. Each morning the sheep were let out and herded back to the downs; in the evenings they returned. The fold was moved daily, or after few days, to manure another part of the fallow, so that each tenant received the benefit in turn. In this way a large part of the nitrogenous waste produced by the grazing sheep was transferred from the downland pasture to the arable, and without labour except for the daily driving of the sheep and the management of the fold. In these southern counties manuring was considered a very important reason for keeping the sheep in later centuries. They were bred for folding, for their ability to climb between field and down, and for dropping manure at night (Kerridge 1959, 54). The importance of folding in the Middle Ages is demonstrated by the villein work-service relating to sheep.

Late sources describe the process very clearly. At Berwick, Sussex, folding regulations were fully stated in 1721. Stints for stocking the laines ('great fields') from September to March were given; those with no land (cottagers) were allowed three ewes. Faggots for hurdles, required each year to make ways for the sheep as well as for the fold, were to be provided by each farmer in proportion to his number of sheep. When finished with, the old hurdles were divided among the farmers. The fold would rest upon the farmer's fallow in proportion to his number of sheep. If, after the fallow folding had been completed, a farmer had not had his share of folding rights, the fold could go on any other of his grounds. Folding in enclosures was not allowed, except by agreement. The shepherd was kept in strict control; if he allowed any overstocking he lost a month's pay. Berwick glebeland had a 'fould tare' in 1619, that is, the right to have the fold in the same the proportion as other tenants. If the fold did not rest on the glebe, the parson could refuse the tenants the aftermath (Cooper 1853).

The duties of the attendant shepherd in 1629 are described for Heale, Wiltshire (Heale House, in Woodford). He was to stop sheep from straying, prevent theft, inform the owner if the sheep became diseased, ensure there was no overstocking, and not allow persons from outside the township to bring sheep in. If a person failed to provide hurdles, then 'the foulde [would] skypp over' the offender's land (Bettey 1982, 2). The seigneurial fold at Amberley, Sussex (see p. 56), was 4½ acres, so could be fitted on nine half-acre strips if the tenants' common fold was the same size as at Heale. This would represent manuring of the strips of nine different tenants at once. It would be quite easy to miss out one of the strips if a tenant had not supplied his share of hurdles (and so the fold would 'skip' over it). In Dorset the wages of the common shepherds were paid in proportion to the owners' stinting level. The owners had to provide hurdles for the fold and bring hay for winter feed. Hurdles were made from hazelwood spars; a lease of the demesne farm of Winterborne Houghton in 1603 included the right to one acre of underwood for making hurdles.

The following examples provide evidence showing (or are in accordance with) the same practice of systematic sheep folding for manuring at earlier dates.

A set of detailed customs and field regulations was made for Winterbourne Stoke, Wiltshire, in 1574 (Goddard 1906). Among them were orders that each yardland owner was to provide four hurdles and 'shores' (stakes) to make a large common fold on the fallow. Elsewhere in the county in the fourteenth century, yardland holders were responsible for moving hurdles at Downton and moving pens across the fields at East Knoyle and Heytesbury. The virgaters of Chilhampton in South Newton sent a man or girl to move the demesne pens (Scott 1959, 22). Sheep were important in the county much earlier: at Winterbourne Monkton villeins washed and sheared the lord's sheep in 1189 and the fleeces were collected by the smith (Jackson 1882, 123).

Villeins at Elsted and Mardon, Sussex, cut branches and thorns for fencing the sheepfold at Drayton in 1253. They could have the old fence and had to move the sheepfold when needed (Fleming 1960, 191). Loventon (Lavant) villeins (1258) gathered branches from the lord's wood to make three hurdles for the sheepfold. They carried wool from Loventon to London, Croydon, or Malling (Redwood and Wilson 1958, 57, 19). The custumal of a yardland at Ratton (1248) included washing and shearing ten sheep (Salzman 1923, 51).

The 1380 custumals of the bishop of Chichester describe work-service relating to sheepfolds at several manors. Villeins at Selsey, Hampshire, mended five hurdles for the fold and if the fold rested on their land they paid the lord five pence (called 'faldpan'). At Sidelsham, Sussex, they gathered stakes, made two hurdles, mended five hurdles, and similarly paid five pence for use of the fold. The villeins of Aldingbourne, also in Sussex, went to the manorial wood, where they collected and split enough stakes to make two hurdles, which were carried to the sheepcote with four fold stakes and four shackles (Peckham 1925, 14, 23–5, 33–4). Aldingbourne's work-services in 1380 were similar to those obtaining there back in 1275.

East Preston had a 300-acre demesne pasture down; tenants could have no foldage on their land unless hurdles were provided (1568) and herds were not to go into the pastures before autumn (Thomas-Stanford 1921, xxvi, 8). At Eastbourne, in 1253, four shepherds kept 600 wethers and 400 ewes on the lord's pasture on the hills. They had to manure the lord's demesne with these sheep from Easter to 21 December and for the remainder of the year they had a fold of their own (*Calendar of Inquisitions Miscellaneous*, i: *1219–1307*, no. 188 (1916)).

Amberley, another of the bishop of Chichester's manors, provides detail of the lord's fold in 1380, which any villein could be chosen as shepherd to look after. One of its purposes was in early summer to obtain sheep's milk for cheese making. The fold occupied 4½ acres and contained 300 of the lord's sheep; the villein shepherd could also put his own sheep in it. The lord provided hurdles, stakes, and hay and expected five weys of cheese, allowances being made for barren ewes. The sheep were milked in about June when there was some meadow service due from the shepherd, but most of his services and rents for half a yardland were remitted (Peckham 1925, 45–6). The lord provided litter for the fold (in winter) when the sheep were on his land and in his sheepcote. He arranged to draw the fold from place to place as needed. The shepherd could have the fold on his own land from St Thomas' Day (21 December) to 5 January but would have to fetch and remove it himself.

The mid thirteenth-century custumals of the Abbey of Bec published by Marjorie Chibnall (1951) illustrate the importance of downland pastures. Of the twenty-one manors described, fifteen operated sheepfolding (according to the villein work-service demanded) and all of them lay on the southern chalk downs except East Wretham in Norfolk, which is near the sandy Breckland. Among the work-service of the tenants was washing and shearing sheep, providing stakes, wattles, and hurdles for the fold, and moving the fold from place to place. At Monxton, Hampshire, it is specified that the land was composted to improve it and that the shepherd could use the fold for ten nights. The six manors where sheepfolding was absent were not in downland locations: Atherstone, Warwickshire, Weedon Bec, Northamptonshire, Lessingham, Norfolk (towards the coast), Blakenham (near Ipswich), Suffolk, Bledlow, Buckinghamshire, and Cottisford, Oxfordshire. Sheepfolding probably occurred as early as 1086, because Domesday records large areas of pasture as part of the manorial resource for most of the southern chalkland parishes.

Away from the south, the Chilterns in Buckinghamshire were used for fuel production as well as sheep grazing. Penn, on the Chiltern summit, had beech woods and the Segrave manor had a hundred acres of demesne woods in 1325. An account of 1371 refers to bundles of underwood cleft and cut into billets. Fuel for London was later the chief source of revenue (Jenkins 1935, 32–41). However, townships near the Chilterns had more sheep than elsewhere in the county. In Essex and south-western Cambridgeshire the chalklands extending from the Chilterns were mainly used for arable. In south-eastern Cambridgeshire, the chalk falls to a sand-capped ridge and sheepfolding was practised where there were sizeable heaths and pastures. West Wratting had a 300-acre sheep walk for 450 animals, and 1,140 sheep were allowed for a Chippenham demesne farm. The foldage rights belonging to each half yardland at Orwell in 1627 are fully described. A seventeenth-century illustration shows a fold placed on the fallow, with new hurdles being made nearby. At Badlingham, in 1306, a half-yardland carried the right of foldage for 140 sheep for its manuring (Spufford 1965, 20).

Folds and sheep in the East Midlands

In Northamptonshire there were very few places with extensive pastures. One such area was Benefield Plain and its adjacent pastures of Deene and Weldon Plains, lying in the woodland of Rockingham Forest, next to Farming Woods, in all about 1,000 acres. In this plain the sheep of the royal manor of Brigstock had right of pasture, and the 'wood of *Ferminwod* beyond the drove of Brigstock' was referred to in *c.*1230 (*Calendar of Close Rolls, Henry III*, i: *1227–1231* (1902), 269). The Plain was approached by way of three great droves cutting through 1,300 yards of woodland coppices. The rights were listed in the seventeenth century (Hall 1995, 214–5 from Northamptonshire CRO, Buccleuch 13/1 in Box X359):

Brigstock men drive theyre sheepe from theyre towne and fields on the wasts of Benefield through the said ridings...Some sheep are att the pleasure of the owners driven to the sayd wasts in the day and brought at night to fold theyr land in the fields of Brigstock and driven again the next morning. Some sheepe are continually kept upon the sayd wasts and are there nightly layed.

These regulations show that there was some night-folding at Brigstock at the will of the tenants, but not the systematic folding organized by the village as in the southern downlands of Sussex and elsewhere.

Sheep were everywhere an important component of farming, as evidenced by the animal stints allowed by right to each yardland. A typical village of 40 yardlands would have 25 sheep for each yardland, hence a potential village flock of 1,000 sheep. The very numerous court rolls and open-field orders that have been studied for Northamptonshire have been checked for their references to sheepfolds. They fall into two types: those that refer to permanent sheepcotes (sometimes called folds) and those that relate to the control of sheep at night to prevent trespass on corn crops and in gardens. There was no folding of the Downs type, even at places like Wollaston that had more meadow than anywhere else upstream from the fen (over 500 acres, with 2,168 acres of arable). An extensive series of field regulations from 1632 to 1733 gives no indication of a fold system. Sheep went into the stubble field six to eight days after the great beasts. They were not to lie unfolded at night, even if the weather was 'unseasonable' (Hall 1977, 144–53).

Sometimes the word 'fold' was used to denote a permanent structure, in other words, a sheepcote, and not a temporary pen made of hurdles. Monastic granges had permanent sheepcotes out in the fields as well as in the farm buildings. Pipewell Abbey, Northamptonshire, had a sheepfold at Cold Ashby called *Suthcote* in *c.*1235 (London, British Library, MS Cotton Caligula A xii, fols. 149–50), and in 1540 (TNA, SC6 Hen VIII 2784 m59d–60d). Pipewell also had a large wooden 'sheepfold' constructed of oaks on four acres of land at Great Oakley (fourteenth century; Northamptonshire CRO, FH 147, fol. 29). Old Warden Abbey, Bedfordshire, similarly had a permanent fold at Eydon, Northamptonshire, in *c.*1200 (Fowler, 1930, 95–6), presumably built on the pasture of the steep scarp.

Manorial sheepcotes are referred to frequently. Kettering paid for eighty-two hurdles used at the seigneurial sheepfold in 1292 (Wise 1899, 29). At Higham Ferrers, in 1314, sixty hurdles were made for the flock fold, and two shepherds were paid for looking after it. Rushden fold contained 200 sheep looked after by one shepherd (Kerr 1925, 55, 61, 66–7). Polebrook manorial expenses included thatching the sheepfold (Northamptonshire CRO, Buccleuch Charters 472 (1365)). The lord at Broughton, in 1459, was ordered to repair the common fold so that animals did not trespass on the sown fields (Northamptonshire CRO, X396, court roll 38 Henry 6). This refers to the permanent structure of the village penfold. Similar orders are abundant. As in most counties the commonest references to folds in Northamptonshire are those to prevent sheep from trespassing on the arable or in the village closes.

As far as can be determined, all the champion parts of the East Midlands used sheepfolds in the same manner as Northamptonshire (excepting Brigstock), that is, to pasture sheep on the meadows, fallow, and commons if there were any, without any intensive management of the fold. In early days lords often demanded preferential foldage, as at Clopton in *c.*1260, where half-yardlanders could have their own folds for composting their own land, but smallholders had to put their sheep with lord's fold (King, 1983, 23–4). Elton in Huntingdonshire was similar in 1307–8 (Ratcliff 1946, 131).

Sheepfolding does not determine the nature of a township field system—it is a management technique. The fields of Wiltshire and Dorset were, for the most part, two- or three-field, regular systems. They differed from East Midland systems in that the arable was a low percentage of the total area; sheepfolding was an efficient method of managing pasture and manuring fallow at the same time. Provided that there was a large area of pasture, sheepfolding could have been applied anywhere, even where the holdings were irregularly dispersed in the fields. The fold could have been erected in turn on whatever few lands were required, leaving out any that belonged to a farmer who was not participating. The case of Brigstock, Northamptonshire, proves that folding could be done privately and individually by any farmer without a highly organized communal village system. There is every possibility that some of the one-field systems found, for example, in the Derbyshire Peak District or Lancashire could have operated the fold system, most likely privately.

References to sheepfolding across the country (excluding East Anglia, which is discussed in Chapter 3 (pp. 69–73)) are as follows.

Bedfordshire: At Edlesborough in *c.*1200, the lord allowed Warden Abbey to pasture 480 sheep on his demesne, provided the abbey's flocks were folded on his land from April to Martinmas (Fowler 1930, 246). Wickers for folding had to be provided at Streatley in *c.*1380 (J. S. Thompson 1990, Bedfordshire Historical Record Society, 69: 88–93).

Buckinghamshire: Taxations of 1332 and 1327 show that there were large numbers of sheep in the Chilterns compared to the champagne lands of the north (J. C. K. Cornwall 1978, *Records of Buckinghamshire*, 20: 57–75).

Cambridgeshire: Villeins had to move the abbot of Ramsey's fold from place to place at Burwell in the thirteenth century (Hart and Lyons 1884, ii, 28). Willingham had a stint of thirty sheep to the half-yardland, and they had to be folded in the lord's fold (Spufford 1974, 99, 129). The rights and obligations of folding in cases like these are to be distinguished from an organized folding system operating to manure the common fallow systematically. At Wood Ditton, in 1290, all sheep had to go in the lord's fold. Linton (1272) had pasture for sheep on the fallow (Tate 1944, *Proceedings of the Cambridge Antiquarian Society*, 40: 59). Bottisham had folding in order to compost the field that was to be sown (Hailstone 1873, 334–9).

Devon: Tenants at Salcombe Regis and Chelson, in 1281, owed sheep-hurdle rent (Morshead 1898, *Transactions of the Devonshire Association*, 30: 132–46). Bishops Clyst manor between 1374 and 1420 received work-service with dung carts for manuring; in addition, there was folding on the fallow for 550 sheep (Alcock 1970, 151–7).

Gloucestershire: The Gloucester Abbey Cartulary records only three manors where washing and shearing were part of the work-service in *c.*1267: Hinton, Buckland, and Boxwell (Hart 1863–7, iii: 56–7, 63, 101). Sheepfolding as a method of fertilizing was used on soils of sand and stonebrash and also on barren lands on hills (Lennard 1932–4, 42). Sheep were an important component of farming in the Cotswolds, where permanent sheepcotes were used and mobile temporary folds are not recorded (Dyer 1995).

Hampshire: Tenants of the bishop of Winchester's manors paid a fine for not putting their sheep in the lord's fold in 1209 (Hall 1903, xxi, xliv–xlvi).

Huntingdonshire: In 1307 Elton villeins were to put their sheep in the lord's fold or else pay foldage if they were kept outside (Ratcliff 1946, 131). Old Weston villeins were to make hurdles for the lord's fold in the thirteenth century (Hart and Lyons 1884, ii, 40).

Lancashire: Youd (1962, 20–34) collected various court orders for most parts of the county. He found that sheep, when on the arable fallow, had to be folded at nights (to prevent trespass into crofts, as in the Midlands).

Yorkshire: The sheepfold belonging to the nuns of Ormsbury, Lincolnshire, at Spaldington (ER, in Bubwith) could be moved from place to place at will in *c.*1250 (Stenton 1922, 18, 62).

Places in the Central Region with low amounts of pasture have fold references that concern either putting sheep in the lord's fold to achieve preferential manuring of the demesne or to control the grazing of the fallow to prevent trespass. Most of the examples listed above are of this type of control and do not prove the existence of any organized sheepfolding system.

3

Field-system types in the Eastern and Western Regions

There is no single type of 'field system' used in the large areas of the country-side lying outside of the Central Region, although in most regions an underlying similarity is that holdings were not distributed throughout a given system. It is possible that a few fields associated with dispersed settlement may not differ in principle from Midland types, except that there were many small townships, no longer necessarily very evident today. However, some fields were quite different from those of the Central Region, with Kent, East Anglia, and the Wash Fenland each displaying a characteristic type, and a fourth to be found in townships with a one-field system. In this chapter, these four types will be described first and then in-field and out-field systems involving convertible husbandry will be discussed.

A. KENT

Many field systems in Kent, although once open and divided into strips, consisted of severalty holdings called 'yokes'. They developed complicated ownership patterns caused by partible inheritance, but there was limited dispersion of arable holdings and no communal crop regulation. The yokes were physical units and not merely a fiscal convenience. Some Kent parishes had irregular fields not called yokes. The existence of detached swine pastures located on the Weald and the effects of partible inheritance caused a fragmented settlement pattern to develop and encouraged early piecemeal enclosure (see the Gazetteer).

Kentish Anglo-Saxon charters make grants of units called 'sulungs' (for example S 1458, S 1511). In the Domesday Survey of 1086, Kent was assessed in 'solini' (sulungs) and 'iuga' (yokes), there being four yokes to the sulung (Round 1895, 103–4 and 108). Feet of fines, manorial extents, and custumals give further information from the late twelfth century. A named yoke at Sheppey called *Stapendun* is recorded in 1197; another at Detling was called *Manesland*

(Churchill et al. 1956, 4–37). At South Acholt, in Sutton-at-Hone, three yokes had their boundaries described in detail in 1203, finishing at the starting point, showing that they lay in a block (Churchill et al. 1956, 31). Rent was paid on each of the twenty-eight yokes in Dartford in 1260 (Larking 1860, 249). Like yardlands they were used to raise taxes in the thirteenth century and had work-service obligations on the lord's demesne farm (Scargill-Bird 1887, 124; Du Boulay 1961b, 82; 1966, 118–22).

Twelve yokes lying at Otford were described in detail in *c*.1425. They each had to contribute to enclosing the 'burgherd' (possibly a ring fence, compare the Lancashire 'ringyards' or the manorial enclosure) and each yoke had a name, as at Sheppey and Detling. The rental states the acreage and yoke size of the holding (some were a quarter, or a yoke-and-a-quarter, etc.), and whether it was in single or multiple ownership. Each yoke seems to have had its own farmhouse, as well as additional dwellings if it had been split up. Earlier deeds referred to some of the yoke names, either as places (farms) or personal names (Ward 1930). Many of the Otford 1425 yokes could be reconstructed using later field names, farm names, and the evidence of very old hedges placed on banks that surrounded them (Hewlett 1973, 106–9). They are thus satisfactorily established as intact blocks of land, as indicated by the 1203 survey of yokes at South Acholt.

Gray (1915, 282–6) discussed a 1447 rental for Gillingham that arranged the tenants' holdings in yokes and described the boundary of each yoke in relation to each other and to roads and other topographical features. It was clear that in principle the yokes were rectangular blocks of land, frequently about twenty-five acres in extent, but could be more. The tenants' holdings were only slightly dispersed, lying in a few, usually adjacent yokes. A similar fifteenth-century survey of Newchurch describes blocks of land called 'doles' of forty to forty-six acres, in which the lands were stated to lie together. Again the tenants' lands were dispersed in a few adjacent doles (Gray 1915, 286–7). From the evidence of earlier rentals for places such as Wye, where information for 1311 can be compared with a fifteenth-century survey, it was found that the names of the yokes were much the same at both dates. The fifteenth-century tenants frequently held several small parcels in a few yokes and there is reference to the former tenants, but each yoke had many tenants (Gray 1915, 287–94). The yokes described in 1447 included meadow, marsh, crofts, and wood as well as arable. A reconstructed plan illustrates the limited dispersion for several holdings. Apart from a concentration of settlement near the church there were many isolated farmsteads lying in twos and threes throughout the area (Baker 1964).

Partible inheritance was a characteristic of Kentish customary law. It is probable that yokes were once held by one or a few people originally, but because of partible inheritance they came into the multiple ownership of many descendants. The fragmentation of yokes added to the already complex settlement pattern, with dwellings being built on detached swine pastures ('denns') as well as on the small parcels within the yokes (see Kent in the Gazetteer). The effects of partitioning in the thirteenth century were examined by Baker. Although it was possible for holdings to become very small, there was some sharing of undivided portions by brothers that reduced the number of pieces Holdings were not dispersed in minute parcels and tended to lie in one area. At Wrotham, land divisions were called 'boroughs', and in 1285 only 5 per cent of the 409 tenants held land in more than one of them. By 1494, 25 per cent held land in more than one borough. At the end of the fifteenth century, most holdings were small, consisting of a messuage and garden with a few small crofts and larger closes lying nearby (Baker 1965a, 157–64). At Bexley in *c.*1214, seventeen yokes were held by forty-seven people. By 1284 there had been much fragmentation and the yokes were held by over 150 persons. Individual holdings were not dispersed; ninety-nine tenants held in one yoke only; forty-three in two yokes, five in three, four in four, and only one person held in five yokes and two in six (Du Boulay 1961a, 21–2).

Baker (1973, 384) concluded that yokes were not individual 'fields' with visible boundaries on the ground, but that they were fiscal units that rendered rents and service. However, at Crundale (TR 08 48) in 1755 the tithe of Cakses Yoke was in dispute. A map of the 132-acre yoke was made showing that it was a block of land then divided into 35 enclosed fields (Northamptonshire CRO, Map 2193). This map confirms the physical reality of a yoke as a single piece of land and its survival as a unit for paying tithe.

There were no communal 'fields' in the county since the yokes and other lands were held in severalty. As late as 1796, Boys wrote to the Board of Agriculture that no part of the county was 'occupied by a community of persons, as in many other counties' (Baker 1965a, 169 quoting the *General View*). There were therefore no common rights on arable for grazing except by private arrangement. No evidence of common gazing on unenclosed arable has been found, but there were a few open meadows that were communally grazed after lotting (Baker 1965b, 25–9).

Parts of the county were probably cultivated in small irregular fields, as in East Anglia. A detailed terrier for Barfrestone described twenty-seven parcels totalling over forty acres in 1236 (Churchill et al. 1956, 130–2); twelve of them lay in 'fields' (furlongs) and six lay in pieces called 'pastures'. Thirteen of them were described as consisting of small groups of furrows. Many fields are described at Limpsfield in *c.*1313 (Scargill-Bird 1887, 139–40). Irregular multiple fields occurred with both 'field' and 'furlong' names at Orpington near Canterbury in 1352 and at Sutton-at-Hone and St Mary Hoo lying near the Thames in the early sixteenth century (Gray 1915, 277–8). A series of fourteenth-century charters for smallholdings at Hadlow near Tonbridge (TQ 62 49) show there were dispersed pieces, some lying in crofts or areas called 'fields'. A number of pieces lay in small blocks and not single lands (Northamptonshire CRO, Westmorland 2.iv.2 C, 1–21).

Farming on the Westminster Abbey demesne at Westerham between 1297 and 1350 was recorded in great detail. This land was not subject to partible inheritance and crop rotation was possible. Crops were grown on a series of 'fields' (furlongs) that were under regular cultivation, each of about 125 acres. Much of this operated on a three-season rotation of husbandry, one furlong having its name missing from the accounts every third year from 1297 to 1350. Another 400–600 acres were sporadically cultivated and so formed an outfield or were part of a system of convertible husbandry. Legumes tended to be planted on the outfields and wheat on the inner area with better soils (Bishop 1938–9). Manors in the Vale of Holmesdale operated similarly between 1272 and 1379 (Baker 1966, 3, quoting A. Smith). Demesnes elsewhere might likewise have been cultivated on a three-tilth system, as at Well (in Ickham) and Luckingdale (in Littlebourne), where 370 acres of arable, 'dry and sandy', could have 100 acres sown each year (*Calendar of Inquisitions Post Mortem, Edward III*, xi (1935): no. 363).

Some use of yokelands spills into Surrey. At Ewell, in the thirteenth century, tenants occupied 'iuga' of thirteen acres, as found in Kent. A field book of 1409 describes lands lying in furlongs, but the holdings, called *tenementa*, had the names of earlier *iuga*. There had been some fragmentation of holdings since the thirteenth century, because parcels in a given *tenementum* belonged to several new owners (Gray 1915, 399–400).

B. EAST ANGLIA

> East Anglian field systems lying on land of low relief with variable soils were mostly irregular. Some had severalty holdings and others were only slightly dispersed, all without communal crop regulation. Open-field strips were once widespread but often had enclosed crofts among them. Large block demesnes were used for sheepfolding in the thirteenth century that developed into extensive fold courses in some areas after the fourteenth century. Elsewhere small irregular holdings encouraged piecemeal enclosure, and open fields began to disappear at an early date.

Case studies of twelve East Anglian parishes showed that common open fields occupied more than half the total area in most places, but were a lower proportion in those parishes that were located in Essex. The remaining parts of each parish consisted mostly of block demesnes or other tenement block holdings (Martin and Satchell 2008, 192). In 1897 Corbett studied Norfolk Elizabethan surveys and noted that the terms 'precincts' and 'fields' were used that had no cropping significance (Corbett 1897, 70–1). Gray (1915, 310–25), using evidence from maps and surveys of similar date, showed that, for a particular holding, scattered open field strips in East Anglia were not located uniformly over a township, but tended to be concentrated near the farmstead. Examples occur at Buxton and Shropham near Norwich (1714–15). There were no great fields, but there were areas called 'precincts', typically from two to five, that were usually bounded by the township roads. Data for the holdings at Castle Acre in 1587 show that although there were three areas called 'fields' the distribution among them was very irregular. A 1575 map of West Lexham was reproduced showing the distribution of all holdings. Again the distribution is irregular; a tenant called Yelverton had his lands entirely in the northern half and another tenant, Lee, nearly all in the southern part. The glebe-holder and tenant Audley had lands in most parts of the township.

Open field occupied about 65 per cent of Weasenham in 1600 and was divided into two precincts, but the holdings were not equally dispersed and many occupied a limited area only. Details of cropping kept by a leasehold tenant in the 1580s show that although his lands were scattered in five disparate parts of the fields (with none in about half of the total area), he ran a three-season cultivation system using some convertible leys to make up the required acreages. The names 'furlong' and 'field' were used indifferently and a given furlong would have more than one crop on different strips in one year.

Cropping these irregular fields had to be fitted in with fold courses in the sixteenth and seventeenth centuries (Allison 1957, 21). Gray (1915, 331) was able to show that at Holkham the unequal distribution of lands among the fields could be taken back to the fifteenth century. With regard to cropping, in September 1392 a smallholding at Holkham had three separate pieces of a few acres, all in the South Field, of which one was ley to be ploughed; another had been cropped with barley and was to be ploughed; the third had been tathed the previous season (i.e. animals had been pastured on the land in order to fertilize it with their dung) and was not to be ploughed. Hence a three-season tilth and a one-in-three-year tathing seem to have operated. Arable fertility was maintained from the dung, urine, and foot action of flocks of sheep. Gray (1915, 331–2) found other places with very unequal distribution in holdings in the fifteenth century: Great Massingham, Ormesby, and Rockland. Three-season cultivation of arable is clear from fourteenth-century demesnes.

Holdings in eastern Norfolk were small during the fourteenth century and dispersed irregularly over extensive arable fields, with common pasture rights over the arable after harvest (known locally as 'shack'). Crop rotation on the demesnes of twenty-three manors in the late thirteenth and the fourteenth centuries showed that fallowing was reduced to a low level (Campbell 1981b, 19–22), commonly below 15 per cent (equivalent to one year in seven). Legumes varied between one-seventh and one-fifth of the area sown. Campbell concluded that there was little whole-year fallow and only shack after harvest, and showed that folding was limited to the demesne areas, but customary tenants had to put their sheep in the lord's fold, unless they paid 'foldage', the right not to put their sheep in the lord's flock. Other tenants paid to have their sheep in the fold, presumably so that their land would be tathed (Campbell 1981b, 22–5).

Before the sixteenth century, north-eastern Norfolk was a champion area with extensive open fields. Six places with sixteenth-century field books provide the earliest detailed data. The extent of old enclosure varied from 23 to 42 per cent of the farmland. There were significant commons, except for Coltishall and Blofield, which were mostly arable. All the fields in the places studied were very irregular with no internal order. Holdings were limited in terms of their distribution among precincts and quarantines. Most were twenty-five to thirty acres, with about 25 per cent less than five acres and 10 per cent over eighty. Neither manor nor parish controlled farming, only the holders, and harvest shack was the sole common right.

The state of holdings in the Middle Ages has been determined from manorial surveys. The Martham survey, dated 1292, was used by Gray and Hudson, and shows that for the 1,057 acres surveyed 56 per cent were in the *Estfeld*, *Westfeld*, and the smaller *Suthfeld*. The remainder lay in areas with furlong-type names. In the 2,122 parcels, with an average area of half an acre, 376 tenants had fewer than six parcels, which amounted to 2.8 acres, and none had more than eighteen acres. In terms of field structure, East Field and West Field are named as well as other places, but there was no equal division of the holdings (called 'eriungs'), among them. A half eriung lay entirely in one field and many of the parcels probably lay in close proximity (Gray 1915, 337–40). More details of the Martham survey were given by Hudson (1917–19), first listing the 107 'former tenants' whose lands were split in 1291 into about 900 small tenancies consisting of near 2,000 strips. The 22 eriungs and 3 acres of the villein lands comprised 266 acres in the detailed listing. All but five of the forty-one villein 'former tenants' were stated to hold a whole or a regular fraction of an eriung, so leaving thirty-six whom Hudson equated with the thirty-six free tenants of 1086 (Hudson 1917–19, 288–9).

A fifteenth-century terrier of holdings (*tenementa*) at Wymondham shows conclusively from its description that they lay, or had lain, substantially in blocks. For Hemsby in 1422, one hundred acres were surveyed. There were no 'feld' names but a hundred 'furlong' type names. The 1,479 parcels had a mean of 0.7 acres, and most were less than 1 acre. Of the 173 tenants the mean was less than 7 parcels amounting to 4.7 acres. Court rolls for other places show that medieval holdings were very small and the distribution of holdings was very restricted in terms of locations (Campbell 1981–3, 15–18). A holding at North Walsham in the late twelfth century was irregularly dispersed. In East Field ('in campo orientali') there were twenty-three acres, *Suthfeld* twelve acres, *Hagene* ten, *Millecroft* nine, with ten acres next to the messuage and six other pieces, in all sixty-six acres. The implication is that all except one were in blocks, since one three-acre piece is specified as 'lying in three parts' (West 1932, 148–9).

Suffolk had irregular fields similar to those of Norfolk. Holdings often consisting of a few pieces lying in small blocks are found in East Suffolk. At Theberton in 1221, 24 acres lay around a messuage called Caldham. A terrier of 11½ acres at Dennington (1230) lay in pieces of 4 acres, 2, 1½, 1, ½, and ½, plus 1 acre of wood, and 1 acre of meadow, most lying near the 'court' (Mortimer 1979, 136). At Darsham, in the mid thirteenth century, 19 acres and 1 rood of arable lay in three pieces: 7 acres in a field called *Le Dune*; 11½ in two pieces lying either side of the land of Robert son of the Deacon; and 3 roods. All three were in blocks and not further dispersed as proved by the abuttals given (Harper-Bill 1981, 135). No 'great fields' are mentioned in monastic cartularies, but there were limited areas of common fields. The demesne of Wade had 125 acres of arable and 70 acres of heath, both 'common' in 1345 (Mortimer 1979, 125).

Postgate (1973, 290–308) gave an account of Suffolk fields and the terminology employed in describing them. There was much variety. Kennett (1670) had three fields, each with twelve furlongs and run in three shifts each. This is similar to a 'normal' three-field system. Lakenheath had a holding equally distributed in three 'fields' in 1774 (Postgate 1962, 85). Tostock, in the sixteenth century had nine fields of widely differing sizes. More commonly land was divided into many locational 'precincts' that had no cropping significance. The field book of Great Barton (1612) had thirty-seven precincts (some with pasture enclosures within them), but in 1566 another survey did not mention precincts. At Norton (1561) mixed terminology was used describing '*stadia*', 'furlongs', and 'fields'. Mildenhall (1618) had forty fields, some called 'furlongs'. Precinct subdivisions of 'stadia' were not always equivalent to furlongs and could contain meadow, but the smallest divisions of 'pecia', were equivalent to selions, being commonly one or two roods.

Bailey (2007, 102–15) further described Suffolk medieval fields. He classified them into three types: irregular common fields with partially regulated cropping, irregular systems with partially regulated cropping, and non-common subdivided fields. The first is found on poor land like the Breckland. It is the most Midland like, and holdings approach equal distribution through the field; shack operated until February. Folding was practised on these fields. The second type occurred on the clays of central, southern, and eastern Suffolk. Individual pecia tended to be larger, with interspersed crofts of arable as well as large pieces of demesne. There was less folding and the common pasture of greens and droves was important. Common rights on arable were restricted. The third type was characterized by enclosed fields of pasture and arable. Holdings consisted of both open and enclosed pieces and had a very limited dispersion. No common rights existed over the closes and only limited rights on the open arable. Cropping on demesnes was intensive, some with more than 75 per cent cultivated each year.

Essex also had irregular fields with no communal regulation. There were many areas called 'fields' or 'crofts', often with small acreages. Richard Britnell studied one such area around Witham. *Frebernescroft* was called a 'field' in the late thirteenth century and fields called *Ravenstoc* and *Halfhide* were in the hands of a single tenant in 1185 (Britnell 1983, 38–55). Gray (1915, 389–90) found that in Essex some places had dispersed strips, such as Wenden in 1207, where six acres lay in twelve named places (Kirk, 1899–1910, 37). At Arkesden, in the early fourteenth century, fifteen acres were scattered as twenty parcels in ten areas described as 'fields', two of which were named 'shotts' (furlongs). He gave further examples and many more have been published since in editions of various cartularies.

Yardland holdings were not as dispersed as in the Central Region, but lay in a few places only. Gray noted a yardland at Dagenham that must have been in a block because its abuttals were simply the four cardinal directions. Another example is Dunmow (1219), where a quarter virgate lay in a field called *Wudelehe* and in half of *Smithscroft* (Gray 1915, 393–4; from Kirk 1899–1910, 52, 32, 257). There are many other examples cited by Kirk showing that virgates lying in a few discrete blocks of land were widespread in the thirteenth century. Britnell (1988, 159–61) gives further examples of holdings consisting of blocks of land at Colchester and elsewhere. At Elmstead, sixteen acres being part of *Audeshide* lay in a piece by the road from Colchester to St Osyth (Essex CRO, D/DU 23/34). At Layer Marney (1207) and Lexden (1288) there is evidence that a 'field' was part of an undispersed hide. Smaller holdings at Colchester also lay in a few parcels or blocks according to fourteenth-century descriptions, and such holdings often acquired a previous owner's name. A late example from the seventeenth century of a virgate lying partly in closes is at Barking, where it consisted of a croft called Whites Three Acres, four closes with twelve acres of arable together, three other closes, plus seven acres of arable land and a four-acre wood (Gray 1915, 392–3).

There is evidence of commoning, some on open field land as well as on commons. At Pitsea numerous tenants had stinted commoning rights in heaths called *Crossihethe* and *Personyshethe* in *c*.1300. Another tenement called *Spakmannys*, consisted of twenty acres of arable and four acres of meadow and had common for forty sheep and four beasts (Fisher 1951, 103–4). The large Tiptree Heath near Witham was used for intercommoning by many adjacent villages. There was also commoning on fallow arable land at Tolleshunt d'Arcy, Maldon, and Colchester in the fourteenth century (Britnell 1983, 53). Maps and surveys (listed in the Gazetteer) show that some open fields survived in north-west Essex until the eighteenth and nineteenth centuries. It is unlikely, however, that anywhere in the county ever developed a simple organized Midland-type field system.

Writtle fields in the fourteenth century have already been discussed in Chapter 1 (pp. 15–16). There was also a demesne cultivated on a three-course tilth.

Cromarty (1966) described and mapped the fields of Saffron Walden, based on a very detailed survey made by Walden Abbey in 1400, and was able to relate it to a 1758 open-field map of the western part of the parish. In this case the term 'field' referred both to a single furlong and to a large area of named furlongs, sometimes called 'shots' or 'crofts'. The demesne crop for any one year lay in several pieces that were not adjacent. Smaller manors were complex. Little Walden had four open fields with woods and pasture. Of the seven sub-manors, two contained villein and demesne lands. Pounces manor (1331) had seven fields that were held in severalty throughout the year and were never opened for folding. Thaxted open fields and parks, reconstructed from a survey of 1393, are described in the Gazetteer.

Sheepfolding in East Anglia

Gray (1915, 325–9) explained Norfolk sheepfolding in the sixteenth century as follows. Townships were divided into a few large areas called 'courses' where the sheep of two or more manors pastured on commons and on the fallow. The courses did not correspond to the precincts (large areas of open field) nor to the arable ownership and tenancies. The tenants and small owners did not have sole pasturing rights over their own fallow unless they participated in one of the fold courses.

Gray used evidence from early maps for Weasenham and for Holkham, which has a map of 1591. This shows three unequal precincts called 'fields' with commons of marsh at the north and ling (heather) at the south. Four mapped folds took in the whole parish but did not correspond at all to the fields: three folds belonged solely to three manors and the fourth was shared by a major landholder and the village farmers. Acquisition by the Holkham Estate of a small manor in 1563 explained the system. The land lay as 25 acres for the manorial site, 235 acres in the large South Field, and 67 and 89 acres in the other two (not a uniform distribution throughout the fields). There were pasturing rights all the year round for this manor, and for all village animals on the commons and over the unsown fields from October to March. In the summer access to fallow arable was restricted to the sheep of the various folds. The animals of each fold were kept within its bounds. Gray (1915, 341–4) concluded that this type of fold-course system, as mapped in the sixteenth century, operated in the thirteenth century on ecclesiastical manors.

Bury St Edmunds Abbey had the right of folding on many of its Suffolk manors in 1086. For example, at Risby all tenants' sheep belonged to the fold except for those of one tenant who had a fold for himself; at Pakenham all the lands of the villeins, freemen, and others belonged to the fold (Williams and Martin 2003, 1236 and 1240–1, 1243). The following examples suggest that, although sheep were important on most manors in the early Middle Ages, the lords' folds were restricted to normal commonable areas with the additional advantage for the lords that tenants were to put their sheep to the lordly fold. In 1278 tenants of manors belonging to the bishop of Ely carried manure to lands in the fields as work-service. There was also service of making wattles and carrying them about for the fold. The tenants' sheep had to fold with those of the lord during the summer. In some manors there was payment of 'foldage' for sheep and cattle when they did not have to lie in the lord's fold (Gray 1915, 341–3).

The abbot of Ramsey dealt with his sheepfolds in two different ways on his Norfolk manors in the thirteenth century. Brancaster folds were let out to eight freeholders and to four tenants holding land that had been 'cotsetland' in the early thirteenth century. Sheep were mentioned on two other holdings. Of special status was one of the tenants, who was a villein but held twenty acres of land and a fold by charter, so mixing up servile status with tenures normally held by freeholders. This was entered on a court roll of 1239 so that the legal status was quite clear and recorded in writing. All the other villeins had to put their sheep in the abbot's fold as part of their work-service. This service was presumably devolved to one of the leased folds, each villein knowing to which one he owed service (Hart and Lyons 1884, iii: 261–5; i: 423; iii: 415).

In contrast, Ringstead had only one fold let out (Hart and Lyons 1884, iii: 266). The other folding rights remained with the abbot, and one villein had custody of the fold with various privileges. He had a three-shilling stipend plus one acre of rye and a rood of barley, with an allowance of bread and cheese for mowing these lands. He was allowed the lord's three best lambs and three good fleeces. He could have his own fold for a total of forty-five days either side of Christmas (Hart and Lyons 1884, i: 408). All the villeins had to carry the fold as part of their work-service—so proving that sheepfolding on a large scale was practised, but probably restricted to the demesne lands only.

Three freeholders of the manor of Bradcar in Shropham in 1298, holding 175 acres in neighbouring places (Great Breckles, Great and Little Hockham, and Snetterton), had fold courses, two of them foldage without number. The pasture of Banham Outwode (1281) was also used by fourteen people from Tibenham who had licence to pasture 1,320 sheep (Hudson 1898–1900). A grant of forty acres of land in three pieces at North Creake made in *c.*1210–15 included the right to erect folds on all the commons of the town, and there was a similar right to set up a fold in South Creake in 1262 (Bedingfeld 1966, 7, 23). In 1350 Creake Abbey also had the run of a fold on all its lands in Quarles, Holkham, and Burnham Thorpe (Bedingfeld 1966, 89). At South Creake, sheepfolding rights as late as 1559 were limited to certain areas on the edge of the waste. Four of the chief landowners were involved and had their own pastures amounting to upwards of half the parish. There were fragmented tenancies within the area, so that arable farming and sheepfolding ran together. Later the fold courses were increased by the lords (Hesse 1998, 87–90).

Castle Acre Priory charters refer to foldage rights at Wesenham in the thirteenth century (quoted by West 1932, 226–7). Thetford priory kept a fold course on the fields of Brettenham and Kilverston and made payments to the tenants (for lands taken in) in 1521 (Dymond 1995, 357, 395). Gimingham court rolls record that the fields had been overburdened by the folds during 1493–9. In 1582 the fold course boundaries were defined and there was to be no enclosure on the field course without agreement of the farmer and tenants, but previous enclosures were exempt and not to be opened (Hoare 1918, 215; 332). Colkirk had three fold courses in the sixteenth century that were not coterminous with the four geographic fields (Rutledge 1995, 18).

The operation of the folding system, with the lord or fold owner grazing other people's lands as well as his own is well illustrated by the activities of Roger Townshend's steward, Thomas Skayman, for the period from 1516 to 1518. During these years Townshend had fold courses in fourteen parishes. There were payments to other farmers and landowners for lands taken into the sheep pasture and rents received for parts of Townsend's sheep pasture that temporarily reverted to tilth and was let out. Thus at Barmer, Skayman went to lay tilth lands out of the pasture for five holdings of 24, 22, 7½, 2, and 4 acres. For which other parts of the holdings were received into the pasture, but as four acres more were laid out than received into it, one of the farmers had to pay a year's rent for these acres. Also at Barmer, Skayman made an exchange by taking out of the pasture and putting into the tilth two furlongs called *Shortmordell* and *Cleypit Furlong* and received into the pasture two furlongs called *Shyplowdyk* and a furlong abutting on *Systern* [Syderstone] field (Moreton and Rutledge 1996, 119).

At Langham in 1593, the lord, Nathaniel Bacon, made an agreement with several owners and tenants about the commons and fold course. It was probably needed to resolve the conflicting interests in a common of 160 acres where the village had pasture throughout the year for great cattle and Bacon had liberty of foldage. It was agreed that the lord was to have part of the common in severalty and part of Langham Heath and all the pasture and closes which he had enclosed. The tenants had right of common over the remainder. Bacon gave up right of shack on some of the tenants' lands and retained it in some of the other fields from October to March with 400 sheep. The tenants could enclose lands over which the lord's right of shack had been given up (Smith and Baker 1990). This kind of agreement regulated the run of sheepfolds and acted as a driving force for piecemeal enclosure by both the lord and other landholders and tenants. Not all owners assented to the foldcourse system and there was payment to the parson of Barmer for 'land kept in tilth to gret hindrance of the pasture at Bircham'. There was also nuisance caused by 'treasure hunting hill diggers at Hedon' (Moreton and Rutledge 1996, 99–132).

Allison (1957, 12–30; 1959, 98–112) showed that during the sixteenth and seventeenth centuries there was an open-field agricultural practice with sheep-grazing alternating with corn over about two-thirds of Norfolk, excluding the fens and broads (maps pp. 13–14). Soils were sandy to varying degrees and there were commons. Arable fertility was maintained by the flocks of sheep tathing on commons and fallow. Large flocks belonged to the manorial lord. The area of open-field and common was allotted to a particular manor and was called a fold course. It included open-field lands of tenants and freeholders as well as the lordly demesne. In the summer, grazing was limited to commons but at other times was increased using the arable after harvest and when fallow. Meadow land was sometimes included in fold courses, to be used after hay was removed. The extent of fold courses varied; at North Holkham, as pointed out by Gray, the whole parish lay in four folds in 1590, but at North Creake, in the early seventeenth century, the folds did not include all the open lands but took in some closes that had to be made available. The fold was made of hurdles to control the herd for tathing a particular area or keeping animals away from sown lands. Further details of flock management during the same period are available. Fold course owners received payment from tenants whose land was dunged by the lord's flock and for the privilege of having a few sheep run with the flock.

Cropping the fields had to be fitted in with the fold courses. When a tenant's land fell in a sown shift, lands were planted, and when fallow, the flock ranged over. Owners of strips that were required to be fallow were compensated by the lord in several ways: by temporary exchange of lands, money payment, reduction of rent, or by allowing an increased number of sheep to go into the lord's flock, for example at Sedgeford and Hindringham. Many tenants were allowed the right to put a few sheep in the lord's flocks, usually relating to the amount of land they held.

Sheepfolding was very extensive in Suffolk from an early date, similar to Norfolk. Relatively large demesne flocks were recorded on the Brecklands in Domesday (Darby 1977, 166). Blythburgh Priory, Suffolk, had right to fold 1,000 sheep on the demesne marshes between Norwich and Yarmouth in 1135–47 (Harper-Bill 1981, 55). In the early thirteenth century the fold course belonging to the abbot of St Edmundsbury at Hardwick lay in one block bounded by a river and roads and included commons and some open fields. Villein work-service included making five hurdles with timber from Hardwick Wood and moving the fold from one field to another after barley had been sown and when manuring began in summer (Tymms 1853, 185). At Hepworth, in *c.*1295, one of the manors had the right of a fold (Corbett and Methold 1900, 40). At Walsham-le-Willows, during the fifteenth century, sheepfolds were erected illegally in various places and pastures were overgrazed (Dymond 1973–6, 205). There is no clear evidence of a fold-course system at that date in this part of the county.

In a study of the fold-course system in west Suffolk by Mark Bailey (1990), it was shown that folds became larger during the fifteenth and sixteenth centuries and that lordly owners increased their grazing rights at the expense of small freeholders and tenants. This led to the highly regulated fold courses of the sixteenth and seventeenth centuries operated by flockmasters as described by Allison for Norfolk. It also necessitated more control of cropping, as detailed in the journal of Thomas Skayman.

References to sheepfolding in Essex are sparse. Hurdles were made as part of work-service at Kirby in 1222 (Hale 1858, 43). A fold system at Saffron Walden, in 1400, was independent and different from the furlong-cropping system, and thus similar to Norfolk examples (Cromarty 1966, maps 3 and 5).

The evidence thus shows that in the early Middle Ages it is likely that East Anglian sheepfolding was limited to manorial demesnes and common pastures. For effective sheepfolding, it was essential to have large areas of pasture, hence less folding occurred in the mainly arable areas of central Suffolk. The oppressive increase in lordly folding found in East Anglia during the sixteenth and seventeenth centuries has no parallel elsewhere.

C. THE WASH SILT FENLANDS

The low basin of Fenland draining into the Wash stretches from near Cambridge in the south, to Peterborough in the west, and north of Boston in Lincolnshire. Peat developed when the basin became waterlogged as a result of rising sea levels and it was partly overlain by various marine deposits from time to time. Close to the Wash and the Lincolnshire coast, a belt of silts accumulated during the Iron Age, which raised the land surface to about twenty feet above Ordnance Datum. The fertile soil provided by this deposit was exploited in the Romano-British period and continued into the early Middle Ages. By the ninth century it was threatened by flooding and was protected by building banks against the fen on the west and the sea on the east (Hall 1996, 186). Holdings were generally held in severalty without any communal crop regulation.

Much of the arable land lying on the Fenland silty deposits was divided into strips called 'dylings' or 'darlands' (Figure 3.1). In Cambridgeshire, strip widths commonly varied from thirteen to twenty-two yards and they were separated by dykes and ploughed flat (from the evidence of the few that survive as earthworks, as well as from numerous pre-1830 maps). Groups of strips were called 'fields' and were surrounded by drainage dykes and banks and sometimes by access roads. The 'field' thus superficially corresponds to a Midland 'furlong', but was often much larger in area. Fields near the earliest villages have short strips, but in reclaimed areas where the ground fell gently towards the peaty fen, massive fields were laid out with strips up to a mile in length. There was

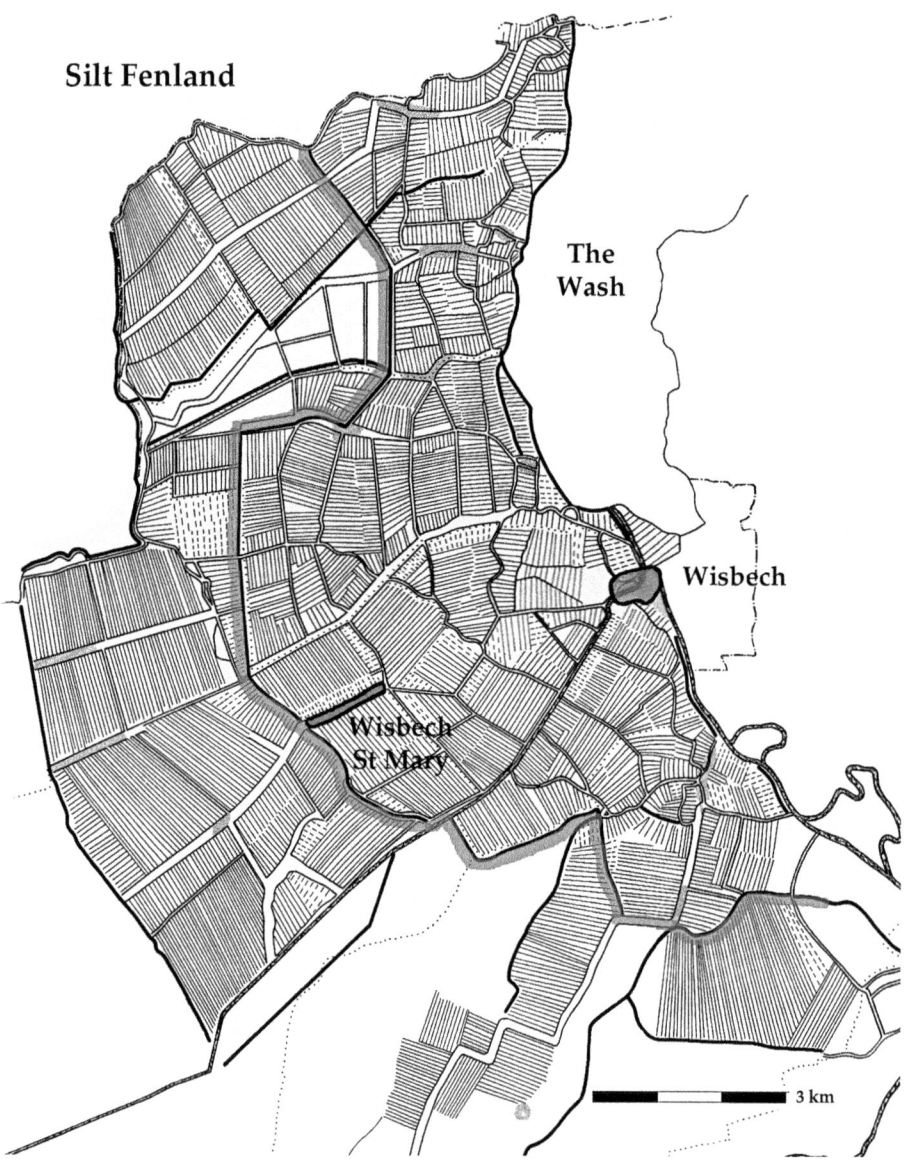

Fig. 3.1. Cambridgeshire Silt Fen ditched fields. Fen banks mark the boundary between inner small fields and large outer fields on the west and south (after Hall and Coles 1994, 147).

grouping into larger areas also called 'fields' that were locational and had no cropping significance.

The pattern is one of small fields near the earliest settlements on the highest ground (*c.*15–20 ft OD) with access droves leading to the grazing resources of the fen. Near the fen, droves become wide and the fields large with very long strips. This phase is associated with reclamation after *c.*1200. Subsidiary settlement was made in the droves, so creating a somewhat dispersed pattern. The drove system is best illustrated by the Marshland villages of Norfolk (Silvester 1988, fig. 124). Nine primary settlements located close to the Wash have wide droves, some of them nearly three miles in length, leading through strip fields to a common fen. Quadring, Lincolnshire, has a few strip fields surviving in earthwork form (Hall and Coles 1994, 156). The complete field systems of the Cambridgeshire and Norfolk Siltland have been reconstructed from historical sources and the general layout is still evident on modern maps and satellite images (Hall 1981b, plan on pp. 44–5; Hall 1996, 187; Silvester 1988, fig. 45). Most of the Cambridgeshire and Lincolnshire parishes have eighteenth- or early nineteenth-century maps and field books.

Hallam (1965, 142–4) has described the Lincolnshire Siltland fields. Late records (acre books or field books) show that the township systems were divided into blocks called 'boundaries' or 'bounders' (exactly equivalent to the Cambridgeshire and Norfolk 'fields') that were separated by dykes, banks, and roads. The sizes varied from a few acres to several hundred. Pinchbeck, in 1647, had 30 fields or bounders extending to 4,517 acres. The number of fields varied according to the whim of the surveyor; in 1611 Pinchbeck was divided into 35 bounders; Spalding had 25 bounders in 1626 describing 3,740 acres, but in 1619 there had been 43 bounders or fields. There is reference to an earlier book for Whaplode in 1491 and that of 1315 for Fleet has been published (Neilson 1920). Hallam showed that the 1315 survey closely resembles one made in 1747. In that year 4,343 acres were accounted for compared to 4,727 acres described in 1315.

Hallam found early evidence for the existence of strips surrounded by dykes at Swineshead and Sutterton in the period 1190 to 1202. A Gosberton selion had a dyke on either side in 1253–74. In 1965, examples of the ditched strips could be seen at Lutton, Long Dutton, Tydd St Mary, and elsewhere in Holland. Old people then called then 'dylings', and an early reference to the name was found for Fishtoft in 1357, when five groups of 'dylyngs' were listed. From Holbeach, oral evidence of about *c.*1820 stated that dylings were also called 'lakes and ollands'; farmers had created them for pasturing sheep on dry land by digging out the lakes and throwing the soil on top to raise the land. This accords with a fourteenth-century record of Spalding Priory stating that manorial lands were dyked and raised. Early records of the name 'olland' were found for pasture at Pinchbeck (1535), where such pastures were called 'holands', and a late twelfth-century charter of Swineshead Abbey referred to a land called *Yelanhouland* with all the dyke on the north side (Hallam 1965, 151–6). The fields were managed using a system of banks to keep out water both from the sea and from the fen. There were six main medieval drainage channels for Cambridgeshire as well as the Old Croft River (formerly the Ouse). These drained by gravity. Water accumulating from rain and seepage into the strip dykes within each embanked field drained into the main channels, and was removed by opening sluices at low tide (Hall 1981b, 43–7).

The villeins of the Ramsey Abbey manor in Walsoken, Norfolk, had holdings ('yardlands') of only nine acres in 1409 (rental held in Wisbech Museum MSS, Walsoken parish box). This is consistent with the high fertility of the soil—a living could be obtained from such a small area. Thus the strips of these holdings must have lain in only a few of the smaller fields; they could not have been in more than one or two of the large fields, where strips of great length could be eight acres in area. Neither were the larger holdings distributed throughout a given township. The land belonging to the Chapel of St Mary by the Sea, at Newton in the Isle, Cambridgeshire, founded early in the fifteenth century, was excluded from the Dissolution in *c.*1540. It can be identified in a field book of 1833 as land belonging 'to the master of the chapel' and has been mapped (Hall 1996, 188, from Wisbech Museum MSS, Newton parish box). There were 17 plots of 2–5 acres irregularly dispersed in only about half of the 1,400 acres of the old strip fields, with 3 larger blocks to the west in the lower peaty ground not drained until the fourteenth century. The holdings at Fleet, Lincolnshire, were more dispersed than at Elm, Cambridgeshire, but never uniformly in all the fields (see the Gazetteer).

The scale of the landscape was large. Elm had about 4,300 acres of fields in its total of 11,378 acres, the remainder being fen ground. Holdings can be studied from a field book made in 1391 (nineteenth-century copy in Elm Box, Wisbech Museum, probably from Ely Diocesan Records A6/1, which is currently unavailable). It was probably made for a tax assessment and valued the fields at different rates, most being two to four pence per acre, with a few of the outer fields worth less than a penny. The names of holders and acreages are listed for 3,273 acres of land lying in 14 fields that varied in size from 59 to 730 acres. Excluded were details of Coldham and Waldersey manors at the south, which from their valuations were about 400 and 600 acres respectively. The 3,273 acres were divided into 311 parcels shared by 184 holders. Six core fields closest to the vill contained 771 acres and 167 parcels ranging in size from 2.7 to 7 acres. Some fields lying farther out had only a few owners (e.g. Lylly Field's 365 acres were held by only four people) and were likely to be late seigneurial intakes.

Of the 184 holders, 73 per cent had a single parcel only and a further 14 per cent had two parcels. Hence there was essentially an undispersed system of holdings and they account for the linear settlement of Elm that is based on a street and drove nearly one mile long with dwellings on the street side of their single strips. Of the twelve holders with more than four parcels, five were seigneurial plus the rector, and three of those had land only in Old Field. This was possibly the first cultivated field of Elm and the pieces very probably had been demesne. Another survey of Elm was made in 1592 (Cambridgeshire CRO, P63/28/2). It, too, omits the southern fields near Coldham Hall but describes 3,823 acres of the main block around the village. Land holding was complex with parcels varying from 0.6 to 600 acres. Most of them can still be identified on the 1840 tithe map, but between 1391 and 1592 there had been much consolidation. The sources quoted show that most holdings in the Wash Siltland were small with little or no dispersion.

The operation of the Siltland fields is nowhere clearly stated. They probably were farmed in severalty by the strip holders, since most holdings consisted of a single strip only. The ditched strips could not easily be opened to common grazing when the ditches were functioning drainage dykes. There are no court orders for communal cropping, which would not be expected in extensive systems of fields where most holdings had very limited distribution. The earliest orders for Walsoken, Norfolk, where there were three manors belonging to Ramsey, Huntingdonshire, Ely, Cambridgeshire, and Lewes, Sussex, only relate to officers and drains. Thus a hog reeve was appointed in 1295, with reeves for the sea dyke. As for drains, a tax was levied for widening the dyke from Welles (Upwell) to Elm (Ault 1972, 76–7). Manuring was probably done from accumulations in winter byres, where the large numbers of animals that grazed the fen in summer were kept. Most of the fen was usable as summer grazing, hence the droves up to 130 yards wide for the movement of animals. This would be essential to accommodate so many animals if a system of manuring by folding was used. Hallam (1965, 162–7) found that in Lincolnshire only the fen grounds were communally regulated, although at Wrangle doles were opened for commoning when the hay was taken off and tithes were paid for hay when the ground was not arable. Pinchbeck and Spalding by-laws of 1422–1743 showed that there was no regulation of arable, only the fen pastures (Hallam 1963). The most likely explanation is that the holdings, which consisted of ditched strips like Elm, were held in severalty. Hallam (1965, 145) referred to selions that seem to be smallholdings similar to those of Elm.

The whole arrangement of Siltland agriculture depended on a great earthwork called the Sea Bank that kept out tidal waters of the Wash. Archaeological evidence has shown conclusively that the Sea Bank was constructed in late Anglo-Saxon times, before 1066 and after the Middle Saxon Period *c.*850 (Crowson et al. 2005, 146, 170–1). The earliest dated field name is 1190, but it is safe to assume the fields were much older than that. Wisbech St Mary is located on the west side of the ancient block of Wisbech fields defined by an inner fenbank. It was given to the bishop of Ely when the see was created in 1109. It must then have been a village with fields and rents or it would have had no value as a temporality providing income. The troubled times of the Norman Conquest in the Fenland are not likely to have been one of seigneurial investment into large-scale wetland reclamation, so the creation of Wisbech St Mary and its fields is likely to have been before 1066. This is consistent with the many vills that were recorded before then, such as Elm in 973, Wisbech in *c.*1016, and Walsoken (in Norfolk) in *c.*983 (Hart 1966, 171, 214, 241). Some fields near to Wisbech St Mary, part of the early block, have strips over 1,100 yards long. West of the inner fen bank, at Parson Drove and Murrow, strips created by the thirteenth century are over a mile long.

The importance of the Siltland strip fields is the chronology of their genesis. They have no relation to the Roman paddock systems that preceded them. The largest area of Siltland Roman ditched fields occurs at Christchurch, Upwell, Cambridgeshire (TL 49 97). They form an approximately rectangular network of 200 acres of cropmarks (Hall and Palmer 1996, 176–7 and fig. 96). The layout is quite different from the strip fields in form and they occur in a low-lying area that reverted to fen. The Lincolnshire Siltland similarly has extensive cropmarks of Roman date. Near Pinchbeck, there are trackways, settlements, enclosures, and linear ditches (Palmer 1995, figs. 2, 8). As with Christchurch, the overall plan does not resemble a medieval field pattern. The strip fields are therefore later than *c.* AD 400.

The Siltland medieval fields could not have existed before the Sea Bank, which, as already mentioned, archaeological evidence shows to be later than 850. However, a large part of the system was in existence at the time of the Domesday Survey in 1086, when all the primary villages were mentioned. Thus the principle of dividing land into strips was adopted in the Late Saxon Period. It was done on a large scale and a great part of the Wash Siltlands must have been embanked against the sea at the same time for otherwise the system could not have been effective. The subsequent process of enlarging the area cultivated, field by field, could have been done piecemeal by each township as it pleased once the Sea Bank had been constructed. It is likely that the major landowners and freeholders would have new ditched strips assigned to them in return for the effort. Such fields are possibly the lands called 'offoldfal' in Lincolnshire that were very carefully measured out (Hallam 1965, 158–60). Thus the expansion of the Siltland field systems closely resembles theories of field development once postulated for Midland systems, taking in a 'furlong' one at a time out of waste and sharing the newly won land. In the case of the Siltlands, the process would not have begun until the tenth century and continued until the thirteenth century, some of the intakes being seigneurial and not shared.

D. SYSTEMS WITH ONE OPEN FIELD

One-field systems may be considered as a variant of irregular systems. Furlongs were not grouped into regular fields, holdings were not dispersed throughout the system, and there was no communal regulation of cropping. The 'half-year' systems of Lancashire are good examples of the type. Records of many northern counties describe the use of 'ringyard' fences that surrounded the arable to prevent trespass by stock. Some irregular fields akin to one-field arrangements are also known in the Central Region.

Many regions of the country, as well as East Anglia, had groups of strips ('furlongs') which were not grouped into any system of 'fields' in the Midland sense, where 'great fields' had a higher hierarchy than 'furlongs' and were cropping units. Instead the terms 'field' and 'furlong' were used interchangeably, both being used for a piece of land that was called a furlong in the Central Region. Many examples are cited in the literature and in the Gazetteer; a few are given below for illustration.

In Staffordshire, a terrier of ten acres at Whitgreave, *c.*1225, referred to a 'field' called *Elfurlong* and two selions in the 'croft' of Hugh Pipart (Parker 1887, 198). At Trentham, in 1357, 11½ acres were dispersed in thirteen pieces, two of them in 'fields' called *Longefurlong* and *Le Cruftes* (Parker 1890, 306). At Manley, Cheshire (1265–91), two-and-a-half selions in *Asponesfurlong* had one of them lying near the land of the abbot of Chester 'in the same field' (Tait 1920–3, 390–3). In parts of Worcestershire parcels of arable were similarly called 'fields' or 'furlongs', such as at Northwick, where in 1299 two demesne *campi* ('fields') were called *Chamberlengesforlong* and *Chamberlengesfeld* (Hollings 1934–50, 1). Wick Episcopi, Hanbury, and Alvechurch have similar examples.

For Cumberland, Gray (1915, 237–40) noted that early terriers of dispersed holdings never grouped furlongs into organized fields. He gave details of a thirteenth-century oxgang at Tallantire that had no indication of fields and had only one furlong in common with another grant in the same township, suggesting that holdings did not have a uniform distribution. The open-field townships of the Gilsland survey of 1602 (Graham 1934) and the 1704 glebe terriers of the Carlisle diocese (Ferguson 1877) likewise had no grouping into great fields.

In many places a group of furlongs was simply called the 'town fields' meaning the 'town furlongs'. Lancashire provides the clearest examples. Youd (1962, 20–34) collected open-field data and court orders for most parts of the county. He concluded that the townfields were cropped as a one-field system. All cattle were removed from the fields on Lady Day or thereabouts. The fences, hedges, or walls around them had to be repaired to make them stockproof. Each tenant was responsible for a proportion of the length. Cattle were not allowed to enter the fields again until the end of harvest, about Michaelmas. In other words there was tillage for six months with spring crops only and for six months fallow, when common grazing was allowed. The lands were otherwise known as 'half-year' lands. Youd cites ten examples of this system with dates from 1496 to 1722. Earlier examples of places with one field in the thirteenth century (consisting of several furlongs) are Allithwaite, Slyne, Halton, Stalmine, Preesall, and *Staynell*. The word 'furlong' was rarely used and strips were called 'acres', 'lands', 'butts', 'doles', etc. Some townships had named 'fields' that were locational only. Salford had five named fields in 1584, but *all* were sown with various combinations of barley, oats, and beans. Each field had variation in the crop planted from year to year. This is in effect a one-field system.

Townships with more than one named field did not necessarily have holdings equally distributed between them. At Bolton le Sands, Lancashire, two tenants had unequal amounts of land in both of two fields, and the lands of twelve tenants were entirely in one field or the other. Other townships had a similarly irregular distribution, such as Leyland and Melling in the Lune Valley (Youd 1962, 19). Another Lancashire township with locational field names, which the evidence suggests was probably run as a single-field system, was Chidwall. Holdings were irregularly dispersed among six named common fields in 1653, all of them small. Grangefield, the largest, had sixty-five acres divided into eleven parcels shared among six tenants. Church Ashfield was entirely arable with twenty-eight acres shared among three tenants. One field had fourteen acres of pasture held by three tenants. The other three fields comprised thirty, twenty-three, and forty acres, and were used for arable and pasture. Of the twenty-three tenants sharing these six fields, four had lands in three of them, one in two, and eighteen in only one field (Dottie 1986, 18–19). The field names are therefore locational only. The mixed land-use suggests convertible husbandry, but not in a communally organized way. Some enclosed fields were used for arable. A wide range of crops was grown, including vetches, and all holdings had small flocks.

In Cheshire, Chapman (1953, 58) thought that 'townfield' was a late term used for remnant open strips after much land had been enclosed, but Sylvester showed that Knutsford townfield was referred to in 1430. A county map of the number of 'field' names used found that there were seventy-seven townships with a single named field, twenty-four with two named fields, six with three, and twenty-one examples with multiple field names (Sylvester 1957, 13, 18). In many cases, however, it is unlikely that the number of names had anything to do with cultivation courses. Most were probably one-field systems with an irregular distribution of holdings. The same likely applies to the fields of Herefordshire that were marked on the county map produced by Sylvester (1969, 220–1). Graham (1910, 132–3) thought that Cumberland fields operated as a one-field system that was continuously in arable and fertilized by manuring.

The William Senior detailed surveys accompanied by maps made for the Chatsworth Estate in the early seventeenth century covered about 49,000 acres of Derbyshire, lying mainly in the centre-north (Fowkes and Potter 1988, xv–xix). Analysis of twenty-three townships that contain some open field showed that field numbers vary from 'none' (only details of furlongs and flats provided) to as many as seven. The field sizes were small and only amount to 22 to 34 per cent of the holdings (and even less as a percentage of the whole townships: 13 to 15). Most field names were presumably used in a locational sense. It is probable that all these Derbyshire townships were cultivated more individually than the Midland system. When open-field arable was supplemented with closes and large areas of common pasture, communal control of the fallow for rough grazing or manuring was not necessary. Arable manuring could have been arranged by a communal fold system (as was described earlier in Sussex and Dorset), but in the absence of any written evidence, it was more likely done by carting dung from private folds and byres. The surveys of Heath, Stainsby, Stony Houghton, Upper Langwith, and Ashford specify that there was a (permanent) fold for nearly every farm.

Wetton, Staffordshire, was also surveyed by Willam Senior for Chatsworth in 1616 (Fowkes and Potter 1988, 142–6). Details are provided for 1,885 acres out of a total of 2,306 for the township. In all, the open field arable amounted to 286 acres (15 per cent) and this was divided among the tenants amounting to *c*.25 per cent of their holdings (the remainder consisting of closes). The tenants' holdings were distributed among many named pieces, collectively called 'arable in all the flatts' or 'arable in all the fields', of which there were six. The holdings were distributed indifferently among them, some being in two, three, four, five, or all six furlongs. Wetton, then, had only a small area of open-field, comprising six furlongs irregularly shared by the holders. It was probably cultivated as a one-field system. Tenants wishing to grow winter crops could do so in their closes.

For one-field systems, it is not necessary for holdings to be uniformly distributed throughout the system. Gray (1915, 355–69, 549) examined many Surrey terriers and found only irregular field systems involving arable closes, and with lands dispersed irregularly in many fields or in just one field. Later terriers and surveys show the same type of complex arrangements with no mention of great fields, as at Merstham in 1522 (Hylton 1907). Putney has a 1498 field book describing six areas called 'fields' (Northamptonshire CRO, Spencer SOX 136). The field sizes vary from 39 to 215 acres (with a total of 581 acres) and the holdings are distributed unevenly among them. For fifteen holdings with 3 acres or more only the two largest and two with 36 and 26½ acres have land in all six fields. The three-acre holding lies in four fields but holdings of 5 and 41 acres lie in only two. It is not possible to group the fields into any combination of equally balanced two or three areas for 'tilths,' so it must be concluded that the 'fields' are locational descriptions only. Another field book of 1617 describes four named fields; two were small and two large. Furlongs, called 'shotts', were open for the most part but there were a few small closes, and some of the meadow pieces were enclosed. Yardlands had only limited dispersion. One holding, held by John Starkey, consisted of thirteen acres in nine pieces lying in only three furlongs, all in Thames field. Another yardland amounted to 12¾ acres, being six pieces in five furlongs: one in Bason Field, one in Park Field, and the others in Thames Field (Northamptonshire CRO, Spencer 7j5).

The structure of the fields of Winnersh and Sonning, Berkshire, in *c*.1600 is illustrated from detailed surveys published by Gray (1915, 553–4, from TNA, LR 2/202). More detail is given in the Gazetteer. Both places have yardlands that include enclosed ground, usually arable, as well as open land lying in many fields and also meadow. Enclosed arable was presumably held in severalty and kept free of common grazing animals. It was possibly used for winter corn, if the open field was run as a one-field of 'half-year's land' like Lancashire, growing spring crops only. Winnersh had an enclosed half-yardland that may represent an 'ancient' half-yardland never dispersed. West Hendred affords an example of a yardland split into only two arable blocks plus a piece of meadow. In a lease of twenty-six acres made by Holy Trinity Priory Wallingford, in 1501, there were ten acres of arable in the 'sowthfeld of Westhenreth bytwene ii ways' and abutting *Spersoldefeld*, and ten acres in the 'north feld' next the crofts of the tenants, with six acres of meadow, and all manner of custom of a yardland (Milne 1940, 71).

Settlement on the Chilterns was dispersed and the fields complicated and irregular. Chalfont St Peter, Buckinghamshire, can be studied through a long series of court rolls from 1308 to 1538. The many small settlements included the assart of Newland and a field called Deanland that was the subject of litigation in 1231. A survey of *c.*1333 lists some virgated land with names that are now identified as farmsteads. Others were called 'quadrentals', defined in 1401 as a messuage, and thirty acres called 'farthing lands'. Enclosure was already occurring in the 1330s, but there were still holdings dispersed in open fields lying far apart. A holding of 1461 consisted of a croft of four acres and another croft adjacent to it with four acres divided in the four fields of Chalfont. Another of 1453 consisted of named properties and three parcels, one in *Hatchesryding* with common in Smith's Heath (Elvey 1961–5, 354).

At East Burnam, in 1208, one yardland was divided into two using 'sun division' to identify the halves. Sun division, or Swedish *solskifte*, is the system of identifying parcels that always lay either towards the east or south of a particular piece, or on the west and north. At Burnham, one half was the 'sun side of odencroft, the shade side of a croft at the head of odencroft, [etc.]…and the shady half of the meadow at Boveney' (Hughes 1940, 33). This is a yardland lying in crofts with very little dispersion.

Cotterstock and Glapthorn, Northamptonshire, lying adjacent, form an example of a one-field system in the East Midlands (Hall 1995, 272–3, and 2009, 36, 94–5) being similar in many respects to the multiple 'fields' found in other parts of the country, especially the north. The use of the word 'field' in this township seems to be for locational purposes only and there was a one-field system in the fifteenth century. It is possible that many of the small fields were taken from woodland by assarting and had never been organized into a formal three-field arrangement.

Hartwell, Northamptonshire, extending to 1,429 acres, has dispersed settlement. Archaeological fieldwork revealed seven small separate medieval settlements yielding thirteenth-century pottery sherds (Hall 1975), three of which where incorporated into a small deer park created in the early sixteenth century (Baker 1834, i: 52–3; Pettit 1968, 14 n.; TNA, E36/179). All are now reduced to single farms or are completely deserted. The present-day village is located in what had been a green on the edge of Salcey Forest and does not have proven medieval origins but was established at the present site by 1727. A map of that date (Northamptonshire CRO, G360) shows the township more than half enclosed. Some enclosures lying adjacent to Salcey Forest are called 'assarts' and some furlongs were called *Stocking*. Groups of enclosures, representing the abandoned settlements, lay in small blocks around isolated farms or groups of houses. One called Chapel Farm had a two-cell Norman church until it was removed to the present village in 1851 (Whellan's *Directory* 1876, 571). The settlements and fields are linked by winding hollow ways and appear to have been created by assarting and never reorganized to form a single compact village.

Hartwell's field system can be studied from a survey made in 1605 (TNA, LR2 201) when the settlements were already shrunken or deserted. In all there were 18 holdings totalling 315 acres of open-field strips, 98 of old enclosure and 23 of meadow, with 17 areas called 'fields' and 8 other pieces. Only 11 fields were referred to more than once and these contained most of the open-field land, 258 acres. The arrangement of each holding in the fields was completely irregular: three holdings were dispersed in two fields, four in three fields, six in four fields, one in five fields, two in six fields, one in seven fields, and one in eight fields. Holdings in each group were irregularly dispersed; thus the four dispersed in four fields had a different selection of four (Plate 7).

Hence, there was not an organized field system of Central Region type in 1605, and no small townships relating to the deserted vills can be established. The multiplicity of field-names at Hartwell were locational, and presumably represent a one-field system of the Lancashire or North Derbyshire type, with no regular arrangement. A simpler field system for the whole township had been imposed by the eighteenth century, when there were three unequal-sized fields occupying only a third of the total area, controlled by standard Midland-type open-field orders (Hall 1995, 291). Hartwell settlements were located in what had probably been woodland assarts, similar to its Buckinghamshire neighbour, Hanslope, where there is much dispersed settlement.

One-field systems and their equivalent 'multi-field' systems are commonly found in regions with large amounts of pasture. A clear example is Sheldon, Derbyshire (see Figure 1.3). Another small-scale system lying in the fells of Westmorland, surrounded by very extensive pastures, was the Sill Field of Preston Patrick, mapped in 1771 (Atkin 1993, 145–7). It has an interesting double-oval enclosure, the larger being a several pasture. The smaller enclosure, called the Great Field, was divided into forty-eight open-field parcels totalling thirty-eight acres, divided amongst four tenants in a completely dispersed arrangement. Each tenant also had two or three adjacent closes (most of them called 'infield') in all 12½ acres, as well as an exact quarter share of the 102 acres of the larger oval pasture. Both enclosures were surrounded by open fell.

Cumwhitton, Cumberland, had a small partially radial strip system surrounded by extensive fell pastures, mapped in the seventeenth century (Roberts 1988, 171), that may have been a one-field system (surveyed in 1603, Graham 1934, 120–32).

North and West Worcestershire did not have regular fields; they were probably 'single' field systems. Glebe terriers refer to multiple fields at Himbleton, Salwarpe, and Rushock in the seventeenth century; Himbleton glebe (forty-seven parcels) was irregularly distributed in six fields (Yelling 1968, 33–4). The Feckenham Forest region of the county, in the sixteenth century, had townships with a nucleus of strip fields surrounded by a fringe of small enclosures held in severalty (Hilton, 1959).

The management of the arable in 'half-year' systems has been described for Lancashire by Yould. An important aspect of their management was to keep animals on the pasture and out the corn during the spring and summer. This was achieved by having a permanent ring fence maintained and repaired by all the tenants. The fence was called the 'ringyard' (with variant name-forms). A tenant of Chatburn was brought to the manorial court for not making his fences, called 'rengyorde', about the town field in 1530 (Farrer 1897, 91). Earlier examples are the ringyards of Clayton-le-Moors and *Hyndefeld* in Huncoat, referred to in a division of lands made in 1376 (Stocks and Tait 1921, 80). Liverpool townfield had to be hedged about in the eighteenth century (Stewart-Brown 1917, 60). Atkin (1985) has drawn attention to oval enclosures in Lancashire, some of which were probably the ringyards of former arable fields and others were once vaccaries. Oval enclosures at Tunley, Lancashire, are illustrated by Roberts and Wrathmell (2002, 95–6).

Other northern counties have records of ringyard fences. The twelve husbandmen at Alnwick, Northumberland, in 1479, had to maintain 'le hege-yard' as part of their work-service. *Daltonfeld* at Dotland had a 'hege-garth' (Raine 1864–5, 4, 11). In North Yorkshire, the Whitby Cartulary refers to an important and onerous work-service called the 'horngarth' or 'horngard'. This appears to refer to the same structure as the ringyard, a fence and ditch surrounding the arable lands to protect them from animals pasturing on the surrounding wastes. Materials for the fence were taken from the abbot's wood. An alternative name for the boundary may be the 'acredike' of which there is reference at Stakesby, Newholm, Leirpelle, Whitbylathes, and Lathegarthe. In 1354 there was right of common pasture within the 'acregarth' of Whitby and nearby places (Atkinson 1878–81, 128–30, 209, 405, 415–29). The field pattern of Crayke, Yorkshire (NR) (formerly Durham detached), has a remarkable inner core that contains all the evidence of ridge-and-furrow ploughing (Adams 1990, 48). Outside of the inner ring lie fields with straight fences reaching to the township boundary (Figure 3.2). This outer area is likely to have been a rough common before enclosure. The inner ring looks like a ringyard fence, although the area enclosed within it is greater than found in most townships (upwards of

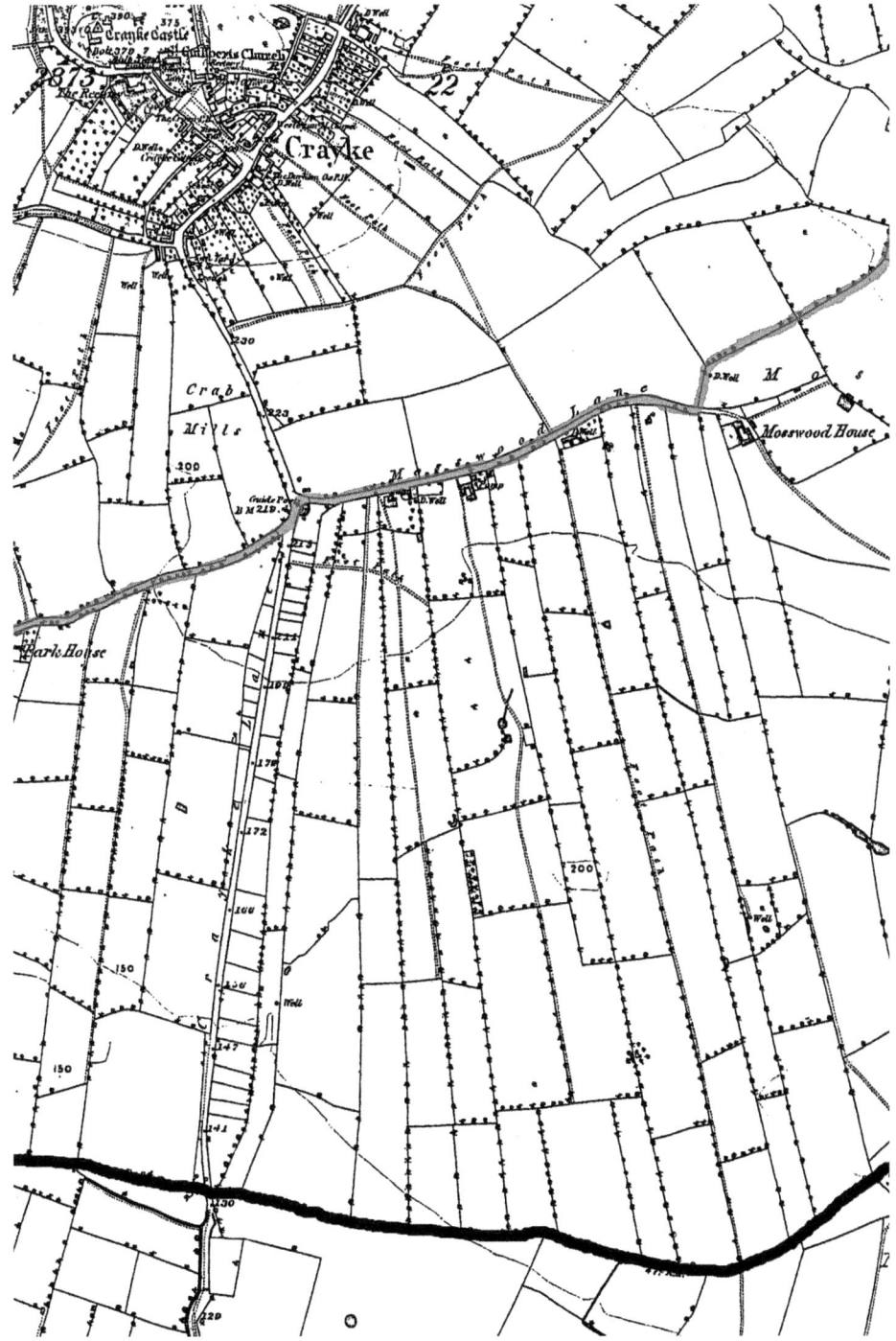

Fig. 3.2. Crayke, Yorkshire (NR). First Edition Ordnance Survey map, 1856, showing part of the inner field boundary (centre) and the parish boundary at the far south.

800 acres). Ring-garths were recorded in Westmorland at Crosby Ravensworth and Helton; that at Crossby was a stone wall five feet high in the sixteenth century. In Cumberland the terms used in the north of the county were 'moordyke' and 'felldyke', with 'ring-garth' used in the south. Preston Richard, in the late twelfth century, had an 'ekergart' (acre-garth) where pasturing within was allowed in winter but beasts had to be outside in the summer (Winchester 1987, 59–60). Several Worcester Anglo-Saxon charters refer to the 'æcer-hege', suggesting they were hedges used to protect arable land (Hooke 1985a, 240). At a very much earlier date, one of Ine's Laws for Wessex, 688–94, refers to a hedge for which it was the duty of tenants to make and maintain between common meadow and other divided land to prevent trespass by animals (Attenborough 1922, quoted on pp. 172–3). This seems very much like an order to maintain a ringyard fence.

E. CONVERTIBLE HUSBANDRY AND INFIELD–OUTFIELD

The infield and outfield system is a method of cultivation in regions with poor quality soils where some land needed periods of rest as pasture before being brought back into cultivation. It is closely related to more informal convertible husbandry regimes where some leys or pastures were taken into arable for a short time. Occasionally this operated in an organized manner, but more often was done at irregular intervals. There is often uncertainty as to exactly which type of system operated because 'outfield' may refer to pasture physically located some distance away from the main core of arable that was occasionally ploughed, but was not necessarily part of a formal infield–outfield arrangement.

Infield and outfield cultivation, as generally defined, was the practice of having a continuously arable inner core of furlongs or fields that were regularly manured, supplemented by occasional ploughing of part of an outer group of fields that were less fertile and left as rough pasture part of the time. After cropping one part of the outer fields, it was left as ley for a few years to regain fertility; different parts of the outer fields were ploughed in successive seasons. They were manured only by grazing animals.

Descriptions of infield–outfield cultivation have been influenced by eighteenth-century Scottish examples (Gray 1915, 158–60; Baker and Butlin 1973, 550–79) that probably have little relevance to medieval England.

Convertible husbandry was a system where some pasture grounds were taken into the arable fields for one or more years and then left to revert to grass for several years before further cultivation. If convertible husbandry were practised on a fixed number of 'fields' of whatever kind (a few furlongs or a few enclosed pieces), then it can be said to be a 'system'. Convertible husbandry is, in reality, an agricultural management technique that can be applied under any circumstances, as required. There is overlap with infield–outfield. Often documents show evidence of convertible husbandry, but without a specific statement (or the availability of a series of cropping data) stating that there was additionally a continuously arable core, then infield–outfield, as defined above, cannot be demonstrated, especially in the Middle Ages.

Clear examples of infield–outfield cultivation are found only in Cumberland. Grainger and Collingwood (1929, 244–9) discussed infield and outfield and other aspects of agriculture at Holme Cultram. The infields, arable closes next to the homesteads, were cropped continuously. The outfields (243 acres in all) were divided into flatts and strips separated by grass balks called 'reines'. Ownership was dispersed. One-third of the area was cultivated for three successive years and left as pasture for six years. Meadows were fenced from the common pasture until mown and carried. A survey of 1538 listed 310 tenants of arable land with 54 cottagers having common rights. Holdings ranged from two to seventy-three acres in size, each with stinted pasture. Coastal townships such as Drigg and St Bees used seaweed to manure the infield. The outfields were only manured by grazing animals and were cropped for limited periods. On the better land, outfields were cropped for five years and left fallow (leys) for five years. Poorer soils were used as arable for three years and then left as leys for nine. The outfield was divided into furlongs or rivings.

Saltmarsh and Darby described a system which they called infield and outfield at the Norfolk manor of West Wretham. A map of 1741 and a field book of 1612 show the township structure. At the south-east was the infield consisting of arable strips with a few closes, and next to them were six brakes called 'outfields', surrounded by an outer ring of heath. At the far north-west was a seventh brake. The infields extended to 226 acres and the brakes to 400 acres. The documents are less clear about the exact cultivation regimes (Saltmarsh and Darby 1935). The statement that each brake was folded with sheep once in seven years may mean that they were cropped once in seven years, being tathed the year before. This seems to be a breck system of convertible husbandry rather than a regular infield–outfield system. There is no information whether the infields were continuously cropped, so an infield–outfield system cannot be confirmed.

The field system of Helperthorpe, Yorkshire (ER), had areas called 'infield' and 'outfield'. In 1772 there were six 'inn field falls' (furlongs) and six outfield falls. Three crops to a fallow were taken from the infield, but in the outfield there was only one crop to a fallow. The infield was also planted with turnips and clover at this date (Harris 1959, 6). This example further illustrates the difficulties of applying definitions. Although the terms 'infield' and 'outfield' are used, the description shows there was an inner four-season cropping system and an outer two-season cropping arrangement. Both had a single year of fallow and there was not a period when either was left as ley. So Helperthorpe was a double-field system, akin to the five-field system (two and three concurrent) described for Newnham, Northamptonshire (see p. 44).

An example of an organized system of convertible open-field husbandry is found in the West Riding of Yorkshire. At Wigtwizzle (SK 250 957), in Bradfied on the edge of the Pennines, there were seven 'fields' in the eighteenth century (but two pairs of these were called upper and lower). Some were divided into two or three parts, making twelve in all, called 'takings'. One of the twelve parts was ploughed each year and cropped for four years, then put down to grass for eight years. So in any one year, only four parts (one-third of the total) were arable and there was a twelve-year cycle. The four owners had strips called 'byerdoles' dispersed throughout all the fields. Two fifths of the whole consisted of thirty-nine strips (hence there were only ninety-eight strips in all). Each byerdole was divided from the next by a 'reyn' or strip (of grass), the north or west reyn belonging to the dole. Beasts and sheep were stinted; beasts were grazed from May Day to Michaelmas and sheep grazed during the winter six months. The term 'byerdole' was also used for strips in nearby Dungworth before they were consolidated and enclosed in 1555 (Innocent 1924, 276–8).

Under such an arrangement the long period of rest allowed the land to re-cover, further helped with manure from grazing animals. Sheep could also have been folded nightly from other daytime pastures on nearby hills. The arable lands were grazed like fallow fields elsewhere after harvest before ploughing, and could also have been manured by carting from winter byres. It is possible that many south-western Yorkshire and northern Derbyshire townships were organized in this way and so appear in documents as 'one-field' systems (like Sheldon). Such a system is not identifiable from surveys and terriers, unless there is an explicit statement. Terriers with multiple 'field' names may be indi-cative of such a convertible system, rather than merely using 'field' as an impre-cise alternative for 'furlong'. The Wigtwizzle system differs from outfield cultivation in that is it convertible husbandry practised within a fixed number of fields which all have exactly the same cultivation in a twelve-year cycle in-volving eight years of ley. Great Longstone, Derbyshire, probably had a similar practice. A 1630 dispute about the enclosure of four acres that had been taken out of the town fields in 1610 states that before enclosure they were 'sometimes used for corn and sometimes for meadow.' Another reference regarding tithes says that land in the fields was converted from arable to meadow, 'soused' for a time, then ploughed again (Wright 1906, 233–4).

The Breckland of north-west Suffolk was run on a convertible husbandry system within a fixed number of 'fields' or shifts. A survey of Eleveden made in 1616 explicitly states that there were six 'shifts and fields'. Each field was, in turn, ploughed and sown with winter corn for one year, then summer corn the second year, and half of it only was sown on the third year. It was then left to lie as ley until all the fields had been cultivated in turn. The sheepfold had continuous feeding on the ley ground (Postgate 1962). So there were always

2½ fields as arable and 3½ in ley, in other words, 42 per cent was arable on a six-year cycle with a 3½ year rest period. Other references to shift systems imply that the same process operated elsewhere in the Breckland. Thetford, Norfolk, had eight shifts in 1338. Tenants at Ingham in the fifteenth century were ordered not to cultivate their lands lying in the lord's fold course. A mid seventeenth-century terrier of Wangford, Suffolk, specifies that the named furlongs were to be sown in three shifts, each of over 400 acres in 1643, 1644, and 1645 (Postgate 1962, 88).

The operation of the Norfolk folding system created a convertible system of agriculture, as described for Barmer in 1516. The fold moved from one tilth course about to revert from ley to arable to the next course that was to be put down as ley (see p. 71, Moreton and Rutledge 1996, 116). This is exactly the same process described in 1599 for Cowpen, Northumberland (Gray 1915, 222–3).

In the Suffolk coastal regions the quality of the soil required a convertible method, not necessarily regularly organized. The demesne of Leiston Priory, in 1345, consisted of 125 acres of arable of low value because it lay in the common field and was so sandy that it could only be sown twice in ten years (Mortimer 1979, 125).

In Sussex an early example of convertible husbandry is found at Barnhorne in Bexhill parish. Detailed cropping returns survive for its demesne, which lay in severalty. Crops growing on twenty-nine fields ('furlongs') for a consecutive sequence of twenty-four years (1397–1420) are recorded. Next to the grange were four fields of highly valued soil ('infields') that were cropped for ten to fifteen years, with only a few years rest (never more than five at once). Twenty fields, some of low value, lying mostly farther away, were cropped from zero to eight times in the twenty-four years with blocks rested commonly for six to twelve years in a single period. This usage represents convertible husbandry with no fixed regular number of rest years (Brandon 1971). The cropping arrangements at Westerham, Kent, during 1297–1350, showed a similar process, but with two crops and a fallow on some of the inner furlongs, as discussed (p. 64; Bishop 1938–9).

Flexible convertible husbandry occurred where, for any particular year, the arable area varied in size and location according to which outer furlongs were temporarily taken in. Under this system 'field' names become primarily locational and have no cropping significance. Groups of furlongs, which varied each year, are the cropping units. The furlongs did not necessarily have to lie in one block if common grazing rights after harvest were limited, or if there was cultivation of spring crops only on the common arable. Where sheepfolding in fold courses occurred, the cultivated areas did have to lie in a block when they were fallow or ley, for the convenient management of the flock.

A guide to the likelihood of the occurrence of any form of convertible or outfield cultivation is the presence of large areas of pasture or waste in a township.

Where there was very little, as in the core of the 'pasture-hungry' Central Region, then convertible husbandry could not be practised. Ploughing up, or removal by enclosure, of even very small areas of common in the East Midlands caused much local strife and sometimes litigation.

Flexible convertible agricultural regimes, often described as outfield, were practised widely in Devon and Cornwall. Fox produced a regional map showing locations of records of unenclosed outfield cultivation in the Middle Ages and later. Outfields were ploughed only occasionally at intervals of twenty, thirty, or even forty years. The main periods when outfield cultivation was practised were the thirteenth and fourteenth centuries, the sixteenth, and from about 1750 to the early nineteenth century (Fox 1973).

In Devon the use of downs for temporary arable is implied in the Arundell Survey of Downinney (Warbstow) by the phrase 'arable land on down'. A survey of 1649 at Climsland Prior (Stoke Climsland) stated that the tenants enclosed part of the down, sowed it for one year, and then threw it open again. In Cornwall commons were still ploughed in the mid nineteenth century at Treligga (St Teath) and Tregurrian (St Mawgan in Pyder). The stitches were bounded by ridges of furze and turf (Fox and Padel 2000, xcv). Probable examples of infield and outfield are recorded at Trebelzue in 1575 and mapped at Bosorne, St Just, in 1696 (discussed on p. 183).

On Exmoor MacDermot (1911, 8) gives similar examples where commons were cropped for two or three years in the seventeenth century (see the Gazetteer). A Kenton (Devon) survey, made in 1787, states that the tenements (holdings) had mark[er]s on the commons which they tilled once in fifteen or twenty years (Gawne 1970, 61). Early evidence from Devon is found in a grant of 958 for Ayshford and Boehill in Burlescombe and Sampford Peverell near Tiverton. It states 'beyond the common pasture here is the paved road. Then there are many hills that man may plough'. Boehill was probably the outfield for Ayshford (Finberg 1971, 9–24; B 1027; S 653). A tenant at Milton Abbot, *c.*1245, had the right to cultivate a portion of Ramsdown and also had pasture for his flock and the right to take fuel outside the cornfields and meadows. A ferling at Blackmorham in Tavistock, also in the thirteenth century, had the right to put cattle to common over the waste of Whiteborough Down when it lay untilled, but the abbot of Tavistock could cultivate it when he wished. There are other similar references to occasional ploughing of the waste of Luscombe Down at Newton in Tavistock in 1310 (Finberg 1951, 33–4).

Examples of mainly convertible husbandry from other counties are summarized below. The brecks of Nottinghamshire and elsewhere are also listed here as these form a variant of convertible husbandry.

Cheshire: The demesne of Frodsham castle consisted of 273 acres in *c.*1346 (Booth and Dodd 1979). One furlong with sowing details for six consecutive years had a three-season cultivation with a fallow. Another had one fallow year in four successive years and may have been on the same system. Other furlongs had mixed crops and sometimes were partly fallow in the same year. This seems to be a convertible system of husbandry.

The demesnes of Wybunbury and Tarvin were listed with acreage and cultivation in 1297 (Sylvester 1959, 183–5). The Wybunbury furlongs were grouped into three seasons of 20½, 53½, and 31½ acres. A piece called *Vicariesruding* was at least 33 acres in size and was split between the three seasons as well as being partly meadow (2½ acres). *Neuruding* was used for 20½ acres of arable, and another 17 acres were in the pasture. The document states that it was not used as pasture when it was 'assarted' (cultivated) and became part of the grain-producing land. *Wombetaggeruding* was similarly used as four acres of arable and one acre of pasture with the same restriction when it was cultivated. Tarvin demesne furlongs were run on a three-season system. One of the furlongs, *Mulnefeld*, was used partly for arable, partly as meadow, and partly as pasture. The mixed land-use indicates that convertible husbandry was practised, furlongs sometimes being used for arable, meadow, leys, or even abandoned as rough pasture. Such practices explain why the three seasons were not of equal size—they are unlikely to represent a regular rotation system but merely ensured that there was a change of crop to maintain good husbandry.

Cornwall: Small closes were farmed on a convertible husbandry system, being cropped for a few years, then left for seven as ley. The system was still used in the eighteenth century and there was no bare fallowing (Fox and Padel 2000, lxxi). See also Bosorne in the Lanhydrock Atlas (p. 183).

Cumberland: Parcels of the glebe of Addingham and Little Salkeld in 1704 are described as being 'arable or pasture' and so were likely subject to convertible or outfield husbandry (Ferguson 1877, 179).

In the sixteenth century, Fingland, on the Solway Firth, had eight tenants who each had twenty-one acres of infield and ten acres of outfield. At Beaumont twelve tenants had twenty-three acres of arable land and two of pasture (Elliot 1960, 98 n. 3). In contrast to Fingland, Beckermet survey of 1587 describes holdings that had twenty-two acres of (infield) land and a large area of outfield of eighty-four acres, plus twenty-two acres of meadow and a share of peat moss (Elliot 1973, 66). The large proportion of outfield presumably reflects that only a small proportion was cultivated in any one year.

Aspatria had 965 acres of open field in 1567 of which 640 were in the outfield and 325 infield. The outfield was divided into eight rivings (four each in the East and West outfields) and protected from stock on the waste by a large boundary wall. Part of the infield (eighty-seven acres) was kept for meadow land and mown for hay. Mockerkin similarly had outfield and infield (Elliot 1959, 92–3; Eliot 1960, 103–8). Edenhall glebe in 1704 was 'part of it in the common-field, and part in the grassing for all the parish', probably meaning outfield (Ferguson 1877, 173).

Devon: Fox described details of convertible husbandry practised between 1500 and 1800 in enclosed severalty farms. The rotation consisted of three successive arable crops followed by seven years ley.

Durham: Stainton had a terrier of a husbandland that described 34 parcels of half a rood to 1½ acres, totalling 14 acres in *c.*1479. The remainder of the husbandland was not terriered because it lay in the waste of the same fields, and 'wherever lies the other portion of Raysland, on the west lies the equal portion of the prior of Hexham's land throughout the fields' (Raine 1864–5, ii: 40–1).

Kent: The terrier of Barfrestone, 1236 (Churchill et al. 1956, 130–2), had six pieces of arable in pastures, three of them called downs *Bestedun, Northdun,* and *Litledun.* These are likely to be part of a convertible husbandry system. Some of Westerham demesne, up to 400 acres, was intermittently cultivated (Bishop 1938–9).

Lancashire: The common field at Deadwinclough in Rossendale was divided among the houses standing on the fold in 1528. The shares were used partly for corn and partly for hay so that animals had to be tethered (Tupling 1927, 103). It is likely that this is convertible husbandry, the arable being rested as pasture in turn. The demesne of Penwortham Priory, in 1543, consisted mostly of arable closes, one of eight acres described as pasture now or lately in cultivation, and is probable evidence of convertible husbandry (Hulton 1853, 112–13).

A grant of land at Tunstall in 1202 referred to five named hills that could be cultivated to whatever extent was wished, the remainder being left as common pasture (Farrer 1899, 15). Four of the five fields of Carnforth in the eighteenth century were cultivated as follows: one had continuous cropping and was never pastured; the other three were cropped for three years and left as pasture for six years. Holdings were dispersed between these three fields so that each proprietor had a crop (Youd 1962, 29–30).

Norfolk: Described on p. 87.

Northumberland: Terriers of two 'husbandlands' (yardlands) at Slayley were made in 1379. One described twenty-three acres lying in seventeen pieces; the remainder of the husbandland lay in diverse acres of fallow upon *Scheldeschaw.* The other husbandland specified eight acres in three pieces and the rest of it similarly was fallow upon *Scheldeschaw* (Raine 1864–5, ii: 27–8). These fallow lands were probably lying in ley indicating convertible husbandry within the main open fields.

Shoreston had land newly broken up from the township common pasture in 1250 (Butlin 1973, 141). Tynemouth tenants, in 1378, had inland and outland that may have been convertible (Dendy 1894, 146). In 1379, if part of the common of Fenwick and Madfen was to be ploughed, the prior and convent of Hexham were to have their share by lot as they had before in the old arable intakes (Dodds 1926, 366). These examples could be regarded as 'communal' assarting, but are more likely to be aspects of convertible husbandry.

Guysance practised convertible husbandry in 1567, when the tenants had every sixteenth rigg in every 'rifte' that was to be made arable ground which had been ley or pasture. Acklington in 1616 similarly practised convertible husbandry. At Morlands, ninety-eight acres detached from the three arable fields around the vill had been recently taken out of the common and converted to arable (*History of Northumberland,* v, 373, plan). It was divided into strips and allocated to the tenants in rotation, one Henry Jackson having the first, twenty-first, and forty-first land. A piece called *Brocks Haining* lying on the edge of South Field next to the East Moor and consisting of thirty-four acres was described as pasture that had been arable but was laid down because it was barren. At Cowpen, in 1599, part of the cornfield 'lay waste' and an equal amount of land taken from the common wastes (Gray 1915, 222–3; Butlin 1973, 117–18, 120).

Nottinghamshire: Townships lying around the edge of the sandy Sherwood Forest supplemented their open-field farming by creating 'brecks' that resembled infield and outfield cultivation. Under this arrangement temporary enclosures were made for three to nine years and used for arable crops, after which the land was opened and returned to a sheep walk. Additional arable was needed because with good land in short supply, each settlement had a core of permanent enclosures around it that included pasture for hay as well as arable. Outside the enclosures was a large area of sheep walk and warren.

In the eighteenth century Budby had 394 out of its 2,096 acres enclosed, and Rufford 1,935 out of 9,910 acres, presumably for the same reason that earlier brecks had been created, namely a shortage of arable land (Fowkes 1977).

On the east of Sherwood brecks belonged to each oxgang (Timson 1973, ii: 223). A mapped example is Carburton, 1615 (2,235 acres). There had been some enclosure of arable but there was still a small common open field next to the village, 175 acres, and lying farther out were several enclosed areas called 'brecks'. The largest, 518 acres, was 'the waste lyinge in brecks' (Beresford and St Joseph 1979, 45–8).

Brecks were used at an early date. At Blyth in *c.*1220 there was a grant of breck and land in the fields (Timson 1973, i: 22). Most townships had brecks in the eighteenth century. Norton had seven different brecks. For individual holdings in Ollerton (1762), a forty-six-acre open-field holding had fifteen acres of breck; another forty-seven-acre open-field holding had ten acres of breck. The brecks were used for spring corn, rye, and sometimes specialist crops such as turnips or improved grass. The area under breck was always about the same but the locations varied. Kirkby-in-Ashfield brecks were called 'fields' and had a nine-year rotation system. Calverton enclosed its brecks for seven years, the expenses being met by the occupiers. Each messuage had one acre of breck, and any unwanted land was let out for the upkeep of the poor. There is no published information on how long brecks were left between the cultivation periods (Fowkes 1977).

Shropshire: Convertible husbandry was probably used at Linley in 1309, where two pieces were brought back into cultivation and assarts were allowed to lie fallow, but would be tithable when they were cultivated again (Rees 1985, 148).

Staffordshire: In 1285 a grant referred to a carucate and five acres of cultivated waste at Rule that may have been outfield (Wedgewood 1911, 189, 259). There are early records of land being newly taken from the waste at Kings Bromley in 1300. Many small assarts of one-half to eleven acres made during the thirteenth and fourteenth centuries were enclosed and used to grow corn for up to six years. Others lay fallow. It is not clear whether this represents convertible cultivation or merely indicates that the assart reverted to long-term pasture (Birrell 1999, 165–7, 207). Cannock in 1419 had holdings with two acres each in three fields and another piece in a 'riding' (Kettle 1979, 56–7). The last was probably an assart and may have been convertible.

Waste in Staffordshire was being cultivated regularly by the sixteenth century. In 1551 all of Ilam Moor had been ploughed, and, in 1597 it was stated that the Aldridge part of the Sutton Coldfield waste had been sown for eight years at a time during the previous half century. In 1666, Little Wyrley in Norton Canes took four crops on the waste and paid tithe to the lord, provided there was visible 'ridge and reign' (ridge and furrow). Similar cultivation occurred in the eighteenth century at Sutton Coldfield for short periods. Great Barr was cultivated for four years in the seventeenth century and then left for seven years. Teddesley Hay was cultivated for five years in fourteen (Kettle 1979, 56–77). Pitt, in 1796, noted 'evident marks of cultivation far more extended than anything known in modern times' on the commons (quoted by Kettle 1979, 49).

Plot (1686, 343–4) recorded that Staffordshire heathland was seldom enclosed except for cultivation. When it was required for tillage, it was never ploughed for more than five years, after which it was thrown open for commons again. Heath was 'stocked up' (uprooted) with mattocks and left fallow for a winter. Moorlands were improved by digging turf (peat) and burning it in May. Lime was applied before Michaelmas ploughing followed by spring sowing. Four crops were obtained, after which the land was laid down for four years.

Sussex: The demesne of Chitlington in 1385 amounted to 313½ acres, of which 48½ could be left as leys (Hudson 1910, 156–7). See Bexhill (p. 89), and Wiston and other places on the South Downs in the Gazetteer.

Westmorland: In Westmorland there were four years of successive cultivation followed by seven in grass in 1797 (quoted by Elliot 1973, 63, from Pringle's *General View*). This may refer to agriculture in enclosed fields.

Yorkshire: Some of the fields on the Wolds and in Holderness used convertible husbandry. Kilham in the eighteenth century was run as two townships, its 7,000 acres divided into the Northside Fields and Southside Fields. A plan made in 1729 showed twenty-nine fields (furlongs) in Northside and eighteen in Southside. By-law men set out a group of fields for sowing on each side and both sides had furlongs that were regarded as outfields. Convertible husbandry was practised in the region, as Harris has shown: Bishop Wilton had ground on the wold sown once in ten or twelve years in 1611, and at Fimber some ground was cultivated once or twice every twenty years (Harris 1959, 7–8).

4

Open-field structure and management

A. THE DEMESNE

The demesne 'farm' was an important element of township field-structure because of its antiquity. Its precise location and extent needs to be established. Study of demesne types across the country show that while both block and dispersed forms occurred widely, the former were more common outside of the Central Region. Although it has often been stated that block demesnes were formed by exchange of dispersed villein strips, no clear evidence has been found for this occurring at any date.

The demesne was the land 'anciently belonging' to the manor-house, in effect the 'home farm'. It remained unchanged for a long time and is well documented since it was usually the main part of a lord's property. The demesne was distinguished from other land held by the lord, such as that of his medieval bond tenants, or in later centuries from land taken in hand by the lord and farmed directly or leased out for terms of years. The antiquity of the demesne makes its precise nature and location important in trying to determine the field structure of townships; it will be further discussed when we consider what might be the origin of organized regular field systems (see p. 188).

Some early authors assumed that all demesnes were originally dispersed amongst the villein lands. Corbett pointed out as early as 1897 that in East Anglia there was variation—some were compact, some dispersed, and others had elements of both (Corbett 1897, 75). For a long time after that, there was doubt where the demesne was in relation to the manor-house. In 1965 Titow still thought that on balance it was probable that most demesnes were intermingled with those of the tenants in the common fields (Titow 1965, 97). Evidence from Northamptonshire, however, showed that there is not a single uniform type of layout applicable to all demesnes, as Corbett had found in East Anglia. A sample of fifty-six Northamptonshire townships where the demesne type is known, eighteen (32 per cent) were compact and thirty-eight (68 per cent) were dispersed (Hall 1995, 72).

When a demesne was identified as being in a block it was sometimes assumed that there had been a 'great unrecorded redistribution of the lands of lord and tenants,' with the implication that it was once dispersed and a compact demesne had been created as a result of exchanges, as at Writtle, Essex (Newton 1970, 33). This statement is often still repeated in the literature without any evidence being provided. As will be seen, such an event seems unlikely since there is no documented example of its occurrence anywhere in the country at any date.

Dispersed demesne

The Central Region Examples of dispersed demesnes occur widely, usually lying in small parcels of contiguous lands. An early recorded example occurs at Braybrooke, Northamptonshire, for which a terrier of 1199 shows that it was distributed in small blocks lying in numerous furlongs, many of which can still be identified on a map of 1767 (Round 1900a, 140–1; Northamptonshire CRO, Map 6251). It is a typical dispersed type, with small blocks of lands in most of the furlongs and a small amount of land, including the Hall Yard, next to the village. There were also demesne woods and a few closes (probably assarts) adjacent to them lying on high marginal ground.

Another early example is Kislingbury, where there were two manors of which the tithe of the demesne had been given to St Andrew's Abbey, Northampton, in the early twelfth century. In the fourteenth century, a terrier of the lands of both manors shows that there were blocks of mainly six to eight half-acre lands dispersed throughout the fields (London, British Library, MS Cotton Vespasian E xvii, fols. 301–3). The lands of the two manors were always adjacent, and there were a few whole furlongs next to the manor-houses called 'inlands'. The two Kislingbury manors were in existence in 1086, but the disposition of the contiguous demesnes suggests that before 1066 there had been a single manor. A reconstructed plan has been published (Hall 1981a, 50–1, 62).

Other reconstructed plans of dispersed demesnes have been given for Brockhall in 1574 and Raunds 1739 (Hall 1995, 66–92). Adjacent to the two Raunds manor-houses was a large block of demesne, but the remainder was dispersed throughout the fields. Gretton, Northamptonshire, affords another example of a dispersed demesne, shown on a map made in 1587 (Plate 8; Northamptonshire CRO, FH 272). As well as the dispersed arable, there were enclosed demesne pastures on the unploughable fractured scarp of the Welland Valley.

Dispersed demesnes existed in many other counties, as summarized in the following list. Examples are widespread, both geographically and in date, and are given in alphabetical order of county.

Bedfordshire: The manorial demesne of Kempston Greys lay in 1341 in nineteen pieces, 302½ acres in all, some large (18 and 34 acres); most were called 'furlongs', two were termed 'crofts', one a 'redynge', and two 'inlond' (Stitt 1952).

Buckinghamshire: At Woburn the demesne was dispersed in many furlongs in 1222, as was the demesne described in a Kimble terrier of twenty acres in *c*.1224. Brill demesne, in 1252, lay dispersed in nine 20-acre pieces plus 6¾ acres, with the name of each furlong given (Salter 1930, 199).

Cambridgeshire: In 1446 Burwell had 5½ acres of demesne dispersed in *Dichefeld*, 2 in *Estefeld*, and in *Bradway* 1½ acres (DeWindt 1976, 314).

Gloucestershire: The demesne of Woolford was dispersed in many furlongs in 1227 (Elrington 2003, 42). Minchinhampton demesne lay mostly in single acres in 1235–45 and at Frampton a half-yardland of demesne was dispersed in twelve separate furlongs grouped into two fields (Watson 1939, 74, 87). The demesne of Rendwyke in 1267–8 (108 acres) lay as small pieces in 32 named furlongs (Hart 1863–5, iii: 44).

Sevenhampton and Stratton demesnes in the late thirteenth century were stated to lie in many furlongs in two fields (Gray 1915, 465).

Huntingdonshire: Most of Wistow demesne furlongs, listed in 1252 (Hart and Lyons 1884, i: 353), can be identified on a map of 1617, showing they were dispersed throughout the township (Patricia Hogan 1988). Cropping details are given in the Gazetteer.

Leicestershire: Kings Norton demesne was dispersed in seventy-one furlongs in 1360 (Hilton, 1954, 157).

Lincolnshire: Two oxgangs at Riby lay in thirteen parcels in the late twelfth century. Great Limber demesne terrier of *c*.1150 similarly had groups of selions lying in twenty-eight furlongs (Stenton 1920, lviii–lix). The dispersion of demesne at Dunsby (*c*.1182) was in four four-acre pieces, showing that although dispersed, it was grouped as contiguous parts of furlongs (Stenton 1920, 83).

Northamptonshire: Dispersed demesnes occur in about 67 per cent of those where the location is known (Hall 1995, 72).

Nottinghamshire: Cossall had a dispersed demesne in the fourteenth century according to terrier evidence (Walker and Gray 1940, 58).

Oxfordshire: At Garsington a hide of inland was said to be dispersed in 1086 (Williams and Martin 2003, 430). In the same county, Dornford, in the mid thirteenth century, had two fields, with the demesne equally dispersed in both of them (VCH, *Oxfordshire*, xi (1983): 274). The demesne of Cuxham, described in 1448, shows it to be dispersed in small blocks lying in several furlongs, most of which can still be identified on a map of 1767 (Harvey 1965, 20).

Warwickshire: The five yardlands of glebe at Hampton Lucy were dispersed in blocks of up to thirty-four ridges lying together (Barratt 1955–71, i: 100). This was probably a grant from a dispersed demesne. Loxley glebe was similar, with several furlong pieces containing up to fifty-two ridges (ibid. ii: 4–5).

Wiltshire: Thirteen of the fifteen demesnes referred to in the Gazetteer are dispersed.

Worcestershire: At White Ladies Aston, demesne pieces lay in many named furlongs. Harltebury demesne was dispersed in 'cultura' with furlong names (Hollings 1934–50, 87–90). Tredington demesne lay in two named fields (Hollings 1934–50, 279).

Yorkshire: Spaldington (ER) had a dispersed demesne lying in several furlongs in the mid twelfth century (Stenton 1922, 62). Walkington demesne of fourteen oxgangs lay as compact pieces dispersed in the flatts of the open fields in 1726 (Harris 1955, 533).

It can be seen that most counties in the Central Region provide evidence of dispersed demesne; many of them date from the twelfth and thirteenth centuries; and some show remarkable stability over several centuries.

Counties of the Eastern and Western Regions Counties away from the Central Region also had examples of dispersed demesne. Moorland, Cumberland, in the thirteenth century, had a demesne of ninety-four acres dispersed in pieces, many called 'flatts' (Prescott 1897, 373, 452). In Devon, a terrier (1324) of 26½ acres of demesne arable belonging to Braunton Gorge's manor was split into 19 dispersed pieces, some of them lying in the Great Field (Slee 1952, 142–9). The 245-acre demesne at Canonsleigh, Devon, in 1323, was dispersed as blocks in several furlongs (Hoskins and Finberg 1952, 274). Uplyme, in 1326, had 213 acres of demesne intermixed with tenants' land lying in 29 furlongs. The arable extended into the hills and coombes on the south and north as well as lying around the vill (Fox 1972, 95). Stoke Fleming demesne, in 1374, consisted of *c.*300 acres dispersed among 44 furlongs, some called 'fields' (Fox 1975, 189). The demesne of Pipe, Hereford, was dispersed in 1694, as identified by a terrier of land free of tithe, formerly belonging to the Priory of St Guthlac, Gloucester (Herefordshire CRO, HD2 (1694)). The pieces described in the terrier are not stated to be demesne but lie next to 'court [manorial] land' and since it was monastic land, this would have been the arrangement before 1540.

Further examples of dispersed demesnes are summarized as follows.

Berkshire: Dispersed demesnes are recorded for Brightwell, Benham Valence, and Sandhurst (see the Gazetteer).

Cambridgeshire: The Chippenham demesne of 1544 was dispersed in large blocks; near the village some were enclosed (Spufford 1965, 17 and end map). The demesne of Kirtling (sixteenth century) and Stow cum Quy were dispersed in pieces and blocks of three to forty acres (VCH, *Cambridgeshire*, v (1973): 69, 238, 380).

Durham: A 1430 terrier of forty acres at Hett, almost certainly demesne, was dispersed in several pieces called 'flats', two of twelve and sixteen acres respectively (Greenwell 1872, 173 n.).

Lancashire: The Bolton-le-Sands demesne (eighty-eight acres arable, eleven acres meadow) was distributed among twenty furlongs in parcels of half an acre to nine acres in *c.*1320 (Farrer 1907, 39). Pennington demesne in 1332 had arable in pieces of two to twelve acres dispersed in eleven named furlongs (mostly called 'flatts' or 'crofts') as well as small pieces of dispersed meadow (Atkinson 1886–7, i: 495–8).

Norfolk: The land of Henry Bedingfield at Cockley Clay in 1722 was distributed over all the open fields in blocks and may be a dispersed demesne (Norfolk CRO, BL 47/1).

Northumberland: Alnwick demesne in 1379 was dispersed in many furlongs (called 'flatts', 'tofts', 'riggs', 'lands') in pieces of four to twenty acres. *Hughe* with a demesne of 169 acres, Echwyk with 88, and Dalton with 150 acres were similar; the demesne of *Hughe* was specified as lying 'in the fields of the vill' (Raine 1864–5, ii: 3, 38–50). A full terrier of Corbridge demesne made in 1635 shows it was dispersed in all four of the township fields in small blocks of rigs (Craster 1914, 124–30).

Somerset: Horsington demesne was dispersed in 1340 among several large furlong pieces lying in two fields, exactly distributed (Whitfield 1981, 21). Bruton Priory demesne in 1390 consisted of four closes and twenty-one dispersed pieces of furlongs, 310 acres in all (Hobhouse et al. 1894, 20–1). Other examples are Puckington in 1525 (Ellerton 1999, *Proceedings of the Somerset Archaeological and Natural History Society*, 142: 339), Stoke sub Hamdon 1615 (J. Batten 1894, ibid. 40: 260–1), and Allerton in 1650 (P. Coleman 1900, ibid. 46: 73–81).

Staffordshire: Wednesbury demesne consisted of groups of two to eleven selions in a large number of furlongs in 1315 (Duigan 1911, 321). Himley demesne in the fifteenth century was dispersed, with an equal number of selions in each field (Birrell 1979, 13). The demesne of Blymhill was dispersed in small blocks of arable in many furlongs in 1585, some called 'flatts' (Bridgeman 1881, 118–20).

Suffolk: Hadleigh demesne of 327 acres in 1305 lay in diverse fields (Pigot 1863, 230), and Elvendon demesne in 1618 was dispersed in all the six fields and shifts (Postgate 1962, 97).

Surrey: At Banstead the demesne in 1325 was dispersed in seven large pieces totalling 348½ acres (Lambert 1912, 62–3).

Sussex: The demesne of Edburton was dispersed in blocks called 'courtland' in 1343 (Salzman 1934, 106–9). A dispersed demesne was recorded at Radmell Beverington in 1588 (Godfrey 1928). It consisted of 139½ acres of arable: a croft of 2 acres, two blocks of 16 and 12 acres, and parcels in 11 furlongs (only five times called furlongs; one was called *South Croft*). The pasture (forty acres) was dispersed, one being a piece (ten acres) called 'croft or feild called paddockfild'. Dichelinge demesne (1624), called 'courtland', consisted of 60 acres of arable and meadow of 40 acres, with pasture (94 acres) dispersed in several fields, and a several sheep pasture on the down for 400 sheep (ibid. 40).

Compact demesne

In contrast to the dispersed type, some manors had nearly all their demesne in a single block lying next to the manor-house. Kettering, Northamptonshire, provides a contemporary map of block demesne, called the Hall Field, consisting of 620 acres, which had already been enclosed before 1587 (Figure 4.1). Other Northamptonshire block demesnes have been mapped: Watford, from a terrier of 1276 (Hall, 1989, 196–204; fig. 11.1), and the Hall Manor at Wollaston, from a terrier of 1430. A block of land lying west of the manor-house site at Wollaston, called the 'inlands', was enclosed in 1583 and mapped in 1789. As well as the inlands block there were other lands dispersed in a few

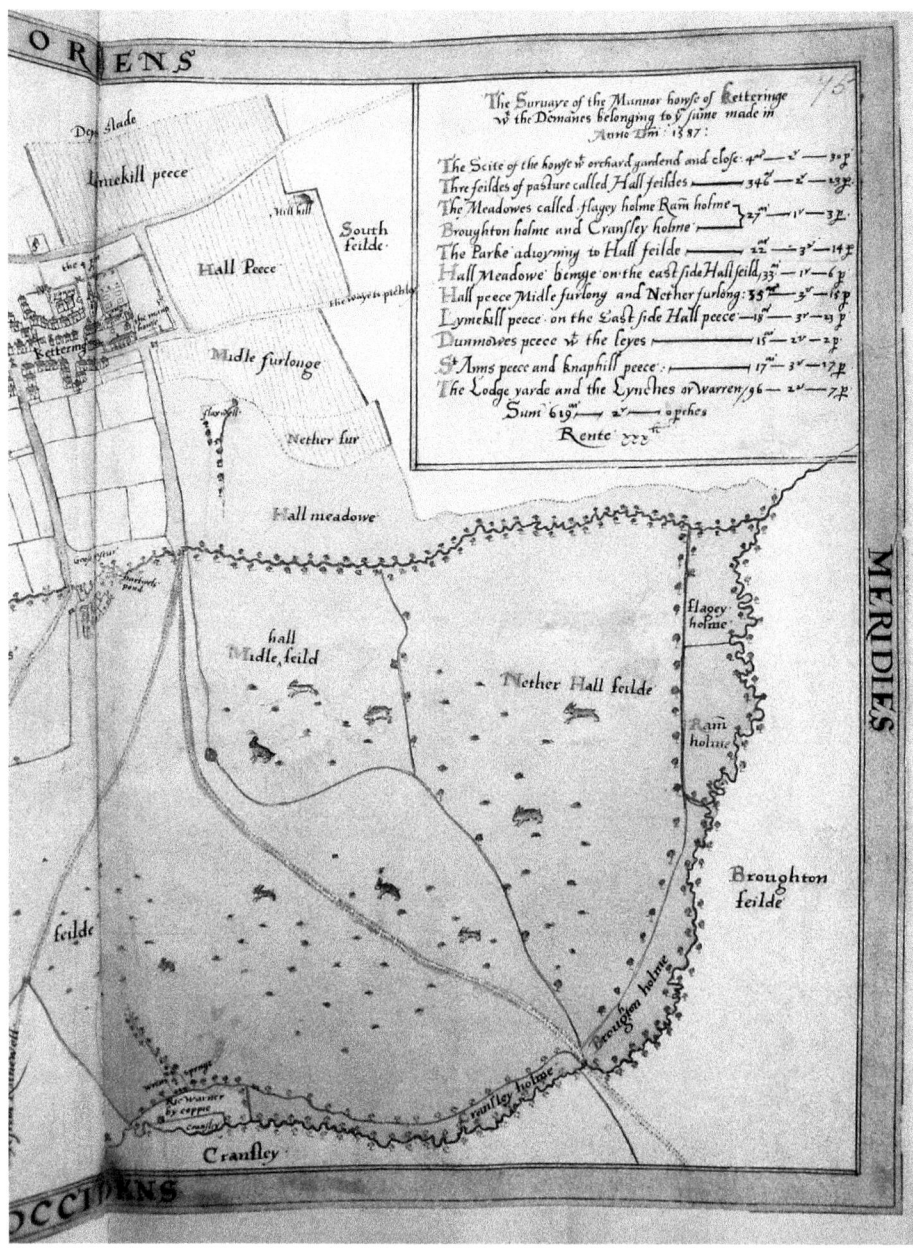

Fig. 4.1. Kettering enclosed block demesne, called the *Hall Feilde*, Northamptonshire, 1587 (Northamptonshire CRO, FH 272; copyright Northamptonshire Record Office).

furlongs and the whole of *Rye Furlong*, lying on free-draining river gravels one mile away to the west (Hall 1983, 119).

The Mears Ashby demesne block was called the Hall Field. It, too, had an isolated furlong away from the rest located on well-drained red ironstone land, probably used for rye. A plan has been reconstructed from a field book of 1577 (Hall 1995, 138, 177–81). The block demesne of Higham Ferrers has been identified and mapped (Hall 1988, 113–14).

The demesne of Stoke Talmage, Oxfordshire, was described in two charters of *c*.1150 and 1180. It lay in a block with two pieces of free land of 20 and 30½ acres and comprised 10 adjacent furlongs totalling 122 acres that were said to be 2 hides and 8 acres (Salter 1947, 96–9).

In East Anglia, block demesnes were the norm in ten out of a sample of twelve detailed case studies, mainly in Suffok and Essex (Martin and Satchell 2008, 40–8, 192–5). Four of them had large central blocks and six had demesnes using resources near the township edge. Most parishes had demesnes that utilized the best agricultural soils.

Many other block demesnes are known throughout the country. Examples of those that have unambiguous descriptions are here listed.

Bedfordshire: Aspley Guise and Oakley had block demesnes in the eighteenth century (Fowler 1936, 7, 24).

Buckinghamshire: Chichele, in 1557, had a block demesne called Bury Field at the north, enclosed as pasture by 1526 (H. Baines 1999, *Records of Buckinghamshire*, 39: 7, 11–17). Sherington demesne of 552 acres lay in a block in 1580 (Chibnall 1965, 260).

Cambridgeshire: The Cambridge property formerly of the Mortimer family transferred to Gonville Hall in 1501 was probably a very ancient block demesne (Maitland 1898, 57, 179).

Devonshire: Braunton Dean manor lay in a block next to the church and was probably demesne (Stanes 1994).

Derbyshire: Six virgates at Wessington (1254) lay together in defined boundaries (Darlington 1945, 461) and were probably block demesne or an assart.

Essex: Writtle block demesnes were surrounded by hedges in 1328 and many small manors also had compact demesnes (Newton 1970, 28, and fig. V). Martin and Satchell (2008, 141, 127, 155) found blocks at Felstead, Great Henney, and Ingatestone. See the Gazetteer for more examples.

Gloucestershire: Bibury had a block demesne in 1299 and later (Dyer 2007, in Bettey 2007, 69–73). Frocester enclosed demesne lay substantially in a block in the sixteenth century (Price 1998, *Transactions of the Bristol and Gloucestershire Archaeological Society*, 116: 16–19).

Hampshire: Foxcotte block demesne of 591 acres was mapped in 1614 (Jones 1985).

Herefordshire: Whitbourne demesne was in a block in 1577 (Williams 1970–2). See the Gazetteer for two more.

Hertfordshire: The 840-acre demesne of Flamstead in 1264 lay as three closes (Roden 1973, 332). Block demesne was found at Ardeley (Martin and Satchell 2008, 163). Codicote demesne lay in eleven named areas (fields); only one was the same as the villeins' fields (Levett 1938, 339–40), hence the demesne probably lay in a block.

Huntingdonshire: Elton demesne, described in 1605, lay in a block divided into three fields (Ratcliff 1946, li).

Kent: Kent demesnes were in blocks at Wrotham, Westwell, and other places in the Vale of Holmesdale (Baker 1966, 3–4, 15–19). Similar was Otford in 1285 (Hewlett 1973, 104); Du Boulay lists the names and acreages of the component furlongs and fields as they were in 1284 and 1516, being 665 acres in all (Du Boulay 1959, 117–18). Other block demesnes were Saltwood, Petham (Du Boulay 1966, 132), Gillingham in 1447 (Baker 1964, fig. 1), and Sibertswold manor (Shepherdswell) in 1616 (Gray 1915, 273).

Lancashire: Prescot enclosed demesne, *c.*1580, lay in a block around Prescot Hall (Bailey 1937, 33–4). Caton enclosed demesne also lay in a block (Youd 1962, 6).

Leicestershire: In 1743 three large closes totalling 368 acres in West Langton belonged to the lord of the manor and were 'seised in his demesne'. The closes were called the Wheat Fields, Bean Field, and Fullers Field, and had common grazing for animals of the other landowners (Hill 1867, 68).

Lincolnshire: Part of the demesne of Brocklesby was in a block of twenty acres in the twelfth century (Stenton 1920, xxxiii, 186).

Norfolk: A block demesne was found at Frenze and Thelveton (Martin and Satchell 2008, 105).

Northamptonshire: In addition to examples that have already been discussed in this section, Helmdon demesne (mapped in Hall 2013, 40–4), 1530, shows there was a block called the 'beryfeld' next to the village.

Northumberland: Barton manor-house had sixty acres lying next to it in a block in 1235 (Page 1893, 208). Bingfield's 238-acre demesne may have been in a block in 1379; some of the named furlong pieces were next to each other. Farendon grange, Silksworth, was in a block defined by detailed bounds (Raine 1864–5, ii: 6, 62). Other block demesnes are identified in the Gazetteer.

Nottinghamshire: The enclosed demesne of East Bridgford lay in a block in 1614 (Ashikaga and Henstock 1996, 84), as did Laxton in 1635 (Orwin and Orwin 1938, 132).

Oxfordshire: The demesne at Stoke Talmage has been described. Deddington demesne, in 1728, was a block split between three manors and lay in about thirty-eight furlongs of arable and leys. In contrast, the copyhold yardlands lay in about forty-eight furlongs, of which only thirteen bore the same or similar names to the demesne furlongs (Northamptonshire CRO, C(A) 3309, pp. 4–29). A four-yardland part of one manor had all of its 25 arable pieces (252 ridges) flanked by the demesnes of the other two manors in 1728. This shows that there had been one single block of demesne before the original manor was split into three in the late twelfth century (VCH, *Oxfordshire*, xi (1983): 91).

Rutland: Whitwell had an enclosed 'Hall Land' in 1684, almost certainly a block demesne (Ryder, 2006, 10, 68).

Shropshire: The reconstructed 1256 demesne of Moortown lay in a block. Ercall demesne was slightly dispersed in large pieces at the same date and similarly in 1746 (Hill 1984, *Transactions of the Shropshire Archaeological and Historical Society*, 62: 48, 66, 70). Myddle demesne lay in two fields of about one hundred acres each, and was enclosed and divided by the freeholders before 1563 (Hey 1974, 29, 31).

Somerset: Stapleton in Martock had a large area called 'the lord's fields' that was probably a block demesne (Bush 1978, 81).

Staffordshire: Keele, in 1331, had a block demesne of 120 acres with another 40 acres dispersed in the fields (Birrell 1979, 15). The enclosed 984-acre demesne of Weston under Lizard was in a block (Bridgeman 1899, 337–45).

Suffolk: Block demesnes were found at various dates at Sutton, Walsham le Willows, and Worlington (Martin and Satchell 2008, 137, 175, 121). The enclosed demesne of Snape Priory, 287 acres, lay in a block next to the priory in *c.*1528. It consisted of wood and closes that were then mostly described as pasture, meadow, and fen (Filmer-Sankey 1984, *Proceedings of the Suffolk Institute of Archaeology*, 35: 213–21). Elveden had a block demesne interspersed with glebe in 1706 (Postgate 1973, 288).

Surrey: Fitznell's manor demesne at Ewell, described in *c.*1350 and 1476, was nearly all of block form lying next to and north of the village in large pieces (Meekings and Shearman 1968, 50–1, 55–61). The block pieces were sometimes called 'fields' and sometimes 'crofts'; *Heecroft*, which contained twenty acres, was called a 'ditched field' in *c.*1350.

Sussex: Laughton on the Weald had a demesne in 1292 consisting of 213½ acres of arable lying in named blocks (none called field or furlong), along with meadow of 78½ acres and 94 acres of pasture, as well as wood and pannage in the park. The reconstructed topography showed that it lay in a block next to the park (Wilson 1961). Other block demesnes are known (see the Gazetteer Table and Bexhill, p. 89).

Worcestershire: Cleve Prior demesne lay in a block and had been enclosed before 1650 (Yelling 1977, 68–9, 124)

Yorkshire: Compact demesne oxgangs at Salton, Yorkshire (NR), were described in *c.*1479 (Raine 1864–5, ii: 72). There were sixteen demesne oxgangs in all, each comprising nine acres of arable land, of which eight lay in four named flatts lying in a block next to Crossow Bridge and were separate. The other eight oxgangs lay discontinuously in diverse parts between the lands of the tenants.

Interpretation of demesne types The Northamptonshire evidence for demesne type has been discussed (Hall 1995, 72–3), and is here summarized. There is insufficient published information for any other county to make a synthesis, other than to observe that both block and dispersed demesnes occur widely in all three regions. It is not clear why there are the two types, dispersed and compact or block demesnes; as will be shown, the block form seems to be 'older'. As mentioned, the literature often refers to blocks being created by consolidation, as quoted for Writtle (pp. 95–6). There is no Northamptonshire evidence that there were ever any large-scale conversions—certainly not of a block becoming dispersed. Muscott in Brockhall provides the best example of scattered demesne strips being exchanged with freeholders in 1433, but the resultant consolidated and enclosed piece was used for pasture at the periphery of the township and not placed next to the manorial residence at Brockhall. This was the poorest quality land and presumably the freeholders were willing to exchange strips there for better quality land nearer to the vill. There was an exchange of dispersed demesne land at East Haddon in *c.*1600 and Great Billing in 1629. In both cases the purpose was to create parkland-type enclosures next to the manor-house and had nothing to do with open-field farming on good quality land (Hall 1995, 196, 282–3).

Many of the dispersed demesnes were lost to sight and let out with other tenants' lands. That of Ashby St Ledgers was only identifiable in the seventeenth and eighteenth centuries because one-third of the tithe was due to the vicar, as it had been since the thirteenth century. It may not be necessary to look for an early uniform type of demesne, either dispersed or compact; there could have been differences from the time that the vills and their fields achieved their final form. This is supported in that some demesnes were assessed, and others excluded, in the 1086 Domesday Survey, thus proving their existence. The block demesnes at Mears Ashby, Kettering, Badby, and Newnham were excluded, taxation being raised on villein yardlands (Hall 1995, 75–6).

Manors with large blocks of demesne often seem to be of some local importance. Higham Ferrers and Dodford were the chief manors of considerable estates in 1086 that survived for many centuries. Higham belonged to William Peverel and became the Northamptonshire centre of the Lancaster estate, to which it still belongs as part of the Duchy of Lancaster. Dodford was part of an Anglo-Saxon estate that became the property of William de Keynes and remained with the Keynes family and its female heirs until 1725. Kettering was a local centre of Peterborough Abbey estate and Watford was the chief settlement of the three townships within its parish. All these places had large compact demesnes.

In some cases block demesnes lie in an area of demonstrable 'old' settlement, as at Hall Manor, Wollaston, where the demesne incorporates a Roman villa and Early and Middle Saxon-period settlement, the villa recalled in the furlong name *Waltonacre* (1430), the 'tun' of the British ('wala'). Middle Saxon structures lie adjacent (partially excavated in 1964 (Hall 1974, 38–40)), and not far away Early Saxon sunken-featured buildings have been discovered in back garden crofts of the medieval vill (Northamptonshire HER, 3250/0/53–4 and 57, from Ashworth and Turner 1999). Higham Ferrers similarly has extensive Roman remains and a remarkable Middle Saxon settlement within the area of its demesne and immediately to the north of the medieval vill and borough (Hardy et al. 2007). It may be that block demesnes represent early separate lordly estates which did not participate in dispersal when villein fields were laid out. An early reference to a block demesne comes from Winterbourne (Dauntsey), Wiltshire, where an Anglo-Saxon charter of *c.*964 distinguished five hides of 'syndrig' land on the west of the village, defined by a boundary, from that of another five hides in the divided land on the east that were not given boundaries (B 1145; S 668).

For a block demesne, the tenants could theoretically have had groups of strips lying together that represented undispersed yardlands. However, Midland demesnes are observed to lie in ridge and furrow strips the same as the rest of the field system, and so dispersed yardlands within the block are to be expected. This is proved to be the case at Brixworth, where nine yardlands lying in a block demesne had full terriers made in 1422. The lands were scattered and intermixed in half-acres and rood pieces, although the tenurial order is not perfectly regular, contrasting with another yardland which was dispersed throughout the township fields with the same neighbours on each side (Hall 1995, 217). It is probable that most block demesnes were laid out in strips quite indistinguishable physically from the tenants' lands because, at manorial level, the demesne was still reckoned in yardlands and held by villeins.

At *Preveta* in Anmore, Hampshire, half of half a hide of demesne was described as a 'virgate' in *c.*1190. It lay in a block either side of the road from Anmore to Hinton, and another half hide lay on the east side of it (Hanna 1988, 33–4). These seem to be examples of undispersed demesne yardlands. In the county list above, eight counties outside the Central Region have recorded dispersed demesnes and twelve have block demesnes. It is probable that block demesnes are more common outside the Central Region than within it; they occur frequently in Essex.

B. OPEN-FIELD LANDHOLDINGS: YARDLANDS, OXGANGS, AND UNDISPERSED HOLDINGS

As already mentioned, tenants' holdings were most commonly called 'yardlands' in Southern England and 'oxgangs' in the North; but there were also many local variants. The size and physical disposition of yardlands in open fields likewise varied across the country. Simple numeric relationships between yardland totals for townships and the 1086 Domesday assessments occur in some counties, and these will be considered further in connection with the origins of field systems. There is some evidence for the existence of undispersed yardlands and hides that also relate to field origins.

The Central Region

The yardland, called the 'virgate' in Medieval Latin records ('oxgang' or 'bovate' in northern counties and other terms such as 'wist' in the south), was the standard open-field farming unit. The name may be derived from the term *yrðland* ('earth-land'); in the eleventh century the use of the term 'gherdelande' at Lutton, Northamptonhire (S 1110; Hart and Lyons 1884, i: 189; dated 1052–65), did mean 'yardland' as shown by its relationship to a hide recorded in 1086 (see the Gazetteer, p. 293).

The yardland comprised a number of arable lands that, in the Central Region, were normally scattered more or less uniformly throughout a field system. Such a distribution ensured that whenever a portion of the furlongs was left fallow there were always lands in the remaining area on which crops could be grown. The yardland also included rights to a proportion of the meadow (and common pasture, if there was any), as well as the right to common (turn out for grazing) a fixed number of cows, horses, and sheep on the meadows and arable after the crops had been removed. Sometimes cows and horses were accounted together as 'great beasts'.

The glebe yardland of Twywell, Northamptonshire, marked on Plate 4, shows a typical distribution. Yardlands are identifiable in the early twelfth-century Survey of Northamptonshire (Round 1902; Raban 2001, 241–57), when they were called 'small virgates'. The area of a yardland, which normally referred to arable and did not often include meadow and pasture, varied a great deal from vill to vill. For most places it remained a stable, unchanging farming unit, on which local taxation was based for many centuries. Northamptonshire yardlands have been discussed (Hall 1995, 77–94, where data and references are given) and the chief items of interest are summarized in what follows.

Yardland sizes, collected from sources in the date range 1250–1788, varied from twelve to eighty acres, although most were between twenty and thirty. There was a close correlation between yardland size and soil quality, taking loams based on river gravels and limestone as 'good' and boulder clay and heathland as 'poor'. The soils formed on intercalated permeable and non-permeable underlying geology on the slopes the Nene, Welland, and Ise Valleys are particularly good and support the greatest density of settlement, many vills being large. An interpretation of this correlation proposes that when a number of yardlands was assigned to a vill for taxation purposes, either the soil quality was taken into account or a standard acreage was assigned, and later when 'waste' ground was brought into cultivation, it was allotted to each yardland, maintaining the same number of yardlands in a vill.

For the bishop of Worcester's estates in 1299, it was noted that the yardland varied according to the nature of the soil. On light land most valued by the Anglo-Saxons, yardlands were small; the land was easily worked and carried heavy crops, but needed feeding (Hollings 1934–50, xviii). Jackson (1882, 144) likewise supposed that yardland acreages varied according to soil quality.

Stenton (1922, Lincolnshire Record Society, 18: xxi) showed that for many places in Lincolnshire the oxgang consisted of twenty acres in the twelfth century. For instance, at Cotes near Ingham, in 1182, two bovates (forty acres) were granted, twenty acres on one side of the village and twenty on the other

(Stenton 1922, 83). He thought that the variation from this found in the sixteenth and seventeenth centuries was caused by taking in additional land. The twenty or so places that afford this evidence occur throughout the county excepting the Fenland (Stenton 1920, xxvii–xxix). In the seventeenth century it was noted for the Berkeley region of Gloucestershire that a yardland was forty acres but in the wolds or hilly parts it was greater (Hilton, 1966, 115), illustrating that poor quality soils had large yardlands.

The variability in the number of yardlands in a hide, as well as the number of acres in a yardland, was commented upon long ago by Seebohm and Neilson. Seebohm listed values for Huntingdonshire from the Hundred Rolls of 1279 and found that there were four, five, six, or eight yardlands to the hide and that the yardland ranged from fifteen to forty-eight acres. However, more unity was found in the area of a hide, which in twelve out of nineteen cases was close to 120 acres. For most townships there were four virgates to the hide, each virgate being approximately thirty acres (Seebohm, 1883, 36–7). Neilson (1898) and Maitland (1897, 455–6) studied Ramsey Abbey holdings. The number of virgates in a hide varied from 4 (most common) to 6.75 and the acreage of the hide from 80 to 256 acres, but 120 acres occurred 7 times out of 31, and was near to the average (Hart and Lyons 1884–93, iii: 208–15).

The total number of yardlands in Northamptonshire townships did not normally change—although for a few places yardlands ceased to be used in later centuries, being replaced by acres. The number of yardlands per vill is variable (21–110) but many places show a close correlation to the number of hides assigned to the vill in the Domesday Survey of 1086. The ratio is not a uniform one, but ten and twelve yardlands to a hide is common, with extremes of five and twenty-four, the underlying factor being that nearly all ratios are five or six or some multiple of them. Thus Watford had twenty-four yardlands to the hide and Ashby St Ledgers ten yardlands. The implication is that the yardlands recorded in thirteenth-century and later documents already existed in 1086 (and therefore 1066) and were used as a basis for taxation (Hall 1995, 82–9).

The finest example illustrating the number of yardlands is for the parish of Clipston. It consists of the two townships of Clipston and Nobold (now a deserted vill site) in which there were four estates of irregular size assessed at a total of 5¼ hides in 1086. The values of the same estates in *c*.1124 add up to eighty-four 'small virgates' (from internal evidence there were sixteen to the hide), and in 1284 the same four were said to be eighty virgates (yardlands). The last total had been rounded to the nearest five for each estate so giving the apparent 'loss' of four virgates. The 1284 document has the advantage of recording how many virgates in each estate were assigned to either Clipston or Nobold, thus establishing that Nobold was in existence in 1086 although not referred to in a written record until this 1284 assessment. Most remarkably, in 1776, on the eve of parliamentary enclosure, the parish was assessed at eighty-four yardlands. So, on four occasions during 700 years, surveyors looked at

Clipston and Nobold townships and recorded that they were exactly the same size in terms of yardlands (references in Hall 1995, 86–7).

Kettering was rated at ten 'cassati' in a royal grant made to Peterborough Abbey in 956, and at ten hides in 1086. A survey of *c*.1130 stated that there were forty yardlands and the same number of yardlands is described in account rolls of 1307–8 and 1309–10 (Raban 2011, 405, 606). Hence there were four yardlands to the hide. A field book of Kettering made in 1739 gives details of the precise structure of the township. All the furlongs had a regular tenurial cycle divided into four parts called 'hides'. Three of them had ten 'yardland parts' (lands, equivalent to yardlands when added up throughout the system). The fourth hide generally had 13½ lands because 3½ lands had been added to it, but the other 10 lands were regular. There was therefore a regular cycle of forty yardlands with a few small items added. The cycle represents forty yardlands in 1739, 1310, and 1130, which if grouped as four-yardland pieces is the same as the ten hides of 1086 and the ten 'cassati' of 956 (Hall 1995, 91). The correlation between the ten 'cassati' of 956 and the later yardlands shows that the yardland assessment of some townships can be taken back to at least the mid tenth century and in this case did not change over eight centuries.

More information about yardlands in other counties is given in the Gazetteer; a summary of some of this material is listed below. Yardlands in all counties are of variable size and there is a widespread correlation that yardlands with large acreages are found in townships that once contained 'marginal' land of wood, heath, or marsh. This accords with the Northamptonshire findings, namely, either that an 'original' (probably pre-Conquest) assessment allowed for land quality being variable, or that later intakes were shared out among already existing yardlands, in contrast to assarts made after 1086. Such a process increased their size, but not their number, in a given township.

The potential addition of extra lands to an oxgang was recorded at Spaldington, Yorkshire (ER), in the mid twelfth century, where it was specifically stated that if the boundaries of the tilled land were to be extended, each oxgang was to be increased by the same amount (Stenton 1920, 62). This is here interpreted to be a memory of what used to be done a century or more previously. The document is written in the past tense; it does not say that increased land was being added to the oxgangs. The statement is therefore seen as evidence that oxgangs and yardlands had been increased in this way previously. It explains the larger size of lands in townships with marginal land being the result of intakes added to a core of yardlands that did not increase in number.

Simple numeric relationships between yardland numbers in a township and the 1086 Domesday assessments have been noted in a few counties. In Bedfordshire, at Aspley Guise, Fowler (1936) drew attention to the meadows being divided into a little more than thirty-nine 'hides', closely relating to the ten Domesday hides at four virgates or yardlands to the hide. Leicestershire hides, carucates, and yardlands have been discussed by Hughes, who has suggested that the fiscal values of 1086 may be correlated with eighteenth-century yard-land assessments (Hughes 1967–8, 19–23), and that yardland sizes may have been increased as a result of assarting.

The oxgang assessments for the East Yorkshire township of Thorpe Bassett in 1563 was sixty-four oxgangs (Harvey 1982, 35–8). This corresponds exactly to the sixteen carucates of 1086 (Williams and Martin 2003, 793, 879) at four bovates per carucate. At Warthill, Yorkshire (NR), Domesday records five caru-cates (Williams and Martin 2003, 798, 806), corresponding at eight oxgangs per carucate to the forty oxgangs recorded at Parliamentary Enclosure in 1756 (Tate and Turner 1978, 294). These relationships suggest that more analysis of Yorkshire oxgangs and their respective Domesday assessments would show similar exact relationships, thus further establishing that there were structured field systems by 1086.

The Eastern and Western Regions

Although the terms 'yardland' and 'oxgang' were used over large areas of Eng-land, there were other terms used for the medieval holding. In Northumber-land, holdings were measured as 'husbandlands' at the manors belonging to Hexham Priory in 1379. In later centuries they were called 'farms' or 'farm-holds' (not to be confused with demesnes that were often called the 'farm' throughout the remainder of the country). The Hexham Priory husbandlands of 1379 varied in size from Dotland's fifteen acres to Kirkheaton's thirty-four acres. Some husbandlands were engrossed—a tenant of Kirkheaton held four. In 1847 evidence was collected from many townships in the county, all en-closed, yet taxes and other dues were still paid proportionally on the number of 'farms' that comprised them in open-field days (Dendy 1894, 138–9). 'Oxgangs' were also used in parts of the county.

'Ferlings' were used in Cornwall and Devon. In Cornwall they were the fourth part of a Cornish acre and varied from four to fifty acres (Gray 1915, 256). Cornish and English acres have a variable relationship. Sometimes 64 or 120 English acres to a Cornish acre were used at different dates (Hull 1971). In the fourteenth century, seven acres English were said to be a quarter-acre fer-ling Cornish (Beresford 1964, 17). In 1706 there were, by Cornish measure, eighteen feet to a yard at St Kew manor, making a Cornish acre equal to thirty-six English statute acres (Maclean 1873–6, ii: 91). Holdings were typically one or half a Cornish acre at Nether Helland in 1569 (ibid. 19–20). Devon ferlings were also of variable acreage. Demesne arable ferlings at Langdon in Wembury contained thirty-two acres in 1283. A thirty-acre ferling at Tavistock, in 1302, lay in only five pieces (Finberg 1952, 275). Other ferlings were Churston Fer-rers, thirty-two acres in 1198, and Paignton, thirty acres in 1567 (Finberg 1951, 39). An Ottery St Mary rental of 1483 shows that one ferling was sixteen acres, with four ferlings to a virgate of sixty-four acres, and four virgates to a hide

(Rose-Troup 1934). Elsewhere ferlings were used to describe a quarter of a yardland, for example at Bargham and Chiltington, Sussex (Wilson 1961, 73; Godman 1911, 177).

In Sussex, the term 'wist' was used as an alternative name for a yardland. Wist areas in the fifteenth century were Berwick twenty acres, East Blatchington eleven, and Alfriston seventeen acres. Fractions of yardlands are given as 'dayworks' in some places such as Bersted near Pagham in 1321, Tangmere in 1258, and Robertsbridge in 1541. At Hamerden in Ticehurst, ten dayworks equalled to a rood, from which a daywork was a tenth of a yardland in area (D'Elboux 1947, xix). In Kent the terms 'yokeland' and 'tenementa' were used, as already described. Surrey was mainly assessed in yardlands, but at Ewell, in the thirteenth century, tenants had not yardlands but yokelands of thirteen acres, as found in Kent. A field book of 1409 describes lands lying in furlongs, but the holdings, called *tenementa*, used the names of earlier yokelands. There had been some partition of lands since the thirteenth century, because parcels in a given *tenementum* belonged to several new owners.

Suffolk and Norfolk holdings were called *tenementa* and 'eriungs'. The term *tenementum* for a holding is found in the Binham Cartulary and in Castle Acre Priory records at *Sudmere* and *Helingeia* before 1148 (quoted by West 1932–3, 224). Wimbotsham had holdings called 'aruinges' of ten acres in the twelfth century (Hart and Lyons 1884–93, iii: 285) but not identifiable in Domesday. The term 'lancetagium' was used at Brancaster, Deepdale, and Burnham for twelve-acre holdings (ibid. iii: 262–3). Villein holdings of St Benet of Holme were called 'landsetti' at Ludham and Potter Heigham (West 1932, 22). Gray described irregularly scattered late medieval holdings consisting of parcels that had once belonged to older holdings called 'tenementa' and owed rent and work-service. In the thirteenth- and fourteenth-century surveys the holdings were listed by the names of the *tenementa* but already they usually belonged to several people. The earliest detailed survey for Martham, made in 1291 has been discussed by Gray (1915, 335–39). From the personal name forms the *tenementa* cannot be dated before the thirteenth century. However, at Martham, it was probable that the 'former tenants' already held eriungs that were subdivided. The evidence is suggested by a fine of 1198 where 3½ acres of Martham land was shared between 5 people, one of whom bore the same surname as a tenant of villein land in 1291 (Hudson 1917–19, 296–8). Suffolk has similar evidence of subdivided *tenementa*.

Below is a summary of data relating to yardlands and oxgangs, also giving local names found for lands and yardlands in other regions.

Bedfordshire: Various early sources show that there were four yardlands to the hide. An estate at Studham consisted of fractions of twelve yardlands lying in named hides which added up to three hides in 1204–5 (Fowler 1926, 123). Yardland data is provided by the Hundred Rolls of 1274–9; at Riseley four hides contained sixteen virgates

(Thompson 1990, Bedfordshire Historical Record Society, 69: 27–63). Yardland sizes from eighteen to forty acres are described in the Gazetteer.

Berkshire: Sonning and Winnersh yardlands have been discussed. Harwell yardland was 20 and 21¼ acres in 1220 and 1310 (Magd. Berks Charters, Harwell 9A and 100).

Buckinghamshire: Lathbury yardlands were eighteen acres in 1241 (Hughes 1940, Buckinghamshire Record Society, 4: 77). A Ramsey Abbey extent of 1244 for Crawley refers to yardlands of forty-eight acres. At Little Brickhill, 1½ yardlands lay in *Cokeshide* in 1197, showing that arable lands were grouped in blocks called 'hides' (Travers 1989, Buckinghamshire Record Society, 10: 102). A hide at Beachampton contained four yardlands in *c.*1200 belonging to four named people (Jenkins 1942, ibid. 9: 27, 40). Leadam (1897, 638) recorded values in 1517 from fifteen to forty acres. In the eighteenth century Akeley had 13 yardlands and Steeple Claydon 80½ (Tate and Turner 1978, 67) which equate with Domesday's 3 and 20 hides (Williams and Martin 2003, 405, 420) at close to four per hide. Grendon Underwood's thirty-five yardlands and three hides correspond closely to twelve yardlands per hide (ibid. 418).

Cambridgeshire: A hide at Haslingfield contained 120 acres in 1279 (Hundred Rolls, ii, 558). Yardlands at Leverington were sixty acres, Tydd St Gyles thirty-two, and Fenton thirty (Walker 1886, *Proceedings of the Cambridge Antiquarian Society*, 5: 101). Ely monastic estate yardlands areas were listed in 1221 and 1277. Among them were Coln fifteen acres, Doddington twelve, and Welles eighty (Pell 1891, ibid. 6: 38). Chippenham yardlands were thirty to thirty-two acres in 1544, Orwell twenty-five to thirty-five in 1607, and Willingham thirty-six to forty-two acres in 1575 (Spufford 1974, 67, 100, 149).

Cheshire: Lands were called 'butts' and the dividing furrows were known as 'reins'. Holdings were measured in oxgangs. At Frodsham manor, in *c.*1350, an oxgang was about sixteen acres (Dodd 1982, *Transactions of the Historic Society of Lancashire and Cheshire*, 131: 24–6).

Cornwall: Ferlings were the normal holding in the Middle Ages, as already described (p. 109). Strips were often called 'stiches' or 'quillets'.

Cumberland: Holdings were measured in oxgangs. In *c.*1200, a Warwick oxgang of eight acres lay in three pieces of five acres, two acres, and one acre (Prescott 1897, 121).

Derbyshire: Holdings were described both as 'yardlands' and 'oxgangs' in the twelfth and thirteenth centuries. Aston on Trent had a holding described in yardlands in 1202 and oxgangs in 1219 (Hart 1885, *Journal of the Derbyshire Archaeological Society*, 7: 200, 213). Willesley was virgated and Finder bovated in 1208. Oxgangs were mostly dispersed, as shown by early terriers and the explicit statement for Alsop in 1287 that twenty acres consisted of 'divers small parts lying in the fields'. Oxgang sizes were variable; at Normanton in 1262 they were sixteen acres, Pentrich approximately twelve acres in 1286, Crich twelve acres, and Alvaston around ten acres (Darlington 1945). Various taxations were raised on the oxgang; at Norton, in 1581, ninety-six oxgangs paid for the repair of the church (S. O. Addy 1922, *Journal of the Derbyshire Archaeological and Natural History Society*, 44: 58–9).

Devon: As with Cornwall, the medieval holding was a 'ferling' (see above). Open field strips were called 'landscores'. The 1566 Dynham survey described customary holdings at Ilsington with parcels lying in the common [field] called *lez londscores* (Chope 1911, *Transactions of the Devonshire Association*, 43: 280). Shorter showed that 'landscore' survived as a field-name, often for small rectangular fields, in tithe and earlier records for many places (Shorter 1940, *Devon & Cornwall Notes and Queries*, 23: 372–80). Landscores were mentioned at Widicombe and Blackslade in 1588 (Gawne 1970, 56).

Dorset: Arable strips were called 'lawns' in the west and south of the county and in the tithe awards for Askerswell in 1848 and Sherborne in 1842 (Kerr 1959, *Proceedings of the Dorset Natural History and Archaeological Society*, 81: 133–42). They were separated by narrow strips of grass called 'walls' or 'lanchets' (Pope 1909). In *c.*1225 a half-virgate lay four acres in one field and four acres in another at Blandford St Mary (Hassall 1946, *Proceedings of the Dorset Natural History and Archaeological Society*, 68: 46–7). Bloxworth yardland, in 1301, contained twenty-two acres (Jones 1942, ibid. 64: 72) and a yardland at Oborne (1515) had forty acres of arable and meadow (J. Fowler 1955, ibid. 77: 157–61).

Durham: The smallest parcel was called a 'rigg' or 'land', about half an acre, and furlongs were commonly called 'flatts'. Holdings were usually 'oxgangs' in the Middle Ages, and occasionally 'husbandlands'. The sizes vary and they were sometimes let out in pairs in the twelfth century. Sizes in 1183 for Boldon, Morton, Warden, and elsewhere ranged from twelve to fifteen acres (Greenwell 1852, 45–8); in 1430, East Rennington was twelve acres, and Eden twenty-four acres (Greenwell 1872, 19–21).

Essex: A Felsted survey of 1572, copying an earlier one of *c.*1462, gives thirty acres for half a yardland (Fisher 1937, *Essex Review*, 46: 64–5). Walton, Kirby, and Thorpe le Soken yardlands were thirty acres in 1222 and there were four to the hide (Hale 1858, 38–43). Hidation and holdings in 1222 have been discussed by Faith (1996, 206–12).

Gloucestershire: There were four yardlands to a hide at Saintbury in 1199, split into one yardland and six half-yardlands belonging to named tenants. At Cirencester one yardland, being half of half a hide consisted of eighteen acres dispersed in ten furlongs (Elrington 2003, 2, 10). A hide at Combe Baskerville consisted of four virgates held by different tenants (Hart 1863–5, i, 238). Yardland size varied and recorded values are twenty-eight to forty-eight acres in the Gloucester Cartulary (Hart 1863–5, iii: p. cix). Large yardlands occurred at Rodmarton: forty-five acres in *c.*1200 (Ross 1964, 344–5). Swindon, near Cheltenham, had a quarter yardland of 15½ acres in *c.*1250 scattered mainly as half-acres (Fullbrook-Leggat 1946–8, 283–4).

Hampshire: Most of the county was virgated, for example at Charlton, Witherington, and East Downton. Barton yardlands were forty acres in *c.*1330 (Hanna 1988).

Herefordshire: The county was assessed in yardlands in the thirteenth century (Bannister 1929). Bosbury yardlands were sixty acres and a quarter was called a 'noke' in 1578 (Butterfield's survey, Herefordshire Record Office, AA 59 A2, fol. 32).

Hertfordshire: Half a yardland at Alswick in the thirteenth century consisted of 17½ acres lying in 10 parts. Terriers of Kensworth, in *c.*1490, described two irregular half-yardlands of around twelve acres consisting of seven and five parcels only and not lying in the same furlongs. Undivided yardlands occurred at Weathampstead in 1211 and at Weston in the mid seventeenth century (Gray 1915, 377–8, 374). Therfield yardlands were sixty-four acres in the thirteenth century (Hart and Lyons 1884, iii: 213). Abbots

Langley yardland was forty acres (1350) and Codicote sixty-four acres in 1331 (Levett 1938, 186).

Huntingdonshire: The size of yardlands, collected from the Ramsey Cartulary, varied from 15½ acres at Abbots Ripton and 16 at St Ives to 34 at Brington and 40 at Knapwell (DeWindt 1976, 376). A Stilton yardland (*c.*1230) was about thirty acres (Godber 1940, Bedfordshire Historical Record Society, 22–3: 206–7). Values given in the Hundred Rolls have been noted earlier (p. 107; Seebohm 1883, 36–7).

Kent: Block holdings were called 'yokes' and varied in size from twelve to fifty-two acres (Du Boulay 1966, 120–1). They were held in severalty and fragmented by partible inheritance. Yardlands were sometimes used, as at Eltham in 1263, which had 28¾ yardlands each of 7½ acres. Domesday assessed the county in 'sulungs' consisting of four yokes. It is frequently recorded that ploughlands and sulungs occur in the ratio of 4:1 (e.g. Hollingbourne, Beckenham, Leeds, etc. (Williams and Martin, 2003, 11–19)), with the implication that a ploughland is equivalent to a yoke.

Lancashire: Lands were often called 'butts'. Some blocks of lands were separated as early as the mid thirteenth century by grass balks called 'ranes'. Holdings were measured in oxgangs during the thirteenth and fourteenth centuries (Atkinson 1886–7, 91, 230). Oxgang sizes in 1346 varied: at Overton an oxgang was twelve acres, at Slyne twenty, at Skerton twenty-four, and at Roshead sixteen (Farrer 1915a, 142–54).

Leicestershire: Bottesford was assessed in oxgangs in 1524. Wigston Magna had ninety-three yardlands in the sixteenth century and ninety-six in 1764. More details of the yardland assessments of several townships are given in the Gazetteer (p. 283). Stathern oxangs were levied at sixpence each by the churchwardens in 1651.

Norfolk: The term 'yardland' was not used, and medieval holdings, often twelve to fourteen acres, were variously called 'eriungs', *tenementa*, or had no name at all (see Martin and Satchell 2008, 18). Many of them had been undispersed pieces split by partible inheritance (see p. 110).

Northamptonshire: Yardlands and the Domesday assessment have been discussed (Hall 1995, 82–94). Whilton was assessed at one hide and two ploughlands in 1086. An inquisition of two-thirds of the manor in 1295 had 9 demesne yardlands and 8½ yardlands of villein land (*Calendar of Inquisitions Post Mortem*, iii: 156–7; Palmer 1984, 66), hence 26¼ for the whole. These descended as 13.445 yardlands each of farmland and townland for the whole village in 1778 (Enclosure Award, Northamptonshire CRO, Inc. 68), a total of 26.9. From this it can be seen that Domesday assessed only 13½ yardlands as one hide, but both this and the equal-sized demesne was assessed as two ploughlands, hence ploughlands seem to measure the overall size.

Northumberland: The smallest unit was called a 'rigg' or 'land', usually about half an acre, and was separated from its neighbour by a furrow, called a 'floor'. Widths varied from four to sixteen yards, often about twelve. Furlongs were called 'flatts', 'furshotts', and 'sheths'. Waste spaces were called 'balks' in 1378; these were probably dividing markers between major holdings. Holdings were sometimes measured in 'husbandlands', as described on p. 109, and some were assessed in 'oxgangs'. Husbandland sizes are recorded in 1378: Alnwick sixteen acres, Dotland fifteen (Raine 1864–5, ii: 3, 10).

Nottinghamshire: Both 'yardlands' (eighteen acres) and 'oxgangs' (mostly twelve acres) were used in the county.

Oxford: Yardland sizes in 1728, taking the 'ridge' of a survey as an average of 0.33 acre, were Deddington twenty-two to twenty-four acres, Clifton sixteen to twenty-five, and Hempton fifteen to eighteen acres (Northamptonshire CRO, C(A) 3309, pp. 18–60). Yarnton and Begbroke hides at 10 and 4.25 in 1086, correspond approximately to the 52 meadow divisions of later centuries at four to the hide. Nether Worton with 24 yardlands in 1279 and 24.75 in the seventeenth century, was assessed at 5 hides in 1086 (VCH, *Oxfordshire*, xi (1983): 285. Great Tew yardlands lay in 'hides' as late as the seventeenth century (ibid. 223).

Rutland: Yardland sizes in the seventeenth and eighteenth centuries were Barrow twenty acres, Cottesmore twenty-five, Oakham twenty-five, Preston approximately forty-four, Wardley twenty-two, and Whissendine thirty-six acres. Wing township contained forty yardlands. Lyddington's 27½ yardlands of 28 acres (Northamptonshire CRO, Box X5208) correspond to the 2 hides of Domesday (Williams and Martin 2003, 595) at near 12 yardlands per hide.

Shropshire: The county was mostly assessed in yardlands (or carucates), as at Walcot in Wellington 1141–7 and Abcott in Clungunford 1195–1222 (Rees 1985, 228, 19). A quarter of a yardland was called a 'noke'. Both yardlands and oxgangs were recorded at Cheswardine and Childs Ercall in 1280 (Fletcher 1908, *Transactions of the Shropshire Archaeological and Historical Society*, 3rd ser., 8: 361–7). Yardland sizes were often high at fifty to sixty acres (Stamper 1989, 33–4). Downton was forty-one acres in 1298; Haughton yardlands sixty-two acres (1327), Aston sixty-four acres (1261) (Rees 1985, 75, 108, 35). Local taxes and rates were levied upon them (Waters 1910, *Transactions of the Shropshire Archaeological and Historical Society*, 10: 60).

Somerset: Milborne, Kingston, and Wyke yardlands were forty acres in 1262 and had commons for fifty sheep, one cow, and two pigs (Whitfield 1981, 22). Cold Ashton yardland in *c.*1150 had commons for eight oxen and fifty sheep (Hunt 1893, 64). The described bounds of a hide at Minehead in *c.*1290 implied that it may be an intact unit (Hunt 1893, 99). Two 'cotsetlands' made half a yardland at East Pennard in 1189 (Jackson 1882, 40).

Staffordshire: Some of the county was virgated, early examples being Slindon, Swinfen, Hampton, Longridge, Chorlton, Stanton, Ackbury, and Woodford in 1199–1206 (*Collections for a History of Staffordshire*, 3 (1882): 167–214). Oxgangs were also used, for example at Marchington in 1227 (ibid. 4 (1883): 227). Baswich yardland was twenty to twenty-two acres in 1298, Wolverhampton and Wigginton sixty acres in 1249, and Wednesbury thirty acres (Anon. 1911, ibid. 144–6, 320). Large yardland sizes are found at places with relatively marginal land.

Suffolk: Carucates were used in the Domesday survey, usually at 120 fiscal acres. The customary holdings were variously called *tenementa*, 'full land', 'landsettagium', or 'sift'. The acreage varied widely, but was usually consistent within one manor or parish; twelve acres was common (Douglas 1927, 27, 37–43). In some vills there was fragmentation caused by partible inheritance (see the Gazetteer, p. 319).

Surrey: Yardlands were studied by John Blair (1991, 71–4). A list of seventeen townships showed that sizes varied from ten to thirty-two acres. Egham yardlands (1378) were 22–37 acres, Epsom 20.5 (1507), Battersea 15 (1545), Chertsey 14–24 (1426) (Gray 1915, 360–3); Bookham and Cobham were 15 acres in the fourteenth century; Banstead (1325) 60 acres (Lambert 1912, 319–20, 324–5). Godalming holdings were called 'cotlands' in the fourteenth century and varied between twelve and twenty-five acres (Woods 1910, *Surrey Archaeological Collections*, 23: 93, 103). Much larger and variable acreages were found on the Weald (Lambert 1912, 53). At Ewell, in the thirteenth century, holdings were described as 'yokelands', as found in Kent.

Sussex: The terms 'wist' for a yardland and 'daywork' for a tenth of a yardland have been explained (p. 110). Battle Abbey hides (*c*.1120) consisted of eight virgates, probably to allow for the poor quality soils in assarted lands on the Weald. Elsewhere four virgates made a hide and four hides a knight's fee. Alciston had 33 yardlands in *c*.1280 and a virgate was 11½ acres. Other values were *Suthram* twelve and Fishbourne fifteen to seventeen acres (thirteenth century). Meeching and Piddinghoe had 22¾ and 32 yardlands in total. Large yardlands occur at Atlingworth, 36 acres; Keymer, 100, 125, 140, and 164 acres; and Cuckfield, 140 acres (Godfrey 1928, 17–40).

Warwickshire: 'Hides' ceased to be mentioned in transactions after 1208 (Welstood 1932, xix, no. 679). Newbold Pacey glebe had four yardlands that consisted of blocks of four lands lying as one acre and Treddington was similar (Barratt 1955–71, ii: 14, 78). These were probably grants of a hide. Haselor yardlands were eighteen acres in 1585, Alveston twenty-four acres in 1240, and Tiddington eighteen acres (VCH, *Warwickshire*, iii (1945): 110, 284–5). Stoneleigh yardlands were thirty acres in 1392 (Hilton 1960, 121). Wasperton's yardland of thirty-five acres consisted of eighty-three lands (Barratt 1955–71, ii: 117). Yardland sizes described in glebe terriers vary from sixteen acres at Weethley to seventy acres at Wolverton, the commonest being about thirty acres (Barratt 1955–71, i: p. xlvii). Dyer (1981, 5) found that the yardland varied from sixteen to thirty-six acres. Exhall's 24-acre yardlands were stinted at 50 sheep, 7 beasts, and 2½ horses in 1585.

Wiltshire: Most yardlands varied in size from seventeen to thirty acres and had stints for large numbers of animals, especially sheep (see table in the Gazetteer).

Westmorland: Bolton near Appleby had a ten-acre oxgang in 1490 (Prescott 1897, 453). Six acres made an oxgang at Penrith in 1307 (demesne); socage holders had the same, but bondhold oxgangs were ten acres. Other sizes are Alston, twelve acres and Langwathby, seven acres (Elliot 1960, 98 n. 3).

Worcestershire: Charlton yardlands were approximately fifteen acres in 1649 (Cave and Wilson 1924, 29). Castlemorton was twenty-four acres in the fourteenth century (Fullbrook-Leggat 1946–8, 286) and Flyford Flavell forty acres in 1584 (Yelling 1969, 31–3). At Acton Beauchamp a quarter-yardland was called a 'ferendell', which was further divided into four 'nokes' in 1594 (Pratt 2000–2, 210).

Yorkshire: Early records of sizes are recorded for Normanby (fifteen acres in the late twelfth century: Farrer 1915b, no. 745), Stapleton near Pontefract (twenty-two acres in *c*.1179) and Skelbrooke (sixteen acres in *c*.1185: Farrer 1916, 292, 231). An oxgang at Ormesby and Caldecote was nineteen acres in the thirteenth century (Brown 1889, 278). Much other information relating to oxgang sizes is available in the literature (Brown 1892, 15–51). Harris (1955) discussed oxgangs in the East Riding that lay in narrow lands and broad lands. The narrow ones were generally half the width of the broad lands, which represented one and two oxgangs respectively.

Undispersed yardlands and hides

Interest in undispersed yardlands and hides, that is, a block of land belonging to one holding and not dispersed as small strips throughout a field system, arises because of its possible relation to Saxon-period severalty farms pre-dating dispersed fields. There are difficulties in interpreting the data of the Central Region before any claim to an 'old' or 'original' undispersed yardland can be made. Some undispersed holdings called yardlands were assarts made in the twelfth and thirteenth centuries. Other blocks called yardlands may have been demesne. Early evidence is often from monastic sources and most of the large religious endowments were from lordly demesnes, whether stated to be so or not. Some demesnes were in block form and could have been let out in smaller intact pieces called yardlands, defined by two or four cardinal boundaries. Such holdings are unlikely to have any direct bearing on early undispersed systems. Possible examples of undispersed or partly dispersed hides and yardlands are collected together, after examples of medieval assarts have first been discussed.

Assarts were generally held in severalty, and most were measured and rented by the acre. On the Ramsey Abbey lands in Bedfordshire some assarts at Cranfield were assessed as yardlands and half-yardlands in the thirteenth century (Hart and Lyons 1884–93, iii: 302–3). These are therefore undispersed yardlands, although of late origin, probably twelfth century, and after 1086. Crofts in the Buckinghamshire Chilterns, probably assarts, in existence by *c*.1300, can be located from their names on the 1843 tithe map of Chesham (Casselden 1987, 145). Roderick (1949, 62–5) provides medieval evidence for assarts and 'stokking' names in Herefordshire at Much Marcle, Wormsley, Clehonger, Stockton, and elsewhere. At Orleton a half-yardland of 21½ acres was an assart. A yardland at Alveley, Shropshire was probably assarted (Rees 1985, 25). Half a yardland consisted of 12½ royal acres (23 statute acres) lying in crofts, at least three of which were adjacent and one contained 12 selions.

The demesne of Reigate priory at Horley, Surrey, in the thirteenth century, consisted of a grant of assarts and dens on the Weald Clay that have been located from detailed field-name studies as five separate areas of closes (Ellaby 2000, 145–55, plan 149). Thorncroft (Leatherhead detached) yardlands on the slopes of the Surrey Downs can be traced from *c*.1265 and have been shown to be compact holdings (Blair 1991, 47; Gardiner, in Christie and Stamper 2011, 103). Virgates were still being added to manors as late as *c*.1220. A Hambledon deed granted lands then recently put together which were stated to be half a yardland. There was often a lack of widespread dispersion (Blair 1991, 76–80).

Southern Berkshire had complicated fields in the Middle Ages. The Arborfield estate of Corpus Christi College, Oxford, included grants of a purpresture called 'hinutherudinge' in 1250 which was described as arable in 1326, and in 1386 consisted of a messuage and five crofts of arable lying in a block. In 1508 there were three crofts of arable called 'bourdlond' containing ten acres. Two blocks of closes were mapped in 1607, one group called 'boorde landes', at 9½ acres (Milne 1942). This is assarting with the land remaining in severalty and being enclosed early.

Assarted yardlands differ from those associated with the ancient vills in that they were marginal land taken in and not part of the original village arable land. Some assarts were divided into strips, as at Field, Staffordshire (thirteenth century), where an assart contained at least fourteen selions (Wrottesley 1906, 321). Two examples of early large intakes on the edge of former woodland are found in Northamptonshire. They are Roade Hyde, lying next to a detached piece of wood in what was to become the royal forest of Salcey. It is identifiable in 1086 as a holding assessed as 'four parts of half a hide' (Williams and Martin 2003, 614) and was called 'hyde' in *c*.1124 (Round 1902, 375, under Ashton, as four small virgates). It had its own two small strip fields separate from those of Roade (mapped in Hall 2013, Map 69M). The other example is *Beanfeild* in Middleton near Rockingham. It was named *Benefeild* in *c*.980 (Mellows 1949, 69, incorrectly assigned it to Benefield near Oundle); it is not recorded in 1086, and is called *Banfeld* in *c*.1124 (Round 1902, 386), assessed as one hide. It became appropriated as an oval-shaped lawn with a lodge in Rockingham Forest, and so never developed any strip fields or became a severalty holding (mapped in Hall 2009, 24, 83). These examples were assessed in hides and were treated as townships. They are to be distinguished from severalty assarts made after 1100.

Next follow examples of early holdings not stated to be assarts that were assessed with core vill settlements. Gray (1915, 265) found evidence for Devon ferlings lying as undispersed in blocks at Coombe and Coleford in 1199. A possible undispersed ferling occurred at Buckland Filleigh, in 1274, where one ferling of land lay between the lands of Richard Shorta and Gilbert Carpenter with all houses upon it (Reichel 1909, 253). Two undispersed oxgangs were recorded at Lee Gallica, Lancashire, in 1233, one at the new corn mill and the other in *Gulhosfeld*, both lying between the lands of named people (Lumby 1936, 33). In *c*.1200 at Warwick, Cumberland, an oxgang of eight acres lay in three pieces of five, two, and one acre (Prescott 1897, 121). A 1455 terrier of lands belonging to Morehouse (south of Warwick), describes several pieces lying in or near a furlong called Halfacres. A possible undispersed oxgang occurred at Melmerby lying next to the land of Daniel son of Henry (ibid., 277, 374–6, 289). Westmorland had undispersed holdings at Hutton Roof (Gazetteer, p. 334).

Yardlands at Hunston and Kepston, Sussex, were partly dispersed in 1253. One had pieces in seven crofts and another lay as fourteen acres in five crofts and two acres in *Hamstede* (Fleming 1960, 187). A hide at Tangmere, in 1222, consisted of 60 acres; a terrier of part of it described 22½ and 18¾ acres both lying in blocks next to roads. A grant of twenty-four acres at Banstead near Croydon, Surrey, had been taken out of a virgate and all lay in *Snithescroft* in

1199, showing that it was an undispersed yardland Gray (1915, 355–69, 549). Eversholt, Bedfordshire, had an undispersed half-yardland lying in a croft in *c*.1200 (Fowler 1926, 130). In southern Berkshire, a yardland at Sonning contained fourteen acres of enclosed land only (Gray 1915, 553–4). This may represent an 'ancient' half-yardland never dispersed.

Medieval holdings that certainly lay in blocks were the yokelands of Kent and the *tenementa* of East Anglia. These have been described (see pp. 62, 110); their sizes are similar to yardlands. Some parishes in Suffolk and particularly Essex had customary and freehold lands that seem to have been undispersed from an early date and were never part of any open-field system, for example Felsted and Ingatestone (Martin and Satchell 2008, 141–51, 154–9). At Ingatestone, some of the block holdings had 'hide' names. Britnell (1988) gives further examples of holdings consisting of blocks of land at Colchester and elsewhere, some of them called 'hides' (described on p. 68).

Severalty small estates that were not necessarily medieval assarts but could be undispersed holdings of pre-eleventh century origin are found in Dorset, where severalty farms were recorded in the late twelfth and thirteenth centuries at Broadwindsor and Thorcombe. The boundary of Willislond in Broadwindsor was described with the comment 'men may live on this land and have commons in Broadwindsor', showing that it lay in a block and was not dispersed. In the same parish Childhay (ST 41 04) and Sandpit (ST 42 04) were similarly bounded, Sandpit having tenements and arable furlongs. Catsley (Catsley Farm, ST 52 03) in Corscombe (mid thirteenth century) and Sadborow (ST 37 02) in Thorncombe (1161–72) were small discrete estates, the property of Forde Abbey (Hobbs 1998, 31–48, 78, 91).

Other holdings likely to be undispersed come from indirect evidence in the Domesday Survey (but there is no certainty that they were not demesne). At Dunwich, Suffolk, a carucate, part of two in 1066, had been submerged into the sea before 1086 (Williams and Martin 2003, 1208). Two hides out of seven at Horton, Dorset had been taken into the royal forest of Wimborne (ibid. 209); unless the two hides formed a separate township, they cannot have been dispersed.

The above evidence shows that medieval assarts, generally held in severalty, can be found anywhere in the country, but undispersed holdings of likely pre-Conquest date assessed as hides or virgates, are mainly confined to the Eastern and Western Regions, with a few in the Central Region where there had once been woodland or relatively marginal land. Early 'hydes' will be further discussed (see pp. 184–5).

C. TENURIAL ARRANGEMENTS IN THE FIELDS

Evidence for a regular order of tenants' holdings in the Central Region was first fully examined by Göransson in 1961, although the closely related occurrence of scattered strips with the same neighbours either side had been known since Seebohm's publication in 1883. More examples are collected here and it is suggested that there was an organized field structure at an early date and sometimes before 1086 in the Central Region, where most of the regular cycles are to be found.

There are numerous pieces of evidence suggesting that many townships in the Central Region had a regular order of tenants' holdings in the fields. Some terriers listing lands of a single holding lying in all the furlongs of a township imply that there was a regular tenurial order with two, or only a few, neighbours on either side throughout the system. The best known is Shipton manor, Winslow, Buckinghamshire, for which Seebohm (1883, 24–6) published a full terrier of one yardland made in 1361. In all there were seventy arable lands and in sixty-two of them one neighbour was the same on one side and only two occur on the other, forty-three and twenty-three times each. There are many other examples of various dates, some noted below or in the Gazetteer.

Much scepticism has been expressed in the past about the interpretation of regular tenurial arrangements, many authors doubting the antiquity and significance of such arrangements. Doubts were driven to some extent by opposition to theories that the Anglo-Saxons introduced highly structured open-field systems in the fifth century, and disbelief that a regular order was a surviving relic of an ordered layout made in the distant past. One of the suggestions was that a dispersed holding with the same neighbours could arise by the splitting of parcels between heirs and need have no relevance to the tenurial arrangement in the remainder of the system.

A holding at Mitcham, Surrey, in 1235, lay on the 'shade side' of six named 'fields' (furlongs). Blair (1991, 85–6) noted that this would 'divide open-field strips along the furrows of their component selions—surely the best explanation for those common and much discussed cases of subdivided holdings where each plot adjoins the same neighbour's land'. Two pieces were named *Inlond* and *Bery*, so they were probably demesne. A Wimbledon holding of 1248 had each of five lands lying next to the same person. These examples are from 'final concords' that frequently deal with estate splitting, and are not random samples of villein land lying in the fields. Such division would give rise to two new holdings with at least one adjacent neighbour the same throughout the system.

The longitudinal splitting of the strips has been suggested as the interpretation of regular neighbours at Chippenham, Cambridgeshire. Land that had been divided between two heirs in the thirteenth century had many scattered strips with the same neighbours. It was commented that 'at first sight, this looks like a much battered survival from the original allotment of strips; in fact it may not be the product of co-aration, but the division of holdings between heirs' (Spufford 1965, 22).

A clear case of splitting lands lengthwise is found at Althorp in Brington, Northamptonshire, in 1302, when a yardland 'in the common fields' was divided equally between two parties, one having the sunny side and the other the shade side (Willis 1916, 74–5). Any division of a single yardland by this 'sun division' method would result in lengthwise splitting of the lands, forming new half-yardlands with the same neighbour throughout the fields. At Ecton, in the same county, in 1608, a deed describes the partition of a land between two heirs by specifying that it was to be done by lengthways splitting (Northamptonshire CRO, E(S) 383b). These are likely to be unusual events, because they are tedious methods of division that involve entering the fields to measure and set out furrows down every land.

Fowler cautioned against interpreting too much about the antiquity of tenurial layout based on the evidence of late maps and field books (Fowler 1936, ii: Oakley, 6). Regular sequences have also been discussed by Joan Thirsk (1964, 28), who doubted if they had much antiquity or would have lasted very long. Field books that have no apparent regular order are most likely to have been subject to the exchange processes noted by Fowler and an example is fully recorded for Lamport, Northamptonshire. A field book of 1583, made for two of its three fields (Northamptonshire CRO, IL 812) betrays little evidence of any regular sequence because of extensive exchanges previously made by the lord, John Isham, with the freeholders in the third field, which he then enclosed (Finch 1956, 15–17).

A regular sequence of holdings would be formed by partible inheritance, especially if the holdings were in large pieces in the first place, as with the yokelands of Kent and the *tenementa* and eriungs of East Anglia. All these types of holding show great fragmentation by the thirteenth century, with many parcels held by people with the same surname. These mechanisms, splitting of holdings and partible inheritance, would explain why a single terrier had parcels with one or both neighbours the same. However, such processes are rare or localized events. Homans (1941, 116, 428) found that evidence for partible inheritance was very limited, only being found in Kent, Suffolk, Norfolk, Hertfordshire, and in some fenland villages in the Wisbech region of Cambridgeshire and Lincolnshire. Some places in County Durham also operated partible inheritance (see the Gazetteer).

Several early authors as well as Seebohm commented upon examples of strips lying in an apparent regular sequence. Vinogradoff (1892, 233–4, 457–8; 1905, 178, 263–7) described the re-allotment of lands at Segenhoe, Bedfordshire, in the twelfth century, which probably created a regular order. All lands were surrendered and re-alloted by six old men who knew to which fees they belonged. Fox (1981, 96–7) used the same source and suggested that it might represent the creation of a regular two-field system (found there in the early thirteenth century) from an irregular arrangement. A 1246 grant of two oxgangs at Gelston in Hough on the Hill, Lincolnshire, stated that they 'lay scattered in the fields always next to the land of Geoffrey le Manner' (Foster 1920, xv, 35 no. 5). Stenton noted that a regular tenurial order was un-usual in Lincolnshire but found a thirteenth-century terrier for Beesby in the Marsh where seventeen acres, dispersed in twenty-two parcels, had the same neighbour on eighteen occasions (Stenton 1920, xlviii). Beresford (1948–51) found clear evidence of a regular order in the sixteenth and seventeenth cen-turies at Langtoft, Wetwang, and Great Givendale, Yorkshire. At Langtoft six oxgangs were described as numbers eleven, twelve, twenty-six, twenty-seven, thirty, and thirty-one from the bank on the south side of the flat throughout the fields.

Göransson (1961) looked for English parallels to the regular systems found in Denmark and Sweden. He referred to the work of Homans (1941, 92), who cited medieval evidence for a regular order, as well as examples where lands or whole oxgangs were described by 'sun division', which was usually associ-ated with a regular layout of a field system. Göransson studied Ecton, North-amptonshire, which has an open-field map dated 1703, further commented upon by Hall (Northamptonshire CRO, Map 2115; redrawn in Hall 1983, 123–5). It shows strips of land called 'hides', which lie in cycles of eleven that repeat themselves throughout the furlongs of the parish. The hides are named and consist of groups of ten lands. Within the hides there was a fixed order of holdings, the neighbours each side always being the same (provided that they held a full yardland). Such an arrangement was in operation from at least the thirteenth until the eighteenth centuries (Hall 1995, 117, 123). The whole township had been set out in a regular manner with a given yard-land always having the same position. The hides have a medieval origin, as shown by charters that record their names (being those of Ecton tenants) in the fifteenth century.

Göransson (1961, 85–7) studied Burton Agnes, Yorkshire (ER). An 1809 open-field plan showed the northern part (published by E. G. Taylor in 1888 and reproduced by Göransson 1961, 87). There was a regular distribution of small blocks of glebe, which consisted of strips lying together, stated to be eight oxgangs in 1685. In 1285 the glebe was described as one carucate. The impli-cation is that there was an ordered system throughout the fields. Whixley, in the West Riding, has a field book dated 1401–11 (Göransson 1961, 93–8). The seventy flatts were divided between the Northside and the Southside, being two townships. There was a regular tenurial cycle with occasional irregular parts, and both townships had their own three fields. Göransson provided examples of the regular sequence in three of the Southside furlongs. One yardland at Harlestone, Northamptonshire, in the early fourteenth century lay throughout the field in half-acres next to the lots of the Dodford fee (Willis 1916, 33). It is unlikely that this could arise by a private arrangement to divide a yardland since it involved two different estates. It is therefore evidence of a regular order at an early date, well before 1300.

Outstanding historical evidence for an ordered field system comes from the abbot of Selby's holding at Foggathorpe, Yorkshire (ER), located a few kilo-metres to the north-east of the abbey. In the fourteenth century six oxgangs consisted of 135 strips lying in 45 flatts that were grouped into three named fields. Each of the forty-five flatts had three strips lying together, and the de-scription goes on to say that in each flatt there were six strips lying together, being three of the abbot and three of the fee of Peter de Mauley. Throughout all the fields the lands lay with the land of Robert de Skrykenbek on one side and the land of John de Pothow (holding of the de Mauley fee) on the other (Fowler 1891–3, ii: 42). This shows there was a regular system at an early date; the regular disposition of the two fees proves that it was a fundamental part of the field structure in the early fourteenth century.

Early examples of regular tenurial arrangements are also found near Durham (Britnell 2004). At Over Haswell an oxgang was described in the fourteenth century as lying divided in the field between the same two neighbours. Regular neighbours are mentioned in two other 1320s terriers of the same place. An oxgang in Great Haswell, part of a larger estate, lay nearest the sun in ten named pieces in 1322. Splitting of estates is indicated in the early fourteenth century, when dispersed land in Upper and Lower Haswell was said to lie next to the grantor's sister and another person. Britnell doubted if sun division or the occurrence of regular neighbours needed to be interpreted as a relict regular order throughout a field system, and suggested, as with earlier authors, that it was caused by estate splitting.

Since there are abundant examples with regular neighbours in thousands of terriers from many parts of the country, the order is unlikely, in most cases, to have been caused by the rare events of splitting between heirs or by part-ible inheritance, which is known in only limited parts of the country. Conclu-sive proof of a regular order is found on the late map of Ecton, which is perhaps unique in the Midlands. The other, more common, proof of the ex-istence of ancient regular sequences of tenurial names is found in field books, such as the case of Whixley, Yorkshire (WR), discussed by Göransson in 1961 as mentioned, yet its important evidence has not hitherto been fully appreci-ated. There are many other examples of field books with regular tenurial sequences.

Field book tenurial analysis

Muscott in Brockhall, Northanmptonshire, has a field book of 1433 describing in detail the great fields, the furlongs, and the individual strip ownership (Hall 1995, 219–21; illustrated in Hall 1982, 14–15). Within the terrier is a 'hidden' tenurial sequence of nineteen names. Some names are missed out from time to time and often the sequence is in reverse order. The regular order must have already been very old to have reached the degree of disorder found in

1433, and the field book surveyor ignored it. The year 1433 was the occasion of further disruption to the ancient tenurial order, since there were exchanges of strips made between the lord and the freeholders so that he could create a consolidated holding of pasture on the edge of the township. Muscott is a fully recorded example of the partial fragmentation of an early ordered arrangement. Field books revealing a regular tenurial sequence for other Northamptonshire townships are Mears Ashby 1577, Hardingstone *c.*1660, and Kettering 1727 (Hall 1995, 177–80, 284–7, 302–3). Sometimes meadow records preserve a regular sequence, a good example being Byfield in 1662 (Northamptonshire CRO, Byfield 59P/91).

Field books hold invaluable data about tenurial arrangements, as well as being essential for reconstructing plans of field systems and for identifying furlong names and the locations of block demesnes and the great fields. If regular cycles of holdings were limited to one area of a field system, then a separate township is implied. Provided a regular tenurial cycle is not too large and not too degenerated, it will be fairly obvious from an inspection of the names, looking for one or a few repeated sequences that can be built up to complete the cycle. Reverse sequences have to be considered also. For more difficult or apparently randomly dispersed examples other information has to be sought. At Ashby St Ledgers, Northamptonshire, the meadows had a better preserved remnant regular order than the arable lands. However, once identified in the meadow, the forty-land cycle could be found in a disintegrated form in the arable (Hall 1995, 142–6). Analysis can be assisted by entering the information into a database.

When there was a regular tenurial order, the owner of a single yardland always had the same neighbours on either side. In the twelfth and thirteenth centuries, yardlands at Newnham and Ecton, Northamptonshire, instead of being described by a terrier furlong by furlong, were defined by stating who the neighbours were, which was quite sufficient, because they belonged to a tenurial sequence, and furlong names were not needed. A Newnham yardland in 1198 was said to lie in the fields next to the yardland of William son of Stephen (Round 1900a, 24, 147). This statement implies that there was a fixed order of holders. Similar statements were made for medieval oxgangs that lay 'everywhere' (in the fields) between the same two neighbours for Ormesby, Yorkshire (Göransson 1961, 42; Brown 1889, 278), for a yardland at Harlestone already referred to (Willis 1916, 33), and for the oxgang at Gelston in Hough on the Hill, Lincolnshire, in 1246.

Most townships do not have field books confirming a regular tenurial order, and terriers of individual holdings have to be studied. The evidence of single yardland terriers is less full than a field book, but if there had been a regular order, then the neighbours would always be the same. It is frequently found, especially with medieval terriers, that the neighbours are the same on either side of the lands in each furlong, thus implying a regular order, as at Shipton,

Buckinghamshire (p. 119). A Mears Ashby field book of 1577 exhibits a decayed regular cycle of forty lands in which positions one, nine, and twenty-six to twenty-eight are invariable, while other positions have various tenants (Hall 1995, 178–80). A test for a regular order of neighbours occurring in a Mears Ashby terrier of 1636 shows that one neighbour held 29 per cent of the total, three people 71 per cent, and five people 88 per cent. Since a regular order is known from the 1577 field book, the 1636 data represent the proportions of neighbours that may be expected, so that, for other townships, where three people occupy 70 per cent or more of the positions, it may be assumed that a regular order had existed and would be identifiable in a field book. In the analyses provided in the Gazetteer, a regular tenurial order has been assumed if three people occupied 60 per cent, or more, of the positions.

Harvey (1982, 28–39) found many examples of regular tenurial order in terriers and furlongs on the Yorkshire Wolds, with holders having the same positions in the furlongs or the same neighbours throughout the field systems in the eighteenth century (such as North Grimston, Sewerby, Rudston, and Walkington). The field books of Butterwick and Thorpe Bassett, both dated 1563, had regular tenurial cycles throughout their flats. Wide lands were sometimes known as two-oxgang lands and narrow ones as one-oxgang (as for Kilham in 1729). For the thirteenth century, at Kirby Grindalythe and Nafferton, small groups of bovates lay together throughout the fields, and at Boynton and Rudston oxgangs were defined as lying between the same two neighbours. Harvey (1980 and 1981, 192–7) also found evidence for a regular ordering of oxgangs through the fields of some Holderness townships (Easington, Ryehill, and Kilnsea). At Barnston it could be traced back to 1473 and at Preston to *c.*1250. An approximate relationship was shown between township areas and the carucate areas of 1086, and was used to suggest that there may be a relationship between taxation and field system area in some cases (Harvey 1978).

There are very many examples of a regular tenurial order in counties of the Central Region, as adjudged from an analysis of neighbours in terriers. A few of the more striking are listed below.

Bedfordshire: A terrier of the South Field of Willington, 1230, reveals a very regular tenurial order, with two out of four people holding 92 per cent of the positions (Godber 1963, Bedfordshire Historical Record Society, 43: 21–3). Other fairly frequent sequences occur at Duloe in Eaton Socon, 1217, where 7.5 acres were dispersed. Of the eighteen neighbours three held 67 per cent of the places (Godber 1940, ibid. 22–3: 39).

Buckinghamshire: The best known example of a regular tenurial order in Buckinghamshire is for the Shipton yardland in 1361. Other regular terriers with few neighbours are found for Steeple Claydon, *c.*1220 (Salter 1935, 215–16), and Maids Morton, 1607 (Reid 1997, 135–8).

Cambridgeshire: Elements of a regular order in the early thirteenth century come from Cambridge, where a terrier of fifteen acres consisting of thirty-six strips had Adam son of Eustace as one of two neighbours in the last thirteen strips (Maitland 1898, 58).

Derbyshire: A terrier of Alsop (thirteenth century), described 3¼ acres lying in 9 parcels in 3 fields, with only 4 people as neighbours, 2 of them occupying 84 per cent of 18 positions (Wrottesley 1887, *Journal of the Derbyshire Archaeological and Natural History Society*, 7: 107). A terrier of five acres in eleven parcels at Willington made in 1252, had six people in twenty-one neighbouring positions, three of them holding 76 per cent of the total (W. H. Hart 1886, ibid. 8: 62).

Durham: Regular neighbours found in the fourteenth century at Upper and Lower Haswell have been described (p. 122; Britnell 2004).

Gloucestershire: A terrier of 5¼ acres in Ham had the same neighbour next to its land in 9 out of 12 cases, 75 per cent (Walker 1998, 111). A Rodmarton yardland, *c.*1200, lay in twenty-nine pieces in two fields with twenty neighbours named, of whom two held sixteen places, 80 per cent (Ross 1964, 344–5).

Leicestershire: Walton and Kimcote had, in the thirteenth century, four acres dispersed as nine pieces lying next to the parson in all places. A terrier of Illston on the Hill, 1364, was regular (Thompson 1933, 247, 265).

Lincolnshire: Regular tenurial order has been noted at Hough on the Hill and in the Senior Surveys. Others are given in the Gazetteer.

Northamptonshire: Many townships show evidence of a regular tenurial order (pp. 122–3 and Hall 1995, 115–22).

Northumberland: The only known examples of townships with a regular tenurial order are those that had newly allotted strip sequences created when field systems were divided into two independent parts: Long Houghton before 1567, Chatton, Charlton 1685, and Acklington, seventeenth century (see the Gazetteer).

Nottinghamshire: Caunton, in the early fourteenth century, had a terrier with regular neighbours (Walker and Gray 1940, 194).

Oxfordshire: Chalford's 1315 terrier had parcels with the same neighbour throughout (Lobel 1935, 90).

Rutland: Oakham had three yardlands of glebe with the positions of individual lands in each furlong numbered (1700). Several furlongs had triplet groups in the following positions: 1–3, 19–21, etc., indicating a regular tenurial order (Northamptonshire CRO, Rutland Glebe Terriers).

Shropshire: Terriers of Downton, *c.*1255, and Newton in 1326 possibly had a remnant regular order (Rees 1985, 84, 164–5).

Surrey: A regular sequence possibly occured at Merstham in 1522, where three neighbours occupy 67 per cent of seventy-nine positions (Hylton 1907, *Surrey Archaeological Collections*, 17: 106–11).

Sussex: A thirteenth-century terrier of Ferles with few neighbours of dispersed strips implies a regular order (Peckham 1946, 97–8).

Warwickshire: Strips at Gaydon in 1324 and Avon Dassett had the same neighbours on each side throughout (Roberts 1973, 223–4). Barford and Brailes glebes of 1714–17 and Binton in 1635 had few neighbours (Barratt 1955–71, i: 23–31, 49). Other places implying a regular order are Exhall, 1585, Oxhill 1585, and Stretton on Fosse, 1616 (Barratt 1955–71, i: 26–9; ii: 66–73).

Wiltshire: Regular orders in terriers have been noted in four townships at dates ranging from 1428 to 1704 (Gazetteer).

Yorkshire: See Harvey (p. 124) for the Wolds and Holderness.

Counties for which no examples of a regular tenurial order have been found are Cheshire, Cornwall, Cumberland, Devon, Essex, Hereford, Lancashire, Somerset, Staffordshire, Westmorland, and Worcester. They lie in regions that were early enclosed and most of them have little evidence of extensive Midland-type fields. The examples found in Durham, Shropshire, Surrey, and Sussex are limited, and regular orders are primarily a characteristic of the Central Region. Göransson (1961, 100) published a map showing where he had found places with evidence of a regular order. Most of them lay in the Central Region.

A different kind of evidence implying a regular order comes from the disposition of the lands of twelfth-century estates in the furlongs where there was more than one manor. The rector of Rushden, Northamptonshire, had the tithe of the 10½ yardlands of villein land that had formerly belonged to the crown estate of Finedon (specified in an enquiry of 1318). Glebe terriers of the eighteenth century describe the lands subject to rectorial tithe. It consisted of blocks, mainly of 10 to 15 adjacent lands (corresponding to 10½ yardlands) lying in all the furlongs of the township (Hall 1985, 64–5). This must have been its disposition in the early twelfth century when the tithes of neighbouring lands were given to Lenton Priory, Nottinghamshire. A regular order is implied in a 1556 description of a three-quarter yardland of bondhold belonging to Richard Boyfield, which lay between the lands of Thomas Launcebon on the south and John Burton on the north (court roll 3 & 4 Philip and Mary, in Northamptonshire CRO, Box X704).

A detailed terrier of Ramsey Abbey land at Hemington, Northamptonshire, was made in 1517 (Northamptonshire CRO, Buccleuch 15/59, in Box X8678). It shows that often Thorney Abbey (Cambridgeshire) or Hinchingbrook Priory (Huntingdonshire) had lands next to Ramsey, indicating that large parts of furlongs had been given to monasteries. Frequently the Ramsey land is at the end of a furlong (proved by abutting the next furlong), so a regular order is likely. This is probably the arrangement in the twelfth century and before. The lands of the monasteries are unlikely to have changed. The great furlong blocks reoccurring regularly imply a regular tenurial order in the fields at an early date.

The regular tenurial order of strips was widespread as has been shown. An early example of an explicit statement relating to the organized structure of the fields is found at Syresham, Northamptonshire. An agreement about the division of the meadows was made in 1258 between the three principal owners, the abbot and chapter of St Mary de Pratis Leicester, the abbot and chapter of Biddlesden, Buckinghamshire, and the prior of the Hospital of Brackley, plus seven free tenants:

that the principal meadow of Syresham shall be divided by hides and virgates as the arable land is divided, so that each tenant may have his own portion marked out by boundaries…with not more than five beasts allowed for each virgate (Magd. charter Syresham 25).

This tells more than that villein yardlands were structured—something that theoretically could be done in the twelfth or thirteenth centuries by the re-arrangement of older irregular fields—but that hides were involved, which points to an earlier date.

The same basic structure of hides is evident in the fields of Whitfield. An early thirteenth-century grant of seventy acres of a dispersed demesne is accompanied by a very detailed terrier that describes the number of strips in each piece and numbers their position in each furlong (Magd. MS 273, fols. 40–1). Nearly all the pieces were called 'acres' and comprised four adjacent selions in almost every case. This demonstrates a highly organized field structure showing that the pieces were 'hides' on the ground like that at Astrop. It was the acres, or hides, that were numbered, not the individual strips, as would have been done later (as in William Senior's surveys of Lincolnshire in the early seventeenth century). The organized structure of the Whitfield fields is mirrored in the furlong layout (see the plan in Hall 2013, map 84M); nearly all the strips lie on the same north-west–south-east alignment throughout the parish. That hides rather than virgates were the structural unit of the fields is made clear in a charter for Astrop in King's Sutton, in the same county, made in about 1310, that granted

a tenement called Selestede, and four yardlands of arable land in the field of Estrop which are called *Selehyde* and which lie between the land of Adam Averey and Adam Godbryne, together with the meadow called *Selesham*' (Magd. charter Brackley A34).

From this we see that the four yardlands lay together in a block called a 'hide' and that they related to a particular tenement and had their own meadow piece. They also lay in a regular order throughout the fields, since only two neighbours were required to define them. These hides and their four component 'yardlands' (virgates) recall the system described in the pages of Domesday in 1086 for most counties, where four virgates were accounted as a hide. The personal name *Sele* is most probably from the Old English 'hall' as in Seal, Kent, and Seale, Surrey (Ekwall 1974, 409) and therefore likely to have been an 'old' name in 1310. The same structure of four adjacent yardlands, on a single occasion called a 'hide', can be found at Astcote in Pattishall (Ross, 1964, ii: 554–60). It was granted to Cirencester Abbey before 1197 and an undated thirteenth-century full terrier shows that the land lay as 'acres' (four roods— four yardlands) with the same two neighbours throughout the field system. Hence, the fields had been structured in hides at an early date—in the twelfth century or before.

Segenho, Bedfordshire, already mentioned (p. 120), provides information about organized field structure. It is not explicitly clear that the document records a first-time ordering of the fields; it may equally be a re-establishment of an order that had existed in the early twelfth century before later illegal occupation of the villein eight hides. What is quite clear is that the villein parts of the estates were laid out in the fields in intermixed strips by the acre and by the hide (Vinogradoff 1892, 457–8). As with the other examples quoted, it was the hides that were thought of as being dispersed in the fields.

It may be concluded therefore that in many parts of the Central Region there was an organized field structure at an early date, and in many cases very probably by 1086.

D. FIELD REGULATIONS AND FARMING METHODS

Manorial regulations, otherwise called 'field orders' were recorded in different parts of the country, and show how the complicated systems of intermixed holdings found in the Central Region were regulated to ensure that agricultural practices operated satisfactorily. Early evidence for open-field operation can be deduced from the work-services listed in manorial custumals. Away from the Central Region, communal control did not include cropping, and principally dealt with maintenance of drains and the stinting of commons.

The complicated system of intermixed holdings required regulation to ensure that it worked satisfactorily. Manorial courts attempted to control overstocking of commons, to prevent trespass by ploughs against a neighbour and to minimize trespass by animals in crops. Fences, ditches, and roads also needed maintenance. Records of regulations varied over the centuries and also varied according to the type of field system. The smaller irregular fields of the Eastern and Western Regions needed less control than the extensive fields of the Central Region. Regulations become detailed in the sixteenth century, probably partly because agriculture was becoming more complicated with changes from two to three, four, or five fields in the case of the Midlands. During these later centuries, Midland courts tightly controlled the agricultural round of which furlongs should be fallow, when stubble ploughing should begin, and when commoning should begin or cease.

Manorial court rolls begin to include orders relating to agriculture and fields at the end of the thirteenth century. Where there was no resident lord or where lordship was divided, the 'homage' or village community regulated the fields themselves. In some late cases this was done from the vestry. Court orders are the commonest source of information and from them the course of husbandry is apparent. Agricultural information can be deduced from work-service dues detailed in manorial extents and custumals. The description of cheese-making from ewes' milk at Amberley, Sussex, in 1358 is particularly detailed. Early work-service details are recorded in a Glastonbury Abbey survey of 1189. Organized control of the field system is evident at Mere, Somerset, where officers were appointed as part of their regular work-service to keep stray animals out of the fields, for which they received half-acres of wheat and oats (Jackson 1882, 29).

Ault (1972) studied the orders recorded in a regional series of court rolls. These mainly relate to the counties of Huntingdon, Bedford, Buckingham, and Oxford in the Central Region. Recorded orders begin in 1270 and those for the next few decades are concerned principally with the theft of grain, the restriction of gleaning to paupers, and trespass of animals in crops. For Newton Longville, Buckinghamshire, in 1291, the lord and community of the town decreed that the ordinances of previous years should be observed (Ault 1972, 84). This demonstrates that a system of regulations existed before then. Regulations also existed by 1312 at Elton, Huntingdonshire (see the Gazetteer). A range of early orders and misdemeanours for Northamptonshire from 1300 cover similar items. At Brampton Ash there was a regular series of numbered orders made before 1434, when John Ruston, a shepherd, broke 'the bylaw...in the neats' pasture...and broke bylaw number 9' (Hall 1995, 12). Not until the sixteenth century are there large numbers of detailed orders about stinting levels and regulations relating to cropping and the dates of ploughing in the great fields, etc. Animal stints per yardland are recorded and sometimes the more limited grazing rights that went with 'ancient' cottages.

One of the finest early set of orders published is that for Wymeswold, Leicestershire, made in *c.*1425 (Historical Manuscripts Commission 1911, 106–9). They were written in a wordy Late Middle English text and are numbered and paraphrased below.

1. Three named neat pastures were to be thrown open on 14 September for one week and any person not putting beasts into them should pay one penny to the church.

2. After those three pastures were grazed, the next was to be the wheat field, for a week, except for one headland used to strew [stack?] corn; fine: one penny to the church for not using the pasture.

3. On Maundy Thursday Eve, the commons of the peafield were to be opened for horses and no other beast. Unfit horses were not allowed to enter. Any beast that strayed in corn or grass would incur for the owner a one penny fine for the church and require damages to be paid to his neighbour.

4. At Whitsun Eve everyone must use their own grass and not tie horses on other people's land, with the same fine for default.

5. Ploughing oxen placed on private grassland were to be securely tied. If they strayed, the owner was to be fined one penny for each beast, payable to the church, and also pay damages to his neighbour.

6. If horses were tied to graze on heads [grass at the ends of strips] or by the side of streams and got into the corn, a penny to the church was paid for each foot in the corn grazed and damages were to be paid to the farmer.

7. Anyone grazing animals at night in corn or grass was to be punished by the law and pay the church fourpence.

8. Those with peas in the field were to gather pods only on their own land. Those with no peas of their own could gather twice weekly, on Wednesday and Friday, by going in the furrows between lands. They were not permitted to gather with sickles but with their hands only. Anyone with peas who gathered on the land of others paid one penny to the church and lost his pods. Those without peasland who gathered more often than allowed or used a sickle were to lose the vessel in which they gathered them as well as the pods, and pay one penny to the church.

9. No one with [animals in] the common herd or with a byherd was to go on the 'wold' after the grass was mowed, until it was made [into hay] and carried away. For each beast in default, one penny was to be paid to the church.

10. Any person who threw sheaves from another's land to tie to his horses [to be taken away] was to make substantial payment to those robbed and pay one penny for the church.

11. Anyone carting or carrying at night, between [ringing of] the bells, had to pay forty pence to the church, except for those with peas who could bind them in the morning.

12. Labourers dwelling in the village who had commons were to do harvest and other hired work, and were not to go to other villages unless they were without work. If they left when there was work, they were to be punished according to the law.

13. No one doing harvest work could carry sheaves home except those that had been given to them. If stolen, a fine of one penny was paid to the church for each sheaf.

14. No one able to work was allowed to glean and no one could glean among the sheaves [before carrying]; if they did, they lost the corn and had to pay one penny to the church.

15. The common herd or a byherd could not go into the wheat or peas fields until the corn and peas were carried. The herds were to go together, on pain of one penny a beast to the church for default.

16. No one was to take his beast from the common herd to go in the wheat field and damage the plants between Michaelmas and Christmas; one penny was to be paid to the church for default.

17. If the hayward impounded a flock he took sixpence for a flock of neat cattle, fourpence for sheep flocks, and one penny for each horse.

18. The 'wold' was to lay separate at Candlemas. If any herder let his beast go on it he had to pay for each animal fourpence to the church each time.

19. Those who had leys within the corn, both the lord and others, were to let them out, and trespassers were to make recompense.

These orders were made by the lords and with the consent of the whole village. They were concerned with management of the crops and grazing, especially prevention of forbidden grazing, animal trespass, and the theft of crops.

Detailed regulations similar to those of Wymeswold are typical of intensively farmed East Midland field systems, and are very similar to those that occur abundantly in later centuries elsewhere in the Central Region. They may, for instance, be compared with orders of 1620 for Aynho in Northamptonshire (Northamptonshire CRO, C(A) 2456), which are paraphrased as follows:

1. The jury was to meet in the common field before 9 May [St Andrew's Day] at 8 a.m. to view ploughing encroachments.

2. After such 'reformation', everyone had to plough one foot of their land so that a balk two feet wide was made between every two ridges.

3. No one after St Martin's day next was to keep above forty sheep for a yardland.

4. Each yardland was to have [a stint of] four beasts and four horses.

5. No one was to put any weaned calf in the commons unless they had abated a cow common.

6. No peas were to be gathered except on their own land unless the owner or someone from his house was present.

7. Foals and sheep were to be taken out of the cornfield before St Luke's Day.

8. Hogs and swine were to be ringed.

9. Both of the town cow pastures were to be kept several and unbroken until 9 May.

10. No sheep were to be placed in the meadow or cow pasture [until after hay time].

11. The meadow and cow pasture were to be hained (kept separate) at Candlemas until 9 May.

12. Pasture beyond the brook [the South Field] was to be hained at Lady Day and broken at May Day.

13. No one was to let a horse or gelding [in the common].

14. Those with no commons could not keep any cattle.

15. No man was to put forth his beasts at night.

16. No gleaners were allowed to glean until the whole furlong was carried.

17. No beast was to go [in the commons] before the herd before 5 a.m., at which time the herdsman would blow his horn.

18. Oxhey ditch was to be cleansed.

19. No horse was to be tied on any balk before the cornfield was carried.

20. No horse was to be tied in the hay meadow before the cocks were removed.

21. No lambs were to be pastured on the common after Holyrood.

22. After All Saints' Day no one was to receive out-town people [in their house].

23. No foreigner's sheep were to be taken in from Michaelmas until May Day [because of the lack of fodder].

24. The jury had to view watercourse encroachments.

Both sets of orders have interesting social items. In the Wymeswold agreement of 1425, labourers with cottage commons (no. 12) were expected to work in the township when required. Presumably this was a remnant survival of work-service, but it was the community that demanded it and fines went to the church. Two centuries later at Aynho, labourers who had been hired for harvest work were to leave the village and no more were be taken on after 1 November (no. 22). Presumably they were unwelcome lest they became a burden on the poor rates if work could no longer be found after the harvest. The orders show that control of pasturing and grazing was important and it was in everyone's interest to see that the regulations were observed, thereby in effect drawing the community together.

The open-field officers at Easton on the Hill, Northamptonshire, in 1644 met and viewed encroachments (to be presented at court), and set out mere markers (boundary markers) in the meadows and fields (Burghley House Muniments, Exeter MSS 21/32). The officers were elected at the manorial court during 1663–86 (Burghley House Muniments, Exeter MSS 15/154–7) and the court decided what new by-laws were to be made and which of the old ones should stand.

Cunningham (1910b) published numerous detailed orders for Cottenham and Stretham, Cambridgeshire. Sets of by-laws for Laxton, Nottinghamshire, from 1686 to 1908 were given by the Orwins (1938, 172–81). Thirsk (1973, 246–62) discussed many regulations for the East Midlands, noting the differences between champion, fen, and forest townships. Kerridge (1967 and 1992) drew primarily upon court orders for his work on agricultural changes and open fields, and supplied a very large number of useful references to original source material. The County Gazetteer provides many further references to sources of open-field orders and farming practices in each county, noting those specific to local areas such as cider-making. The Central Region has much relating to the regulation of crops and open-field grazing. Detail of medieval demesne farming has been given by Campbell (2000).

The preceding discussion has provided information on agricultural practices revealed by open-field regulations. Further data are given by the numerous records of monastic and crown estates, but records of farming at a lower status level giving details of stock and crops are relatively few. However, a taxation of farm stock made in 1297 for Bedfordshire provides information about small lay farms of freeholders of non-manorial status and of the villeins (Gaydon 1959, xxvii–xxxii). Oxen were used less by the villeins than on manorial demesnes. The usual villein draught animal was the affer, a small type of horse. Not all had two affers, the number likely to have been needed to draw a plough. Most had at least one milking cow, and all had sheep. Farmers with sheep and no other animals were presumably shepherds, though seventy sheep at Caddington belonged to a Dunstable wool merchant. It seems possible that many commons' stints were not fully taken up. Wheat was the main corn crop, followed by barley and oats. Beans and peas were grown on clay soils, and rye was a marginal crop, grown on sandy soils in the Ampthill and Sandy regions. Similar details of village farms of all tenurial types are available for the Suffolk Hundred of Blackbourne in 1283, listing crops and animals. Wheat could not be grown on the sandy soils of the west at Barnham, Euston, Rishworth, Weststowe, and Wordwell. It was replaced by rye at Barnham and in the other settlements barley was the main crop (Powell 1910, xxiii and tables). Farming in early seventeenth-century Berkshire was described by Loder (see Berkshire in the Gazetteer, p. 222). Farming details elsewhere can be found in the Gazetteer.

5

Early evidence for settlement and fields

The description of the different types of field system found in England along with details of their internal structure and management reveals that there are a variety of parameters that need explanation to understand how such arrangements may have originated and developed. This chapter collects together additional types of physical and historical evidence that have a bearing on field origins.

The main themes fall into four rather disparate types:

1) archaeological evidence relating to the distribution and dating of settlements associated with early medieval (Saxon-period) occupation;

2) supplementary data provided by alluvium and pollen indicating that large-scale changes in the management of the countryside took place, and also information made available by archaeological surveys of open-field earthwork remains;

3) possible antecedents of the field forms that come into view by the twelfth century;

4) the early dating of intermixed and dispersed holdings, including items relevant to field systems found in Anglo-Saxon charters and other early historical sources.

A. ARCHAEOLOGICAL EVIDENCE RELATING TO SAXON-PERIOD SETTLEMENTS AND FIELDS

Archaeology has made a significant contribution to the nature of early medieval settlement and to the known extent of field systems consisting of intermixed strips. Evidence for Early to Middle Saxon-period settlement was reported during the 1960s and continues. The original published data is quite limited, however, and has been quoted and applied to parts of the country where there is no pottery evidence. Settlement studies, drawing upon the known existence of only a few of these small deserted Saxon-period settlements lying away from the present village sites, in some townships, concluded that 'by the eleventh century...the pattern of settlement...lay in a curious pattern of hamlets and farmsteads[, some]...scattered across the landscape' and 'few if any [villages] existed by 1000' (Taylor 1983, 124–5). This has led to statements such as 'fieldwalking in central and eastern Northamptonshire and south-eastern Leicestershire, has indicated a plethora of small hamlets scattered across the landscape' (Lewis et al. 1997, 92). The currently known sites in the north-eastern quarter of Northamptonshire, from the Welland Valley across to the Nene Valley, is thirty (Plate 9), and thirty-four sites have been found in the adjacent part of Leicestershire by Richard Knox and Peter Liddle (Bowman and Liddle 2004, 97). It is almost necessary to 'start again' with the evidence, each county publishing a township map showing where there are Saxon-period sites lying out in the fields, where they have been discovered under medieval vills known to be have been in existence in the eleventh and twelfth centuries, and where there are Roman sites with Saxon-period remains.

'Saxon-period pottery' in these pages refers to handmade material dating from the Early and Middle Saxon era (*c.* AD 400–850) and not wheel-made Saxo-Norman pottery dating from the Late Saxon period into the High Middle Ages (*c.*850–1150). There is no implication of ethnicity in the use of the term; it is employed to describe the ceramic tradition. It is not assumed that settlements were occupied by descendants of Northern Germanic origin, nor that settlements possibly founded by Anglian and Jutish settlers are excluded. In the East Midlands most Saxon-period sherds found by fieldwalking or excavation have hard fabrics and often contain temper of sand, mica, and crushed igneous rock. They are therefore resistant to frost and survive in ploughsoil; they are also distinguishable from handmade Iron Age pottery, which is not often sandy. There are, of course, many local and regional variations in the fabrics of Iron Age and Saxon-period pottery that can cause difficulties with identification.

Dating is always difficult. Early sherds tend to be thick in proportion to the vessel size, with an occasional stamped decoration. In East Anglia and adjacent counties, Middle Saxon Ipswich Ware (*c.*700–900) occurs, helping to refine the dating. Another type of Middle Saxon pottery called Maxey Ware is found in the East Midlands and Lincolnshire. It, too, is hard and handmade, but is distinguished by having shelly inclusions. Away from the East Midlands and East Anglia, Saxon-period pottery is often vegetable-tempered, frequently difficult to distinguish from Iron Age sherds, and does not survive well in plough soil. Additional difficulties are met in areas such as Dorset, where twelfth-century sherds are coarse and sandy, degrading to small pieces that look similar to East Midland Saxon-period material. Very little pottery has been identified in the Western Region, or the western part of the Central Region, such as Herefordshire and Worcestershire.

Early to Middle Saxon-period pottery was recognized in Northamptonshire at development sites within present villages from 1964 onwards at Grendon, Irchester, Higham Ferrers, and Wollaston (Hall and Hutchings 1972, 6). More pottery was discovered at Weston Favell in 1972 and at Raunds in 1976. Previously, sherds had been found at Castor near the church in 1958 (Green 1972).

The first site investigated in Northamptonshire in modern arable fields was at Maxey (Addyman 1964). Other pottery scatters lying far distant from later medieval and present-day settlement was discovered in 1971 at Newton Willows (SP 868 845) (Hall 1973), extending to about five acres. A detailed study of Brixworth, made by Paul Martin during 1973–4, yielded a remarkable series of small sites (Hall and Martin 1979). Glenn Foard made a similar field-work survey of Doddington in 1976 that revealed seven Saxon-period sites located out in the fields, one on a Roman site, none lying on clay (Foard 1978). The plans of Brixworth and Doddington were reproduced in 1983 (Taylor 1983, 113) and their evidence has since been widely quoted to indicate that a dispersed Saxon-period settlement pattern occurred throughout the country.

Countywide fieldwork in Northamptonshire has revealed many more Saxon-period sites, most of them discovered by Paul Martin. A brief account and location plan of those in the Peterborough region was published in 1980 (Hall and Martin 1980; reproduced in Hooke 1988, 103). The total number of sites found in modern arable during the county survey is indicated on Plate 9, distinguishing townships that were not searched in detail for remains other than medieval fields.[1] In all, 129 sites have been discovered of which 29 lie on Romano-British settlements. Additionally (not marked on Plate 9) there are 65 find spots with 1–4 Saxon sherds plus 50 Romano-British sites with 1–4 Saxon sherds, in all, 244 'activity areas' and sites. During the open-field survey, farmland in each township was visited once, so that searches for artefacts were not as intensive as at Brixworth. For most townships only a small proportion of the total area was 'line-walked' for artefacts; the areas studied intensively were recorded, but the implications have not been included in the following analysis. Therefore, almost certainly there are many more sites hidden under fields that were in pasture or were not in a suitable condition for discovering surface finds at the time of the single visit. However, a general distribution is clear within the 239 townships studied in sufficient detail to discover some Saxon-period sites. Firstly, most of the 129 sites lie on light soils of river gravel, limestone, or Northampton Sand and Ironstone (only 8 small sites lie on clay, excluding ephemeral sherds found on Roman sites). Secondly, there are twenty-five townships that have more than one Saxon-period site out in the fields (seventeen townships with two sites, three with three, three with four, one with six, and one with fifteen sites). Currently, fifty-three townships have one known site only (see p. 186 for further analysis of the Saxon activity with additional information). A few sites are large (Castor, Dingley, Helpston, Newton Willows, Warmington), many are small. Such is the rareness of Saxon-period sherds that

---

[1] Plate 9 presents the site information found by D. Hall and P. Martin during the Northamptonshire Field Survey undertaken during 1961–2011. Some data have been made available previously (Foard et al. 2009, 46–54; Williamson et al. 2013, pl. 15a) and used with secondary information from other sources. Plate 9 is the currently complete version from a single source (2013) and includes additional fieldwork data not previously available, checked with pottery finds and site notes.

just a few are reckoned to be significant, an opinion held by Peter Wade-Martins (1980b, 5). For example, in 2009 fieldwork near Brampton, Huntingdonshire, identified three Saxon-period sherds. Subsequent 'grid-walking' produced nine sherds, and partial excavation revealed a sunken-featured building and other settlement remains (Evans and Standring 2012, 84–8, Site 5, Area B1; site located by Hall). Saxon-period sites rarely give cropmarks, as demonstrated by recent work examining many aerial photographs taken of sites where sherd-scatters have been discovered (Deegan and Foard 2007, 125–35). An exception to the norm is a site lying west of Polebrook (TL 06 87) discovered by Stephen Upex. Excavation of a cropmark area revealed ditches and postholes relating to large hall buildings that had footings cut deeply into limestone (Upex 2003 and 2008).

In other counties, parish fieldwork studies were made by Barry Cunliffe at Chalton, Hampshire (Cunliffe 1972, 1–12), where a settlement site, subsequently excavated, was located by the presence of sherds lying in ploughsoil. The desertion was thought to have occurred by the ninth century. In Wiltshire, Saxon-period pottery was recovered by Peter Fowler in the 1960s from Wellhead, Westbury (ST 873 503), and from Round Hill Down, Ogbourne St George (SU 215 755) (Fowler 1966, 31–7). Both sites had Roman material and are presumably early.

In the 1960s, John Owles discovered Saxon-period pottery at Witton, Norfolk (Wade 1983, 50, 70). Some of these sites were excavated and proved to have structures under them. Wade-Martins (1980b) walked the areas around isolated churches, greens, and moats in fourteen Norfolk parishes during 1967–70, finding Middle Saxon sherds, often in locations different from the later medieval villages. Andrew Rogerson's studies of Barton Bendish, Norfolk, during 1983–7 showed a pattern of Early and Middle Saxon-period settlement as well as later remains (Rogerson 1997, 19–23). The largest sherd scatters of Early Saxon pottery were at three areas designated 'sites', one lying nearly two kilometres east of the vill (eighty sherds) and the others just north and west of the present village of Barton (thirty and eighteen sherds). Middle Saxon Ipswich Ware sherds (133 in total) were nearly all concentrated around the vill. Fieldwork at Fransham has revealed a similar pattern (Rogerson 1995).

North Elmham has a Middle Saxon-period site near the vill (Wade-Martins 1980b, 37–73), and another Middle Saxon settlement was found in a field near Sedgeford (Cox et al. 1999–2001, 174–6, 351). Hargham revealed a few sherds of Early and Middle Saxon-period pottery 800 yards south of the village, and a settlement with 120 surface sherds of mainly Ipswich Ware immediately south-west of the vill site (Davison and Cushion 1998–2001, 260–1). West Acre had Early Saxon-period sherds away from the village centre on the east, the site of a cemetery, and also at the south-east at Custhorpe. No concentration of Middle Saxon sherds was recovered except for nine sherds around the vill, suggesting that material of this date lies under it (Davison 2003, 212–13).

In the Siltlands of the Wash Fens, Middle Saxon-period remains have been found at the southern end of West Walton by Silvester (1995, *Norfolk Archaeology*, 39: 107–8). Detailed fieldwalking by Tom Williamson of twenty-eight square kilometres of the north-west of Essex, near Saffron Walden, during the 1970s revealed thirteen Saxon-period sherd scatters on or near to Roman sites, and ten on or near Saxo-Norman pottery scatters. The sherd-level was low, not more than five per site (Williamson 1982, 126, 131). In the same county, a grid-walked survey of a block of 1,156 hectares, revealed 7 Saxon-period sites (among many others of different date) at Stansted (Medlycott and Germany 1995).

In summary, Early to Middle Saxon-period settlement sites have been found in modern arable fields lying distant from present-day settlements in all of Eastern England from Essex to Yorkshire, including Suffolk, Norfolk, Cambridgeshire, Huntingdonshire, Bedfordshire, Buckinghamshire, Northamptonshire, Rutland, Leicestershire, and Lincolnshire. Some sites are known in Berkshire, Gloucestershire, Hampshire, and Wiltshire, as adjudged from the literature and the county HERs. The reason for the strong eastern component of the distribution is partly because of the existence of hard durable pottery, as mentioned, and partly because these counties have had by far the most fieldwork studies undertaken during the past forty-five years. Other sites, mostly found by more recent commercial excavations, are recorded in annual summaries in *Medieval Archaeology* and county journals, and online. The major excavated settlement sites are listed and discussed by Sam Lucy in the report of Carlton Colville, Suffolk (Lucy et al. 2009, 1, 7, 428–34). A good discussion and synthesis of settlement and building forms is given by Helena Hamerow (2012), using much information from commercial excavations not yet fully published. A national map of the sites studied (ibid. 4–5) confirms the largely eastern distribution, with significant sites also around Oxford and in Hampshire. There is a riverine location for many settlements.

It is assumed that pottery scatters represent settlement and this has been proved in many cases. Where excavation has failed to discover structures, the most likely interpretation is that ephemeral posthole and other remains have been ploughed away (as at Courteenhall, Northamptonshire (Jones et al. 2006, 28–31)). The interpretation that scatters represent manuring is very unlikely, since pottery was not much used. At Yarnton some Saxon-period features were only visible by hand trowelling (Hey 2004, 15). Other excavations revealing evidence of structures on the sites with known Saxon-period pottery scatters in Northamptonshire have occurred at Brixworth and Upton (Ford 1995; Shaw 1993–4), Higham Ferrers (Oxford Archaeology 1991; Hardy et al. 2007), Warmington (Shaw 1993), and at the edge of Wollaston in 1964 (Hall 1974). Villages in the Whittlewood region sampled by one-metre 'test-pits' have produced Saxon-period sherds (Jones and Page 2006, 84–92). The programme of making test pits in many East Anglian villages has revealed Saxon-period remains (Lewis, in Higham and Ryan 2010, 81–105; Lewis 2011). In other counties, a summary of the information available is given below. Apart from the eastern swathe of counties referred to above, nearly all Saxon-period pottery has been revealed by excavation.

Some of the most significant excavations of Saxon-period sites underlying medieval furrows were made at Great Linford, Buckinghamshire, and Eye Kettleby, Leicestershire, with an important site at Brough, Nottinghamshire, found in 2004. The Great Linford site at Pennylands excavated by Robert Williams revealed enclosures, thirteen sunken-floored buildings, and three timber hall buildings, with wells and other features dating to *c.*500–750. The site was completely sealed by later ridge and furrow, which had no relationship to it in plan. The furlongs overlying were called *Long* and *Short Dunstead*, which probably reflects knowledge that the site had existed by those who lay strip fields over it (Williams 1993). A site underlying ridge and furrow with Middle Saxon-period features has been excavated at Eye Kettleby, Melton Mowbray, Leicestershire (Knox, in Bowman and Liddle 2004, 99). Previous fieldwork discovered 144 sherds, and concentrations coincided with excavated features. At Brough, Nottinghamshire, furrows were found on the same alignment as buildings.

To these should be added sites in Bedfordshire (Clapham), Buckinghamshire (Chichele and Fenny Stratford), Huntingdonshire (Orton), Lincolnshire (Quarrington), and Northumberland (Shotton), where excavation has proved that furrows overlie Saxon-period settlement in all cases (see the list on p. 142). The important Saxon-period settlement sites at Yarnton, Oxfordshire, showed several phases of occupation and were finally covered by furrows (immediately) after the tenth century. Pottery sherds in the furrows dated from the eleventh to fourteenth centuries (Hey 2004, 207–9, fig. 11.5). The plan of the Saxon-period settlement at nearby Cassington shows that it lay under medieval furrows. The site was deserted after the eighth century (Hawkes and Gray 1969).

Of earlier work on settlements, the Saxon-period village at Sutton Courtney, Berkshire, was the first excavation of a site made under controlled conditions (Leeds 1923–7 and 1947). It was discovered by gravel workings and there is no record that pottery was found on the surface beforehand. The same applies to other well known older excavations of which the largest were Old Down Farm, Andover, and Cowdery's Down, Basingstoke, Hampshire (Millet and James 1983), Catholme, Staffordshire (Losco-Bradley and Kinsley 2002), West Stow, Suffolk, Mucking, Essex (Hamerow 1993), Cassington, Oxfordshire (Hawkes and Gray 1969), and Bishopstone, Sussex (references in Lucy et al. 2009, 1).

Various excavations showed that some nucleated settlements existed, with desertions taking place in the seventh and eighth centuries. The excavation plans of Chalton, Catholme, and Stow are unlike later villages, with little organized plan and no streets. New Wintles Farm, Cassington, Oxfordshire, is very dispersed. In contrast, the Yorkshire Saxon-period site at Heslerton studied by Dominic Powlesland (2003) was laid out in a planned, regular form and extended over many acres. Similarly the Saxon-period settlement at Carlton Colville, Suffolk (AD 500–700), had an organized plan and, although it changed in detail, remained in a fixed location (Lucy et al. 2009, 429–30).

Hamerow (1991 and 1993) reviewed the then-known evidence for settlement plans. Mucking shifted in three phases over a considerable area during the fifth to seventh centuries, and West Stow was similar. Some of the small rural sites discovered by fieldwork probably did not move until deserted; if they did, some of them can represent little more than one or two farms 'moving about'. On the Continent, large scale excavations give clearer evidence that settlement did relocate in Holland, Jutland, and north-west Germany during the seventh to ninth centuries (Hamerow 2002, 104–5, 110–24). In view of the very few complete excavations of large Saxon-period sites in England (only thirteen, two unpublished, Lucy et al. 2009, 1) it is difficult to put forward a valid view of a sedentary or shifting settlement pattern as a statement applicable to a region or to the whole country. In the absence of excavations that have extremely good dating evidence, it is not possible to say whether closely adjacent Saxon-period sites discovered by fieldwork were shifting; pottery is not closely datable outside of East Anglia within a span of four centuries. A site-distribution map alone provides no information about shifting settlement. Saxon-period house- and site-plans have been revealed, largely resulting from commercial investigations (Hamerow 2012, 67–119). It is clear that many sites had an organized structure, with houses and plots set out in a rectilinear manner during the eighth to tenth centuries, as at Cottenham, Cambridgeshire (Mortimer 2000), and Quarrendon, Lincolnshire (Taylor 2003). There were also examples of reorganization of site layout, notably at Flixborough, Lincolnshire, during the ninth century (Loveluck and Atkinson 2007, 72, 82).

The chief interest for open-field studies is to determine when Middle Saxon-period sites were deserted. Since some excavated settlement sites are known lying away from present-day villages, it is obvious that the occupants must have moved elsewhere when they were abandoned. In a discussion of the relative locations of pagan Saxon cemeteries and villages with early place-names, it was concluded that new settlements were formed with the 'desertion of the majority of known early Anglo-Saxon settlements, and the development of new centres in the seventh and eighth centuries' (Arnold and Wardle 1981).

Subsequent mechanisms proposed for village formation have followed the conclusion of Arnold and Wardle, resulting in statements that villages are not old and did not exist much before the eleventh century, as quoted above. Arnold and Wardle were making a general statement to account for settlement and cemetery locations then known. It has been extended to models of the Brixworth type that have several small deserted Saxon-period settlements, proposing that the occupants moved to *new* sites in a better location, which became the established villages. It had already been known in 1972 that Brixworth,

Northamptonshire, had Middle Saxon-period occupation remains near the church (Everson 1977, 71–5), before the discovery of sites in the fields in 1973–4. Sherds (240) were recovered from a grid twenty by fifteen metres contemporary with a ditch of eighth-century date. Other examples of villages in the same county with known Saxon material within them and sites out in the fields are Cogenhoe, Cold Higham, Pattishall, and Preston Deanery. In Buckinghamshire, three small Saxon-period settlements were found in the fields of Leckhampstead and occupation of the same date existed within the village (Jones and Page 2006, 85–6).

There are several unlikely aspects about the 'new site' interpretation. Saxon-period farmers were unlikely to choose poor agricultural locations and leaving the best areas unoccupied. A more likely development is that, in town-ships of the Brixworth type, the occupants of the several small settlements eventually moved to the largest or most important site that was located at the optimum agricultural location with good quality soil and near to water. In other words, according to this model, there should be Middle Saxon-period remains lying under many villages.

Saxon-period remains under present day settlements in Northamptonshire were found in the 1960s, as described (see p. 135), and continue to be dis-covered throughout the Eastern Region. In Cambridgeshire sites have been found at Cottenham, Cherryhinton, Long Stanton, Madingley, Willingham, and in Cambridge around Castle Hill (excavations by the Cambridge Archaeological Unit). Barton Bendish, Norfolk (see p. 137), has both Early and Middle Saxon material very close to the village. Excavations of the deserted medieval village sites at North Coningsby, Lincolnshire, and Shotton, North-umberland, have revealed Saxon-period settlements underneath or nearby (see pp. 144–5).

A view of some villages in AD 973 can be seen in the confirmation of the es-tates by King Edgar to Thorney Abbey (S 792; Hart 1966, 165–86). Bishop Æthelwold of Winchester was buying and exchanging estates in the later tenth century with the intent of endowing the fenland monasteries of Ely, Peterbor-ough, Ramsey, and Thorney. There seems to have been a composite record of all the estates acquired before it was decided exactly what each monastery would receive. Two different extracts were made, one in the Thorney Cartulary and a different one in a Peterborough cartulary, each pertaining to the re-spective monasteries' own estates. The record shows purchase of many manors, vills, and parts of vills that were already well-established places. A similar view of the Ely estates during 963–84 gives details of the abbot's transactions in many villages, some small, with sometimes several purchases made in the same vill, for example in Cambridgeshire at Chippenham, Horningsea, and Wilburton (Hart 1966, 213–30; Blake 1962, 82–117). These documents reveal an active land market occurring in known vills, similar to dealings familiar in the thir-teenth century and later. They are concerned with places of substance and having value suitable for the endowment of various monastic houses, not just a few scattered huts in the fields.

It seems likely that most of the large manors mentioned in Domesday in 1086 were at vills already in existence by the Middle Saxon period, if not earlier. In the central village belt, many of the places that were not listed in 1086 were likely to have been in existence, but assessed with another vill. The deserted medieval vill of Nobold in Clipston has been shown from taxation records to have existed in 1086, and Muscott (in Brockhall) and Onley (in Barby) have each produced a few sherds of Middle Saxon-period pottery, as have five other Northamptonshire deserted village sites.

A summary of significant information about some of the Saxon settlement remains found in each county is given next. Information on cemeteries is not included. Many sites have been discovered during development work during the last fifteen years, and the results can be consulted in the county HERs, with much more available online. The term 'Saxon' refers to the ceramic material as defined above. It will be seen that evidence for Saxon-period pottery is sparse in many counties, especially in the west, namely Cheshire, Cornwall, Derby, Devon, Dorset, Hereford, Lancashire, Somerset, and Worcester. In some counties there is more data than can be conveniently included, so that a selection has had to be made, with the emphasis on those lying under medieval fields for providing dating evidence.

Bedfordshire: About fifteen Saxon pottery scatters have been discovered by fieldwork on arable land (HER), including a large one at Bletsoe (Hall and Hutchings 1972, 9). Early to Middle Saxon features were discovered at Oakley Road, Clapham (TL 0220 5280), partly overlain by medieval furrows (Edmondson et al. 2002, *Medieval Archaeology*, 46: 153–4). Saxon settlement sites have been excavated on river gravels at Harrold and near Lower Caldecote (Ingham and Shotliff 2012, 46–51; 2003, *Medieval Archaeology*, 47: 220).

Berkshire: For the Sutton Courtney Saxon site, see p. 139 (Leeds 1923–7 and 1947). Fieldwalking in the Loddon Valley (SU 780 778) recovered eleven sherds believed to represent a site (Ford. 1994–7, *Berkshire Archaeological Journal*, 75: 1–8). Various sites have been found at building development excavations near existing settlements, such as Abingdon (SU 5055 9732) (Keevill 1992, *Oxoniensia*, 57: 55–79).

Buckinghamshire: About fifteen sites have been discovered, some by development near existing settlements, and some found by fieldwork on modern arable. Sites at Chichele (Middle Saxon sherds with Maxey type fabric (Farley 1980, *Records of Buckinghamshire*, 22: 92–101)) and Fenny Stratford (sunken floored buildings; Ford and Taylor 2001, ibid. 41: 80) were sealed by ridge and furrow. Middle Saxon sherds were found by fieldwork at Lillingstone Dayrell (2002, ibid. 42: 152). Leckhampstead has produced Saxon pottery under the village and in modern arable fields (Jones and Page 2006, 85–6). The important site at Pennylands in Great Linford has been discussed (p. 139).

Cambridgeshire: Saxon sites have been excavated at Cambridge (1997, *Proceedings of the Cambridge Antiquarian Society*, 86: 173, and 2004, ibid. 93: 112–19), Cherry Hinton

(2000, ibid. 89: 92), Chesterton (2004, ibid. 93, 128), Cottenham (Mortimer 2000), and Fordham (1997, *Proceedings of the Cambridge Antiquarian Society*, 86: 178). At Gamlingay, on the east of the village, a settlement and an early Christian cemetery of fifth to seventh century date were discovered (Murray and MacDonald 2005). Saxon-period sites were excavated at Hinxton and Harston (1997, *Proceedings of the Cambridge Antiquarian Society*, 86: 111), Madingley and Melbourn (2003, ibid. 92: 120–1). Sites are known from fieldwork at Waterbeach (Hall 1996; Robinson and Gutman 1996, *Proceedings of the Cambridge Antiquarian Society*, 84: 172) and from excavations at Willingham (1997, ibid. 86: 187), Ely (2002, ibid. 91: 152). Ipswich Ware occurs on most sites, enabling the Early and Middle Saxon periods to be distinguished.

Cheshire: Small quantities of Saxon pottery have been found during development work (HER).

Cornwall: Some Dark Age pottery has been identified, but it has not been discovered on ploughed fields (HER).

Derbyshire: Saxon remains are very sparse, even from recent development work. House sites have been discovered at Willington (HER).

Devon: Pottery occurs in towns from the mid tenth century (Exeter, Totnes, Lydford, and Barnstaple) but cannot often be distinguished from later material. A few Saxon sherds have been found at Stockland, but none is known from the fields (HER (J. Allen) 2011).

Dorset: Only one Saxon site is recorded (Mepham 1993, *Proceedings of the Dorset Natural History and Archaeological Society*, 114: 115), but there are problems with pottery fabric identification.

Durham: Saxon pottery has been found as scatters on some arable fields (HER).

Essex: The best-known site is Mucking with 256 identified buildings excavated (Jones 1974 and Hamerow 1993). Settlement along the Chelmer–Lower Blackwater valley in central Essex shows early Saxon sites located on the fertile river gravels with a move further inland to more clayey soils during the Middle Saxon period (Sue Tyler 2011 and pers. comm.). Within present settlements, pottery and structures have been found at Clacton (Letch 2004, *Medieval Archaeology*, 48: 259). For sites near Saffron Walden and Stansted see p. 138 (Williamson and Meddlycott). Rippon (1996) reviewed evidence for the state of Saxon Essex. See below for Saxon sherds on late Roman sites (Morris 2005).

Gloucestershire: Saxon pottery scatters have been found away from present village sites at Somerford Keynes (twenty-eight sherds, Barclay et al. 1995, *Transactions of the Bristol and Gloucestershire Archaeological Society*, 113: 43), west of Roel (Aldred and Dyer 1991, 142), and at Hazelton one scatter, Hidcote one scatter, and Roel two scatters (Dyer 2002, 13). Excavations have produced Saxon sherds at many locations: on a Roman site at Matson (1997, *Transactions of the Bristol and Gloucestershire Archaeological Society*, 115: 290) and eighteen sherds at Bishop's Cleve (1998, ibid. 116: 124). West of Fairford twenty sherds were found at Southrop (2000, ibid. 118: 223, 213). Early Saxon buildings have been excavated at Lechlade (2003, ibid. 121: 58–63). Middle Saxon sherds were found near the church at Churchdown (2005, Nichols, ibid. 123: 90–2) and at Lower Slaughter a Middle Saxon ditched enclosure lying near the church was interpreted as a manor-house site (Kenyon and Watts 2006, ibid. 124: 73–110).

Hampshire: Several sites are known, including those discovered by fieldwork at Old Down Farm, Chalton (Cunliffe 1972), and at Andover (Davies 1979, *Proceedings of the Hampshire Field Club*, 36: 161–74). At Cowdery's Down, Basingstoke, there was a break in site layout on a Roman site that was superseded by Saxon settlement of the fifth and sixth century (M. Millett and S. James 1983, 192–257). Sherds were discovered in the Middle Avon Valley (Schofield and Shennan 1994, *Proceedings of the Hampshire Field Club*, 50: 51), at Bentley Green Farm (SU 780 434) (S. Ford 1997, ibid. 53: 74–5), and at Southampton (Correll 2004, *Medieval Archaeology*, 48: 270–1).

Herefordshire: Vegetable-tempered pottery was excavated from features at Ripple (SO 866 391) (Barber et al. 2004, *Medieval Archaeology*, 48: 273; Russell et al. 2002, ibid. 46: 173).

Hertfordshire: Only a few Early and Middle Saxon sites are known. Pottery survival is very poor, bring predominantly made in a grass-tempered fabric (HER).

Huntingdonshire: Several Saxon sites are known out in the fields as at Abbots Ripton (TL 2275 8000) (D. Hall unpublished). Saxon occupation on Roman settlement was found at Haddon (Upex 2002), and at Orton Hall Farm a fourth-century Roman site was succeeded by Saxon buildings in the fifth century, and later cut by medieval furrows (Mackreth 1996, 42, 85–91).

Kent: Saxon pottery has been identified at seventy-four sites, some from excavations, with a good site at Springhead (HER 2001; Wessex Archaeology online). Middle Saxon settlement lies south of Lyminge church (Thomas site report 2008; copy online and HER). Welch has summarized the evidence for Anglo-Saxon settlement (in Williams 2007, 201–9).

Lancashire: Small quantities of Saxon pottery have been found during development work only.

Leicestershire: Saxon pottery has been discovered by fieldwork at about 120 sites. Some are on the edges of present villages and others out in arable fields that were overlain by ridge and furrow. The distribution has been published (Knox, in Bowman and Liddle 2004, 96–7). Only a few Ipswich Ware sherds have been found, but handmade Middle Saxon material is plentiful on some sites; for example Stonton Wyville produced 200 sherds on gravelly and sandy soil. Eye Kettleby has been discussed (p. 139). Middle Saxon sherds have been excavated at Kirby Bellars churchyard (Hurst 1967–8, 10–12) and at Saxby village (SK 822 199) (Thomas, 2002, *Medieval Archaeology*, 46: 183).

Lincolnshire: Many Early and Middle Saxon pottery scatters have been found in the south of the county, both on upland and the Siltland fen ground near Spalding (Hilary Healey, in Hayes and Lane 1992, 249–52). Fillingham has Saxon remains at Chapel Road (Buckberry and Hadley 2001). A settlement site near Quarrington, Sleaford, produced 2,330 pottery sherds but no ninth- or tenth-century Saxo-Norman fabrics. The site lay on limestone and was cut by medieval furrows twelve yards wide (Gary Taylor 2003, 249, fig. 11, 276). The site of North Coningsby deserted village, near

Flixborough, produced Saxon and medieval occupation from the late seventh century with site replanning in the ninth century (Loveluck and Atkinson 2007, 6, 96).

Middlesex: Sherds and features came from Winslow Road, Hammersmith (TQ 3233 7790) (Jamieson 2002, *Medieval Archaeology*, 46: 166) and Westminster, Floral Street (Taylor and Humphrey 2002, ibid. 171). More results are provided by Rackham et al. 2004.

Norfolk: Saxon settlements as known in 1994 were seven early sites and several Middle Saxon sites. It is not clear how many of the latter have been found within village confines, or out in the fields (Penn and Rogerson, in Wade-Martins 1994, 37–9). Sherds and features were excavated at Snetterton (Birks 2004, *Medieval Archaeology*, 48: 280), Great Ryburgh Manor House (Emery 2002, ibid. 46: 193), and Norwich, Wensum Street (Taylor 2002, ibid. 195). Developer-funded excavations in West Norfolk showed that 62 per cent of Romano-British sites have fifth–sixth century occupation on or within 500 yards of them (Rippon et al. 2012, 63–4). Finds of 1960s–1980s have been described (see p. 137).

Northamptonshire: Early and Middle Saxon handmade pottery, much of it tempered with crushed igneous rock or sand, survives in plough soil and has been found in many parts of the county by fieldwork. Its distribution lies predominantly on light soil and few sites have been discovered on clay except those associated with Roman settlement (Plate 9). Some of these have been excavated and proved to relate to structures underneath them at Brixworth (Ford 1995, identified by fieldwork in 1973–4) and Courteenhall (Jones et al. 2006, 28–31; identified in 1983).

Northumberland: A settlement dating to the sixth to eighth centuries, has been excavated next to Shotton deserted village site, west of Cramlington (NZ 228 778) (HER report, December 2010). The site underlay ridge and furrow.

Nottinghamshire: Near Brough, Saxon pottery, sunken-pit-features, and postholes forming a rectangular building were found at SK 8380 5890, with furrows of medieval fields superimposed (HER, M 18429, May 2001).

Oxfordshire: Excavations in 1964 revealed early Saxon pottery at Clanfield near the church (Pocock 1968, 89). Fieldwork at North Stoke produced 17 Saxon sherds initially and 796 sherds when intensively searched in 5 by 5 m squares. Subsoil features were found by trial trenching over the site (Ford and Hazel 1990). Maureen Mellor (1994, *Oxoniensia* 59: 36–7) has reviewed Saxon pottery finds. Yarnton excavations revealed substantial settlement remains in the fields that were overlain by furrows in the tenth to eleventh centuries (see p. 139; Hey 2004, 55). The HER has other records of Saxon finds.

Rutland: Hard pottery with igneous temper of Early and Middle Saxon date is found in the fields and within village confines. Occupation sites have been recorded at Hambleton, Manton, Oakham, Tickencote, and Tixover (Leicestershire HER 2001). Uppingham produced sherds of Ipswich ware (Field 2000, *Transactions of the Leicestershire Archaeological and Historical Society*, 74: 259). A remarkable late Saxon site at Ketton (SK 969 056) had a Saxon church and hall buildings. The only pottery associated with it was Stamford ware, *c.*900–1100. Furrows of open-field strips ten to twelve yards apart lay next to the site, but did not overlie it, showing they were contemporary, or respected it as an older site (Meadows 1999, ibid. 73: 119–21; Meadows 2002, ibid. 76: 127). This is probably the deserted site of *Soulthorpe*, known to lie in the west of Ketton parish (Cox 1994, 151–2).

Somerset: Pottery does not appear to have been used much in the county (Aston 1988, 69) and no Early or Middle Saxon pottery sites could be identified in the HER in 2000. A few sherds, dated by radiocarbon to the tenth century were recovered from Cheddar (McSloy 2006, *Proceedings of the Somerset Archaeological and Natural History Society*, 149: 114). Mick Aston (1994) has reviewed settlement studies, including the problem of Saxon pottery finds; see also the Shapwick Project Report (Gerard and Aston 2007).

Staffordshire: Saxon pottery from the county is friable and not often discovered in plough soil. The excavated site at Catholme had some trackways, and many buildings dated early seventh to ninth century. Pottery was found on the surface of nearby fields (Buteux and Chapman 2009, 150–8; Losco-Bradley and Kinsley 2002).

Suffolk: The early Saxon sites excavated at West Stow by Stanley West (1990) lay on sand and Early Saxon-period sites generally avoided claylands, but some Ipswich Ware sherds occur on clay. Middle Saxon sites are known at Brandon, Butley, and Culpho, and about ten were discovered in the south-east Suffolk survey area (Dymond and Martin 1999, 45; Hardy and Martin 1988, *Proceedings of the Suffolk Institute of Archaeology*, 36: 44–50, 145–6).

Surrey: Saxon pottery in the county is mostly vegetable-tempered and does not normally survive in plough soil. It has been found during development, both within medieval settlements and out in what had been open land such as Battersea ((McCracken) 1980, *Surrey Archaeological Collections*, 72: 250), Bletchingley (Poulton 1990, ibid. 80: 213), and Goldaming (Jones 1998, ibid. 85, 194–5). It has also been discovered during development in some cases on Roman sites. At Kingston, Tolworth, pottery was found in an enclosure ditch (2004, *Medieval Archaeology*, 48: 266), and other Saxon finds have been made at Kingston (Hawkins 2004, *Surrey Archaeological Collections*, 87: 220).

Sussex: Saxon pottery has been excavated at Bishopstone (Bell 1977, *Sussex Archaeological Collections*, 115: 227–35) and at North Marden (Foster 1986, ibid. 124: 110–16). Early settlement was discovered at Itford Farm, Beddingham (James 2002, *Sussex Archaeological Collections*, 140: 41–7).

Warwickshire: Saxon sherds were found on the site of a probable Saxon Palace at Hatton Rock (S. Hirst and P. Rahtz 1973, *Transactions of the Birmingham and Warwickshire Archaeological Society*, 85: 160–72), and on the surface near where a sunken-floored feature was excavated at Baginton (Wilkins 1975, ibid. 87: 124–7). Saxon pottery has been recorded on a few plough-soil sites in the county: Idlicote in 1986 (R. C. Hingley 1986, HER 6027), and from Bidford on Avon (Hingley 1987, HER 5101). An early Saxon settlement on gravel lying out in the fields of Salford Priors, south of Alcester, was excavated in 1999 (Palmer *Transactions of the Birmingham and Warwickshire Archaeological Society*, 103: 97–110).

Wiltshire: Saxon sites have been identified during development work (HER). The finds from Westbury and Ogbourne St George have been described (p. 137). Draper (in Higham and Ryan 2011, fig. 5.1) provides a map showing the distribution of Saxon-period sites in the county.

Worcestershire: Saxon sunken-featured buildings have been discovered during commercial developments, for example at Ripple in 2002, and produced vegetable-tempered pottery (Barber and Watts 2003, *Worcestershire Recorder*, 64: 12–13). The site also produced hard gritty ware, some of it associated with early dwellings dated late sixth to late seventh century (Barber and Watts 2008, 54, 78). Other sherds are known from Kemmerton and Upwich. No sites have been discovered by fieldwork. Nine rural Saxon sites are known (HER).

Yorkshire: An early Saxon settlement was discovered at Wykeham, NR (Moore 1966, *Yorkshire Archaeological Journal*, 41: 403–44). The large-scale excavations at Heslerton on the south side of the Vale of Pickering by Dominic Powlesland revealed an early Saxon cemetery (fourth–seventh centuries) and the related settlement extending to twenty-two acres. It continued from the Roman period, in about AD 400, and lasted until about 850. It was laid out in a planned manner, with areas for housing, industry, and animal husbandry, justifying the term 'village'. There were in all 220 buildings including some with sunken pits and rectangular timber buildings (Powlesland 2003, 32–44). One Romano-British site at Butterwick (SE 9881 7099) yielded three sherds of Saxon-period pottery (D. Hall and P Martin, Figure 2.1, site 6).

Alluvium and pollen

Archaeological evidence of a completely different type is provided by river alluvium. It is the latest deposit found in river flood plains and post-dates many prehistoric and Roman sites, demonstrating that it is relatively recent. It was formed by run-off from arable land when extensive areas were left as bare fallow. The lowest levels of the alluvium therefore give information about the date of large-scale arable ploughing that might be associated with open fields. River valley alluvium is very extensive, sometimes deep, and there are no equivalent extensive deposits associated with the Roman or earlier periods. In the past, not many excavations paid attention to alluvium, more interest being taken in the prehistoric remains buried underneath, but a few sites have now provided dates.

In the central Nene Valley of Northamptonshire, the work of the Raunds Area Project showed that the latest infill of a channel beneath the river alluvium was dated to AD 740–880 (HAR-8563). Thereafter alluviation continued into the Middle Ages on neighbouring riverine sites (Brown 2006, 25). Beneath the alluvium was an extensive plough soil containing Roman artefacts and no evidence of Roman alluviation was observed (Mark Robinson, in Parry 2006, 34–5). Saxon-period occupation of the nearby Stanwick Roman villa was buried by alluvium after the sixth century (Kevel, in Needham and Macklin 1992, 183). Robinson (in ibid. 197) noted that 'overbank alluviation with deposition of fine sediments was not a feature of these rivers [the Nene and Thames] until comparatively recently'. This evidence tells us that in the Raunds region, and upstream, the river meadows are of the Late Saxon period and after, and that therefore these are the dates when there was extensive ploughing with bare arable on the river valley slopes and beyond. At Yarnton, Oxfordshire, in the Thames Valley, the main floodplain alluviation was dated to Cal. AD 790–1050 (Hey 2004, 54). The onset of minerogenic alluvium deposition near Panborough, Glastonbury, was found to be between the eighth and tenth centuries (Cal. 1165–982 BP, Aalbersberg and Brown 2011, 143).

It is worth emphasizing the importance of the dating of riverine alluvium beds, which are found in all major river valleys. The often extensive tracts of this deposit, on which lie the valuable meadow resources referred to in Domesday and in manorial extents thereafter, did not form until the Late Saxon period. Development of alluviated meadows represents a major change in agricultural resources, complementing extensive areas of bare fallow systematically used in routine agricultural procedures.

The Somerset Levels provide an important type of evidence relating to antecedent fields. The Roman landscape was drowned by flooding, engulfing all settlements and fields with 0.6 m of marine deposits. The Saxon-period settlements that developed on top of them before 1086 had no precursors—yet both dispersed settlement and relatively large villages with common fields are to be found. This appears to be the result of lordship, since the village settlement belonged to Glastonbury Abbey, whereas other owners were content to accept or allow dispersed settlement to develop (Rippon 2008, 61–105). Few peat deposits of the Late Saxon period now survive, but studies of changes in the pollen record of those that do reveal significant agricultural changes. A series of eleven small bogs lying around the edge of Exmoor with intact medieval deposits has been studied by Ralph Fyfe, Tony Brown, and Stephen Rippon. The pollen record shows that the landscape was opened up (after the Bronze Age) and was mainly pastoral with a low level of cereal cultivation, probably not very near the bogs. Tree pollen was also low, consistent with its limited survival on the steep edges of coombes. The pollen profile continued without any significant variation from the Late Iron Age through the fifth and sixth centuries, suggesting there was no great change in farming methods, in other words, there was no settlement expansion or desertion. Between the later eighth century and the tenth (depending on the altitude of the bog, those located higher being later) there was a considerable increase in the level of cereal pollen that then continued through the Middle Ages. Pastoral activities continued alongside the arable and the evidence is consistent with what would be expected from the system of convertible husbandry practised in the region, recorded from the fourteenth century onwards (Fyfe et al. 2003; Fyfe and Rippon 2004, 38–9).

Pollen evidence does not characterize the adoption of any particular agricultural regime, but does show that much more countryside was being opened up. A similar sequence was found in the pollen record from Amberley, Hampshire, which implied clearance of the countryside, with increasing herb, grass, and cereal pollen deposited from the seventh century onwards (Waton, in Bell and

Limbrey 1982, 85). The dating accords well with the alluvium evidence. In East Anglia, at Oakley near Scole, an increase in cereal pollen in a palaeochannel was recorded in the eighth century (Cal. 670–820) with viticulture and hemp cultivation (Wiltshire unpublished, quoted by Rippon, in Higham and Ryan 2010, 58). At Micklemere near Pakenham there was a marked increase in cereal pollen dated 588–972, with an influx of minerogenic sediment implying increased soil erosion (Murphy, in Rackham 1994, 29).

Open-field furlongs

Evidence for the extent and nature of open-field patterns draws upon three main sources: open-field maps, aerial photographs, and archaeological survey. The use of maps to study open fields is fairly straightforward—though care must be taken to avoid confusion with estate maps marking enclosures filled with 'artistic' strips that have no physical meaning. Relatively few townships have contemporary plans and even fewer have been published, partly because when reduced they are apt to be unreadable. Accuracy and detail vary—some have schematic furlong outlines only.

Ridge and furrow often preserves details of open fields on the ground and can be effectively recorded by aerial photography. Among the earliest photographs are those of Calstone in Calne Without, Wiltshire, published by Crawford and Keiller in 1928, and Crimscote, Warwickshire, first published by Venn in 1933, already referred to (p. 26). In the 1940s the RAF made vertical surveys of the whole country, some of which are of outstanding quality. From the 1970s most county councils commissioned vertical surveys every decade and now the whole country can be viewed on websites. The surviving ridge and furrow in the Midland counties in 1996 has been discussed (Hall 2001).

There is the question of dating. An early modern historical reference to ridge and furrow was made in 1860 by the Rev. J. Wilkinson (*Wiltshire Archaeological and Natural History Magazine*, 6: 29). Writing about Broughton Gifford, Wiltshire, he states 'the appearance of our pasture, in ridge and furrow, the ancient mode of carrying off the surface water, tells the tale of the land having been once under the plough'. In 1925 Clark studied the field system of Marston near Oxford, of which there is a 1605 map made for Corpus Christi College. He noted that on the ground 'ridges correspond closely with the Corpus map' (Clark 1925, 12). This is the earliest known comparison of ridge and furrow with map evidence. Shortly after, Crawford and Keiller (1928, pls. 28 and 29, pp. 166–8) photographed the terraces of 'typical medieval cultivation' on the downs of Calstone in Calne Without, Wiltshire. An extract of an open-field map (dated 1713–32) of the same area shows an almost identical plan. A map of Soulbury, Buckinghamshire, dated 1769, was compared by Mead (1954) with ridge and furrow plotted from aerial photographs. The agreement between the furlong boundaries and the map is good. Beresford and

St Joseph (1979, 25–37) gave an elegant visual proof that Midland ridge and furrow equated to open fields by illustrating maps and aerial photographs of surviving ridge and furrow showing identical areas. Exact parallels are evident for Broxholme, Lincolnshire, 1600; Crimscote, Warwickshire, 1842; Ilmington, Warwickshire, 1778; and Padbury, Buckinghamshire, 1591. In all cases there is detailed and precise agreement.

Ridge-and-furrow evidence must be used with care and not confused by post-enclosure ridging. Kerridge (1951 and 1955) cited a range of agricultural authors and sources to show that in some areas land was gathered into ridges after enclosure. These are often straight, sometimes narrow, and fit within a hedged or walled 'modern' fields (Hall 1995, 38). Difficulties can occur with those dating to the seventeenth or eighteenth centuries, which are curved. Remains of this type are best referred to as 'cultivation ridges'. They were the result of ploughing techniques used by severalty farmers, mostly before the era of widespread underdraining (from 1820 onwards). They have no relevance to open fields, yet have been published as open-field ridge and furrow—for example Northamptonshire, at SP 885 924 (RCHM 1979, *Northamptonshire*, ii, 61 and pl. 29; north at the bottom), which shows mainly nineteenth-century cultivation ridges on a broken scarp, the top of which, although ridged, was part of Rockingham Forest woodland until 1835.

RAF vertical photographs taken during 1945–50 are the most useful, since they recorded the countryside before it was assaulted by intensive arable agriculture and urban growth. They have been used to prepare sketch plans at the 1:10,560 scale of Leicestershire and Rutland (HER; some published by Hartley 1984, 1987, 1989; see Leicestershire in the Gazetteer, p. 285), Oxfordshire (in the HER), and parts of Warwickshire (N. Palmer and HER). For most places it is difficult to map a complete furlong pattern with precise boundaries, even from the best photographs, and ground survey is necessary to fill gaps.

The formation of linear soil banks along furlong boundaries produced as a result of ploughing has been explained previously (Hall 1972; 1982, 25–33; 1995, 39–48). The resulting earthworks can be mapped by archaeological survey and used to prepare furlong plans, to which may be added data from aerial photographs (and from a contemporary plan, if there is one). Crawford (1937) was the first to identify furlong earthworks, and they were observed in the Welland Valley in 1960 (RCHM, 1960, 32). I independently recorded them at Wollaston, Northamptonshire, in 1961 and prepared the first such maps for the county in 1964 (Hall 1972, 59). Arthur Pocock (1968), working at Clanfield, Oxfordshire, realized that the soil banks in the fields were pre-enclosure furlong boundaries (he called them balks) and reconstructed a plan of the whole parish in 1963. Most of these earthwork furlong boundary studies were made in areas where there had been ridged strips. The earthworks have nothing to do with the ridging of strips, but were caused by the continued turning of ploughs in the same place for many centuries with deposition of slight amounts of soil. The survey techniques are therefore applicable in most parts of the country, irrespective of whether there was ever ridge and furrow ploughing. Baker drew attention to the 'relatively neglected field form' in 1973; outside of the Central Region they remain neglected forty years later.

Furlong plans are helpful when used with field books to elucidate the field structure of a township. The survey technique is applicable to much of the country, as has been explained, and plans of townships in Yorkshire, Leicestershire, Norfolk, Suffolk, Essex, Sussex, Oxfordshire, Berkshire, and Dorset, as well as in Northamptonshire, have been prepared and several used in this volume (Plates 3, 7, 10, 11; Figures 5.3, App.2, App.3, App.4). For the townships of Stourpaine, Dorset, and Iford, Sussex, the furlong patterns correspond exactly to open-field furlongs marked on tithe maps.

Apart from individual studies to identify field structure and demesne types, etc., and to illustrate specific points, furlong maps have more general applications. Open-field maps of Yorkshire are striking with their long lines. Strips of such length are not apparent in the Midlands until the plans of townships are considered. Places in areas with relatively planar pieces of countryside have furlongs with strips in the same alignment, suggesting that lands had been lain out in great lengths initially and subsequently broken down to the familiar a chequerboard pattern. This has been previously pointed out for Raunds and Wollaston (Hall 1982, 50; 1983, 119). Publication of the north-eastern quarter of the county shows that such alignments are widespread (Hall 2009, 73–101); their significance is discussed later (p. 190). More alignments can be found in the remainder of the county (Hall 2013, 37).

At Fen Ditton, Cambridgeshire, the furlong pattern affords dating evidence. The fields sit astride the High Dyke, a northern extension of the Fleam Dyke, one of the four dykes that divide the county south-west of Cambridge into blocks. Various dates have been suggested for the dykes, with the Saxon era thought to be the most probable. The Fleam Dyke, from which Ditton (*dic-tun*) takes its name, is first recorded in *c.*995 (Reaney 1943, 35, 142). High Dyke at Fen Ditton became a central trackway in the Middle Ages, forming the boundary between two of the great fields. On the east, High Ditch Field is separated from the other two by a narrow belt of fen and marsh. The form of the furlong pattern shows that the fields are later than the dyke, especially in High Ditch Field, where furlongs and strips slope in radial curves either side of the dyke to drain into the surrounding low ground (Cambridge University Library, MS Plan R.B.9 (1796); also Cambridgeshire CRO, TR626/P1, *c.*1730). No furlong boundary in the mainly rectangular network of the other two fields crosses the dyke. The main Fleam Dyke has been dated to Cal. AD 330–510 for the first phase, and 450–620 for the last phase (Tim Malim 1997 and 2000, bringing together previous work). If the High Dyke is of the same date, it follows that the furlongs were laid out after the Middle Saxon period.

There have been several good detailed archaeological parish surveys in, for example, Norfolk, Suffolk, Northamptonshire, and Somerset to study settlement history by the systematic collection of pottery sherds in ploughsoil (Rogerson 1997; Davison and Cushion 1998–2001; Hardy and Martin 1986; Foard 1978; Parry 2006; Gerard and Aston 2007). The technique has been applied to identify early medieval site locations and more recently to ascertain the extent of open fields at a given date, assuming that the sherd scatter measures the area of arable which received manure containing domestic refuse (Jones and Page 2006, 92–4; Jones and Hooke 2011, 39). The last application is an interesting approach, but the intensity of fieldwork necessary limits sample sizes to a few townships, and the method is not applicable to early medieval fields because so little pottery was used or survives. The assertion that 'it was a universal practice to manure ploughlands with an admixture of animal dung and domestic refuse' (Jones and Page 2006, 92) requires comment. For the Saxon period no such practice can be proved or justifiably assumed. Manuring by sheepfolding in southern England did not create dung in byres that might acquire sherds, and even where manure was collected from byres and permanent folds, it cannot be assumed that domestic refuse containing pottery was consistently added to it. Additionally, if there were parts of a field system that were not manured (such as outfield), then mapping sherd scatter cannot be a satisfactory method of measuring the extent of arable.

Reconstructed maps: mapping using field books and terriers

The Rev. H. E. Salter reconstructed the open fields of Oxford in 1939. The base was a map of 1828 made before enclosure, which shows furlong shapes and sizes, although by that date there was much old enclosure. Salter superimposed on the nineteenth-century map information from a field book of 1382–6 that describes open land and from which he reconstructed a fourteenth-century map of Oxford's fields. With the caution of an historian, he stated that the 'map...is imaginary' (Stevenson and Salter 1939, 502).

In 1965 Chibnall reconstructed maps of Sherington, Buckinghamshire, for 1580 and 1300, using in the first instance detailed terriers of 1770 that related scattered strips to neighbouring furlongs. The block demesne had been split into parcels and closes by 1580 that were then fully described in relation to each other. Rectangular pieces of card cut to scale had abuttals written on the appropriate edge, and by adjusting them in relation to each other and to known roads and woods, etc., it was possible to achieve a reconstructed map. This was used to plot the distribution of lands described in medieval terriers to confirm the locations of the great fields (Chibnall 1965, 259–76). The method was developed and used to reconstruct from field books the field systems of the Northamptonshire parishes of Ashby St Ledgers, East Haddon, Raunds, and elsewhere (Hall 1995, 36–50). Field books give the name of each furlong,

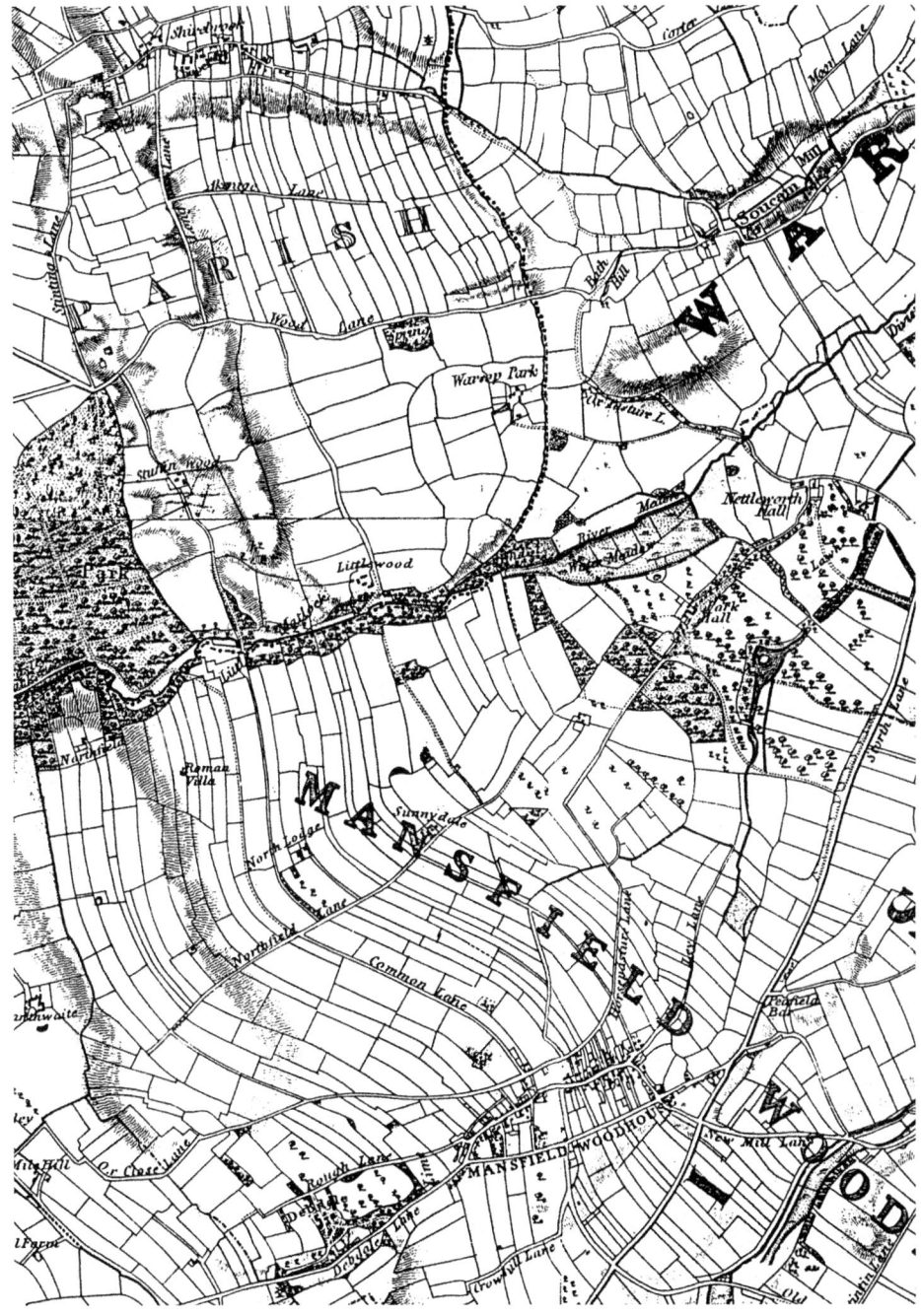

Fig. 5.1. Curved hedge patterns at Mansfield Woodhouse, Nottinghamshire (after G. Sanderson, *To the Nobility, Clergy, Gentry…, This Map of the Country Twenty Miles round Mansfield…*(Mansfield: Privately Printed, 1835)).

describing its topographical location and abuttals with neighbouring furlongs, and listing all the lands and owners. In the absence of a field book, it is possible to work from a parliamentary enclosure commissioners' quality book. A quality book is a valuation of a township made prior to enclosure, and was needed to value the new allotments as well as the old open-field property. Each furlong is listed as in a field book, but groups of lands are taken together as a single entry (without ownership details) and valued in shillings. More detail has been provided for the reconstruction of Newton Bromswold fields by this method (Hall 1995, 46–8). For townships without field books or quality books, precision will generally be much less. Success depends on good survival of furlong names as modern field names, and matters are easier to work out if the township is small with fairly large furlongs. Grendon, Northamptonshire, is an example where nearly all the furlongs can be identified from modern field names.

The 'reconstructed' furlong maps are considerably better as a data source than implied by the term 'imaginary map' used by Salter for his plan of Oxford fields. They are based primarily on mapped physical evidence surviving on the ground. They have *not* been constructed by retrospective historical geographical techniques, using data recorded on nineteenth-century or older maps. They therefore tell of the maximum extent of medieval ploughing and can be brought to life by historical documents that reveal topographical names and tell how they were managed, and can be used to determine the detailed structure of a township. I have completed a survey for the whole of Northamptonshire (1961–2011) and the results can be viewed in the published atlases (Hall 2009, 72–101, and 2013, maps M1–M86). Some of the data have been used by Williamson et al. (2013, pl. 15).

Many counties had operational open fields surviving into the twentieth century and arable open-field strips can still be viewed in Lincolnshire, Devon, and Cornwall. More information can be found in the Gazetteer and via the index. Many hedge patterns recorded on maps of the nineteenth century reflect former open fields, as at Mansfield Woodhouse, Nottinghamshire (Figure 5.1). More examples are noted by county in the Gazetteer.

B. ANTECEDENTS OF MEDIEVAL FIELDS

There have been many attempts to relate Romano-British villas and their presumed estates with medieval manors and fields. The historian Ephraim Lipson (1937, ch. 1) discussing the origin of the manor, pointed out that there is little evidence of any direct connection of medieval estates with Roman villa estates. There are no relevant Celtic or Latin words in the language and no elements of Roman law. Evidence for archaeological continuity has been sought in several counties from time to time. For example around Cirencester, Gloucestershire, there was one villa known per medieval parish (Reece 1983, 13). In eastern

Cambridgeshire, it has been suggested that there is an unexpected degree of coincidence of pagan Anglo-Saxon presence with major Roman sites that might indicate continuity of administration (Oosthuizen 1998, 104–5). However, Draper (2009) has shown that the Wiltshire evidence for continuity of Roman estates and Saxon-period estates or hundreds is unsatisfactory.

The intensive work of the Raunds Project concluded that 'the widespread and almost total dislocation of the settlement pattern with further change at the end of the middle Saxon-period…shows that there is no simple development from Roman villa to medieval manor' (Parry 2006, 95). The Raunds studies also showed that there was no Roman alluviation of the river flood plain and therefore, although there are many Roman sites surrounded by sherd scatters possibly indicative of manuring, there cannot have been at any one time an extensive area of arable with bare fallow. Probably convertible husbandry and possibly horticulture were practised, as stated by the classical Roman authors for Italy. The map of Roman sherd distribution around all the sites away from the river shows a sherd spread covering near half the total area (Parry 2006, 82, fig. 4.13), but in agricultural terms this is not evidence of a large area of arable at any one time—the outer edges of the scatters represent a density of one sherd in forty hectares. Allowing for 300 or so years of site life, this displays low-intensity activity.

Near Hardwick, Cambridgeshire (TL365 599), excavation revealed an Iron Age and Roman site on what was then open grassland, later overlain by furrows seven to nine metres apart (Abrams and Ingham 2008, fig 1.11, Scotland Farm site 7, and p. 97; see also fig 1.6). On the claylands near the Ouse Valley at St Neots, Huntingdonshire, excavation at Love's Farm revealed droveways and paddocks dating from the late Iron Age to the end of the Roman period. The site lay in open grassland grazed by sheep, cattle, and horses and was later also overlain by open-field furrows that ignored it (for photographs see Jessica Mills and Rog Palmer 2007, 11, 13). These sites, and many others, demonstrate pastoralism and the absence of extensive arable in the pre-Saxon era.

Archaeological evidence: ancient fields and cropmarks

It is not possible to know what tenurial and practical arrangements operated in the fields belonging to prehistoric and Romano-British farmers in England, but archaeological evidence gives some information of their physical remains. On high ground, such as chalk downland, there are networks of small rectangular fields defined by low earthen banks. They were called 'lynchets' by early authors (and are not to be confused with medieval strip lynchets). Curwen (1927) described and illustrated exceptional lynchets occurring in various parts of the country. Those forming rectangular paddock systems, often associated with Iron Age or Roman settlement earthworks, were called 'Celtic fields'.

Crawford and Keiller (1928) and Bowen (1961, pls. Ia, III, V) published further examples.

References to the occurrence of lynchets are given by county in the Gazetteer. Some of the networks of lynchets are impressive in extent, yet few of the largest cover more than a hundred acres. It may be possible that Saxon-period fields originated by breaking lynchet systems into strips. However, there are problems in relating them to furlong patterns, even though attempts have been made to do so. None of the many plans published by Bowen and Fowler (1978) provides a likely precursor in plan form. Fowler (2002, 127–60) has further discussed first-millennium field forms, but no obvious 'medieval precursors' of the Central Region type can be found in the examples given. Although some areas of earthworks and cropmarks on chalk downland parishes in Berkshire and Wiltshire are considerable (Hooke 1998, 130–1), and there are around six square kilometres of Celtic Fields near the River Avon on Salisbury Plain (McOmish et al. 2002, 62), generally cropmark areas of prehistoric and Roman fields are not nearly so extensive as medieval fields. This is demonstrated on the Yorkshire Wolds (Stoertz 1997) and Cambridgeshire Silt Fens (Hall and Palmer 1996, figs. 95, 102), both of which regions have soils very responsive to cropmark formation and the published plans can be regarded as substantially complete.

Many surviving Celtic fields lie on marginal land and it may be that there were once equivalent but larger systems on the lowland, adjacent to Roman sites, but which are now obliterated by medieval and modern arable agriculture. Unless the boundaries were associated with ditches that give cropmarks, no trace of such fields would be discoverable once the low banks of the Celtic fields had been ploughed away. Even if there were at one time Roman fields of this type on the lowland, it does not remove the problem that the pattern is very different from medieval fields. Lowland evidence for early fields comes mainly from cropmarks. Roman and older sites lying on permeable soils yield cropmarks of fields or paddocks around them. They are not often extensive, but exceptional are the Roman ditched fields of the Fenlands at Elm, March, and Christchurch in Upwell, Cambridgeshire, already mentioned (p. 78). The larger systems cover 200–300 acres, but again, they do not form a furlong-type pattern and did not develop into strip fields of either medieval fenland or Midland type; the most extensive examples lie in what became medieval fen and certainly had no influence on the subsequent strip fields. They were most likely paddocks used primarily for cattle rearing and relate to saltern sites in the region, since they lie near a major transport route, the Fen Causeway, which links the Midlands and East Anglia (Hall and Palmer 1996, 176–7 and figs. 84, 102).

Near to Colchester at Gosbecks are forty acres of cropmarks laid out in a regular six-cell grid, although not quite rectangular (Crummy 1978, 81). At Fordington, Dorset, a map of 1779 shows an extensive furlong system forming an approximate rectangular arrangement fitting in with roads approaching

Dorchester. The furlongs completely ignore the extensive prehistoric and Roman cropmarks lying north of Maiden Castle (Peter Woodward, in Sharples 1991, 18–19). Martin (in Martin and Satchell 2008, 7–10) has reviewed the archaeological evidence for early fields in East Anglia extending to South Essex and the Thames Valley. Generally fields or paddocks were small and did not extend over very large areas compared to medieval fields.

The long strip fields of the Yorkshire Wolds provide very clear cases where medieval fields can be shown to have little relationship to earlier agricultural arrangements. Published cropmarks of Butterwick show a series of sites picked out by four paddock systems (Stoertz 1997, Map 1, in the grid square SE 98 71 area; detailed on fig. 44). Iron Age and Roman pottery was discovered at the cropmark locations during the field survey by Hall and Martin made to prepare Figure 2.1 of this volume in 1981 (Stoertz 1997, sites nine and ten). A remarkable linear 'ladder' settlement running for over a kilometre above Northdale (site two) yielded Roman sherds towards the west end. Cropmarks south of the village (Stoertz 1997, fig. 30, no. 1) also yielded Roman sherds and three handmade Saxon-period sherds (SE 9881 7099). The long medieval strips completely ignore the alignments of the cropmarks. If comparison of a contemporarily mapped field system is preferred, rather than a reconstructed one, then the linear and paddock cropmarks at West Field, Kilham (Stoertz 1997, Map 2, TA 02 63 area), bear no relation to the published strip fields of 1729 (Harris 1959, 7–8). Other cropmarks in the region similarly have little or no relationship to the strip fields, all of which were of the simple structural type with long lands.

Furlong boundaries overlying cropmarks of all dates have been published for the Welland Valley at Maxey, Northamptonshire (RCHM 1960, fig. 6, pl. 12). The boundaries can be dated to the end of the Middle Saxon period or later, since they ignore the Saxon site excavated by Addyman (1964). The RCHM noted (1960, 37) that the 'ridges' (furlong boundaries) paid no attention to buried ditches or other features. Striking, vertical summer photographs of the area taken in 1976 and 1977 show multiperiod cropmarks and extensive blocks of curved parallel ditches underlying the furrows of the medieval strips, which were cut into gravel bedrock. South of Northborough it is clear that the strips (as cropmarked furrows) do not relate to any aspect of the previous landscape, but they do stop at a (medieval) droveway running along the meadow-edge south of Maxey, called the Nunton Outgang (vertical photographs CUCAP, RC8-DE 14; for more photographs of the Welland Valley, including Helpston and Glinton, Northamptonshire, and West Deeping, Lincolnshire, see CUCAP, RC8-BO 212–240; and 1976 vertical aerial photographs of the Welland Valley held at Peterborough Museum).

Examples of ancient ditches underlying furlong boundaries have been described (Bowen and Fowler 1978, 159 n. 43). Upex (2002, 84–7) has excavated ditches on a Roman site at Haddon, Huntingdonshire, that underlie four adjacent furlong boundaries and has suggested that some small furlongs lying on or near to Roman sites may represent a continuity of field systems. The sites also produced some Early Saxon-period pottery, suggesting there was activity until the sixth century. A Roman site at nearby Lutton, Northamptonshire (TL 1119 8700) (D. Hall unpublished), also lies under a network of small furlongs. The terrain is almost flat and the topography does not require complicated small furlongs unless there was something to interfere with the large-scale layout of strips normally found in the locality. Plate 10 shows that the cropmarks and the small furlongs coincide. In four cases the ditched cropmarks underlie furlong boundaries exactly and are in close alignment on a fifth. This would seem to prove that the furlong layout was influenced by the presence of surviving Roman earthworks, which affected the drainage and needed to be dealt with. It does not necessarily prove that there is any agricultural or physical continuity in field systems—merely that there were banks or ditches still surviving and that these impeded the plough. The boulder clay upland plain of Lutton would not normally be a preferred site for Saxon-period settlement. For the most part, Roman sites in Northamptonshire do not have small furlongs associated with them and the many cropmark sites have no similarity to a furlong pattern (Deegan and Foard 2007, 106–7,112, 115), indicating that few surviving earthworks offered the plough any difficulty. An example of excavated Romano-British ditches that align with the hedge pattern as mapped in the 1880s is at Bishops Cleve, Gloucestershire (Rippon et al. 2012, 62). Again there is no proof that an extensive complete ditch network exists, nor that there is any more continuity other than that the Roman features represented an obstacle to ploughing strips at a much later date (if the ditches can be shown to relate to furlong boundaries).

Late Iron Age ditches at West Stow, Suffolk, were overlain by the Saxon-period settlement, which was overploughed in the Middle Ages on an alignment that had no relation to either of the earlier land divisions (West 1990, figs. 10, 17, 33). At Carlton Colville, in the same county, a Saxon-period settlement and cemetery dated AD 600–680, was sited on top of a Roman trackway and enclosures after a 200-year gap with no activity in-between (Lucy et al. 2009, 428). One of the finest examples of a furlong boundary overlying an ancient feature is at Ashton Keynes, Wiltshire (SU 034 960), where a double-row pit alignment, of likely Iron Age date, lies along the curved strip terminations at the headland between two furlongs (Figure 5.2). A similar boundary with curved furrows approaching it on both sides, overlies a ditch at Hardwick, Cambridgeshire (Oosthuizen 2003, 54), and a furlong boundary at nearby Caldecote followed the line of a late prehistoric trackway (Scott Kenney, in Mills and Palmer 2007, 125, 128–9).

Oosthuizen (2010, 380–3) has given a useful review of examples of ditches and features that may represent Saxon-period or older field systems. This interpretation is very probably correct for parts of the Eastern and Western Regions. However, on the whole, ditches and cropmark features underlying furlongs of

Fig. 5.2. A headland at Ashton Keynes, Wiltshire (SU 03 96), marked by two sets of crop-marked furrows lying at right-angles (centre). Remarkably, the furrows can be seen re-cut towards the left at their ends, as a reverse-S shape developed. Underlying the headland is a row of prehistoric postholes that were probably once buried by an earthwork (18 June 1999, NMR 18338/10; copyright English Heritage).

the Central Region are so few that they cannot account for complete furlong patterns. Probably, at the time when fields were being created, the ditches were associated with surviving earthworks that would present an obstacle to the plough and interfere with the drainage of furrows if overploughed. If in a suitable location, an earthwork could be conveniently used as a turning place and so develop into a furlong boundary. The Aston Keynes, Wiltshire, pit alignment (Figure 5.2) was probably buried under an earthwork; pits have been revealed under linear earthworks on the Yorkshire Wolds (Stoertz 1997, fig. 21, p. 43). Some ditches have been reported under furlong boundaries in the Whittlewood region of Northamptonshire, but no plan or details of the number or extent are provided (Jones and Page 2006, 94).

Many of the attempts to relate pre-Saxon fields to 'medieval fields' have been to find precursors relating to the extensive fields of the Central Region. It will be shown that most of these fields are dated to about the ninth century and later, so that the expectation of a simple relationship between systems five centuries and more apart should not be high. We shall return to the theme of dating small dispersed 'hyde', 'huish', and other settlements of the Eastern and Western regions (see p. 185).

Present-day field patterns

A different approach to seeking antecedents for medieval fields has been made by the analysis and interpretation of field patterns preserved in enclosed fields recorded on the First Edition Ordnance Survey 1:10,500 maps (dates varying from *c*.1865 to 1890). The earliest use of them made by antiquarians was to try to discover patterns of roads and hedges in the English countryside by dividing the landscape into a network of regular squares that may have derived from a system of Roman 'centuriation'. Remarkable areas of countryside in Italy, Tunisia, and Yugoslavia are divided in this manner (Bradford 1957, pls. 42–3, 46–9). Many of the English attempts were fanciful. Haverfield (1918) commented upon 'the old controversy, as to the continuity between Roman Britain and Saxon England', and doubted if there were any proven examples of centuriation in England; he referred to a 'treatise' by a mid nineteenth-century author (H. C. Coote) in which 'ingenuity and ignorance are about equally characteristic'. Haverfield drew attention to parallel roads in Essex: one from Great Dunmow to White Roding (the current B184 road, a five-mile straight length) and the other, seven miles away, running through Braintree to Little Waltham (now the A131), which might have been due to Roman planning. The roads are now accepted as Roman roads, but a case study at Felsted, lying between them, found no evidence of centuriation (Martin, in Martin and Satchell 2008, 214).

The hedge patterns of Ripe, Sussex, form a remarkable series of very elongated rectangles with sinuous boundaries divided by short straight hedges into smaller rectangular fields (Margery 1940). Margery interpreted the pattern to be evidence of Roman centuriation, but it really looks like a series of sinuous furlong boundaries subdivided by short straight hedges that do not reflect the reverse-S of strips. Finberg (1972, 92–3) superimposed a grid on the 1803 open-field plan of Much Wymondley, Hertfordshire, in an attempt to show a relationship between the sizes of furlongs and Roman measurements. The connection between them is not very apparent. A grid-system of field boundaries at Brancaster, Norfolk, was suggested to have a Roman origin (Ward 1932–4), but it was later proposed that they were more likely to have been set out under the influence of Ramsey Abbey, as lord, after AD 969 (Moore 1966, 8–9).

Field systems of hedges that form approximately rectangular networks over large areas of the countryside have attracted more recent attention. Studies in East Anglia and Essex have proposed that patterns of hedges in the old enclosed countryside of the region date from the Iron Age or Roman periods. In south-east Essex, a network of rectangular hedged field systems, lying on London Clay at Dengie, was commented upon by Drury and Rodwell (1978, 34–7) and Oliver Rackham (1986). Drury and Rodwell (in Buckley 1980, 59–75) assigned a prehistoric or Roman date to them. They were further discussed by Stephen Rippon, who illustrated the plans of Orsett and the Southend area (Rippon 1991, figs. 4, 5). He showed that a Late Saxon date, between the eighth and ninth centuries is more likely than a Roman one. When examined on large-scale maps, the rectangular systems do not run smoothly over considerable areas of countryside, but have discontinuities at parish boundaries (Rippon 1991, 57). These field patterns are very different from most of Essex and are located on low-lying land. They look similar to rectangular furlong patterns found in lowland parts of the East Midlands, as near Glinton, north of Peterborough (Plate 11), and a field survey of the north-east part of Orsett shows that there are linear banks of furlong boundaries. Northill and Southill, Bedfordshire (near Upper Caldecote), have regular roads and paths that make up a rectangular pattern (interpreted by Bigmore (1979, 43) as centuriation). They most likely represent a simple layout of open fields or even parliamentary enclosures made in 1780 and 1797.

Williamson (1987) proposed that the 'coaxial' hedged field systems on the claylands of the Scole–Dickleburgh area of southern Norfolk had a prehistoric origin, because the pattern was cut by a Roman road that crossed them. This was criticized by David Hinton (1997), who suggested that the field patterns shown on early nineteenth-century maps were post-medieval. In a reconsideration, Williamson (1998) modified some of his earlier statements, but still thought that the basic structure of the system (trackways) had late prehistoric elements. More detailed study of some of the areas suggests that the apparent alignments are made up of several components. The Roman road need not be considered to pre-date the Scole–Dickleburgh network, but more likely forms a non-comfortable element in later field systems that were laid out according to drainage requirements. The neighbouring South Elmham–Ilketshall system is likely to represent the fields of several parishes making up a Saxon estate. The St Michael, South Elmham Ordnance Survey map of 1891 shows a large area of narrow fields with long curved boundaries (Martin and Satchell 2008, 97). The exact similarity to Yorkshire medieval fields is readily apparent (see Butterwick and Towton, Figures 2.1, 2.5, and online vertical images of Pickering). Fieldwork at St Michael has revealed the presence of furlong boundaries underlying the hedges (Figure 5.3), and therefore a late medieval date is the earliest that can be assigned to the field pattern preserved on nineteenth-century maps. The Norfolk example at Dickleburgh is similar, with furlong boundaries underlying hedges or visible where hedges have been removed. Again, the hedges cannot be much older than the late Middle Ages (observations based on a site visit in 2012).

Bull (1995) studied central and north Buckinghamshire. The alignment of selected roads, trackways, and some parish boundaries, as depicted on Ordnance Survey Maps, was noted as forming an approximately rectangular system, especially on the Chiltern Ridge. Again the proposal is that the system is planned and dates substantially from the Iron Age. Was the Iron Age population really so dense that it occupied the whole of the landscape and

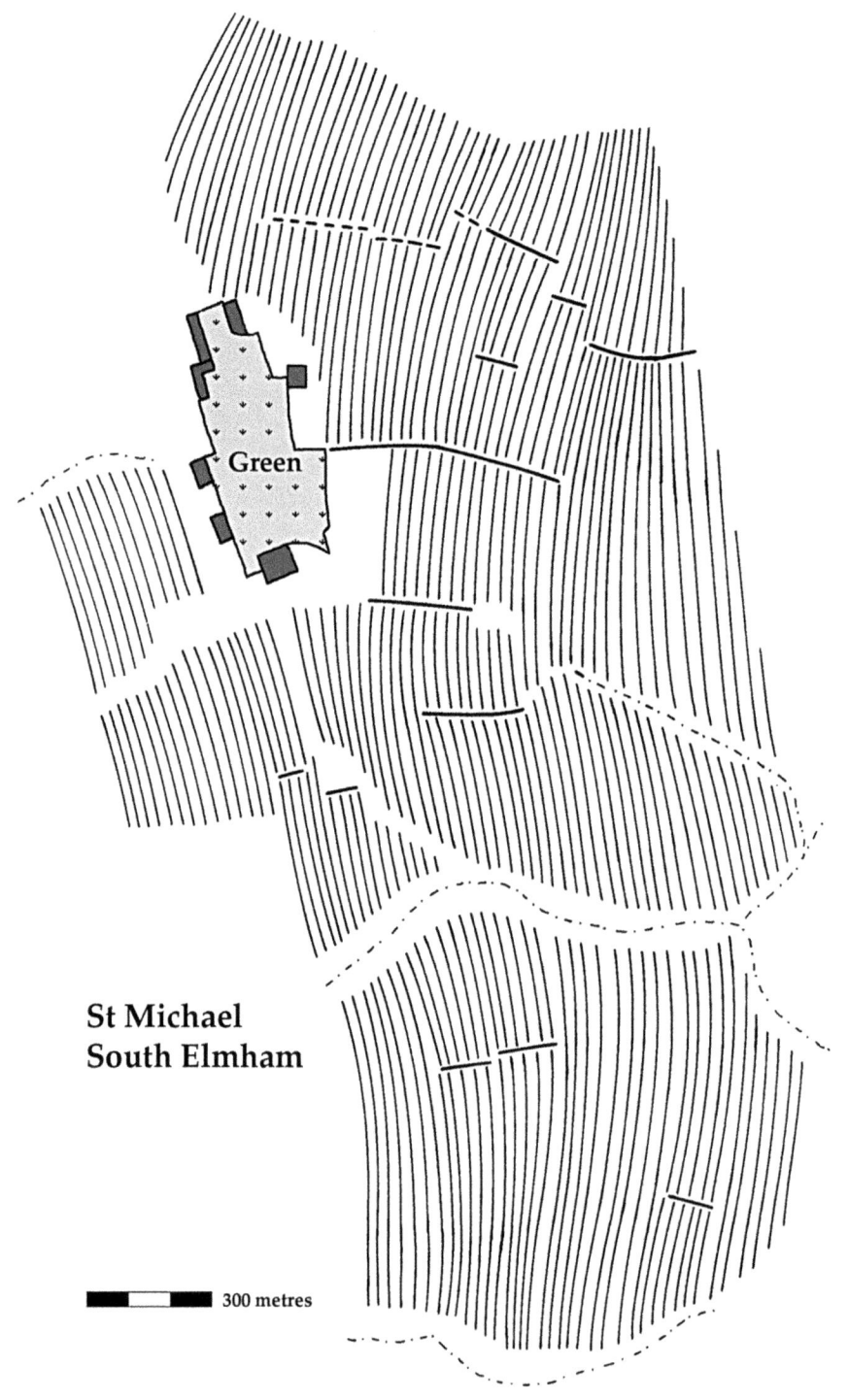

Green

St Michael
South Elmham

300 metres

Fig. 5.3. Furlong boundaries and strip fields at South Elmham, St Michael, Suffolk (field survey by D. Hall, 2011).

needed to divide it with permanent and planned boundaries and roads? If so, then a population similar to that of the Middle Ages is implied. It is incumbent upon authors to prove that the boundaries date from before the Middle Ages, or to show convincingly that the medieval landscape has Iron Age origins. On the south of Bull's map, on the chalk ridge, there is much more rectangularity than in the north. Again this is immediately attributable to open-field boundaries set out on the large scale. Soil ridges of furlong boundaries are readily visible in the similar terrain in south-east Cambridgeshire, where they have been photographed from the air and recorded over large areas. In this region the furlong boundaries are very large, and form cropmarks and soilmarks, because considerable depths of topsoil have accumulated in the banks contrasting with the shallow soil of the chalk slopes. The soil banks were identified as remnants of open fields by Crawford in the 1930s at the Cambridgeshire parish of Litlington (Crawford 1937). A draft enclosure map of 1804 has all the furlongs mapped, and their boundaries corresponded exactly with the soilmarks on Crawford's photographs.

It has been suggested that part of the furlong pattern of Caxton, Cambridgeshire, based on an eighteenth-century map, has earlier prehistoric origins since the pattern is bisected by a Roman road (Oosthuizen 1997). An alternative explanation is that the furlongs are aligned on the Bourn Brook, which the road crosses at an angle, the drainage of strips across the heavy clay towards the brook being more important than the presence of an older road that was ignored as a planning element. An early date has been proposed at Harlton, Cambridgeshire, where a Roman road interrupts furlong patterns (Oosthuizen 2003, 57). Again, drainage may have dictated such an alignment, with strips running across the contours. The main dating evidence for rectangular and linear hedge systems comes from the observation that Roman roads cut across the field pattern which, it is argued, must therefore be older. However, the apparent bisection of field systems is not sufficient evidence alone; stratigraphic data based on excavation are required. Medieval fields were laid out according to local topography to achieve natural drainage by alignment across contours, and would approach and continue beyond a Roman road obliquely if it suited the lie of the land, as suggested for Caxton and by Martin and Satchell for Dickleborough, Norfolk.

Apart from looking like medieval furlong patterns, there is a difficulty with the extensive systems of coaxial hedged fields in terms of intensity of land-use. Many occur on claylands and their management implies that a considerable population was required to farm them—and that there ought to be an appreciable density of Iron Age and Roman settlements equivalent to that of the Middle Ages. The extent of ancient settlement in the areas of published coaxial fields is not readily available. As recorded in 1994 the density of Iron Age settlement in Norfolk is sparse, there being one site in Dickleburgh and one in Scole (Green, in Wade-Martins 1994, 32–3). If coaxial fields had a pastoral function, it is not clear why they were enclosed and why so many paddocks were required, unless they represent severalty farms, implying a considerable population. Which brings us full circle, because when these coaxial and rectangular fields are first viewed in late medieval surveys or on early modern maps, they *are* the enclosures of severalty farms!

There is a further difficulty in trying to relate ancient fields to furlong patterns. Whereas Roman paddock systems and Celtic Fields make up a coherent network of enclosures, in contrast, furlongs, especially in the Central Region, do not. Furlong boundaries are formed by linear banks that represent convenient divisions for turning the plough on an open landscape; they are not enclosure boundaries. Plate 12 shows the furlong-boundary pattern of Rothersthorpe and Kislingbury, which does not make a pattern of enclosed areas. The apparent chequerboard network shown on many open-field plans is made up of 'political' boundaries between different named furlongs, but the strips often have the same alignment either side and there was no physical division between them. The planned nature of many fields (see p. 190) shows that some boundaries are secondary to the original layout, which was much more akin to those of Eastern Yorkshire with its long strips. It is therefore not likely that there will often be 'ancient' boundaries underlying furlong boundaries since they represent divisions created subsequent to the initial layout.

This account has tended to take the view that in the Central Region there is very little antecedent landscape, somewhat against the trend of current research, which is looking for various types of continuity that might have been overlooked previously (Oosthuizen 2013). In the other two regions there are examples of modern features following ancient ditches that have been proved by excavation. At Saltwood, Kent, a track and a road (in separate locations) follow the lines of linear ditches (pers. comm. J. H. Williams 2010). It is perhaps in cases of this type and in some of the irregular hedged fields of Essex that there are features in the modern (or recent) landscape that have a genuine origin in the pre-Saxon period. A good example of excavated Late Iron Age, Romano-British, and Early Saxon-period features lying in the same general alignment as irregular hedges mapped in 1881 can be found at Hunts Hill Farm, Upminster, Essex (Rippon et al. 2012, 63; 2013, 40). This is the type of settlement that might be expected in the 'hydes' and block demesnes of Essex and elsewhere in the Eastern and Western Regions.

Having failed to discover any convincing evidence of large-scale extensive ancient field systems that could readily be transformed into the type of medieval fields found in the Central Region, it is necessary to examine Early and Middle Saxon-period settlements for any evidence of associated fields. Unfortunately, sites of these periods present very few cropmarks, and a recent study of Northamptonshire yielded little evidence for 'fields' (Deegan and Foard 2007, 125–35). The Polebrook site is probably alone in being discovered by the photography of cropmarks as representing ditched paddock systems. These do not differ much in form from the tofts of medieval vills and do not help to reveal any extensive arable lands that may have existed around and beyond them. Few large Saxon-period settlements have been excavated; those that have, do not exhibit extensive ditched field systems. Any traces of unditched open fields would have long been destroyed, unless subsumed into later furlongs.

Historical records tell nothing of English agriculture in the fifth and sixth centuries and the archaeological record does not provide any satisfactory antecedents for the strip-field systems found in the Middle Ages. For this early period we therefore have to engage in speculation.

C. INTERMIXED STRIP HOLDINGS

The most characteristic feature of medieval arable holdings when they first appear fully described in detailed terriers of the twelfth century is their subdivided or intermixed nature. A single farm is scattered in small strips of one to two roods lying among similar pieces belonging to the other holdings. This is true of most of the country at core settlements (not considering peripheral assarts), except in Kent and parts of East Anglia. How old is this type of arrangement? The earliest direct documentary evidence for all counties comes from the twelfth century. A map and field book of 1758 locates furlongs at the extremities of Great Corringham township, Lincolnshire, described in 1200 (Beckwith 1967). A terrier of Cranford St John, Northamptonshire, dated 1154–69, shows that the strips are scattered over most of the township when plotted on a map of 1748 (Hall 1995, 3). At Rushden the dispersed rectorial land granted in 1114 can be identified in eighteenth-century terriers as already discussed (p. 126) (Hall 1995, 340; plotted in Hall 1985, 64–5). Indirect eleventh-century evidence is found at Yardley Hastings and Denton, Northamptonshire. The manorial structure of Denton was unusual in that one hide belonged to Yardley Hastings manor in the Domesday Survey of 1086 (Williams and Martin 2003, 600) and in the Northamptonshire Survey of *c.*1124 (Round 1902, 376). By analogy with Bernwood Forest, Oxfordshire, this hide was very probably given to the keeper of Yardley Woods or the whole of Salcey Forest. The Bernwood forester was granted a hide of arable land containing sixty acres, called the 'derhyde', referred to 1252 and identified as lying in the fields of Borstall (Salter 1930, 157–9, 169, 177, 195).

Terriers of the Yardley Hastings part of Denton township dating from the seventeenth century and later (Compton Muniments, 770B) can be plotted on an open-field map of 1760 (Compton Muniments, 1348; see Hall 2013, 44–5, for a plan). The land was still called the Yardley Hide in 1704 and paid tithe to Yardley church. It consisted of blocks of lands dispersed in all the furlongs, most of them consisting of eight to twelve lands (Northamptonshire CRO, Yardley glebe terrier). This hide therefore suggests that there was a regular disposition of lands by 1086, and that the field system was fully in existence. This is fairly reliable historical evidence that most of Denton was arable in the 1086 (excluding its woodland). The dispersion of the Yardley Hide in Denton shows that Denton lands were likewise dispersed. These can be traced back to a grant made to Ramsey Abbey before 1029 (Hart and Lyons 1884-93, iii: 167); there is not then any internal evidence that they were dispersed. An interpretation that in 1086 the Yardley Hide lay in a small arable area that increased in size as the field system expanded to occupy most of the township (Williamson et al. 2013, 123) seems unlikely, because intakes of the twelfth century and later always seem to be severalty assarts, as has already been discussed (p. 177). The mechanism is even more implausible at Clipston, where yardlands comprising four estates of variable sizes in 1284 were stated to be intermixed in a complicated way between the two townships of Clipston and Nobold. The same four estates are precisely identifiable in 1086; the likelihood of their being intermixed in two small embryonic field systems in 1086 and then all four continuing to remain intermixed in exactly the same proportion, as furlongs were added piecemeal to two separate field systems during the next two centuries, must be very remote.

The evidence of late Anglo-Saxon charters

The earliest documents giving information about land are late Anglo-Saxon charters. Often they record grants of a township or groups of townships and many describe the boundaries of the estate. In some cases there are references to arable and other information relevant to the nature of the holding, although specific information about fields and furlongs is usually slight and incidental. Kemble (1839–48) and De Gray Birch (1885–93) published most Anglo-Saxon charters and they have been discussed and assessed for authenticity by a large number of scholars, including Peter Sawyer (1968), Cyril Hart (1966 and 1975), H. P. R. Finberg (1961 and 1964), and more recently in the series published jointly by the British Academy and the Royal Historical Society, 1973–2013. Topographical items listed as markers in the boundary descriptions have long been studied by local and national authors to interpret land-use and landscape features, as well as to locate the exact boundaries.

Several early writers studied Anglo-Saxon charters in order to seek evidence of dispersed strip fields because of the similarity of terminology to that of the High Middle Ages. Seebohm (1883, 106–8) quoted the bounds of Hardwell, Berkshire, to show the occurrence of ploughing terms. Vinogradoff (1908, 274–85) also used evidence from them, and Gray (1915, 51–62) discussed several charters, reviewing critically interpretations made by Nasse and Seebohm. Later authors have drawn upon the same significant material and discussed examples of 'shared land' (Finberg 1972, 487–97). Gray listed extracts from nineteen charters which he thought suggested the existence of intermixed strips, although some of them are open to other interpretations. Below are the most relevant items in chronological order (translations have been verified with the original charters as printed by Kemble and Birch).

Denchworth, Berkshire, had some undispersed hides in 947. A charter describing the boundary of the five hides notes that three of the five at the north were not divided (S 529; B 833).

At Charlton by Wantage in 956, a five-hide estate had 'no clear boundary, for ploughlands are parted by adjoining ploughlands' (S 634; B 925; K 1207). The same five hides were later granted in another charter of 982, stating that 'the boundary is divided [and] unimportant, for adjacent ploughlands are intimately connected' (S 839; K 1278).

Nottinghamshire townships belonging to Southwell, near the Trent, had intermixed holdings in 958. The Archbishop of York had at Halam every sixth acre, and at Normanton-on-Trent every third acre (S 659; Farrer 1914, 7, 9; B 1029).

An Ardington (in Hungerford, Berkshire) charter of 961 granted nine hides, stating that the open pasture ('feld') lay common, the meadow is common, and the arable ('yrðland') is common (S 691; B 1079; K 1234).

A charter granting three hides at Hendred, in 962, says the boundaries (of all three) were common because the lands 'lay every acre under acre' (S 700; B 1095; K 1240).

At Avon, Wiltshire, in 963 (probably Avon Farm in Stratford sub Castle), three hides lay as 'single lands mixed in the common field here and there in dispersed places' (S 719; B 1120; K 503; and previously noted by Jones (1870)).

Land at Upper Stratford, on the divided hide, had every other acre given, and at *Fachanleage* every third acre of 'feldlandes' in 966 (S 1310; Robertson 1956, 88–9; B 1182).

At *Cudinclea*, Worcestershire (Cudley in St Martin's-Without, Worcester), thirty acres were said to lie 'in the shared lands of two fields' in 974, the only example where fields are mentioned (S 1329; B 1298; K 586).

Intermixed acres occurred at Kingston (Bagpuise), Berkshire, in *c*.977, where the boundary description refers to land lying 'acre under acre' (S 828; K 1276).

A parcel of divided land seems to have been called a 'sticca'. At Oxney, Peterborough, sixty 'sticca lands' amounted to thirty acres in *c*.985 (S 1448a; B 1130), and thus were half-acre lands.

A division of two hides at *Uppthrop*, Worcestershire (otherwise called Moreton, now Morton Farm in Lower Westmancote in Bredon), in 990, between two brothers was made so that in all places the elder brother had three acres and the younger the fourth, in the inner and in the outer parts of the estate (S 1363; K 674). This is clearly not a description of two blocks of land lying in severalty.

Dumbleton, Gloucestershire, had 2½ 'mansae' located 'in the common land' in 995 (S 886; K 692).

At Aston Somerville (Gloucestershire; Worcestershire since 1931), land was held 'sorte communes' in 1002, 'shared out by the local community' (S 901; K 1295).

From the above, it is seen that items relating to dispersed arable occur in charters dating to the second half of the tenth century. Earlier information is found in one of Ine's Laws that refers to divided land held in common *c*.688–94 (discussed on pp. 172–3). Most of these Anglo-Saxon references relate to property in southern England, nearly all in Wessex, apart from one for Southwell, Nottinghamshire. There is less detailed evidence for subdivided fields in the north of the country before the twelfth century (Hart 1975, 112–30), but the regularly ordered field systems of Eastern Yorkshire almost certainly date from before then. The boundaries of Howden (ER), are described in 959 and its dependent hamlets listed (S 681; B 1052; Farrer 1914, 12–13).

The perambulations described in some charters can be analysed with reference to township boundaries recorded on maps of the sixteenth century and later. Berkshire has seventy-four charters, which have been discussed by Margaret Gelling (1968, 1976). Della Hooke (1987, and 1988, 125–38) has further considered the boundaries, land-use, and topography of the Vale of the White Horse. Uffington and Kingston Lisle have open-field maps of 1785 and some of the right-angled parish boundary bends caused by negotiating furlongs are identifiable with 'steps' described in the Anglo-Saxon charters. There is generally a close correlation between the angled boundaries of the charters and the later open-field system. Prehistoric and Roman fields (considered as possible precursors) in the Lambourn (Berkshire) and Wylye (Wiltshire) region do not have a furlong-like plan nor do they relate to township boundaries (Hooke 1988, 130–1, with plans).

The boundaries of Hardwell (SU 28 87) in Compton Beachamp, Berkshire have right-angle bends that can be identified in a charter boundary of 903 (S 369; Gelling 1976, 684–5, CIII, map). It is assumed that arable lay on both sides of boundaries where the terms 'furlang' and 'hēafod' were used. A field survey of the Hardwell area, made in 2004, provided a complete furlong pattern from which a new interpretation of the course of the boundary can be determined (Figure 5.4). Apart from being a long narrow township, taking in resources of the chalk downs and heavy clays of the Vale, the field survey reveals a normal chequerboard pattern of furlongs. There is a break in the pattern on the steepest slopes, which were likely left as unploughed pasture and woodland. The right-angled bends of the township do indeed go around furlongs. The Ridge Way is older than the furlongs, according to discontinuities in the pattern on either side of it, consistent with its being named in the 903 charter. Its seems, therefore, reasonable to assume that the pattern revealed by Figure 5.4

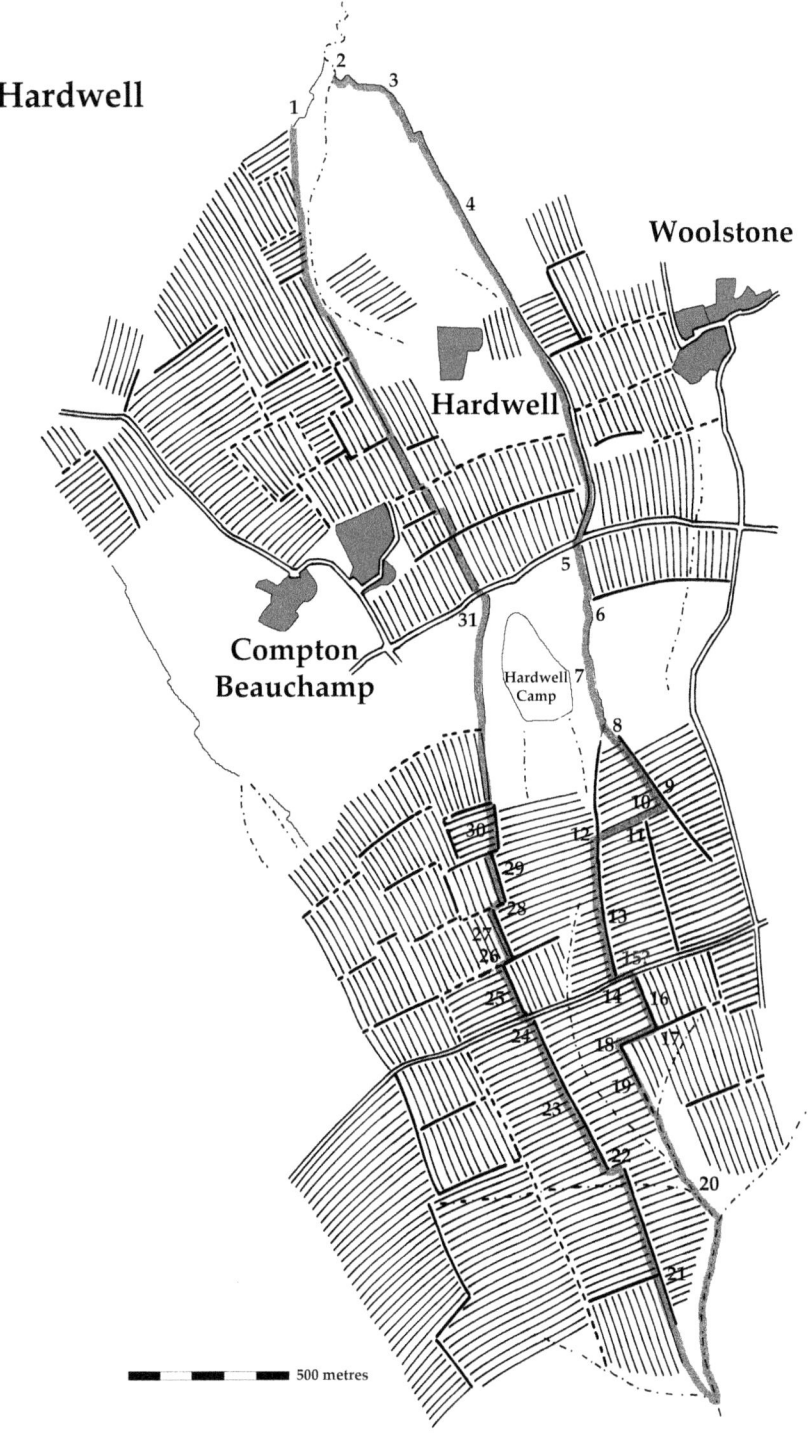

Fig. 5.4. Furlongs at Hardwell and neighbourhood, Berkshire. Boundary mark numbers after Gelling 1976, 684–6 (field survey by D. Hall, 2003).

was already in existence by 903, and that the landscape of Hardwell and the adjacent townships had been opened up as an extensive area of arable before then, in the ninth century. Further details of the charter boundary description and its relationship to the furlong pattern are given in the Berkshire Gazetteer (pp. 223–4).

Hooke has made several studies of the late Anglo-Saxon landscape based on charter evidence and published evidence for open-field agriculture in the West Midlands, Gloucestershire, Warwickshire, and Worcester in 1981. Many charters in the region have boundary clauses (plan p. 41). The analysis was further developed in 1985 (Hooke 1985a). Worcestershire had many settlements on the boundaries of the areas being described (Jones and Hooke, in Christie and Stamper 2011, 36–7). There are maps of various name-elements that indicate the state of the countryside at the time. The agricultural terms 'forierð' or 'heafodæcer' probably mean the same as the later 'headland' used to describe a boundary running between the contiguous arable lands of two adjacent townships (see the details of the Hardwell boundary for the precise meaning of these terms there, pp. 223–4). 'Garæcre' and 'hlinc' (gore acre and lynch) may not necessarily refer to arable but just to triangular pieces of land and a hillside or scarp. The use of the term 'feld' seems to refer to land lying near woodland that had been opened up. At the Upton on Severn boundary an item of 'feld' was stated be 'always by the wood where plough and scythe are wont to go', possibly indicating its use as outfield. In the Vale of Evesham there are references to 'old open land', as though it had long been cultivated (Hooke 1985a, 184). Finberg also considered these agricultural terms in charters from various southern counties. Caution is advised in interpreting Anglo-Saxon agricultural terms as relating exactly to similar medieval ones. Thus 'hēafod' may not always mean headland but merely the top of a valley or other topographical feature. The plural form most probably does refer, however, to 'headlands'. 'Land', occurring as 'oatland' or 'earthland' refers to arable but when linked as 'ham londe', it was meadow. The use of 'furh-lang' was originally the length of a furrow, but had the medieval sense in the boundary of Adlestrop, Gloucestershire, where '*Rahulfes furlung* which is in the field of Evenlode' was the neighbouring vill (S 1548; Kemble 1367). 'Furh', furrow, indicates that there was cultivated arable land on both sides. At Broadway a large open field called Shear Field extended to the parish boundary to meet the open field of neighbouring Willersey, separated by a 'furh', where the parish boundary goes through a 'step' (mapped in Hooke 1981, 43–4).

A map of agricultural terms shows that the main open-field areas were north-east Gloucestershire, south-east Worcestershire, and possibly the mid Avon (Hooke 1985a, 191–5). Hooke (1981, 46–53) also studied in detail the topography of north-east Gloucestershire and the Vale of Evesham. The distribution of Roman sites and Anglo-Saxon agricultural terms correspond closely, showing that the farmers of both periods used the best agricultural soil. It compliments fairly exactly with the regions where the charters refer to woodland (Hooke 1981, figs. 7, 13, 14).

The following is a summary by county of items relating to field systems.

Bedfordshire: An Aspley Guise charter of 969 refers to a headland at the boundary of Aspley and Wavenden, Buckinghamshire, showing that the arable fields of the townships were touching (S 772; B1229). An open-field map of 1747 (partly reproduced by Bigmore (1979, 66–8) shows three large aligned furlongs and other cross furlongs, which look planned.

Berkshire: The many Berkshire charters have been discussed (p. 168).

Buckinghamshire: The boundary charter of Winslow, made *c.*948 (S 1546a), had a 'head acre' along the bank of the arable field belonging to Winslow and Swanbourne, where there is now ridge and furrow, suggesting that the arable lands of both townships touched (Bull and Hunt 1998, *Records of Buckinghamshire*, 38: 90–108).

Cornwall: Charters with boundary descriptions, for example St Buryan, Tywarnhayle, and Lanow (S 450; S 684; S 810), follow natural features and do not contain information relating to arable agriculture (Hooke 1994 and 1999, 95–104). Padel has studied Cornish place-names. Celtic place-name forms predominate in the county, except for the far north-east around Stratton. Thus Old Cornish *tre* 'hamlet, farmstead' as a first element is densely distributed, contrasting with only two found in Devon near Dartmoor, whereas Old English *tun* 'farmstead, settlement' is not common except in the Tamar Valley next to Devon (Padel 1985 and 1999, 88–94; Maps 13.1, 13.3).

Devonshire: Devon charters contain little information relating to arable agriculture since most boundaries follow natural features (Hooke 1994, 95–104).

Dorset: A headland at Handley is mentioned in 956 (S 630; Grundy 1936, *Proceedings of the Dorset Natural History and Archaeological Society*, 58: 103) and the Hinton St Mary charter of 944 refers to an acre headland at the north of the parish (ibid. 119; S 502).

Gloucestershire: Charters of 779, 816, and 840 describe the boundaries of Donnington, Hawling, and Willersey (S 115; S 179; S 203), partly following arable headlands and boundaries making right-angled turns still visible in the parish bounds. This indicates that arable had reached the edge of the township. Several small places have boundary descriptions, such as Harford in Naunton, Pegglesworth in Dowdeswell, Upton, Aston Magna, and Ditchford in Blockley, proving their early existence as townships (Finberg 1964). Dumbleton has been mentioned (p. 168). The Anglo-Saxon landscape of the county has been studied by Hooke (1985a).

Hampshire: Arable land is referred to in some Hampshire Anglo-Saxon boundary charters, for example Stoneham near Southampton (S 1012; Currie 1994, *Proceedings of the Hampshire Field Club*, 50: 103–21, at 123).

Herefordshire: An Acton Beauchamp (now Worcestershire) Anglo-Saxon boundary charter of 972 has only three of the twenty-one reference points that refer to agricultural features, suggesting a landscape not much opened up (S 786; Pratt 1997, *Transactions of the Woolhope Naturalists' Field Club*, 49: 33–46).

Nottinghamshire: Townships belonging to Southwell had intermixed holdings in 959 (S 659; Farrer 1914, 9).

Oxfordshire: Ardley had a 'land boundary' on its eastern boundary in 995 and a head-land boundary is mentioned at Benson (S 887; Grundy 1933, Oxfordshire Record Society, 15: 1, 7). In 904 the boundary of Water Eaton (S 361) ran 'diagonally over a furlong' and in 1002 the boundary between the two Haseley townships ran along the headland of the acres (S 902; Blair 1994, 128).

Shropshire: Tenth-century charter bounds of Church Aston, Plaish, and Aston near Wellington were studied by Finberg (S 723; S 802; 1957–60, *Transactions of the Shropshire Archaeological and Historical Society*, 56: 28–33). None of them refers to any agricultural marker.

Staffordshire: Pre-Conquest charters were published in 1916 (Bridgeman, *Collections for a History of Staffordshire*, 1916: 67–137), Sawyer (1979), and further studied by Hooke (1983). Boundary clauses in the twenty-three charters refer to much woodland but very little arable. Arable probably lay near to the vills at Hatherton and Rodbaston. The 'feld' of Wolverhampton was cleared land.

Warwickshire: Late Anglo-Saxon charter boundaries of places in the Avon Valley indicate that arable land lay at the extremities of townships (with reference to furlongs, headlands, and furrows). No boundary settlements are referred to, implying that settlements were located in the central parts (Hooke 1985b, 126, 134). Likely open fields reaching estate boundaries are recorded at Bishopton in Stratford, Longdon, and Blackwell in Tredington (formerly Worcestershire detached).

Wiltshire: A charter of 921 refers to a headland and three strips of ploughland at Aughton in Collingbourne Kingston (S 921; Bonney 1969, *Wiltshire Archaeological and Natural History Magazine*, 64: 60–3). At Stockton an Anglo-Saxon boundary charter of 901 refers to a mere furrow and landshare linch (S 362; Jones 1870, ibid. 12: 216). Other examples of arable indicators in the Wylye Valley are given by Hooke (1988, 129).

Worcestershire: Morton in Bredon has been discussed (p. 172). The Anglo-Saxon landscape has been studied by Hooke (1985a). Charters show that some of the areas being defined had boundary settlements, indicative of the dispersion that was still characteristic of the region in the Middle Ages (mapped by Jones and Hooke, in Christie and Stamper 2011, 36–7). There was much wood in the county according to the charters (Hooke 2011b, in Higham and Ryan 2011, 153–4).

The above evidence shows that by the late Saxon period, in the tenth century, there were many places with dispersed arable land held in common. The arable lands of several adjacent townships in the Central Region were touching. Hardwell demonstrates that a fully developed furlong pattern had been established by the time of its 903 charter; therefore the field systems had been formed during the ninth century or earlier. The single early historical evidence is one of Ine's Laws, which are dated from internal evidence to 688–94, and the best surviving text was copied about 925 (Attenborough 1922). The main item relating to intermixed fields is clause 42, much quoted by various authors, which states:

If commoners have a common meadow or other partible land ['gedalland', or 'gafol-land'] to fence, and some have fenced their portion and some have not, [and cattle get in] and eat up their common crops or their grass, then those who are responsible for the opening shall go and pay compensation for the damage which has been done to the others, who have enclosed their portion. They [the latter] shall demand from [the owners of] the cattle such amends as are fitting.

Note. If, however, any beast breaks hedges and wanders at large within, since its owner will not or cannot keep it under control, he who finds it on his cornland shall take it and kill it. The owner [of the beast] shall take its hide and flesh and suffer the loss of the remainder.

This, taken with the reference to cornland in the note, seems to be referring to intermixed holdings of meadow and arable held in common. The 'fence' is most easily understood if it is interpreted as a ring fence, the 'ringyard,' that surrounded small open-field systems in Lancashire and Cumberland in later centuries (see p. 84). This is the earliest information that English documents contain, and we are left to presume that the Anglo-Saxons brought with them, or developed, a semi-communal form of landholding with intermixed strips.

Earlier Continental information is given by Tacitus in his account of the agricultural system used by German tribes in the late first century AD. The problems of interpreting Tacitus were discussed by Cunningham (1910a, 36–44). The original Latin text states (Peterson and Warmington 1970, *Germania* XXVI, 168–71):

agri pro numero cultorum ab universis in vices occupantur, quos mox inter se secundum dignationem partiuntur; facilitatem partiendi camporum spatia praestant. arva per annos mutant, et superest ager.... sola terrae seges imperatur.

This may be translated:

The lands are occupied by the whole community in turn, in proportion to the number of cultivators; these they afterwards divide among themselves according to rank; the partition is easily carried out because of the extensive plains. They change the sown fields annually and there is land remaining…only corn is required from the land.

On the one hand this might mean that the Germans were semi-nomadic, moving about a wide, relatively unoccupied, landscape. But moving about does not go well with agriculture. Tacitus says that the sown fields were changed, not that new lands were taken in as they moved on. Since arable land was taken out of plains, not from woodland or waste, it follows that the plain (*campus*, 'field' in the later English place-name sense) must have been grazed, otherwise it would revert to woodland. I think Tacitus is describing a system of convertible husbandry, with a central core of arable land surrounded by pasture that was grazed regularly. It is explicit that the land was divided into strips, and since they were allotted by rank there is the implication that there was something of a regular order of strips, since the ranking of a given community would not often vary from one year to another. The final comment quoted above, that only corn was grown, is very reminiscent of the large areas of strips sown for corn in the Middle Ages. Tacitus was contrasting this monoculture with the orchards, enclosed meadows, and watered gardens used by the Romans.

Also in the first century AD, Pliny commented on the corn crops grown by the German races, presumably to the exclusion of much else (Pliny, *Natural History*, ed. and trans. Rackham 1961, 283):

Barley also degenerates into oats, in such a way that the oat itself counts as a kind of corn, inasmuch as the races of Germany grow crops of it and live entirely on oatmeal porridge.

This concludes the assembling of many differing types of evidence relating to various aspects of open fields. The information will be used to discuss possible explanations for the origin and development of the diverse field systems found in all parts of the country.

6

Open-field beginnings

This Chapter begins with a review of previous accounts of open-field origins. The wide variety of evidence collected in earlier chapters is then used to discuss, firstly, the fields of the Eastern and Western regions, and then those of the Central Region. The chapter concludes with a summary and suggestions as to where future research is needed.

In regions with dispersed settlement there was Saxon-period occupation of Roman sites as well as formation of new ones. Some of them may survive intact as farms called 'hydes'. The fields associated with these settlements had limited dispersion, becoming complicated where there was partible inheritance. Communal cropping arrangements were unnecessary. Fields expanded by developing further small settlements around greens and new severalty farms in waste. Enclosure was relatively simple and open fields began to disappear at an early date.

For the Central Region, the complicated disposition of demesne and villein land provides a possible mechanism for explaining the change from small fields to extensive fields. A model suggested envisages a large-scale ploughing of commons at various dates from the ninth century until the Conquest or shortly after. A reassessment is made of the extent of arable recorded in Domesday.

The origin of field systems consisting of holdings intermixed in dispersed strips has been the subject of much debate for a long time. Before beginning any analysis of how such systems might have arisen, previous theories considering how dispersed fields might have developed are briefly reviewed. A wide variety of proposed mechanisms for the formation of fully developed open-field systems has occupied many printed pages since the late nineteenth century. Most discussions are limited to the extensive Midland type and do not mention the fields found in other regions. It is convenient to review these early interpretations before returning to field systems of the Eastern and Western Regions, and consider what might have happened in the eighth to the eleventh centuries, taking into account the recent archaeological evidence and the diverse regional structures of field systems as revealed in the earlier chapters.

A. PREVIOUS THEORIES ABOUT THE FORMATION OF INTERMIXED FIELDS

Only a brief outline of early theories will be given, since they are well known and have been summarized by Robert Dodgshon (1980, 1–28), who has provided an analytical comment and review, and also by Williamson (2003, 1–13). Many of the suggestions made by early authors draw upon the description of German agriculture given by Tacitus.

Seebohm was the first historian to describe dispersed open fields in 1883 (Seebohm. 1883, 1–7). He used the example of Hitchin, Hertfordshire, in 1816, lying on the south-eastern fringe of the Central Region of intensive open-field cultivation. He assumed (ibid. 40–70) that the same system as found at Hitchin occurred throughout the country (see Hertfordshire in the Gazetteer (p. 272)). Although stated to be three-tilth in the nineteenth century, it is unlikely that Hitchin ever exemplified a 'standard' Midland system. His interpretation passed into the literature as a (universal) model, attempts often being made to apply it to counties where it was irrelevant. Gray (1915, 18–23) criticized the choice of model, proposing instead the 1841 open-field tithe map of Chalgrove, Oxfordshire, as an example of a three-field system. The plan shows that there were eight fields in all (two of them small). The copyholders of Magdalen College had lands distributed fairly equally in five fields lying at the north. Gray does not explain the other three fields in the south—whether there was another township, or whether it was a block demesne. So this does not seem a perfect 'type-site' either!

A better model than either Hitchin or Chalgrove can be found. Twywell, Northamptonshire (see Plate 4), is presented as a late unambiguous example of a simple three-field system in one township and one parish with one nuclear settlement, not distorted by partial enclosure that may have involved exchanges of land.

Seebohm (1883, 117–25) thought that co-aration, the sharing of ploughs by tenants, might have been the reason that strip fields were created, a land being assigned to each tenant in return for providing an animal or part of the plough team. He drew upon Welsh laws for evidence and also thought that the system might have had Roman origins. Vinogradoff (1892, 235–8, 244) attributed dispersed holdings to a system of equalizing shares of good and poor quality land that interweaved everyone's rights and made the rights of the community superior to that of the individual. Since taxes were raised on hides and fractions of them, dispersion removed the difficulties of assessing holdings held in severalty, where soil quality, accessibility, etc., would have to be allowed for by a complex calculation. Vinogradoff made a detailed study of Great Tew, Oxfordshire, using it as an example of shareholding made to equalize rights that continued for a long time. Yardlands were used to apportion work-service and rents in the thirteenth century, and were still used for animal stints and other open-field rights in the eighteenth century (Fisher 1928, i: 139–48, 286–96). Maitland accepted that intermixed fields had a share-holding character, but in a study of Aston and Cote, Oxfordshire, concluded that the village or township as a community had no proprietary or legal rights, and that each landowner held his land privately (Fisher 1911, ii: 337–63).

Gray's detailed study (1915) classified the field systems of England into several main types. The two- and three-field arrangement belonged to the Midlands, with other more complex forms elsewhere. Different again were the fields of East Anglia and Kent, where holdings were compact or localized in their distribution in the fields and cropping regimes were variable. He also described the 'Celtic System', where fields were small and had infield and outfield cultivation. The Midland system was seen as having been brought from the Anglo-Saxon homeland. The fields of East Anglia, Kent, and the Celtic areas were subdivided by partible inheritance. Those of Kent and East Anglia he suggested were inherited from the Romans (ibid. 413–16). He saw outfield cultivation as possibly developing into a two-field system.

Bishop (1935–6) considered that intermixed fields may have originated through assarting. Utilizing charter evidence, he studied Yorkshire, much of which had been laid waste in 1069. He assumed that any open-field development thereafter would have been made from new. This is questionable since it is more likely that there would have been strip fields visible and ready for reoccupation, although their previous tenurial arrangements might have been lost. Bishop investigated the evidence of assarting. Grants of land often referred to the shares of assarts and there were assarts next to the open-field in many places, some of which had personal names, likely belonging to those who first opened up the land. Bishop proposed a mechanism of assarts being created and then split up to later become part of the township open fields during the twelfth and thirteenth centuries. He does not, however, give clear evidence that land lying in assarts or furlongs bearing personal names was part of, or became, the regular core of villein holdings. He concluded with the remark that his account was 'in part a hypothetical one of the rural development of the Yorkshire plain after the devastation' (ibid. 29).

Yet his suggestions have been quoted many times since as a proven mechanism for the formation of open fields. It is certainly a good theoretical mechanism, but he did not prove that it operated after 1066. In the twelfth and thirteenth centuries assarted land in the Central Region seems to have been kept outside of the oxgang or virgated land because it did not pay royal taxation, as is specifically stated for Ramsey Abbey estates in the thirteenth century. The assarts at Cranfield, Bedfordshire, and adjacent Crawley, Buckinghamshire, were assessed at 10 hides for crown taxation purposes, and at 11 hides 1½ virgates by Ramsey Abbey (Hart and Lyons 1884, iii: 3); the crown was to receive dues based on the 1086 assessment and not the assarted land, which was 'outside the hide'. In other words, it was *not* added to the regular fields. Dodgshon

(1980, 86–7) gives many other examples of assarts and new land lying outside the assessed villein land. Hilton (1960, l) found that in the Stoneleigh area, Warwickshire, 'assarted wood and waste were often kept in severalty and even enclosed....arable severalties [were] found in considerable numbers from the twelfth century onwards'. In northern Oxfordshire, assarts at Wootton next to Wychwood Forest were not taken into the open fields and were held by people from nearby townships (Crossley 1983, VCH, *Oxfordshire*, xi (1983): 271–2).

Examples apparently supporting Bishop's hypothesis may have other explanations. The twelfth-century charter of Spaldington (Yorkshire (ER), in Bubwith) states that any increase in the area of arable *was* to be added to each oxgang; this has been interpreted as being a retrospective statement (see p. 108). In Northumberland, in 1359, if part of the common of Fenwick and Matfen was to be ploughed, the prior and convent of Hexham were to have their share by lot as they had before in the old arable intakes (Dodds 1926, 366). This could be regarded as 'communal' assarting taking in waste, but much more likely represents temporary convertible husbandry. At Martham, Norfolk, in 1292, small blocks of land that once had formed a 'villein' holding were being split up. Superficially this looks like an example that supports Bishop's theory, but there were family connections between tenants, and the subdivisions were caused by inheritance and partition (see p. 110). This is a different process, being the breakdown of intact holdings, not the creation of strip fields by the sharing of assarts. The same process occurred by partible inheritance of the 'yoke' block holdings in Kent.

The Orwins (1938, 37–40) approached open fields from a practical viewpoint, seeing strips as being the physical result of using a plough with a fixed mouldboard throwing soil to one side. They attributed the development of a two-field system to population pressure on a system of convertible husbandry. A community was assumed to have ploughed pieces of land until fertility was exhausted, when they would plough other pieces, eventually returning to the first ones. As the population increased, more and more land would be needed until half of the cultivatable land was being left as fallow or reverting to grass in a two-field system. The scattered parcels they saw as a direct result of communal ploughing, a strip being allotted to each person who had contributed to the team.

Since about 1950 there has been considerable interest in field systems, with more evidence becoming available from regional studies as well as new theoretical ideas being presented. Thirsk (1964) defined four essential elements as the characteristics of mature fully developed common-field systems: subdivision of arable and meadow, rights of common pasturing over arable and meadow when crops were gathered, a common waste, and a village assembly (in the form of a manorial court) to control the process. It was considered unlikely that such an arrangement could have been created *ab initio*. Partible

Plate 1. Ridge and furrow in 1999 at Braunston, Northamptonshire (SP 53 66), preserved when enclosed in 1776 (Cambridge University Collection of Aerial Photographs, Z-KnHN 177; copyright reserved).

Plate 2. Ridge and furrow illustrating furlong boundary types. *Upper:* a headland at Castle Ashby (SP 867 609); *lower:* a joint at Wollaston, Northamptonshire (SP 903 634) (D. Hall, 1970).

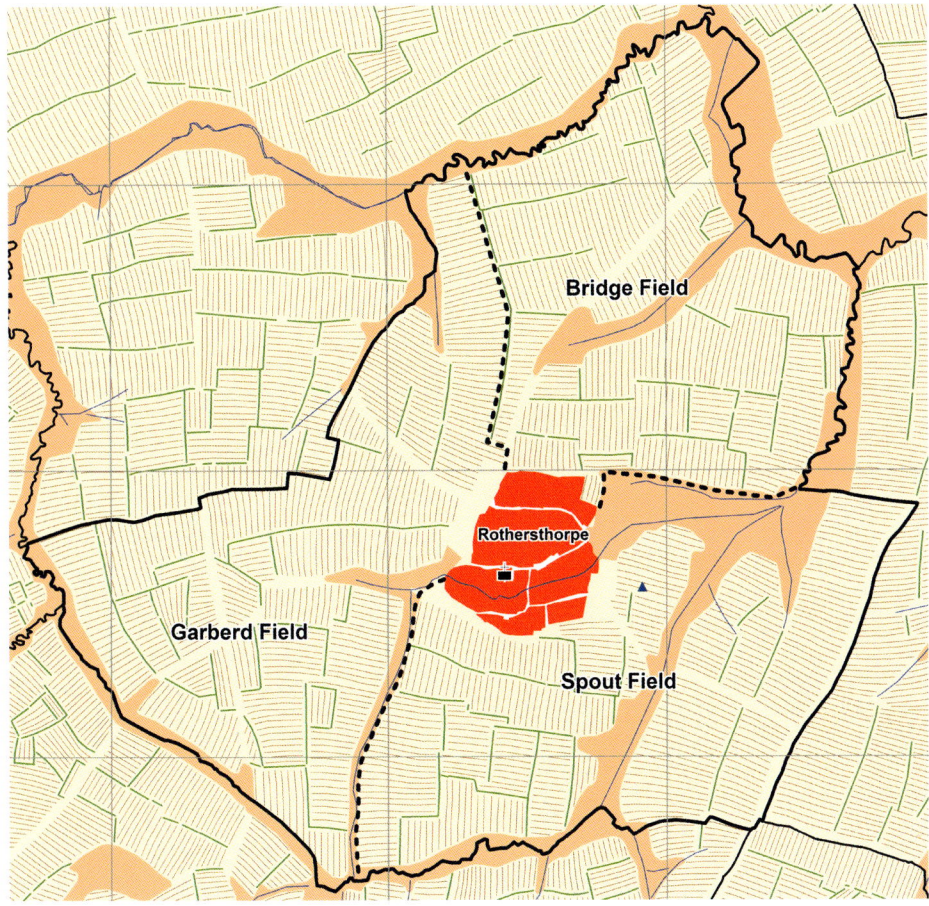

Plate 3. Rothersthorpe open fields, Northamptonshire (SP 71 56), with field names of 1803. A Saxon-period site lies under the strips, east of the village. The grid lines are at 1 kilometre intervals (Northamptonshire Historic Data-base; field survey by D. Hall, 1977).

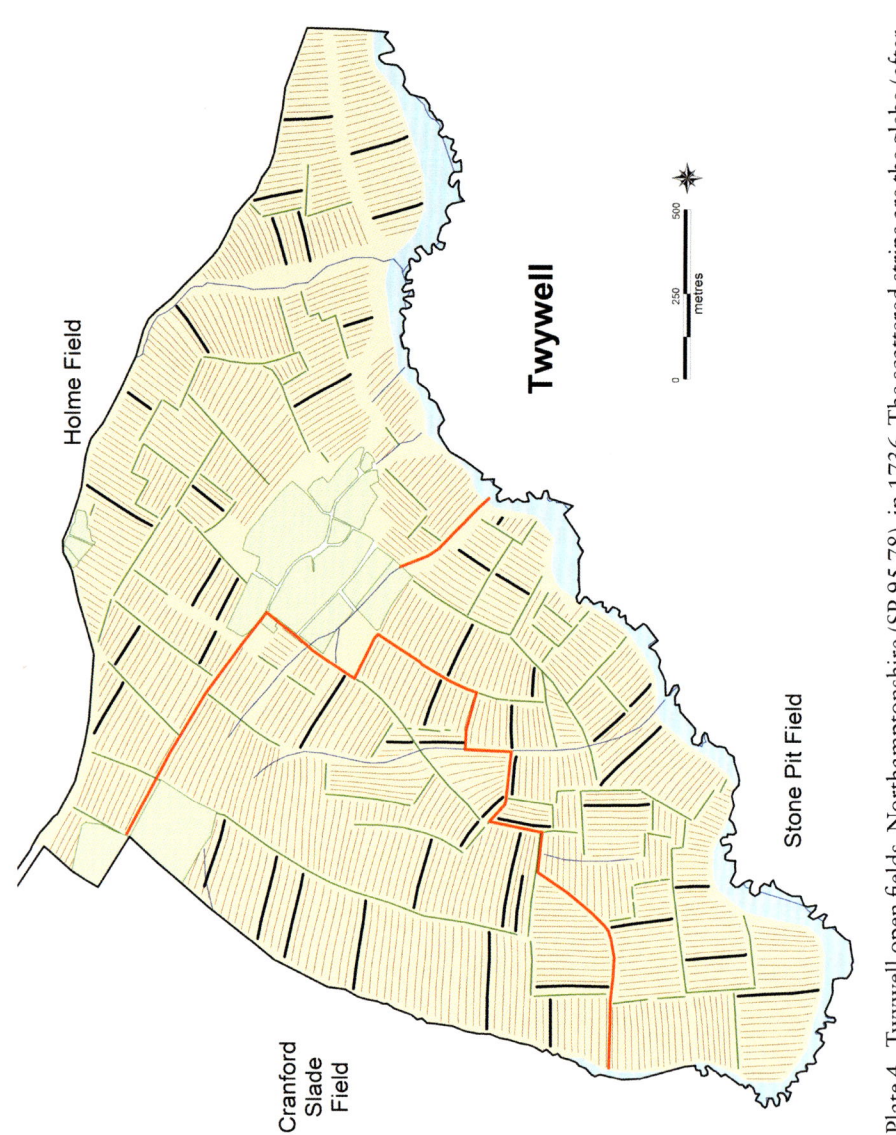

Holme Field

Twywell

Cranford
Slade
Field

Stone Pit Field

0   250   500
metres

Plate 4. Twywell open fields, Northamptonshire (SP 95 78), in 1736. The scattered strips are the glebe (after Northamptonshire CRO, Maps 1409 and 4323).

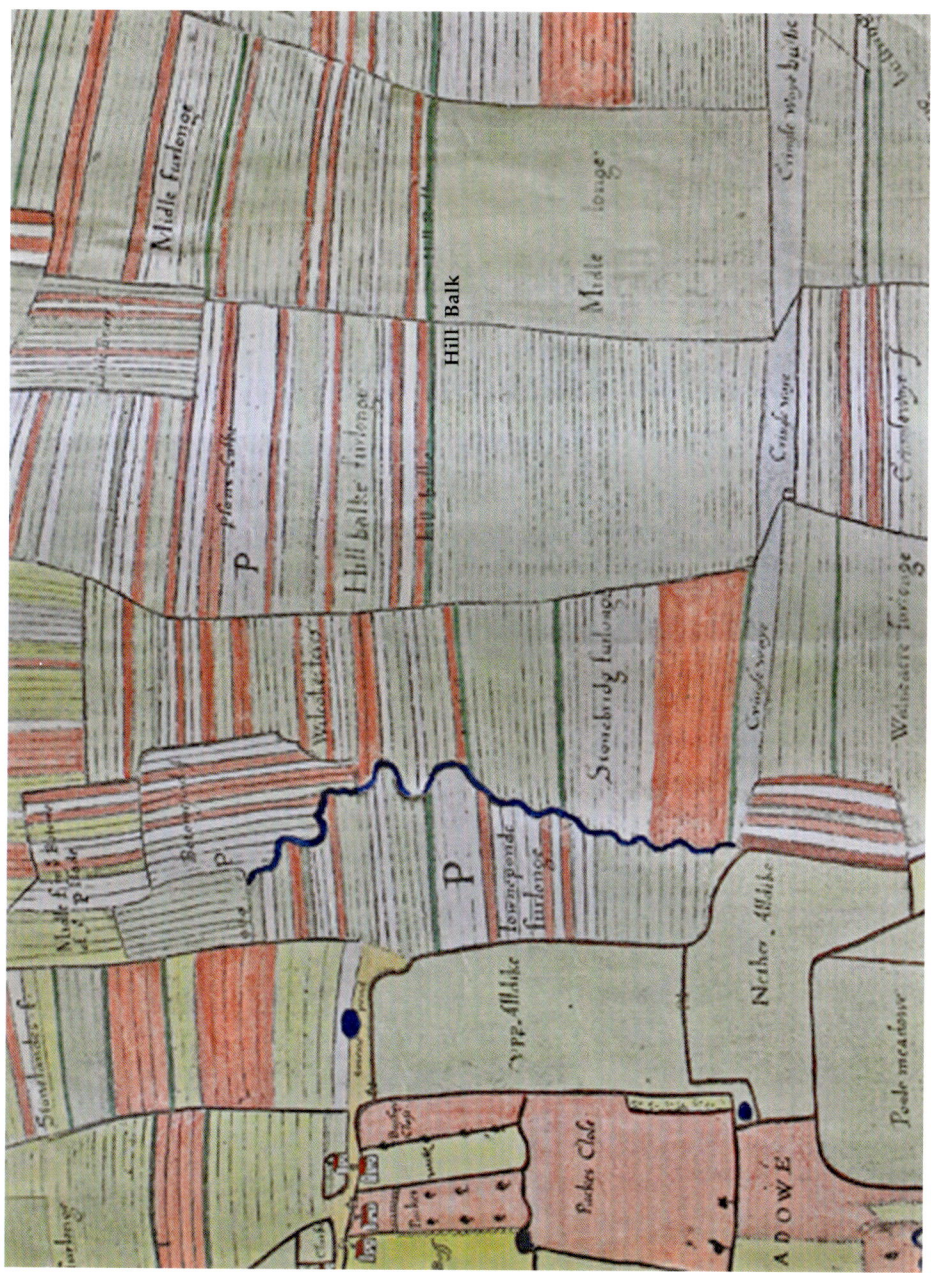

Plate 5. A balk at Strixton, Northamptonshire, in 1595, at SP 899 619. The balk is in the top-centre (after a map by Ralph Treswell, Northamptonshire CRO, Map 2993).

Spratton

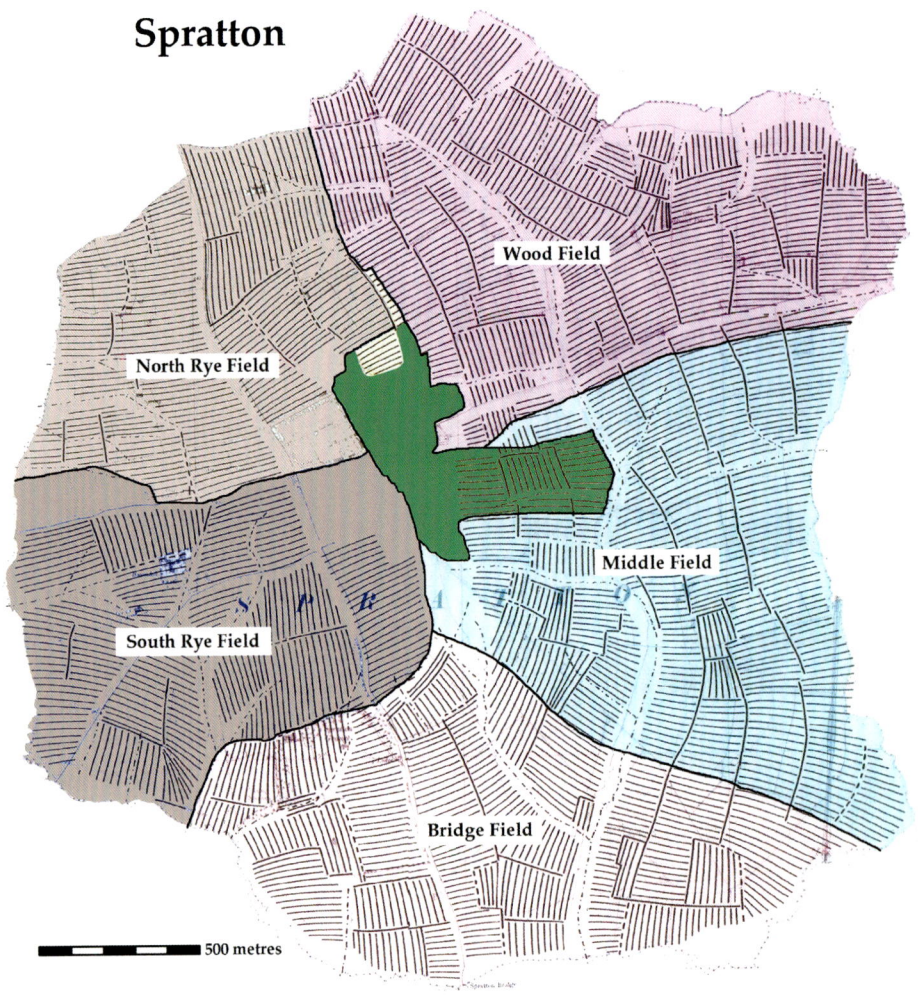

Wood Field

North Rye Field

Middle Field

South Rye Field

Bridge Field

500 metres

Plate 6. Spratton, Northamptonshire (SP 72 70), showing five fields in 1766 (after Heaton 2009, 8–9; field survey by D. Hall 1975).

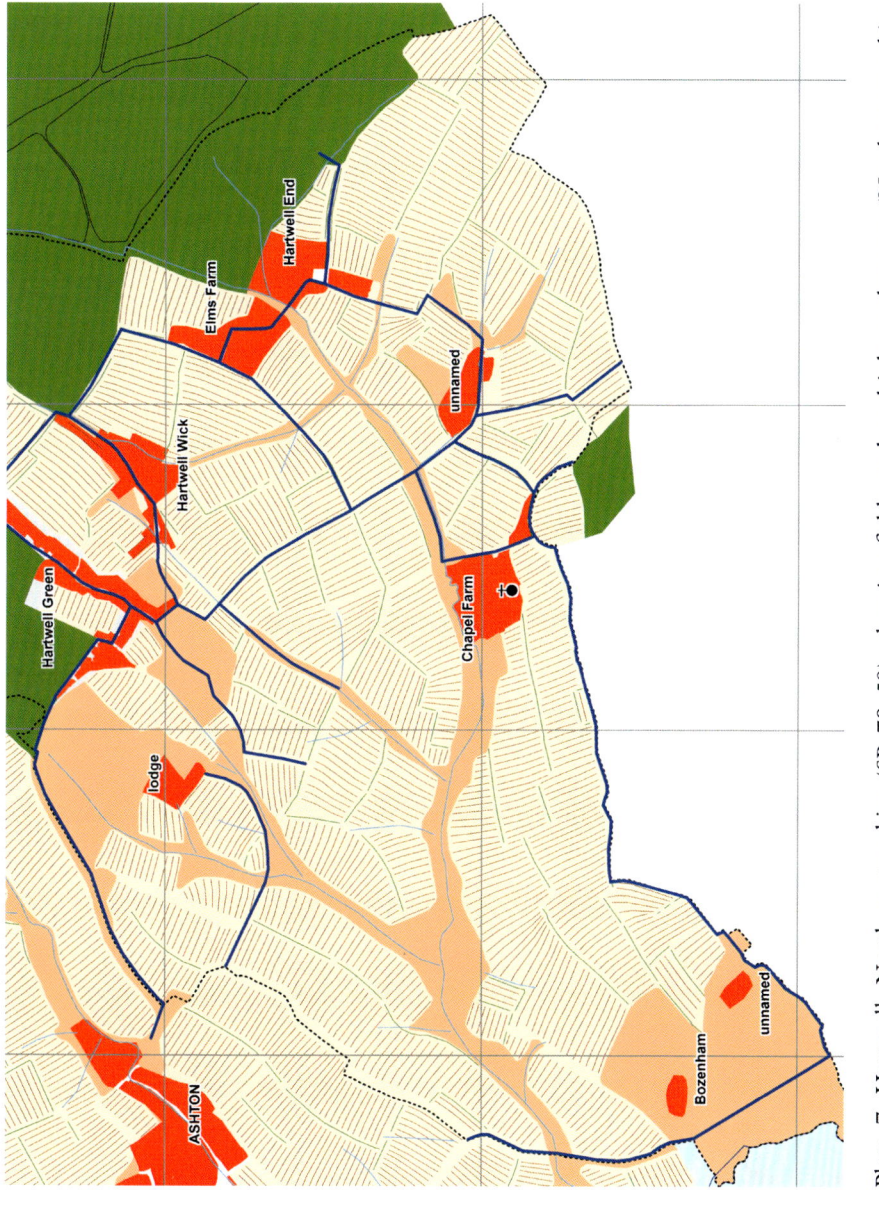

Plate 7. Hartwell, Northamptonshire (SP 78 50), showing fields and multiple settlements (Northamptonshire Historic Data-base).

Gretton

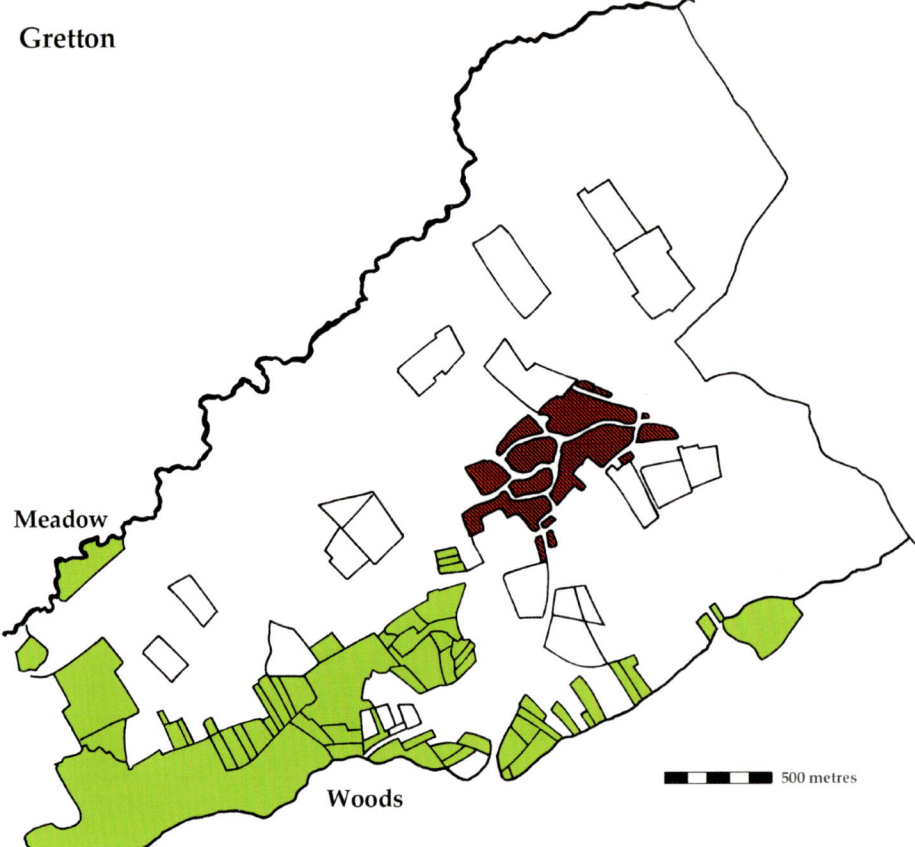

Meadow

Woods

500 metres

Plate 8. Gretton, Northamptonshire, in 1587. The open areas are blocks of dispersed demesne strips; at the south lie demesne pasture closes on a scarp next to woodland (after Treswell survey, Northamptonshire CRO, FH 272).

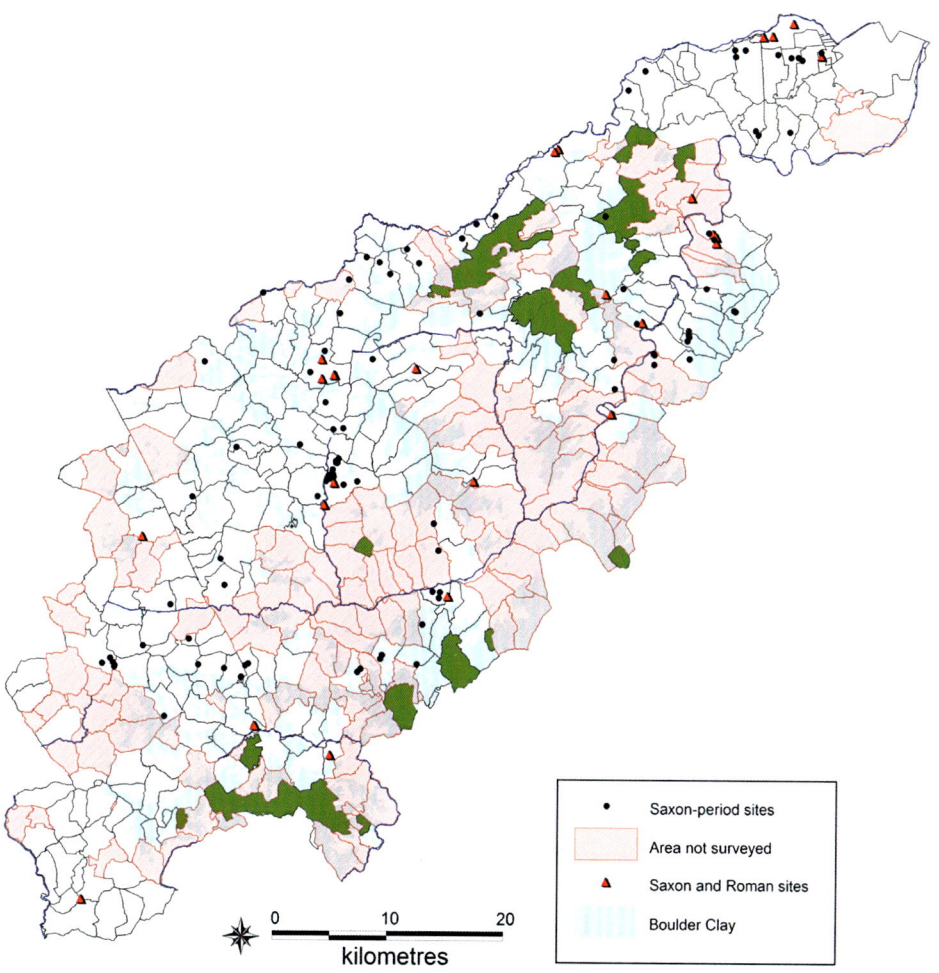

Legend:
- ● Saxon-period sites
- Area not surveyed
- ▲ Saxon and Roman sites
- Boulder Clay

0   10   20
kilometres

Plate 9. Saxon-period sites in Northamptonshire, showing the area searched for pottery scatters (D. Hall and P. Martin, unpublished survey).

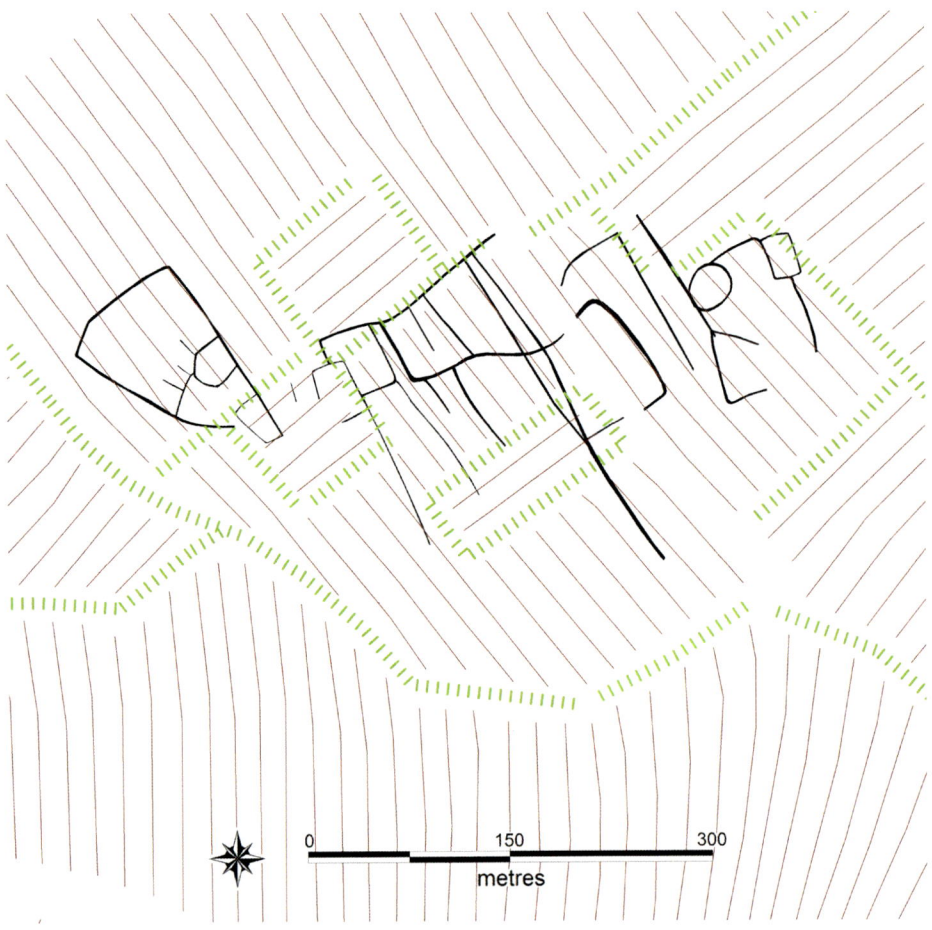

Plate 10. Incidence of cropmarks (solid lines) and furlong boundaries (hachured lines) at Lutton, Northamptonshire (TL 11 87) (Northamptonshire Historic Data-base).

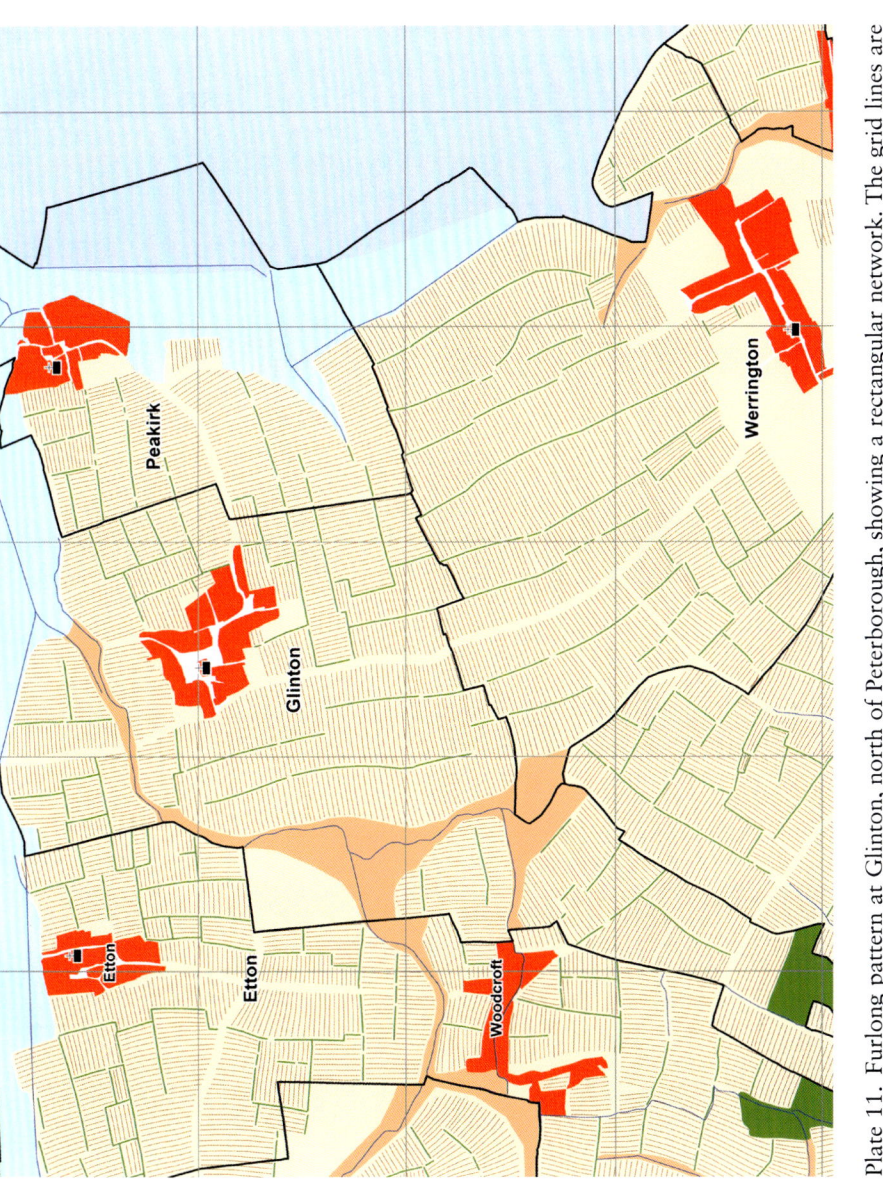

Plate 11. Furlong pattern at Glinton, north of Peterborough, showing a rectangular network. The grid lines are at 1 kilometre intervals (Northamptonshire Historic Data-base).

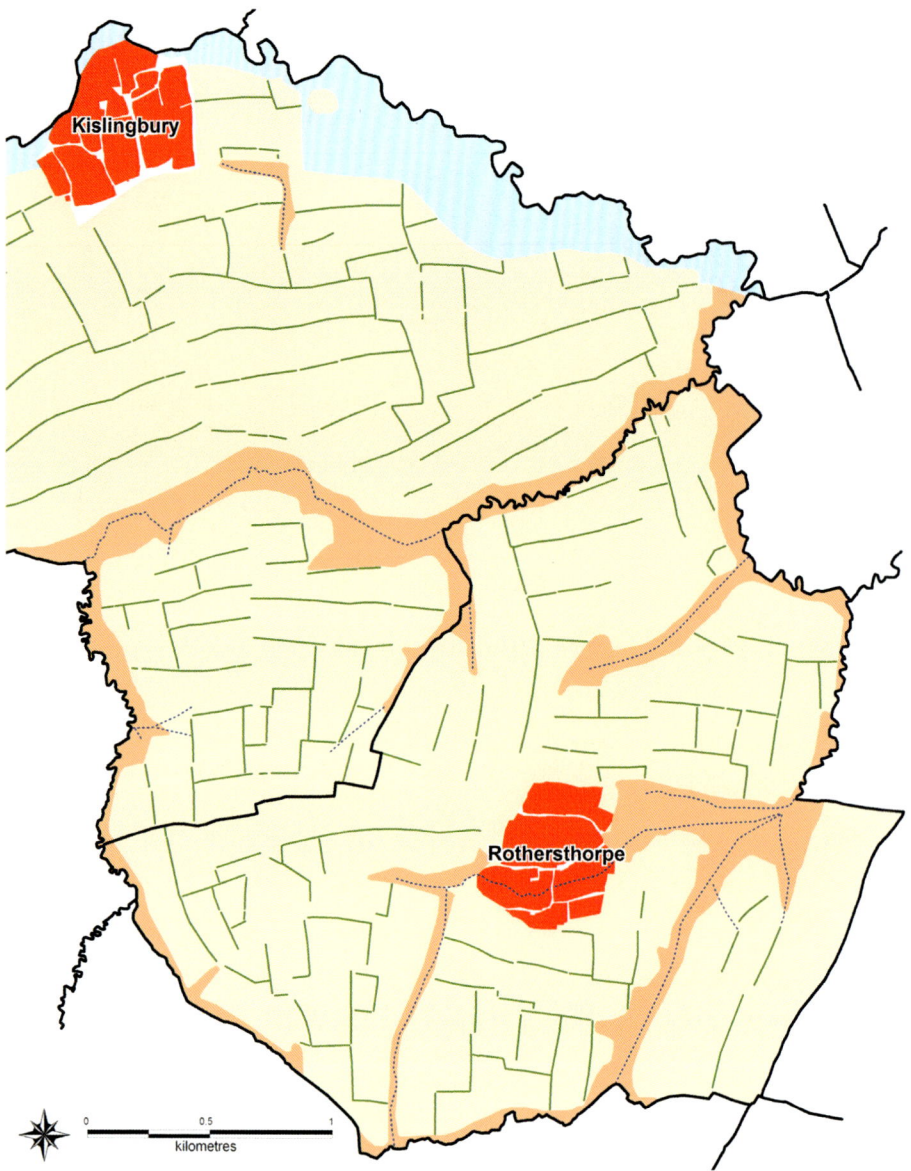

Plate 12. Furlong boundary map of Rothersthorpe and Kislingbury, Northamptonshire.
(Northamptonshire Historic Data-base).

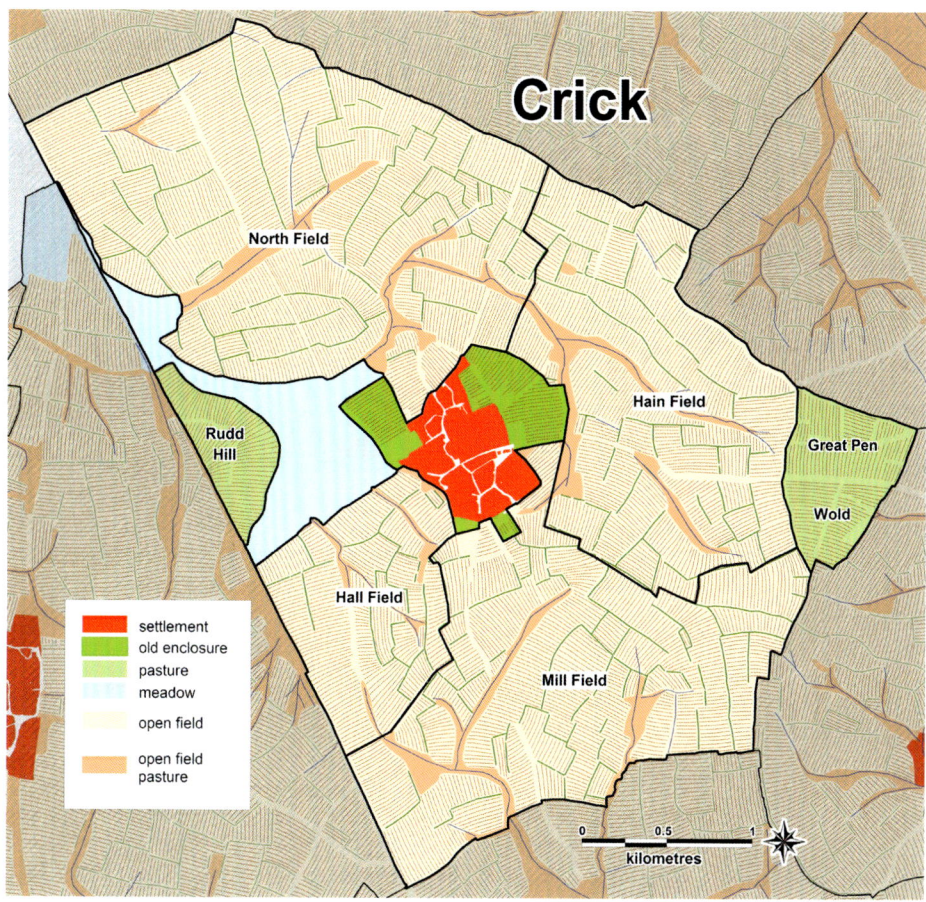

Plate 13. The four fields of Crick, Northamptonshire (SP 59 72), showing the land-use in the seventeenth century (Northamptonshire Historic Data-base).

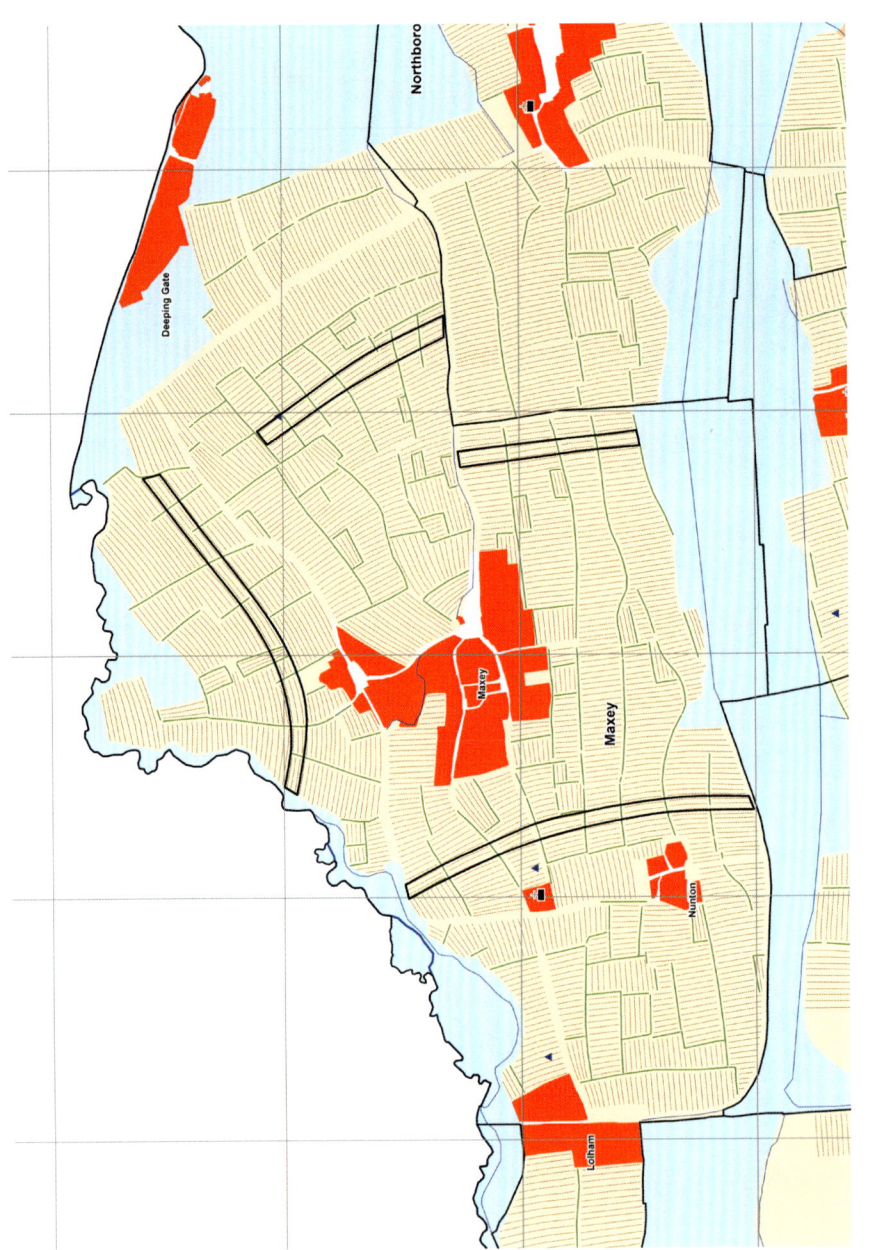

Plate 14. Fields at Maxey, Northamptonshire (TF 13 08), emphasizing three long alignments, two of them curved. The triangles locate Saxon-period sites. The grid lines are at 1 kilometre intervals (Northamptonshire Historic Data-base).

Plate 15. Strips with grass balks at Forrabury near Boscastle, Cornwall, in 2002 (D. Hall).

Plate 16. Upcott Barton, Thorverton, Devon, in 2004. In the foreground are strips cut by a leet (B. J. Coles).

inheritance, the division of land equally between sons and partners, was seen to be a prime factor in causing subdivision of what had once been a sizeable piece of ground or assart. A scheme of division following the ideas suggested by Bishop was proposed as an explanation for intermixed holdings. Complete, mature systems were suggested by Thirsk to be as late as the thirteenth century, and created by rearrangement of earlier irregular parts.

Titow (1965, 92–8) disagreed with Thirsk's theories. He pointed out that in the areas where extensive common fields predominated, there was no evidence of partible inheritance which could be invoked as a mechanism of subdivision amongst coparceners. He did not think that regular ordering of lands was a late development, for then it would have been known from historical sources; he pointed out that there was plenty of evidence for an early equal distribution of lands between cropping units.

Finberg (1972, 491–3) considered charter evidence that might support a Danish origin for intermixed lands. But this is at variance with the earlier dating of shared land apparently described by Ine's Laws (see pp. 172–3).

Baker and Butlin edited a series of important essays in 1973 which showed the variety and complexity of British field systems. There are pockets of subdivided fields, even in the Midland area, and whole regions, like East Anglia and Kent, where holdings were not scattered around the township, but concentrated in a particular part. More detail for areas in the North had become available, and Gray's classification and extent of field types needed revision. The editors confined themselves to a discussion of the various problems of dealing with such an enormous variety of field systems, and did not offer much comment on possible origins. They thought that infield–outfield was likely to be an early form of field system (ibid. 655).

Dodgshon gave an account of the origin of British fields in 1980. He put forward township splitting as a possible mechanism that explained the creation of regulated fields. Maitland had pointed out that in most parts of England there are examples of two or more adjacent settlements with the same name and distinguished from each other by prefixes such as East and West, Great and Little, etc. He noted that in many cases the Domesday Survey of 1086 did not recognize the existence of two settlements, or, if it did, it referred to one such as Addington and then another estate was said to be in 'the other Addington'. He interpreted this to mean that two settlements had recently been formed, or if only one was mentioned, then one of the two later settlements had not yet been founded in 1086. He also suggested that in the south, groups of places such as the Winterbournes in Dorset were once a single territory, perhaps forming a small hundred (Maitland, in Fisher 1911, ii: 84–95). Dodgshon (1980, 108–19) re-examined the evidence for dating some of these settlements, pointing out the dangers of assuming that a place did not exist because it was not mentioned. This was often the case for central manors belonging to one person, where subordinate places can be proved (from Anglo-Saxon charter evidence) to exist earlier. Dodgshon showed that in some cases it seemed likely that settlements were split after 1086, citing Northumberland evidence as a late example of the act of splitting in the sixteenth and seventeenth centuries. Dodgshon further considered what effect splitting might have had on the field systems of a township (1980, 137–50). The formation of a regular layout was likely and it was also an opportunity to create a two- or three-field system that closely related to cropping arrangements, possibly for the first time. Since, however, split townships evidenced by place-names are widespread throughout the country, the splitting alone does not account for the creation of two- and three-field systems without other elements playing a part.

Trevor Rowley edited a volume of papers specifically considering possible origins of open fields in 1981. The contributions to the volume of some authors (Dodgshon and Harvey) have already been discussed (p. 46). Hooke (1981, 39–63) examined evidence from pre-Conquest charters. A plan of the area of the West Midlands employing charter evidence was given, with a discussion of the areas where there was appreciable arable cultivation. Fox (1981, 64–111) reviewed the Midland system and pointed out that an essential feature was the equal distribution of arable holdings between the great fields, and the regular laying down of one field to fallow. He showed that such arrangements were widespread in lowland England by about 1300, with substantial evidence for the twelfth century, and that they may have already been present in the tenth century. He agreed with Thirsk that an organized mature system probably had involved a reordering, necessary when arable had encroached on pasture to such an extent that the only way of obtaining sufficient grazing was to use the fallow. However, the date for such changes needed to be put back to before the Conquest. He noted that the Midland region was surrounded by more complex field systems that were not organized in the same regular way, and that many such townships had large areas of pasture, contrasting with many Midland places that had little or no pasture. These places did not require regulation of grazing on fallow arable. The reorganization phase that produced regular two- and three-field systems had ceased before the middle of the twelfth century, since documents later than that, often quoted as field reorganization, do not describe the fundamental changes needed to create a regular system of the Midland type. Fox assigned the formation of regular systems to the Late Saxon period, perhaps related to the break-up of multiple estates.

Campbell (1981, 112–29) noted that in the thirteenth and fourteenth centuries the Midland system accounted for no more than a third of England. Elsewhere holdings were fragmented and in many there were high levels of wealth and population. He doubted if population growth was the sole cause of the imposition of the Midland system, and suggested that irregular fields could have developed their agriculture through technological innovation, such as the substitution of fodder crops for bare fallow. He further supposed that a period of falling population would make it easier for changes to be made in an irregular system in order to convert it to a regular one. He also noted that strong central lordship was common in the Midland counties and may have had a significant influence.

Roberts, in Rowley (1981, 152–61) discussed the example of Cockfield, Durham, which lies in the highland zone of the county in an area with small fields that seem to have been taken in as cultivated land after 1183. The village plan is a small two-row settlement with an oval enclosure on the south and long tofts on the north side. A series of plans suggests the settlement development, beginning with the oval (twenty-five acres) as the arable nucleus of a very small settlement, which then developed with the addition of tofts. Harvey (1981) described field systems in Holderness, showing that they were very different from the chequerboard Midland types. The consensus opinion from these observations, together with a paper by Gelling describing the place-name data and Hooke's reassessment of evidence from Anglo-Saxon charters, was that the Midland system of subdivided fields seemed to be the result of a planned operation, taking place in some regions around the ninth century.

Dahlman (1981, 31–4, 143–4) considered fields from a theoretical viewpoint using published information. He reviewed previous theories and pointed out the partible inheritance theory of a mechanism of reducing private assarts into scattered parcels, as suggested by Bishop and developed by Thirsk, would, if continued indefinitely, lead to infinite division, whereas parcels were, in fact, stable in size. Dahlman argued that at the time lands became intermixed it was likely to be the best method of achieving what was desired, namely, mixed farming with the production of arable crops alongside animal husbandry. The scattering of lands could not then have been considered inefficient or it would not have occurred.

Nitz (1988a and 1988b) discussed and illustrated the field systems of Central Europe in Carolingian and Ottonian times. These have relevance to England in cases where there were long strip fields comparable to those in Yorkshire. The nature and history of open fields for the whole of Europe over a wide date-range, with a bibliography, have been summarized recently by Renes (2010).

The above outlines previous thinking about the origins of open fields, much of it relating to the extensive or Midland-type systems. Some of the suggestions are conflicting; many are purely theoretical. No attempt has been made here to deal with all the points raised by early authors; some of their arguments cease to be relevant when all the evidence is brought together and collated with archaeological information.

B. OPEN-FIELD BEGINNINGS AND DEVELOPMENT

There can be no historically satisfactory account or proof of the origin of open fields—their creation is lost in time without written records. It is, however, useful to consider various plausible mechanisms and investigate what evidence can be found to support and date them. The fundamental principle of subdivided arable holdings with some communal interest in their disposition seems to be a characteristic of Germanic agriculture in the first century AD according to Tacitus. It is found in Anglo-Saxon England in the seventh century as shown by the laws of Ine. It would seem that this method of agriculture was bought to England during the migration period and that the meagre early English historical evidence cannot explain how it arose, long before the seventh century. It may be speculated that intermixed holdings were the result of cooperative ploughing and the desire to equalize land quality, as postulated by Seebohm. Fully developed extensive fields of East Midland type, as found in the twelfth and thirteenth centuries, occupying most of a township, cannot be associated with the arrival of the Anglo-Saxons in the fifth century. It is implausible because of the number of people required to operate extensive fields was lacking and the evidence of dispersed Early and Middle Saxon-period sites lying underneath some of the fields proves that they are not as old as the fifth century. A small area of arable land is likely to have been cultivated by early settlers, whether they were located on new sites or were utilizing Roman farms and villas. It seems, therefore, best to remove from the discussion at this stage the extensive systems later found in the Central Region and first to consider the fields of the Eastern and Western Regions.

Fields of the Eastern and Western Regions

Field systems of the Eastern and Western Regions (excepting Kent and the Wash Fenlands) have similarities in that they consist of irregular fields, sometimes of small extent in regions where ground was unsuitable for arable. They are associated with dispersed settlement and could have severalty holdings lying among them. Few examples of these field systems have been mapped because they were enclosed and disappeared early. Three late examples lying on relatively marginal ground demonstrate some aspects of how early field systems may have looked and operated: Sheldon, Derbyshire (1617), Preston Patrick, Westmorland (1771), and Creney, Cornwall (1696).

Sheldon in Derbyshire lies on the fringe of the Western Region and the arable was partially enclosed by 1617, but it illustrates the principles adequately (Figure 1.3). Sill Field in the fells of Preston Patrick, Westmorland, was mapped in 1771 (Atkin 1993, 145–7) and has been described under one-field systems (p. 83). The arable, lying in an oval, was surrounded by a fence and divided into dispersed open-field parcels divided amongst four tenants. They had an exact quarter share of an adjacent enclosed pasture. Similar, but enclosed and run as a single farm in 1696, was Creney, in Lanlivery, Cornwall (SX 073 617) (Pounds 1945, 23; Holden et al. 2010, 213). An oval enclosure had a farm in a small square, the former town-place, called a 'park', surrounded by four arable closes and one pasture close. The oval enclosure had one additional exterior enclosure on the west and the whole was encircled by open pasture and moor. Creney may once have been a small settlement with intermixed holdings.

Some of these examples have intermixed ownership of strips of less than an acre, as with most open-field holdings of the twelfth century for which detailed terriers are available. As a small early settlement increased in size, either by population growth or by new arrivals from elsewhere, more land would be taken in. If it were done by cooperative effort, then it may well be that it was shared between the inhabitants, especially if their land was already held in intermixed strips. Partible inheritance could also have caused fragmentation of holdings, but it would not account for the dispersion of strips unless the pieces were already scattered. The arable area surrounding a small settlement could have been increased according to the mechanism of communal assarting and the land subsequently taken into the fields, as suggested by Bishop (1935–6). This process would explain some dispersion of arable lands. The method of cultivation of such an early system could have been as a one-field system with spring sowing, like later Lancashire examples. There would be no cropping regulations with 'great fields', and holdings did not need to be equally dispersed over all the system—as found with one-field arrangements from the thirteenth century onwards. Animal stinting would be necessary if common pasture was threatened by overgrazing. The fields belonging to settlements would not occupy a very large proportion of the township and the settlement would be in the centre or adjacent to the best land, leaving as common pasture poorer ground on hills, fen, or heath. If there was further population pressure, the original core of the single-field could be supplemented by adding intermittently cultivated outfield pieces. Another mechanism would be to take land from the less fertile ground into the arable fields and then run all the land on a convertible system, leaving agreed pieces to grass over for several years.

Examples of small communities apportioning resources can be found in Cornwall. Many hamlets had holdings that were equal shares of the whole, and the number of shares was varied according to the number of tenants over time, the lands being re-allotted periodically. The exactly similar terriers of three holdings at Trebelzue in St Columb Minor attest to this (Fox and Padel 2000, 215–19). Two equal shares described in 1575 had 25½ acres of open-field lying in 45 stitches ('lands', with an average size of 0.57 acre) and 28.625 acres of enclosed pasture. A half-acre curtilage and two small closes (0.75 acre and 20 perches) were held in severalty; 12 acres of the pasture lay in a common heathy down with no division and 15½ acres were dispersed in parcels in two closes. These last were most likely the outfields; since the dispersed pieces were 'known' land, when converted to arable they would be seen to belong to the appropriate farm. Enclosure of the arable stitches would create a system like that mapped for Bosorne in St Just and elsewhere in 1696, reproduced by Pounds (1945, 21, 23; Holden et al. 2010, 81). There was an inner belt of arable closes (possibly continuously cropped) surrounded by a ring of pasture closes ('outfield'). More examples of equal shares of holdings and rents are given in the Cornwall Gazetteer (p. 236). The Cornish settlements are not of Saxon-period origin, but those that can be demonstrated to date from the fifth century may represent what the Anglo-Saxons found in much of lowland England (but nothing can assumed about intermixed strips in Cornwall at that date).

Much of lowland England found by the Anglo-Saxons was not marginal land, but consisted of villas interspersed with Romano-British settlements, often small. The Anglo-Saxons might be expected to continue dwelling at some of these villas and farmsteads, and create new settlements of their own. If this were the case, then the archaeological record should reveal dispersed Saxon settlements lying among Roman sites which have evidence of Saxon occupation. Saxon-period remains identified in parts of Essex reveal such a pattern. Early Saxon-period activity has been found on Roman sites in the Chelmer–Blackwater Valley (Tyler 2011), and finds of the Roman and Saxon periods have been found in the north near Saffron Walden (Williamson 1982, 126, 131) and Stansted (Medlycott and Germany 1995), as previously noted (p. 143). Bryn Morris (2005, 40) found that about eighty-five of the Roman sites occupied in the fourth century (34 per cent) yielded Saxon sherds, as adjudged from the results of online archaeological reports for parts of Essex. Similar finds have been made in Hampshire in the Meon Valley (Hughes, quoted by Hamerow 1991, 15), and the national county list of Saxon-period site discoveries (see p. 142) notes that many lie on Roman sites. Developer-funded excavations in West Norfolk showed that 62 per cent of Romano-British sites have fifth–sixth century occupation on or within 500 metres of them (Rippon et al. 2012, 63–4; 2013, 48–9).

Parts of the field-system plans of the Eastern and Western Regions might be expected to contain relict elements of Early and Middle Saxon arrangements, surviving as small compact farms still visible in the landscape. Estates may have existed in isolation until the land between them was taken in or assigned to new lords and farmers. It has been noted in the search for undispersed medieval holdings that the place-name element 'hyde' is 'old' and hence it may be a useful guide to identify early settlements. Two counties provide detailed evidence. Herefordshire, in the Western Region, has twenty-five 'hide' and 'hyde' names recorded in local place-names (HER, online 'Field-name' search, 2010). Most of them lie on the more wooded land east of Hereford towards Malvern, with a few either side of the Lugg and Wye Valleys. It is possible that some of these (block) hides were Early or Middle Saxon-period holdings; others will perhaps be Late Saxon intakes. They date the small estates to the eleventh century and earlier—such names would not be used after the eleventh century, when intakes were called assarts. By the time the 'hydes' are mentioned in documents they are usually held in severalty, but this may not always have been so for the larger ones. In the Eastern Region, the Essex Place-Name Project

(Essex Record Office online, 2007) has identified forty-five parishes containing the 'hyde' element. They are distributed throughout the county except for the south-east. Case-studies by Martin and Satchell (2008, 126–7, 155, 159) have noted many 'hide' names at Great Henney and Ingatestone, lying in a landscape with complex tenurial arrangements. Some of these are small and may be very ancient severalty holdings.

Ros Faith (1997, 137–40) has discussed early 'hydes'. They often lie on the edge of parishes and are not related to any nucleated settlements. In Somerset, Michael Costen (1992, 93–5) suggests that settlements and farms of various sizes with 'huish' names (from Old English *hiwisc*) were independent Saxon-period farms, pre-dating the open-field system. Many can still be identified as block units, and are 200 acres and more in size. The larger ones lying in the west of the county included tracts of rough grazing. Faith (2006, 2007) has also drawn attention to enclosures called 'worthy' located around Dartmoor, often coupled with an Old English personal name. Roberts and Wrathmell (2002, fig. 7.2) plotted the occurrence of 'worthy' place-names nationally. Not all of them will necessarily refer to one-time enclosures of pre-Conquest origin, but the map emphasizes that they lie primarily in the Western and Eastern Regions, consistent with their relating to an earlier settlement phase than the planned landscape of the Central Region.

That some of these small estates represent undispersed hides is clear from a Worcestershire example at Himbleton. A grant of one hide in 977, to be given to Worcester church, had its bounds described, suggesting it was a block unit. There was also a separate wood, with bounds described, that belonged to it and one-fifth of a partible meadow (the whole estate consisted of five hides: S 1373; Robertson 1956, 116–17, 362–3). Other examples of 'hyde' in Worcestershire (thirteenth–fourteenth century) are at Mathon, Pershore, Rashwood, and Stoke Bliss (Mawer and Stenton 1927), and in Staffordshire at Brewood in Horsebrook and Castle Church (Oakden 1984, 41, 75). Mawer and Stenton (1934) identified eight examples of 'hyde' farms in Surrey, six of them of thirteenth- or fourteenth-century date.

Further evidence of small dispersed 'estates' is provided by those parts of the Western Region that have late Anglo-Saxon charters which refer to boundary settlements, such as Worcestershire (Hooke, in Christie and Stamper 2011, 36–7) and elsewhere. The dispersed medieval settlement pattern of Chittlehampton, Devon, may be the result of similar Saxon colonization. There is no charter or pottery evidence to confirm this, but most settlements were referred to in medieval sources and some were recorded in 1086. Kent yokes are recorded in Anglo-Saxon charters. Small estates with bounds described in the twelfth and early thirteenth centuries have been referred to: Creacombe in Holbeton, Devon, 1185 (Weaver 1909, 158), and in Dorset the boundary of Willislond in Broadwindsor (p. 118; Hobbs 1998, 31–48, 78, 91). These small settlements and their associated lands are likely to be of pre-Conquest date, similar to 'hydes'.

In the Eastern Region, early block holdings are found in some of the fields of Kent and parts of East Anglia which were held in severalty by family groups. These are perhaps to be seen as villein pieces akin to manorial block demesnes, being an alternative to holdings allotted as intermixed strips.

It seems that in the Eastern and Western Regions, Saxon-period settlements and fields began by the continued usage of dispersed Roman settlements and at sites founded in new locations. Some of the new settlements may have been in oval form, as discussed by Oosthuizen (2010, 387–8), like the late examples described above. Most of them remained small, but expanded until their boundaries touched, or alternatively, new intakes and farms were made in pasture and woodland lying between them. Some early settlements were later assessed as a single hide for taxation and acquired the name 'hyde'. Greens used for pasture also attracted dispersed settlement. The result would be a landscape of small settlements sometimes reducing to single farms over the centuries.

The fields associated with dispersed settlements remained largely in severalty or had limited dispersion, becoming complicated where there was partible inheritance. Communal cropping arrangements were unnecessary. Enclosure was relatively simple with this type of ownership and open fields disappeared at an early date, making them more difficult to study than late surviving Central Region fields. The search for relict field systems dating from before the early Middle Ages, in primarily the Eastern and Western Regions, has made good progress with the 'Fields of Britannia Project' (Rippon et al. 2012 and 2013). The project has drawn extensively upon the considerable quantity of unpublished data arising from developer-funded commercial excavations.

Field systems of the Central Region

Like the Eastern Region, the Central Region has evidence of small dispersed Saxon settlements lying away from present-day villages; currently, Leicestershire and Northamptonshire have the most extensive record. In Northamptonshire, of the 129 sites identified, yielding over 5,000 sherds, 29 lie on Romano-British sites and there are a further 50 Romano-British sites with Saxon activity (1–4 sherds), including 12 lying on clayland (Hall and Martin, unpublished field survey, not marked on Plate 9). Additionally there are sixty-five find spots with fewer than five sherds that may prove to be settlements if further studied. These results demonstrate that there was widespread initial settlement, with near one-third of the findspots sites (seventy-nine, 32 per cent) being directly associated with Roman sites. If the data are analysed by township numbers, then there are seventy-eight with Saxon-period sites and fifty-four with other Saxon activity (a few sherds out in the fields and on Roman sites). Counted up in terms of townships, 132 (55 per cent) of the 239 townships searched for pottery finds have yielded Saxon-period material, not including cemeteries or finds within present-day villages. Most substantial Saxon sites lie on well-drained soils suitable for agriculture, based on gravel, limestone, and ironstone; only eight lie on clay (excluding twelve ephemeral small sites associated with clayland Roman settlements). This early occupation pattern is similar to that of the other two regions, showing that the Central Region did not differ from them in the first phase of Anglo-Saxon settlement.

The historical evidence has outlined field-system types and their structure and operation in all parts of the country. The Central Region has been defined by the occurrence of regular two and three fields at an early date. It has been shown that convertible husbandry and infield–outfield occur outside the Central Region proper, except for parts of Yorkshire. Demesnes lying in a block rather than being dispersed are more common outside the Central Region than within it. Tenurial cycles are primarily a characteristic of the Central Region. Göransson's 1961 study found that most examples occurred there, and the list (p. 126) notes only a few cases that lie outside of the Central Region, in Durham, Shropshire, Surrey, and Sussex. Open-field orders and farming practices in each county are given in the Gazetteer. The Central Region has many records relating to the regulation of crops and open-field grazing. Ault's 1972 study was made in Central Region counties, where some places had a numbered list of orders dating from the thirteenth and fourteenth centuries that regulated cropping. Such detailed controls are not found outside of the Central Region.

Another type of evidence that needs to be explained is the location of counties where furlongs with strips lying on the same alignment suggest that there might have been a planned arrangement. Examples are noted in the Gazetteer and are found in Bedfordshire, Berkshire, Buckinghamshire, Cambridgeshire, Lincolnshire, Northamptonshire, Nottinghamshire, Oxfordshire, Rutland, Warwickshire, and Yorkshire. All these are evidences from ridge and furrow. Norfolk has strip alignments recorded on open-field maps of North Creake and Cockley Clay. Additionally there are long hedge lines known at Dickleburgh and the Elmhams in Suffolk.

As already emphasized, a striking feature of many East Midland field systems is the high proportion of arable and the shortage of pasture, as shown in the examples of Rothersthorpe and Twywell. Many East Midland townships were of this form, although it is not always immediately obvious from surviving open-field maps how great the extent of arable had been in the Middle Ages because of the masking effects of early modern partial enclosure and the formation of cow pastures. Medieval field patterns of the whole of Northamptonshire are now available for study (Hall 2009, 74–101, and 2013, Maps 1M–86M). The regional differences in land-use are made clear in Plates 3 and 4, and Figures 1.2 and 1.3, arable varying from as low as 9 per cent at Sheldon to 87 per cent at Twywell. How could such an arrangement come about? One of the chief differences between the Central Region and others is that it is mainly lowland and largely fertile. It is nearly all capable of being ploughed. It was therefore possible for an early small settlement to continue creating arable lands out of its commons. But how and why was this effected? It cannot be that, in early days, waste was slowly taken in, piece by piece, until most ploughable land was converted to arable and the farmers were 'suddenly' confronted with a serious and 'unexpected' shortage of meadow and pasture.

A clue to a possible mechanism comes from the disposition of demesnes. It has already been shown that some of them existed in blocks, called the lord's land, commonly the Bury Field or the Hall Field. This was the lord's private farm, usually sited next to the manor-house on good quality land. Most townships with block demesnes had the villein yardlands dispersed throughout the rest of the fields. Crick, Northamptonshire (Plate 13), had a block demesne, and another discrete block of land called the Hain Field (the 'villeins' field'). Additionally there were two other open fields, making four in all, but they were run on a three-year cycle. There were two types of yardland, called Hall Land and Hain Land, in which the lands were restricted to only three of the four fields. The North and Mill Fields were common to both types of yardland, and the third field was either the Hall Field or Hain Field, according to the type. Hall Land was, or had been, the demesne. The following examples, for which there are early modern terriers, illustrate the arrangement (terrier references in Hall 1995, 243–5).

The last column gives the number of parcels valued in the parliamentary enclosure quality-survey of 1777, which shows that the fields were not equal sized: the Hall and Hain fields were similar, but both were smaller than the other two. This is also evident when the fields are plotted (Plate 13). The approximate acreages are 360, Hall Field; 450, Hain Field; 390, meadow and slade pasture; this leaves about 2,000 acres shared between the other two fields. In 1249 a third of the manor belonged to William de Essebi and consisted of 6¼ yardlands of demesne, and 15 yardlands held by customary tenants (*Calendar of Inquisitions Post Mortem*, i: *Henry III*, iii (1904): no. 123). Another view of the demesne is provided by an inquisition made after the death of John de Mounbray in 1327. He had four yardlands for his two-thirds of one-third of the manor (St John's College, Oxford, Mun. VII. 48). The 1249 survey is in

Table 6.1 Types of yardland found at Crick, Northamptonshire.

Date of terrier	1598	1605	1632	1648	1680	1698	*c.*1750	1777
Field	North	North	Dockame	North	North	North	North	110
Names	—	Heyne	Craxhill	Hayne	Craxhill	—	—	68
	Mill	Mill	Mill	Mill	Mill	Mill	Mill	126
	Hall	—	—	—	—	Hall	Hall	57
Yardland type	Hall land	Hain land	Hain land	Hain land	*not stated*	Hall land	*not stated*	—

remarkable agreement with a different third of the manor that had descended to Thomas, Marquis of Dorset, which, in 1526, was assessed at 14 messuages, 1 house, 1 cottage, 18 closes, 6 yardlands of Hall land, 15 yardlands of Hain land, and 22¼ acres of meadow (TNA, E36/179).

From these three descriptions it follows that there were in total during the Middle Ages 18 or 18¾ demesne yardlands and 45 villein yardlands. It is also clear that the Hall lands of later centuries were the demesne, as would be expected, and the villein lands were called Hain lands by 1526 (probably from the Middle English *hean*, 'poor or wretched', referring to the one-time servile status of the villeins). It is postulated that the Hall and Hain Fields represent very early separate field systems for the lord and his villeins, the rest of the land being common waste. So the plan is very similar to some field systems of the Eastern and Western Regions with more than half the township area left as a common. Both Crick fields may have operated as 'all years' land', being manured by animals returning from the large area of common, as well as by using accumulations in byres. Sheepfolding could have been practised as an intensive method of manuring.

So how did Crick fields develop into the array of furlongs that later occupied the whole of the ploughable land? In most places, block demesnes remained as a fixed number of yardlands confined to their own area, but with common rights extending throughout the township; the tenants' yardlands occupied the remainder. Unusually, at Crick the demesne interests expanded into all the additional arable in the same way as the tenants' yardlands did. This points to a lordly decision, agreed communally, that pasture should be converted to arable and taken in by both fields at the same time. Otherwise, had there been independent intakes of one furlong at a time, each field would have expanded its own discrete area and the parish would have become two independent townships and not the complicated interlocking system found. However unlikely such a decision may seem, there is supporting evidence from the charter for Aston Somerville, Gloucestershire, already quoted, where land was described as 'shared out by the local community' in 1002 (S 901; K 1295), presumably meaning that common land had been taken in, divided, and allotted among the owners and tenants.

There are several fundamental implications resulting from such a drastic, and at first sight, unlikely, action. Firstly, the large and greatly increased area of arable could not be manured from the limited quantities of dung available in springtime from over-wintering in cattleyards. Neither could animals be brought to the arable from the common pasture in the manner of sheepfolding, because in some cases there was eventually almost no pasture left, apart from limited areas of meadow. The solution was to develop the system long known: leave part of the arable fields fallow and use them for rough pasture each autumn. Very likely a two-field system was created, with half the arable left as fallow for a year. This partly compensated for the removal of the permanent commons; there was space for flocks to graze from late July to December. Hay would then have to be used until early April, when animals were turned out on to what limited areas of pasture were left alongside minor brooks and to any part of the meadow that was not to be used for hay. Stubble grounds provided quite an amount of fodder. Medieval oxen and horses were not strong enough to plough hard dry ground until it was softened by autumn rains. During this time, grass, weeds, and dropped corn grew rapidly and provided fodder until frosts arrived in January. This vegetation growth would have been a significant resource, as evidenced by tens of thousands of manorial court orders made over the centuries regulating and stinting animal fallow grazing.

A second consequence of ploughing the commons is that every holding would have to be split up. Equal portions would need to be dispersed throughout the system, so that no matter which part of the new great field was fallow, there would elsewhere be productive land. There is no theoretical reason why scattered lands could not be limited to a few pieces each of several acres. However, smaller parcels are found in terriers from the twelfth century onwards, consisting of rood or half-acre lands scattered in a dispersed arrangement throughout any particular field system. Greater dispersion ensured an equal sharing of differing soil qualities. If such a division of the new strip fields were done systematically it would account for the regular tenurial cycles found in some field books. The other option of apportioning by lot would be impracticable for a large field system—maps and notes would be required. Marked pegs could have been used, as was done for some meadow strips that, in later times, were assigned by lottery, but for large arable fields several thousand markers would be needed. An agreement to take in pasture possibly involved a large area—it seems unlikely that there would be multiple agreements, taken over several centuries, whenever intakes of furlongs were required by either field system (as at Crick). If a large area of pasture was converted to arable and divided into strips and allotted, then an element of planning might be expected, potentially visible in the field pattern. Very possibly whole areas were laid out in long strips where the topography was not too broken by brooks and springs. If soils were well drained, then such a large-scale planned arrangement would survive, as can be seen in several areas of Yorkshire (Figures 2.1–2.5). In those parts of the country where the landscape was undulating and had significant amounts of clay-based soils, the planned units would have to be broken up immediately to allow natural drainage, and so a chequerboard furlong pattern would result, dictated by the drainage system. However, in many parts of the Central Region elements of an original planned layout can still be discerned in the pattern, with alignments of strips continuing through many furlongs. Previously published examples are Doddington (Hall 1995, 134; reproduced in Stamper 1999, 259) and Wollaston, Northamptonshire. At Wollaston, 14 adjacent furlongs were aligned in part of its chequerboard pattern, making an original strip length of 1.6 miles (2.58 km; Hall 1983, 119; Hall 2013, map 48M, approximately along northing 64). The evidence comes from a large-scale 1774 open-field map that marks the strip alignment with great accuracy (Northamptonshire CRO, Map 4447), showing that the fourteen 'conventional' furlongs were taken out of a very large 'furlong', similar to those of Yorkshire. It is probable that the name 'furlong' came from the cross-division of long strips, marking them off in convenient furlong units, being a day's ploughing work. The word 'furlong' is not much used in the Eastern or Western Regions, where the cross-cutting of long strips would not often be the mechanism for furlong formation. The Gazetteer refers to maps for many counties where several furlongs are in alignment, suggesting an older underlying planning stage. They are mostly restricted to the Central Region. Many partial alignments have been published long ago, such as the plan of Crimscote, Warwickshire (Roberts 1973, 196).

A furlong plan for the whole of north-east Northamptonshire has mapped evidence showing large-scale planning. Easton on the Hill had seven furlongs with strips forming an arc 1,000 yards long. There is no satisfactory explanation for this pattern other than that the furlongs were cut out of an extensive block of curved strips that had been previously marked out in a single operation (Hall 2009, 31–3; maps pp. 97–100). Another example is Maxey, lying north of Peterborough (Plate 14). Groups of 4–8 furlongs lying in alignment are evident, two of them forming extended curves up to 1,600 yards in length. The tithe map of Barrowden, Rutland (1844), shows six furlongs lying in a curved alignment 1,400 yards long (Northamptonshire CRO, T286). More examples can be seen elsewhere in Northamptonshire (Hall 2013, 35–7); it is the *curved* alignments that are particularly relevant, as they can only have been achieved by an initial setting out. The suggestion that curved alignments were created piecemeal (Williamson et al. 2013, 105) is implausible, not least because such furlong patterns do not often occur outside of the Central Region. At Maxey excavated evidence proved that furrows continued underneath a headland without a break (Pryor 1985, 15).

On the Continent there are many examples of field systems with long strips forming a simple plan, where, for example, surviving open fields south-east of Vienna lie in alignment (see satellite images). Nitz (1988a and 1988b) has studied East German fields in Saxony. One series of townships had linear row settlements lying in the midst of their territory, with arable land divided into two 'furlongs' of long strips that reached either side of the vill to the boundary, nearly a mile in length. The settlements are associated with colonization during the eastern expansion of the Carolingian empire in the eighth and ninth centuries.The vills and fields were in existence before 780, by which time their tithes had been given to the royal abbey of Hersfeld. Hence there is Continental evidence of large-scale planned field systems comparable to those found in parts of England.

Intake of pasture commons with an expansion of arable, as postulated for Crick, explains the high percentage of arable in the Midland system. It also explains the intense interest in common grazing rights—the farmers had sacrificed permanent common pastures for poor quality grazing on fallow in return for more arable. Any encroachment on that common right by enclosure or trespass by animals from outside was a serious threat to the precarious balance of the farming economy, and was vigorously resisted over the centuries, as numerous litigations testify.

Some system akin to a two-field one with a fallow needed to be set up at the time of the postulated great commons ploughing. Animals had to feed somewhere and the small meadows were quite inadequate. The rearrangement may be considered as a change in farming technique, rather than a desire for increased arable production. If the arable core had been intensively manured by a shifting folding system requiring the transfer of sheep from the commons every night in the summer, then after the rearrangement this would cease—there would be insufficient commons. The new system was a less intensive method of managing the arable, at the expense of the pasture. The explanation is agricultural: replacement of an inner core of every-year's land by a larger system of arable land with one year's fallow rest. Such an arrangement could not occur in regions where the pasture and commons lay on hillsides or were relatively infertile. In those areas, a small core of arable remained and could be manured from winter byres and by summer sheep folding.

If the core fields were already assessed at a fixed number of yardlands, then they would increase in size in proportion to the area of pasture ploughed. For places lying on the edge of woodland and other waste where large areas of common had developed, yardlands would become larger than in those townships with good quality soil whose arable boundaries touched or were close to their neighbours. Yardland sizes recorded from the twelfth century conform to this variation, which is explicable if there had been a major single event (or a few events) of size increase, when commons of varying size were ploughed and incorporated into the yardlands.

The names of furlongs recorded in the Middle Ages, such as 'breach', 'stockings', 'stibbings', 'stubbs', etc., that were an integral part of the open fields attest to former clearances. The distribution of Northamptonshire furlong names shows that the forms 'breche' and 'stubb', occur throughout the county. 'Stibbing' and 'dibbing' are restricted to the northern parts in what became Rockingham Forest. All these are Old English terms for woodland or scrub clearance and are found as furlong names incorporated into regular field systems in the Middle Ages. In contrast, the Norman French terms 'assart' and 'sart' are restricted to the three areas of the medieval royal forests of Rockingham, Salcey, and Whittlewood. 'Sarts' were severalty intakes, not part of the open fields (mapped with discussion in Hall 2013, 18–22).

Theories of arable expansion and the formation of subdivided fields (Bishop 1935–6) assumed there had been communal assarting and division of the assart into strips that were assigned to each participant to make a furlong, so increasing the yardland size. The problem with such a mechanism is that recorded assarts made since the twelfth century were held in severalty, although sometimes called (undispersed) yardlands. The distinction between the old arable yardlands, recorded as hides in 1086, and the new assarts was made in the Ramsey Abbey surveys of the thirteenth century, already referred to (p. 177), when a township was recorded as so many hides for the king and a greater number for the abbot. There would therefore seem to be a difference between early clearances of land for conversion into arable before 1066 and those made afterwards. The former became part of the township field system whereas later assarts remained in severalty. The distinction was likely made to avoid increased taxation. A township was taxed on the number of yardlands or oxgangs (grouped as hides or carucates) in the eleventh and twelfth centuries. Having had the number of yardlands or oxgangs fixed, and forever recorded in the Domesday Survey, there would rarely be an increase in yardland number after 1086, and new land was normally added to a township by keeping assarts in severalty and assigning them, or parts of them, to particular tenants.

There are other fields similar to Crick. Aynho, in 1617, consisted of two overlapping field systems. Lying on the east side of the village and next to it were the Cotman fields. East of the Cotman Fields were the Town Fields. There were also the West Fields and the South Fields which were shared by both the Cotman and Town Fields. As with Crick, there were two types of yardland. Those called Cotman Yardlands lay in the Cotman, South, and West Fields, with no land at all in the Upper Fields. The Town Yardlands lay also in the South and West Fields with strips in the Upper Fields but had no land in the Cotman Fields. The two types of yardland are of great antiquity because two half-yardlands in Cotman Field were mentioned in a charter of 1314 and a grant of half a yardland made in *c.*1190 was said to lie between the lands of the 'cottagers' (Hall 2006). This presumably refers to Cotman yardlands; 'cottagers' of the usual labouring type would never have held a half-yardland.

The model of open-field expansion proposed above is the addition of arable lands on a large scale to a core of older arable, and represents a significant change in agricultural procedure. It is a technological change, not a political one. It could have been arranged by any community and did not have to be driven by a lordly management policy. The lord would have to agree, if resident, and if it affected the location of his demesne, but otherwise the village could arrange matters itself.

Since it was first noted that East Midland open fields overlay Middle Saxon-period sites (Hall 1979), the settlement-change implications of the desertion of Middle Saxon sites has been called the 'great replanning' and has been associated with the fields—but the two did not necessarily have to be changed at the

same time. The field expansion suggested from the Crick evidence is 'new planning' rather than 're-planning'. The field patterns for Northamptonshire show
that large areas of aligned furlongs exist over all parts of the county where the
topography permits (Hall 2009 and 2013, 37), consistent with the large-scale
ploughing of commons. The furlong patterns do not readily reveal a central
core of old arable. Even at Crick, with its identified Hain and Hall Fields, the
early fields are not obvious in the field pattern alone. It may be that they had
jagged edges representing earlier piecemeal intakes, in which case they would
be difficult to identify. What would be expected is some survival of ring fences—
similar to the field pattern found at Crayke, Yorkshire (NR), with its central
core of older fields. But none has so far been identified. It may be possible to
distinguish an inner field additional to a block demesne from the tenurial
details of a field book.

The absence of physical evidence of early ring fences may hint at a complete
replanning of fields, as well as expansion. Archaeological evidence of this can
be found at Papworth Everard, Cambridgeshire, where open-field furrows
overlie a Romano-British settlement and a subsequent Late Saxon-period one
dated to the tenth or eleventh century (Patten 2012, 124–6, 129–30). Here
there must have been a fresh beginning adjacent to the village, even if it did not
involve the whole field system. The excavated Late Saxon-period settlement at
Ketton, Rutland is an example of strip fields being laid out around or at the
same time as a settlement. The site is very probably the deserted Soulthorpe,
which offers a chronology additional to that of the pottery, being presumably
a Danish settlement of the tenth or eleventh century.

Dispersed Saxon-period sites underlying some Northamptonshire fields
(Plate 9) were possibly associated with greens, as seen in East Anglia and elsewhere (Figure 1.4, Semley, Wiltshire), yet none has been observed in the field
patterns, and there must indeed have been a 'great planning' of the fields, rather
than an infilling. In contrast, fields of much of the East and West were based
more on pre-existing Roman sites, Saxon clearances, and small intakes that
often remained unchanged into the Middle Ages in terms of general location
and structure. It is clear that the irregular field types of Eastern and West Regions are quite different to those of the Central Region. Any infilling of waste
or common land once lying between early block demesnes or droves would not
achieve a Central Region field pattern (compare Figures 1.1 and 1.4 with Figure
2.5 and Plate 11).

Some planned field systems occur outside of the Central Region. The hedged
fields of the Suffolk and Norfolk claylands are likely to represent open-field
strips and show that there had been substantial areas of planned fields. The
dating of the St Michael, South Elmham field systems would appear to be eleventh or twelfth century according to the archaeological evidence of settlement
remains. They could have been newly laid out from the Saxon-period centre at
Flixton when new settlements were created (Martin and Satchell 2008, 102–3).

The final result of long aligned strips broken into smaller furlong-type pieces is exactly similar to Towton, Yorkshire (Figure 2.5). South-eastern Essex has a planned (enclosed) landscape that is likely to have a medieval origin, as discussed by Rippon (1991, 57; Wilkinson 1988, 126 and fig. 95). Fieldwork at Orsett has shown that furlong boundaries do occur in Essex. It seems likely that these Essex fields represent a planning process similar to that proposed for the Central Region. It is significant that they do not have any of the 'hyde' place-names, believed to relate to Saxon severalty holdings that are so characteristic of the remainder of Essex. Both these and the central East Anglian planned fields need studies of their medieval records of operation and holding distribution in order to be able to explain why, since they seem to be planned, they did not develop like Midland fields.

A common feature found in all field systems with surviving long lines (coaxial) is that they lie on flat or nearly planar landscapes, for example east Northamptonshire, parts of Yorkshire, Saltfleetby St Peter, Lincolnshire, the later intakes of the Siltlands of the Wash Fens, and the hedged landscapes of the Suffolk Elmham type. There are also open-field furlong alignments, for instance on the 1722 plan of Cockley Clay, Norfolk. In general, field patterns, both open and enclosed, show some evidence for large-scale planning of fields in limited areas outside the Central Region. Further examples of furlong alignments occur at Braunton, Devon, Broadstone in Corvedale, Shopshire, and Great Bookham, Surrey. Anywhere with fertile soils and a gentle terrain could have developed planned fields—they are not specific to the Central Region.

The model, then, for the Central Region is for a large-scale ploughing of commons at various dates from about the eighth century until the Conquest or shortly after. The implications of such an operation are that it would tend to create a planned and expanded field system (still discernible where not masked by variable topography and changes necessitated by drainage difficulties); it would need holdings to be uniformly dispersed over the new fields; and it would require an overall communal system of farming to control, amongst other things, the stinting of animals over the fallow, which had further implications for cropping arrangements. All of these elements are observable in the field systems of the Central Region from the twelfth century onwards and therefore the model seems to be a satisfactory interpretation.

It is not suggested that this model is a monocausal explanation for what can be observed in the parameters that define all Midland fields, or that all places in the Central Region undertook such an expansion and rearrangement at the same time, or even at all. Nor need the extent of ploughing commons have reached the maximum possible as a single operation, but could occur in more than one stage. In some townships considerable areas of pasture did survive and remained unploughed, as is evident in the west of Northamptonshire, where there were steep stony hillsides rising to near 700 feet (Hall 2013—see Edgcote, map 72M), and at the clayland Whinning Green, lying alongside the medieval woods west and south-west of Oundle from Stoke Doyle to Wadenhoe (Hall 2009, 86). Medieval drainage slades and steep hillside pasture in the Badby and Byfield areas have been illustrated elsewhere (Williamson et al. 2013, pls. 7, 64); their apparent similarity to Norfolk greens mapped in the eighteenth century is illusory (ibid. pl. 63), since Norfolk had high greens that have no parallel in the East Midlands. Some places in the Central Region never undertook such a rearrangement and continued to have dispersed settlement and complex fields, such as Glapthorn and Hartwell, Northamptonshire (p. 82). Conversely, some townships outside of the region may have adopted the 'Midland system' if their soils and topography allowed it.

It may be doubted if there were enough people to undertake such large-scale ploughing before 1100, or even before 1200. The Clipston yardland evidence suggests extensive arable had developed by 1086, but further supporting data would be required before one could conclude that appreciable areas of the country had extensive arable at this date. Anglo-Saxon charters—and especially that for Hardwell, Berkshire (pp. 223–4; S 369)—indicate widespread arable in the early tenth century. Yet, at Clipston, there are only 41 people recorded for all of the four component estates of 5¼ hides (Williams and Martin 2003, 591, 594, 599, 611) in a parish of 2,844 acres. Elsewhere in Northamptonshire, Domesday records low numbers of population and low hidages for estates and vills. Since there is a widespread correspondence between 1086 hidage values and the later number of yardlands recorded in a vill (Hall 1995, 82–9), there is the implication that considerable areas of many townships were under the plough by that date, and already divided into the villein holdings familiar in the High Middle Ages. Compton Beauchamp, Berkshire, the township bordering Hardwell, was assessed at five hides with ten people in 1086 (Williams and Martin 2003, 141). This is comparable to Clipston, with even fewer recorded people, yet the Hardwell boundary description shows that the landscape was extensively arable in the early tenth century. The evidence for population levels and the nature of the hidage assessment in the eleventh century therefore needs a full and careful reassessment, which is summarized briefly.

The evidence of Domesday relating to the extent of arable land

The history of Domesday studies and the state of current research and interpretation has been summarized and discussed in a masterly volume by David Roffe (2007). There are still many problems with its interpretation. One purpose of the Survey seems to be a geld or tax return—to record all properties expected to pay royal dues. Many of the assessment units are rounded numbers and are therefore artifial, presumably fiscal, presenting problems of how to relate them to actual estate sizes and values.

The hide: The hide was the Anglo-Saxon unit of assessment and taxation used until the twelfth century. It was discussed by Maitland (1897, 416–21) and seems to have been a notional 120 acres, as stated in an Essex will dated 946–51 (S 1483; Whitelock 1930, 8–9), although several Cambridgeshire vills had a 'double hide' of 240 acres in the tenth century (Blake 1962, 104, 106, 108). However, Round (1895, 36–42) showed that, for fiscal purposes in 1086, the 120-acre hide made up of four 30-acre virgates was used in Cambridgeshire (see Ryan 2011 for a recent historiography of the 'hide').

Hides recorded in Domesday are fiscal, as demonstrated by the prevalence of the five-hide unit, first noted by Round (1895, 44–7, 55–69). A statistical analysis of ten sample counties showed that in nine of them a five-hide unit occurred in 40–62 per cent of the estates (Leaver 1988). The antiquity of the five-hide unit is demonstrated in Somerset by the estates of Anglo-Saxon abbeys such as Glastonbury, where grants made in the tenth century consisted of five, ten, and twenty hides (Loud 1989, 9). If manors and vills are grouped together as they occur geographically, then most of Somerset lay in twenty-hide units in 1086 (Bates 1899, and see the Oxfordshire charter evidence below, p. 203).

For some Cambridgeshire hundreds, Round (1895, 50–1) found a hidage reduction of 20 per cent was applied in 1086. It was presumably a fiscal adjustment intended to provide a 'tax reduction'; it can hardly mean that there was a reduction in the area cultivated since it was applied to selected Hundreds. Hart (1974, 26–31) discussed the Cambridgeshire hidage reduction and showed that it was temporary, restored in the twelfth century according to Pipe Roll evidence.

The ploughland: The ploughland is a difficult Domesday statistic, and the history of its interpretation and likely meaning has been discussed by Roffe (2007, 203–9). It was only used in 1086 so there are no other medieval assessments to compare with it directly. Previous interpretations supposed that it was related to land in some way (as the name suggests), or that it measured the potential resources of an estate. Roffe (2007, 205, 207) suggested it was a measure of the total size of an estate irrespective of how much geld it paid and how much land was exempt because it was demesne or for other reasons.

The example of Whilton, Northamptonshire (p. 113), would appear to conform to this interpretation, where the Domesday assessment of hides and ploughlands agrees closely with the number of yardlands of demesne and the former customary land in the eighteenth century. Moore (1964, 129–30) argued the same interpretation for ploughlands (which he called teamlands) and showed for the Sussex manors of Laughton and Stockington that the twenty-four ploughlands of Domesday could be equated with the total estate size of twenty-four hides and one yardland in the High Middle Ages.

When the relationship between hides and ploughlands is considerd, several interesting points emerge. For Northamptonshire, Round (1900b; 1902, VCH, *Northamptonshire*, i: 263–5) noted that the ratio of hides and ploughlands for villages lying in hundreds at the south-west had a constant ratio of 2:5. Plough-lands for individual vills also occurred in multiples of five, exactly like the hides of Cambridgeshire and elsewhere. He suggested that these ploughlands were therefore nothing to do with ploughs, but were an earlier hidage assessment, and the theory was further developed by Hart (1970, 26–31, 38). If correct, it means that the south-west of the county was reduced by 60 per cent; hence places once assessed at ten hides were recorded as four hides and ten plough-lands. Another example is Aynho, where Domesday records 3.2 hides, 8 plough-lands (2:5 ratio), 23 villeins, and 9 bordars (a sum total of 32 people; Williams and Martin 2003, 616). In 1617 there were thirty-two town yardlands (North-amptonshire CRO, C(A)5112), the same as the number of 'tenants' in 1086 and four times eight ploughlands.

Further information on hides and ploughlands can be obtained for the East Anglian estates of Ramsey Abbey. In 1086 Lawshall, Suffolk, was assessed at eight carucates with two ploughs in demesne (Williams and Martin 2003, 1253). Brancaster, Norfolk, had three ploughs in demesne and seven belonging to the men (ibid. 1133). The steward of Ramsey assessed the estates in the thir-teenth century, as Lawshall ten hides (Hart and Lyons 1884, i: 278 and iii: 213) and Brancaster, also ten hides, broken down as two hides in the abbot's de-mesne, five hides held by named freemen, and the remaining three held by vil-leins (ibid. i: 412). From this it is evident that Ramsey thought that hides, carucates, and ploughlands were exactly equivalent, which gives credence to the interpretation of Northamptonshire 'ploughlands' as the total hidage as-sessment. The 1222 survey of St Paul's, London, treats hides and carucates as equivalent (Hale 1858, cxxii). As noted above, the Kent section of Domesday often records sulungs and ploughlands that are in the ratio of 1:4, suggesting that ploughlands and yokes are the same unit.

Counties other than Northamptonshire have simple numeric relationships between the Domesday hide/carucate value and the number of yardland/oxgangs of later centuries. Three of those in the table below also have twelve yardlands to the hide, suggesting the same kind of 'reduction' as found in Northamptonshire.

Domesday population: The estimate of population levels from Domesday Book is also fraught with interpretational difficulties (Roffe 2007, 229–33). Clifford Darby (1977, 56) noted that Domesday was an 'imperfect record of population'. For example at Avebury, Wiltshire, there was an estate of two hides with a church, but no people were recorded.

Instructive information is supplied by a 'Domesday satellite' document for Cambridgeshire, the *Inquisitio Comitatus Cantabrigiensis* (*ICC*), which is a twelfth-century copy of one of the original surveys used to compile the Domesday entries for that county (fully described by Round (1895), 3–27; printed in 1938, VCH, *Cambridgeshire*, i: 400–27). The *ICC* provides more detail than Domesday. In some cases there were enough villeins to occupy the

Table 6.2 Domesday hide/carucate values and number of later yardlands/oxgangs.

Place	County	DB hides or carucates	Yardlands (YL)	Ratio YL/DB	Date of yardlands	Source
Aspley Guise	Bedfordshire	10	39	c.4	1745	Fowler 1936, 29
Studham	Bedfordshire	3	12	4	1204	Fowler 1926, Bedfordshire Historical Records Society, 10: 123
Akeley	Buckinghamshire	3	13	c.4	1794	Tate and Turner 1978, 67
Steeple Claydon	Buckinghamshire	20	80½	c.4	1795	Tate and Turner 1978, 67
Grendon Underwood	Buckinghamshire	3	35	c.12	1769	Tate and Turner 1978, 67
Farmington	Gloucestershire	4	47¼	c.12	1432	See Gloucester in Gazetteer (p. 263)
Lyddington	Rutland	2	27½	c.12	eighteenth century	Northamptonshire CRO, X5208
Snitterfield	Warwickshire	4	48	12	1616	Barratt 1955–71, ii: 56, 198
Ettington	Warwickshire	3	36	12	1616	Barratt 1955–71, ii: 56, 198
Lighthorne	Warkwickshire	5	20	4	1723	Tate and Turner 1978, 256
Thorpe Bassett	Yorkshire (ER)	16 carucates	64 oxgangs	4	1563	M. Harvey 1982, 35–8
Warthill	Yorkshire (NR)	5 carucates	40 oxgangs	8	1756	Tate and Turner 1978, 294

number of virgates implied by the hidage at four to a hide. An example is the Pampisford Ely manor, which had 2 hides, 3½ virgates and 12 villeins (ibid. 410). The size of many villain holdings at Ickleton is spelled out, for example eleven of them held half a hide each. On the other hand, estates at Weston Colville, Fulbourn, and Foxton had no population recorded (ibid. 406–7, 412), although they were not said to be 'waste'.

One must conclude that a population census was not the main concern of the hundredal juries which conducted the survey. Sometimes the population was recorded and at other times not—perhaps it depended whether the jury knew the answer. Each jury had only eight members. If the jurors had not come supplied with written details of occupancy for all of the many estates in each hundred, they could hardly be expected to know or remember them all. Thus, in the brief, highly abstracted returns of the Domesday Book for the East Midlands and elsewhere, population figures cannot be relied upon to represent the true number of village inhabitants.

That the correspondence between the number of Domesday virgates and the population is a ratio of 1:1 has been noted for Huntingdonshire. One villein often held one virgate as shown for many of the twenty-eight entries in which the demesne was exempt or its hidage given. Coppingford had four hides, with half a hide in demesne. The remaining 3½ hides (14 virgates) were held by 14 villeins (Roffe 1989, 11). Maitland (1897, 144) listed the hides and virgate components of the large manor of Staines, Middlesex, along with the number of holders. There were more than enough people to occupy each virgate. Other Middlesex manors give some detail of villein holdings that approach one villein per virgate, such as Harlesden (Williams and Martin 2003, 360).

Roberts and Wrathmell (2002, 69) have provided a national map of ploughlands. There is a high density in the East and South Midlands, similar to East Anglia, so there were certainly enough population to plough large areas as adjudged by this interpretation. Allowing for caveats that Domesday does not record every settlement and that woodland such as the Weald was noted as an asset of major manors located some distance away, the ploughland map reveals useful information regarding the *relative* population levels: the Wash Fens are largely blank; there is a thin crescent along the East Anglian brecklands; the Weald has few ploughs; and there is none on Exmoor and Dartmoor.

The preceding discussion leads one to the conclusion that Domesday Book is not a reliable record of either population or ploughs and that there may well have been far more people and much more arable land in the East Midlands than is suggested by a literal interpretation of the data provided.

Since the antiquity of the Central Region field systems argued here draws upon the Domesday data for Northamptonshire, we need to consider the information further. The county had an extensive hidage reduction as shown by the overall figures for 1086 when compared with earlier county assessments. The assessments for many vills and estates have fractions of hides. Like the remainder of the hidated part of the country, the reduced hides were assessed at four virgates to the hide. At the vill level this works out at very low hidages compared to the vills of other counties. Environmental evidence suggests that a literal interpretation of Domesday is very dubious for Northamptonshire. The Central Nene Valley floor and valley sides had been opened up at an early date long before the Saxon period; 'The pollen, sediments and soils record from Raunds suggests that the Raunds landscape, along with most of the Nene valley, was one of the earliest and most completely transformed of British [pre-historic] cultural landscapes with ritual, agriculture and rural industry all playing a role' (Brown 2006, 31).

The Domesday Survey may be better understood by comparison with an early twelfth-century document, the Northamptonshire Survey (published by Round in English (1902, 357–92) and in the original Latin by Sandra Raban (2001, 241–57)). The Survey used the hides of 1086 as the basis of measurement, but more detail was given for some holdings by the use of 'small virgates' as opposed to 'great virgates.' When compared with Domesday, the 'great virgates' are seen to be the quarter hides of 1086 and the 'small virgates', were one-tenth or one-twelfth of a Domesday hide in many cases, and for Clipston one-sixteenth of a hide. These can be shown to be the familiar yardlands of later centuries, which must therefore have been present, although 'hidden' in the 1086 values. Their presence explains the many fractions found in the reduced hides of Domesday. The small virgates solve many problems of the Northamptonshire Domesday. If there were ten yardlands to the Domesday hide, then vills assessed at four hides contained forty yardlands, which at four yardlands to the hide would, in other counties, have been assessed at ten hides—which is the 'ploughland' value assigned in Domesday. Such examples are not occasional numerical accidents. Four-hide and ten-ploughland vills are found, for example at Ashby St Ledgers, Newnham (*Celverdescote*) and Badby. Significantly, these places all had regular tenurial cycles of forty lands throughout their fields on the ground (Hall 1995, 142–6, 320, 185–6), which lords and farmers bought, sold, cultivated, and paid appropriate taxes for until they were obliterated in the eighteenth century. Therefore the field structures are very 'real' and relate to the Domesday assessment and are a fundamental part of the fields, already present in 1086 and therefore in 1066.

Various pieces of evidence show that, like other counties, four virgates or yardlands to the hide were known and used in Northamptonshire 'on the ground' and not the unusual reduced units of 1086. The earliest is an Anglo-Saxon writ of Edward the Confessor, 1052–65 (S 1110; Hart and Lyons 1884, i: 188–93), recording the holdings of the abbot of Peterborough at Lutton:

Leofrik abbot of Burgh 9 gherdelandes at Lydyngtone [*Lodintoniam* in Latin] of Seynt Petres soknelande of Burghe, Huntyngseshide by name, and Godriches twa gherde the Denske [Godrici Dani], and Brandes gherde, and Leofgares gherde and Alfwynes gherde the blake [Alwini Nigri].

From this one can see that nine yardlands equal one hide plus five yardlands, and therefore one hide contains four yardlands. Later charter evidence demonstrates that there were four yardlands to the hide at Astcote before 1197 (see p. 127) and at Astrop in 1310. It would seem that Clipston and the champion country west of Northampton was substantially open land, if not arable, in 1086, since there is no record of woodland. The 1086 distribution of woods is very closely similar to that recorded in the later Middle Ages, as seen by comparing the woodland still surviving in 1826 on Bryant's county map (reproduced in Hall 2001, 27) and that for the Rockingham Forest region (Hall 2009, 15, 34) with the Domesday county woodland map given by Darby and Terrett (1971, 405, 438). On the national scale, most of the Central Region had fewer woodlands than elsewhere in 1086, and similarly fewer pre-Conquest place-names indicating woodland (Roberts and Wrathmell 2002, 19, 22). The work of Rippon et al. (2012, 58–9) has shown from the environmental evidence for tree and shrub pollen that the Central Region and East Anglia were very open in both the Roman period and the early Middle Ages.

If the Northamptonshire champion lands were not ploughed in 1086, then reference to pastures might be expected in the absence of recorded woodland, but there is none. This absence of pastures might be because Domesday never recorded them, but elsewhere pastures are recorded; East Anglian sheep pastures are specified on the Essex coastal marshes (Darby 1977, 157–8). The chalklands of the south, where there were extensive downland pastures in the Middle Ages, such as Chilfrome, Dorset, has recorded pasture seventeen by seventeen furlongs (Williams and Martin 2003, 217). In Wiltshire (Williams and Martin, 2003, 163–4), Winterbourne Stoke had pasture two leagues square. Devon also had large areas of pasture recorded in 1086. In Yorkshire, Pocklington Soke had wood pasture four by four furlongs and Hemingborough had wood pasture half a league by half a league (ibid. 788). So the large pasture areas of the downs and elsewhere were assessed in 1086 as a resource and not ignored or subsumed as part of open-field rights. In some areas where there were large areas of pasture, marsh, or other 'waste' in the Middle Ages, Domesday takes notice of them in different ways. Places in Sherwood Forest had substantial areas called 'wood pasture' at Mansfield, Edwinstowe, and Arnold by Nottingham. Many other areas now heathy were still wooded in 1086, such as Cannock and Rugeley, Staffordshire. It seems safe, therefore, to assume that where no woodland and no pastures were recorded in the East Midlands a considerable proportion was already arable in 1086, as at Clipston, so agreeing with the presence of extensive fields at an early date.

When was this field structure set up? Before 1066 in many cases according to the Domesday and yardland evidence, and after the Middle Saxon period according to the archaeology, so at some time during the ninth to eleventh centuries. A date before 1066 would seem to be required for much of the Central Region. If it were not before 1086, then it is difficult to know when it did occur.

It is unrealistic to attribute all the changes to the short time from 1086 to 1150 (Williamson et al. 2013, 124–5), because hundreds or thousands of township field systems would have undergone change and expansion, yet were not noted in the historical record, even though documents become relatively numerous. If many field changes had occurred in the twelfth century they likely would have left some record, as at Segenhoe (when given to Dunstable Priory soon after 1189 (VCH, *Bedfordshire*, iii (1912): 321)), but there is no written evidence for widespread changes. A date after 1066 also requires the Normans to have been significant agricultural innovators, but limited to the Central Region, which seems unlikely since the major magnates had estates throughout the country (see Somerset in the Gazetteer for an example (p. 148) where Glastonbury Abbey *did* influence field-system types (Rippon 2008, 61–4)). General dating for both extensive fields and for the villages associated with them as recorded in 1086 comes from the distribution of pagan Anglo-Saxon cemeteries. Few of them are later than the seventh century and many of them lie under strip fields (see Hall 2013, maps 1M–86M, where both cemeteries and settlements are mapped in relation to medieval fields at the 1:25000 scale). Christian cemeteries are not found underneath strip fields, presumably because they were established in the vills, or in some suitable locally central position in those regions with dispersed settlement.

A small irregular one-field system and its surrounding commons could theoretically be converted to create a large regulated field system at any time before the fourteenth century. It might be expected that such a rearrangement of field systems would be found on the periphery of the Midland region, where it could have occurred late enough to be recorded. None has been found. It is likely that the 'fashion' for the creation of extensive field systems had passed before or soon after the Conquest. Perhaps most of the Central Region had undergone the change at an early date and the problems of too much ploughing and the chronic shortage of pasture and hay were already apparent.

The organized structure of Central Region field systems appears to reflect a Late Saxon central administration. It seems that a hidage was imposed on groups of villages to form a Hundred, part of which was passed on to each vill or township. The artificial multiples of five-hide units have a long history. Original Oxfordshire charters dated 681–969, thirteen in all, show that most vills had multiples of five hides (Blair 1994, 78; see p. 197 for Somerset five-hide evidence). The general antiquity of estate hidages is shown by Gloucestershire charters in the Bristol region, where in eighteen out of twenty-seven examples the charter hidage and Domesday agree (Taylor 1893–4). The hidage for taxation purposes needed to be imposed on the new expanded field structure, and they were laid out accordingly. Divided hides laid out as repeating groups of four strips occurred in the fields of Syresham and other Northamptonshire places in the twelfth and thirteenth centuries (see pp. 126–7). At Little Brickhill,

Buckinghamshire, 1½ yardlands lay in *Cokeshide* in 1197 (Travers 1989, 102). An estate at Studham, Bedfordshire, had yardlands lying in named hides in 1204–5 (Fowler 1926, 123).

References in tenth-century Anglo-Saxon charters specifying 'divided hides' previously noted for Denchworth, Berkshire, and Upper Stratford, Gloucestershire (p. 167), suggest that the process of hide division was ongoing in the tenth century. Local place-names containing the element 'hyde' and 'huish' indicate the presence of undivided hides surviving into the High Middle Ages. These are found predominantly outside of the Central Region, consistent with the absence of extensive common fields. That is, there were blocks of land assessed as hides, but never dispersed or reorganized.

Regarding the earlier theories of open-field formation, most of which related to Gray's Midland system, there is little that needs reiterating. No satisfactory antecedent field structure has emerged and is unlikely to do so, in view of the great gap of time between the Roman period and the development of Late Saxon-period fields. There are clear enough examples of fields being influenced by major obstacles such as hill forts (see the ridge and furrow lying within the ramparts of Willingham, Cambridgeshire (Hall 1996, 143), or the remarkable 1757 open-field map of Goodworth Clatford, Hampshire (Hampshire CRO, Map 117M93/1), with strips inside a hill fort). However, major networks of pre-furlong systems have not been forthcoming, in spite of many reports of ditches lying under furlong boundaries. More pre-medieval ditches have been found during the last thirty years, but they have not been shown to be part of systems covering a significant area. Outside of the Central Region it is probable that some of the block holdings have Early Saxon-period or even Roman origins, and the countryside between them became occupied during the seventh to eleventh centuries, possibly associated with relatively small areas of 'planned' strips in some cases. In these regions ditches underlying field boundaries may well prove to be old.

The assarting theory has been dealt with, and is not seen to be a major factor in the Central Region. Another proposal favoured in the 1960s was that the processes of partible inheritance might account for field origins. Baker thought it was a possible mechanism leading to the creation of strip fields in Kent, although he showed that it did not lead to minute subdivision—there being reaggregation and some holdings consisted of 'undivided shares' (Baker 1964). It is likely that Kent yokes were always divided into strips for agricultural convenience, irrespective of ownership (as with the 1422 demesne at Brixworth, Northamptonshire). A carucate at Barfrestone, already referred to (p. 63), was made up of twenty-seven dispersed parcels in 1236 (Churchill et al. 1956, 130–2). Thirteen parcels, although in single ownership, were stated to include small groups of furrows. The individual strips lying between the furrows served no purpose in terms of the 1256 ownership, but they would when let out to tenants.

Campbell studied the fragmentation of holdings at Martham and Coltishall, Norfolk. Partible inheritance was balanced by the opposing process of consolidation by purchase and exchange during the thirteenth and fourteenth centuries. He concluded that although holdings remained small and dispersed, there was no evidence of any attempt to reorganize the fields into a regulated and ordered Midland type. Therefore the fragmentation of holdings into small parcels by partible inheritance was unlikely to be a precursor of a regular communal field system. The scatter of strips was not increased to develop into a communally regulated arrangement as proposed by Thirsk (Campbell 1980). The required mechanism of partible inheritance is strikingly absent in the Central Region.

It is likely that setting up a two-field system was a fundamental part of the field organization. Two fields are characteristic of the Central Region at an early date and are absent elsewhere. Some counties lying at the extremities of the Central Region have relatively extensive fields on their lowlands but do not have any records of two fields. Among them are Durham (with three-field townships recorded in 1325 and from 1550 onwards), Shropshire (three fields from the thirteenth century), and Staffordshire (three fields from the fourteenth century onwards). It is likely that these counties adopted a change from irregular one- or multi-field systems by converting directly to three fields without a two-field stage. It was probably a grouping together of irregular fields into blocks for cropping and did not involve any physical changes to strips on the ground. Early regular two-field systems are therefore to be seen as an identifier for planned fields of the Central Region type.

The preceding pages have dealt with many types of evidence relating to the fields of the Central Region. The following relevant items have been noted:

There was no extensive river-valley alluvium before about the eighth century.

Many Middle Saxon-period sites were deserted and overploughed by the ninth century.

Late Anglo-Saxon charter evidence shows there were dispersed strip holdings before the eleventh century.

The developed furlong pattern of Hardwell, Berkshire, accounts for its stepped township boundary as early as 903.

Furlong patterns showing elements of long 'planned' alignments are found in many parts of the Central Region.

Field systems exhibit regular tenurial cycles.

There are simple proportional relationships between the number of medieval customary holdings and Domesday assessments.

From the evidence discussed, it is seen that Central Region field systems were organized and extensive, and were developed during the Late Saxon period. The geographical distribution of the fields seems to be largely topographical, with climate and soil fertility being important determinants. That is, they developed on landscapes that were relatively planar, had fertile soils, and a climate dry enough to favour cereal production. Such field systems could develop any-where where the landscape and rainfall were suitable. Within the Central Region, where the physical conditions were not suitable for extensive arable, for instance in areas with sand or wooded clayland, dispersed settlement and irregular fields can be found.

In contrast, the Eastern and Western Regions remained largely unchanged in the Late Saxon period in terms of general structure. They had few large villages and there were many dispersed settlements with assarted severalty holdings and greens. Field systems were small and irregular, but not static. Fyfe and Rippon have shown that in the South-West there were agricultural changes in the late first millennium, with evidence consistent with convertible husbandry beginning in the ninth century (Fyfe et al. 2003; Fyfe and Rippon 2004, 38–9; Rippon 2008). Additional farms and assarts were made in the twelfth century, as has been shown. The occurrence of Middle Saxon-period sites lying under the field sys-tems of the Central Region prove that once, in many townships, there was a more dispersed settlement pattern similar to the Eastern and Western Regions, but abandoned in favour of larger-scale operations. The Early Saxon-period site distribution is further discussed in the final paragraphs that follow.

It is evident, from the discussion above, based on a wide range of historical and archaeological evidence from all parts of the country, that there cannot be any single process or mechanism describing the origin of all English field sys-tems. There were at least two types of development: that associated with dis-persed settlement and irregular fields, contrasting with extensive and physically planned systems often found in the Central Region. There can be elements of both found anywhere in the country, where the landscape, climate, and pos-sibly seigneurial owners favoured them.

C. SUMMARY AND FUTURE RESEARCH

English field systems, their characteristics and their possible origins, which have been described and analysed in this chapter and those that precede it, will now be summarized and then followed by concluding remarks that will suggest where further research might proceed.

The first four chapters of this volume examined a range of historical sources which describe the varied types of open-field systems found in England, with an account of their internal structure and operation. The Gazetteer provides more detail for each county.

Chapter 5 discussed diverse types of evidence that have bearing upon the beginnings of settlements and fields during the first millennium. The main find-ings are as follows. Archaeological survey and excavation have identified many Early and Middle Saxon-period sites, showing that there was occupation of some Romano-British sites and foundation of new settlements, some of which developed into medieval villages, while others were deserted. Environmental evidence indicates an increase in the extent of arable agriculture in the Middle Saxon period, as shown by pollen species and by the development of extensive alluviation in major river valleys caused by the introduction of large areas of bare fallow into agricultural regimes. Former headlands and furlong boundaries, still surviving as linear soil banks on the ground, allow the maximum extent of medieval fields to be mapped in most regions.

Although many studies have been made of cropmarks and earthworks left by Romano-British fields, they do not seem to form satisfactory antecedents to large-scale medieval fields. Rarely are datable pre-fifth-century fields very extensive and their plan forms are very different to those of later medieval furlongs. Surviving field patterns (as mapped in the nineteenth century) have been studied to identify elements of Roman or Saxon-period fields. Outside of the Central Region there are some examples, but considerable archaeological input is required to date them.

Intermixed holdings have a long history, being fully described in historical sources of the twelfth century and satisfactorily deducible for the eleventh century. Evidence provided by Anglo-Saxon charters, mainly dating from the tenth century, shows there were intermixed strips and that arable land extended to township boundaries in some cases. Intermixed strips are likewise implied in the seventh-century Laws of Ine, and on the Continent the description of German agriculture given by Tacitus in the first century AD also indicates the presence of intermixed strips.

These pieces of evidence have been used in the discussion in this present chapter to try to elucidate what may have been the sequence of events that led to the complex and varied field systems found in England before the fourteenth century. Interpretations and suggestions bearing upon possible mechanisms have been introduced. After a brief account of previous theories of open-field origins put forward by historians since 1883, the discussion deals with fields of the Eastern and Western Regions taken together, and concludes with those of the Central Region. The analysis has been based on the information provided or referred to in this volume. It is stressed that the data gathered has originated from manuscript material, or from printed sources 'one stage up'—such as cartularies and calendars published by learned societies or articles examining themes which are directly based upon manuscript sources.

Fields of the Eastern and Western Regions

As adjudged by the occurrence of regular two- and three-field systems, the country falls into three parts: a Central Region, where extensive fields associated with nuclear villages are found, and Eastern and Western Regions that are separated by it (Figure 0.1). Apart from the rather different field systems found in Kent, parts of East Anglia, and the Wash Fenlands, the Eastern and Western regions are broadly similar, consisting of irregular fields, often of small extent, in regions with high ground or other 'wastes' unsuitable for arable. They are associated with dispersed settlement and sometimes had block demesnes and severalty holdings lying among them.

Few examples of these field systems have been mapped because they were enclosed and disappeared at an early date (frequently by the fifteenth century), being converted to severalty farms which themselves were further changed over succeeding centuries. Three late case studies lying on relatively marginal ground demonstrate some aspects of how early field systems may have looked and operated: Sheldon, Derbyshire (1617), Preston Patrick, Cumberland (1771), and Creney, Cornwall (1696). In two of these late examples land was held in intermixed ownership and the community therefore had some interest in the agricultural process. The Cornish example, although held in severalty, illustrates a mapped oval enclosure. Other Cornish townships such as Trebelzue in St Columb Minor had intermixed holdings in the sixteenth century, and rentals show that shared ownership occurred in the Middle Ages. Elsewhere in the two regions, terriers describe intermixed strip holdings that were characteristic of the Middle Ages. Anglo-Saxon charters surviving for the Western Region provide some evidence of this in the tenth century. Agricultural arrangements in these field types could have been similar to those described in later one-field systems with spring crops. Such fields could have been expanded by assarting.

Much of lowland England as found by the Anglo-Saxons was not marginal land, having many Romano-British settlements, often small. The Anglo-Saxons would be expected to continue dwelling at some existing villas and farmsteads, and create new settlements of their own. If this were the case, then the archaeological record should reveal dispersed Saxon settlements lying among Roman sites that have evidence of Saxon occupation. Such a pattern of settlement remains is now being identified in Essex, and excavation reports of sites dating from the late Roman period indicate that a quarter of them have yielded a few sherds of Saxon pottery. The dispersed medieval settlement pattern of Chittlehampton, Devon, may be the result of similar Saxon colonization.

The systematic identification of Roman and Saxon sites by fieldwork over a large part of a county is a slow process. Another approach to identifying small Saxon 'townships' is to investigate settlements with 'hyde' and 'huish' names in the Eastern and Western Regions. From their name forms they are likely to pre-date the twelfth century and may well represent early settlements surrounded by their fields, settlements either of Roman origin or arising out of new Anglo-Saxon dwellings. The best county-wide evidence for 'hydes' currently available is from Essex and Herefordshire, but a significant number of examples occur in Surrey and elsewhere. 'Huish' names are found in Somerset where they appear as severalty holdings of early date, similar to 'hydes' elsewhere. To these sites may be added 'worthy' names and block demesnes, but identification of demesnes in most areas other than East Anglia is at an early stage. Further evidence is provided by those parts of the Western Region that have late Anglo-Saxon charters referring to boundary settlements, such as Worcestershire, so emphasizing the existence of dispersed farms or hamlets. Some fields of Kent and parts of East Anglia were held in severalty by family groups. These are perhaps to be seen as undispersed holdings similar to manorial block demesnes, but belonging to freemen or tenants, being an alternative to holdings comprising strips lying in intermixed ownership. Elsewhere, small block estates with bounds described in the twelfth and early thirteenth centuries are possibly of pre-Conquest date, similar to 'hydes'.

The settlement patterns of the Eastern and Western Regions are envisaged as being the continuation of Roman sites and the formation of new ones located between them. Field systems would be small and have intermixed strips with some communal interest in their regulation. Such settlements could expand by 'assarting' or by taking in pasture grounds. New holdings added to them may have had strips near to the dwellings and not be completely dispersed. Assarting as suggested by Bishop would give a pattern of small irregular fields. Greens could be left as commons amongst the expanding arable. In lowland areas where the soils were sufficiently fertile, the whole landscape would be largely filled by such a pattern, leaving locally less fertile parts for woodland or pasture. In hilly regions arable field systems would form a low proportion of a township area.

The Central Region

Like the Eastern Region, the Central Region has evidence of small dispersed Saxon settlements lying away from present-day villages, and there are many Romano-British sites with Saxon activity, even those lying on claylands. Larger Saxon sites nearly all lie on well-drained soils suitable for agriculture and many present-day villages yield Saxon remains underneath them. This early occupation pattern is similar to that of the other two regions in the first phase of Anglo-Saxon settlement. In contrast, from the twelfth century onwards, Central Region field systems as defined by the extent of regular two- and three-field arrangements, had characteristics not found in the outer regions, which require explanation. The extent of arable was a high proportion of a township and holdings were uniformly dispersed throughout, often in a regular tenurial order, with the same neighbours lying either side of a given set of strips. The whole system had limited amounts of pasture, and agriculture was controlled by communal regulations, restricting the number of animals that could be kept and determining the times for rough grazing on stubble after harvest. There were few block demesnes. Physical evidence shows that arable land was laid out in large blocks divided into parallel strips. Eastern Yorkshire provides the best examples, but long strip alignments are found in other counties, occurring where the terrain was substantially planar and blocks of arable land were not much interrupted by natural drainage networks.

The difficulty in the past has been to explain satisfactorily how such systems developed. Theories of a Germanic 'transplant' can be dismissed, as proved by the occurrence of Middle Saxon-period sites underlying the fields in some places. One might also ask why such a postulated transfer of an organized field system from the Continent in the fifth century did not take place in the Eastern Region, which was nearer to it. Another difficulty is that Continental extensive field systems are not as old as the fifth century. The following mechanism is proposed as a possible model for the organic growth of a small field system into an extensive one.

Significant evidence is afforded by the structure of some manorial demesnes. Crick, Northamptonshire, has a very unusual field system involving a block demesne, from which it is possible to deduce how fields of the Central Region may have formed. Terriers of Crick holdings taken during the sixteenth to eighteenth centuries show that there were two types of yardland dispersed in its four fields, and that there were two overlapping three-field systems. This arrangement can be associated with development from the two smaller fields of the four, called the Hall Field and the Hain Field. Statements listing the manorial and villein yardlands distinguish demesne (Hall) yardlands and villein (Hain) yardlands in the thirteenth century. It is proposed that these two small fields, at a much earlier date, represent the separate arable lands of the lord and the villeins set amongst a large area of pasture. It is further suggested that the fields expanded into the township pastures by agreement between the lord and the tenants, each maintaining their own arable core and sharing the new lands in what became two new great fields. The intake of pastures probably occurred in one or a few operations; if it did not, there would have been piecemeal intakes resulting in the development of two independent townships, each with its own separate field system.

There are several consequences resulting from the postulated large-scale intake of pastures. The most obvious was the removal of much permanent grazing, which necessitated the use of rough pasturing on stubble. This would have necessitated the limiting of the number of animals allowed for each holding and the regulating of grazing periods so as not to interfere with sowing needs. Furthermore, this would have required the regulation of the agricultural seasons and of the area over which crops were to be sown. Probably a two-field system was set up as part of the operation, which would satisfy the grazing and sowing requirements at the same time. An early two-field system would need to have holdings dispersed throughout to ensure there was cultivated land available every year for any particular holding. If a large area were ploughed in a single operation and allotted to tenants, it is possible that it might still be visible in a landscape as linear arrangements of simple form. Yorkshire fields have evidence of large-scale planned layout, which are paralleled on the Continent with settlements made in the eighth and ninth centuries.

If a township's core arable area was assessed for taxation as a fixed number of hides or yardlands, then, when it was expanded, there was the opportunity to either increase the number of yardlands of fixed size, or to keep the number the same, each yardland becoming larger. The first action would have invited (unwanted) higher tax payments. The second would account for the variable size that is found in the yardland from township to township, and seems therefore to be what was adopted. The actual yardland size would depend on the amount of extra land available. Yardlands were generally small in regions with good quality soils, where townships would be smaller than those lying on less preferred soils with more 'waste' in early times.

These features are characteristics exhibited by field systems of the Central Region, and are therefore consistent with township pastures being converted to arable on the large scale, as postulated. The enlarged planned field systems are envisaged as a change in agricultural technique: stubble fallow being manured by its use as rough grazing, which was an alternative method to daily sheep-folding from adjacent pastures. It is not necessary to suppose there was need for increased grain production from the newly won arable which acted as a driving force for its creation.

Expansion of a small core of fields into pastures might be expected to leave the primary core visible as an inner ring, as in Crayke, Yorkshire. None has so far been identified. Townships with deserted Middle Saxon-period sites may have been associated with greens similar to East Anglian settlements, but again no former 'potential green' has been observed in furlong patterns. These two pieces of evidence may indicate a large-scale replanning that ignored earlier cores. However, at Crick the two postulated early fields, although located by early modern terriers, are not readily visible in the furlong pattern—there may be other similar ones that would be revealed by analysis of field structure.

It is not supposed that a planned intake of pastures occurred in Central Region townships at one particular time. It probably took place from about the eighth to the eleventh century. Some townships may never have changed and continued with irregular fields, as can be found near Sherwood Forest, Nottinghamshire. Nor is the proposed pasture intake exclusively confined to the Central Region. Any place in the other two regions with good quality soil and a climate suitable for corn could undertake it.

Direct dating evidence from historical sources shows that regular two-field systems can be identified in the twelfth century. The relationship between the number of yardlands recorded from the thirteenth century and the fiscal assessments of Domesday, 1086, found in several counties, suggests that sizeable field systems were already established by that date. Huntingdonshire had many places with a recorded villein population to virgate (yardland) ratio of 1:1, implying that the Domesday hides represent a fully occupied agricultural landscape. However, the recorded population for much of the Central Region at that time is low. The interpretation of Domesday assessments has been discussed in relation to this apparent difficulty, as well as the very low 'reduced' assessments of hides given in the Northamptonshire folios. It is concluded that most Domesday assessments were primarily fiscal and only indirectly related to land areas, but that there were extensive field systems by the eleventh century. In Northamptonshire, excepting three areas that developed into medieval royal forests, no large areas of woodland or of pasture were recorded in 1086, which is consistent with the presence of extensive areas of arable.

With a mechanism for the formation of extensive fields now provided, it is opportune to consider their dating and origin, bearing in mind that the views put forward will mostly relate to fields of the Central Region, which represents less than a third of England. Fields have now been shown from the distribution of Saxon-period settlement sites to have developed *before* the mid ninth century, when wheel-made pottery was introduced, but *after* the seventh century, because abandoned pagan cemeteries and Middle Saxon-period sites were overploughed by strip fields.

The lack of any satisfactory Romano-British antecedents of similar plan form to extensive medieval fields and the gap of several centuries before the latter were laid out preclude any likely direct relationship between Romano-British and Saxon-period fields. The open fields represent a new beginning. Clearance of land and breaking up of pasture to create such fields is indicated by furlong name-forms. This process seems to have ceased by the eleventh century since assarts recorded in the twelfth and thirteenth centuries were held in severalty and did not form additions to open fields. Partible inheritance is not seen as a formative precursor to dispersed and regulated fields because it was not practised in the Central Region. Where partible inheritance did occur, in Kent and East Anglia, regulated fields never developed, as shown by the detailed study of Martham and Coltishall, Norfolk.

The previous discussion, based on a wide range of historical and archaeological evidence from all parts of the country, shows that there cannot be any single mechanism that will explain the origin of all English field systems. There were at least two types of development. That which was associated with dispersed settlement and irregular fields contrasted with extensive and physically planned systems, the latter often found in the Central Region. The distribution of extensive fields indicates that they occurred where there were good quality fertile soils, where the topography made it possible to utilize large areas for arable, and where the climate was suitable for corn production. They were excluded from the wetter and hilly parts of the Western Region and also from the poor quality soils in the Eastern Region, which were better suited to a regime with limited arable and a significant amount of pastoral husbandry. These combined factors seem to explain what is found and it is not necessary to invoke any ethnic explanation, either in terms of the initial Anglo-Saxon settlement, or of any later Danish settlement in parts of the Eastern Region.

The evidence shows that planned fields occur widely in the Central Region, and that many fields overlie Saxon-period settlements and cemeteries. It therefore follows that many townships underwent major changes in both settlement and field structure. The sequence of events—whether both processes occurred together as a single operation, or whether some small settlements became deserted, leaving 'space' which was then used to lay out large-scale field systems at a later stage—still remains the subject of debate and further work. Both processes may seem somewhat improbable—the choice is to decide if there was one major upheaval in landscape management or whether it was followed by a second. The results of such cadastral changes lie in the present-day Central Region countryside for all to see—relatively large villages with deserted Saxon-period sites out in the surrounding fields.

The amount of Saxon-period material being identified is now considerable in Essex, Leicestershire, Norfolk, and Northamptonshire. The Northamptonshire pottery evidence needs careful interpretation, having due regard to the circumstances of its discovery (*contra* Williamson et al. 2013, 57), in particular to take account of the limited areas that have been studied sufficiently intensively to discover sherds in ploughsoil. Of the 239 townships that have had some detailed survey, only one (Brixworth) was visited on more than one occasion. Thus, although Plate 9 presents a substantial bulk of new information that took a long time to collect during the county survey, it is nevertheless no more than a preliminary statement. To reiterate, seventy-eight townships yielded one or more sites (for the details see p. 136), and fifty-four revealed areas with a few sherds (one to four), most of which with further work would likely have produced more material and have been classified as 'sites'. So the total number of townships with Saxon activity is 132 (55 per cent), excluding those with no data other than Anglo-Saxon cemeteries or pottery sherds found within the village settlement confines. There are therefore 107 townships (45 per cent) with no Saxon-period remains identified during the field survey that represent deserted settlements lying out in the fields. This should not be interpreted as permitting the conclusion that there is none to be found in these places; it is very probable that further survey would identify more. If townships with Anglo-Saxon cemeteries and Saxon-period sherds found within the settlement be added (as well as sites found commercially and by other surveys, see HER), the township number with activity increases appreciably. Therefore the desertion of Saxon-period sites is concluded to be a widespread phenomenon.

Future research needs to proceed on various fronts, especially to refine the dating for the beginnings of extensive fields. Firstly, with respect to settlements, the earliest archaeological dating needs to be determined for existing and deserted medieval vills so as to distinguish those with Saxon origins from those later in date. Saxon-period sites left deserted under medieval fields require more study. It cannot be proved from fieldwork collections that in a given township Saxon-period sites were all deserted on a single occasion for the purposes of laying out new strip fields. Dating is required to discover how long before *c*.850 they were abandoned. The reason for desertion remains speculative; the occupants either died out or moved to an existing village, which by default or by design left space in which new fields were created. The block 'hyde' settlements of Essex, Herefordshire, and elsewhere seem to be particularly important features, as are 'worthys', and the 'yokes' of Kent, where some of the farms belonging to them are likely to have Anglo-Saxon origins. Yokes recorded as having pasture as well as arable could have functioned as independent farms or townships like the large Somerset 'huishes'.

Secondly, it is desirable to establish the extent of medieval field systems over large areas of the country by field survey before the physical evidence is destroyed. Some regions such as East Anglia have received much historical attention and landscape discussion, based on hedge patterns, but the ground evidence of furlong boundaries has not been recorded and incorporated into analyses.

Thirdly, many more detailed historical studies are needed on township field systems to ascertain their precise structure. The importance of such studies has been stressed by Fox and Padel (2000, lxix) and by Roberts and Wrathmell (2002, 192). This is particularly true for the Eastern and Western Regions, where open-fields were removed by enclosure at an early date, leading to fewer records than can be found for the Central Region. Since all the relevant documents are in Latin, they have so far received limited study. It is necessary to select places with detailed medieval records to establish the exact locations of dwellings and the extent of their holdings, and relate the results to fieldwork survey and selective excavation. East Anglia would provide suitable places for such a combined historical and archaeological study because of the survival of datable types of pre-tenth-century ceramics. The number of tenant holdings, the nature of the demesne, and the relationship of both to Domesday and other fiscal assessments are essential research tools. Settlements and fields are intimately related and cannot be studied in isolation. There are many sources such as township field books of various dates and the Elizabethan and early seventeenth-century surveys used by Gray. (A specific example of the latter is the detailed survey of the Hereford bishopric estates made *c*.1578.) Outstanding medieval and later records are preserved relating to the estates of early established Oxford and Cambridge Colleges.

Much more local detail of this type is required to build up an accurate statement for the whole country, otherwise regional and national interpretations of landscape development and history risk becoming divorced from credible supporting data.

Gazetteer of field systems

This Gazetteer gives a brief summary of field systems and related information for each county. The counties are the 'historic' ones defined in the 1880s before the Local Government Act of 1894, and are similar to those used by the Tithe Commissioners following the Tithe Commutation Act of 1836, which are mapped by Kain and Oliver (1995, 4).

Information has been collected from published literature and from analyses of original manuscripts. Observations of ridge and furrow and reconstructed furlong plans have been made as a result of fieldwork in many counties. The Gazetteer does not incorporate much of the extensive and useful data provided by the economic sections of recent volumes of the Victoria County History. The reasons are in part because this would be an enormous task and in part because, without further work on the sources given, it is not always certain what type of field system is being referred to from the summary information provided. These volumes and many other documents held by various public and private archive offices can be consulted online to provide information for more detailed local studies.

The Gazetteer entries do not include archaeological evidence for Saxon settlements or the historical evidence given by Saxon charters, which are both discussed in the text. Neither do they reiterate details of items such as the types of fields or demesnes found in a given county if they have already been discussed (but there are linking cross-references). A fuller view of each county field system may be obtained by use of the index to locate discussion in the text that amplifies information in the Gazetteer entry.

The structure of each county entry has the same general format, beginning with a brief description of its topography and a summary statement of its field-system type. Sources are provided to locate published open-field maps and relevant literature, and evidence for regular and irregular field systems is collected together, sometimes in tabular format. Comments are made on demesne types, yardland or oxgang sizes, and regular tenurial structure, if any, followed by work-service, farming, assarts, and any items unique to a particular county. The account concludes with a description of the physical remains of ridge and furrow and strip lynchets.

Citations appear in the bibliographies at the end of the county sections, the main Bibliography at the end of the book, or, in the case of minor references, can be identified in full by using the abbreviations list at the end of each county section.

1. BEDFORDSHIRE

Bedfordshire had woodland located on the northern clay belt and on much of the sandy area south of the Clay Vale towards the chalk hills of the Chilterns. Both regions have dispersed settlement; the remainder of the county had regular two- and three-field arrangements of Midland type.

Fowler (1936) published maps of Aspley Guise, *c*.1745, Overshot, 1764, Oakley, 1795, and Renhold, 1781. A list of maps in *A Guide to the Bedfordshire Record Office* (Bedford: County Council, 1957), 114–35, includes many that show open fields. Copies are on open access at the CRO and some reveal strip alignments in several furlongs, for example Barton le Clay up to three furlongs, Biddenham five, and Eaton Bray in seven furlongs. Tithe maps, 1838–52, show open-field lands at fourteen places (Kain and Oliver 1995). The HER has published several parish booklets and has archive files with copies of maps. Bedfordshire fields have been described by Gray (1915) and Fowler (1936), who published many medieval texts. Roden's 1973 account is part of his Chiltern study.

Field numbers

Gray (1915, 450–2) provided examples of two- and three-field systems. Among the fourteen two-field townships of medieval date, five were based on demesne data; four of the seven three-field examples are demesne extents. The Table below gives eleven two-field and five three-field examples additional to those cited by Gray. The twelfth-century rearrangement of Segenhoe fields has been described.

Irregular fields are known. A terrier of Wilden of *c*.1250 describes the grant of a messuage and 4½ selions in a croft adjacent with land irregularly distributed in four areas called 'fields': *Stanilond*, eighteen selions plus half a headland with two pieces of hedge; *Malestoch*, five selions (one on *Snakehegfurlong*); *Sortland*, three selions; and *Uvere-croft*, containing three headlands and a hedge. All of which implies woodland and assarts, but there is a regular order of neighbours throughout the sixteen parcels. Of the twenty-four named neighbour positions, half were occupied by Widow Beatrice, who was also a neighbour of the messuage (J. Godber 1940, BHRS 22–3: 136–7).

In *c*.1300, an estate at Ravensden consisting of 12 roods and 43½ selions had 8 selions in a field called *Schortecrofte*, an 11-selion block in a field called *Benefeld* and 6 selions next to *Wildenestrate* with a ditch and hedge, and 6 selions in a croft called *Benecrofte* with heads and a hedge and ditch extending to *Benecrofteyerd* (Godber 1963, BHRS 43: 198). Three other lands also abutted *Benecrofteyerd* and had heads, so this holding was not very dispersed and there were enclosures.

Demesne: Totternhoe demesne was dispersed in the fields in *c*.1175, as was Toddington in *c*.1200, two furlongs of which were called 'ridings' (Fowler 1926, BHRS 10: 78, 88). Kempston Grey's demesne has been described (p. 96); those of Higham Gobion and Streatley were also dispersed in *c*.1380 (J. S. Thompson 1990, BHRS 69: 64–123). In 1387, Chalgrave's 144-acre demesne was dispersed in 24 pieces (M. K. Dale 1950, BHRS 28: xxv). Aspley Guise and Oakley had block demesnes in the eighteenth century (Fowler 1936, 7, 24).

Yardland sizes varied between eighteen acres at Radwell in 1266 (Fowler 1919, BHRS 5: 239) to thirty-four at Harrold in *c*.1254 (Fowler 1935, BHRS 17: 118–19) and forty acres in 1311 at Stagsden (Fowler 1930, BHRS 13: 284). Radwell is riverine and the other two had woodland. Cranfield, also wooded, was forty-eight acres in the thirteenth century and Barton le Clay on the chalk scarp was twenty-three to twenty-eight acres. These yardland acreages accord with low values for good quality soil and high for places near woodland. Other values are Ravensden, mid thirteenth century, thirty acres, and Renhold with Salpho, twenty-three acres (Godber 1963, 206–7). The regular dispersion of strips at Willington and Duloe has been given (p. 124).

Work-service

Work-service for Studham was described in 1200–25 (Fowler 1926, BHRS 10: 122) and at Chalgrave in 1267 and 1303 (Dale 1950, BHRS 28: xxv, 46). Ramsey Cartulary gives full details of work-service for its manors in the thirteenth century. Barton le Clay included carriage to Cambridge, Ramsey, or London (Hart and Lyons 1884, i, 475). Shillington services, additional to the usual ploughing, etc., were to ditch, help enclose the park, cut and carry wood (except oaks and ash) and thorns for repairs to the enclosed field, and take pigs to the lord's wood (ibid. 460–4).

Kempston work-service was fully listed in 1341. The cottars held one or two acres of land in addition to their cottages. There were many tenants of small pieces of 'forland', possibly ground lying by assarts (F. B. Stitt 1952, BHRS 28: 85–9). The Higham Gobion and Streatley account roll for 1379–82 gives details of works; some were sold to tenants who took on dispersed demesne land. Sheep were kept in the lord's fold and forty wicker hurdles were bought to maintain it. There were four cart horses, four affers, and twelve oxen on the grange. Wheat and malt was sold in London; bushes growing (on the steep slopes) at *Berdecoumbe* were sold (J. S. Thompson 1990, BHRS 69: 64–123).

The management of meadows and balks has been given (pp. 24, 27).

Assarts

Assarts at Cranfield have been described (p. 216). The woods and assarts of Lidlington were used for common pasture in 1204 (Fowler 1919, BHRS 6: 31). A hedge was set around a wood at Luton in 1227 to prevent commoning on a pasture recently made next to it (Fowler 1916, BHRS 3: 13). Part of Harrold Wood was given to the nuns of Harrold to be assarted in *c.*1140 (Fowler 1927, BHRS 11: 48). A seven-acre piece (of assart) called *Hastingstoching* at Clapham lay next to the wood of Turstan Basset (Fowler 1930, BHRS 13: 113). Assarts are also recorded at Cople at *Moxworth* (*c.*1200), Sandy (*Maneswude*), and elsewhere (Fowler 1930, BHRS 13: 218–19, 244). A wood, assart, and park are recorded at Eaton Socon between 1198 and 1217 (Godber 1940, BHRS 22: 30–7). In 1200, at Toddington, fourteen acres of assart at Northwode adjacent to Wadlow had another nine acres between an assart and a homestead, implying that messuages were being built away from the main settlements (Fowler 1926, BHRS 10: 218). A spinney at Flitwick was to be assarted in *c.*1200 and there was woodland lying next to a croft, and pasture and arable at *Buckokesleie* (Fowler 1930, BHRS 13: 76–9).

In a dispute between the prior of Chicksand and the Knights Templar at Little Staughton in 1244, the bounds of the parish were described in detail. They refer to the 'moor', 'Grescroft', to groves, hedges' to Pertenhall Wood and an assart under it, as well as the balks of the field of Keysoe (G. H. Fowler 1919, BHRS 6: 130–1). Little Staughton Wood and land in *Leverunes Croft*, *Almareshey*, and a croft of 4½ acres with hedges and ditches were mentioned (*c.*1243) in a full terrier where the lands had many neighbours (Godber 1940, BHRS 22–3: 99–101). At Astwick in *c.*1220, a messuage had a

Table App.1. Bedfordshire field systems.

Place	Demesne type	Field no.	Date	Source
Astwick		2	c.1215	G. H. Fowler 1930, BHRS 13: 52–5; four acres in East Field and five acres in the West Field
Bushmead		2?	c.1500	Godber 1940, BHRS 22–3: 256–7. Terrier grouped as 297 and 294 acres
Chalgrave	dispersed	2	c.1225	M. K. Dale 1950, BHRS 28: xxv; G. H. Fowler 1926, BHRS 10: no. 444; three acres in each field Gray. Demesne 1387 (from London, British Library, MS Harley 1885)
Clifton		2	c.1221	G. H. Fowler 1930, BHRS 13: 203
Edworth		2	c.1185	Lees 1935, 223
Flitwick		2	c.1236	G. H. Fowler 1926, BHRS 10: no. 494; half an acre in each field
Hare in Toddington		2	c.1225	G. H. Fowler 1926, BHRS 10: no. 183; five acres in each field
Harrold		2	1263	G. H. Fowler 1935, BHRS 17: 121. Terrier two acres in West Field and two acres in Wood Field
Hinxworth		2	1200	G. H. Fowler 1930, BHRS 13: 52–3; five-acre terrier
Holme and Clifton		2	c.1220	G. H. Fowler 1930, BHRS 13: 243
Milton Bryan		2	c.1216	G. H. Fowler 1926, BHRS 10: 193–6; four acres in each
Stanbridge		2	1259	G. H. Fowler 1919, BHRS 5: 220. Demesne forty-eight acres in one field, forty-four acres in another
Tebworth		2	c.1225	G. H. Fowler 1926, BHRS 10: 31–2; six acres in each field
Wadlow in Toddington		2	c.1215	G. H. Fowler 1926, BHRS 10: nos. 718–19; equal amounts
Bedford, Newnham		3	1506	B. Crook 1949, BHRS 25: 82–94

Deane	dispersed	3	1649	Cave and Wilson 1924, 222
Elstow		3	1600	G. H. Fowler 1917, BHRS 4: 17–18. Terrier with a few neighbours only
Farndish		3	1607	D. N. Hall and J. B. Hutchings 1976, *BAJ* 11: 44, 49
Goldington		3	1200	G. H. Fowler 1930, BHRS 13: 166
Houghton Regis		3	c.1235	G. H. Fowler 1926, BHRS 10: no. 528
Hulcote		3	1575	Oxford, All Souls College, terrier, TM 286/No. 2
Oakley		3	c.1635	Northamptonshire CRO, Map 2991 fols. 17 and 65
Podington		3	1543	Y. Nicholls 1985, BHRS 64: 29
Willington		3	1230	J. Godber 1963, BHRS 43: 21–3. Yardland terrier

croft containing three selions with a garden and pasture at the edge of the strips, being an example of settlement expanding on to arable lands (Fowler 1930, 13, 172).

Ridge and furrow

Ridge and furrow occurred widely north of the Greensand ridge on the claylands. It is often about four yards in width, narrower than the East Midland norm of five yards or greater. Reconstructed maps based on a field survey have been published for Farndish (Hall and Hutchings 1976; *BAJ* 11: 43–50), Barton le Clay, Edworth, Higham Gobion, Sundon (D. Hall 1991; *BAJ* 19: 51–6), and Milton Ernest (Hutchings 1969; *BAJ* 4: 69–78). Strip lynchets occur on the steep scarps of the Chilterns. Barton le Clay, Streatley, and Totternhoe have the best examples, which are recorded on open-field maps.

REFERENCES

BAJ   Bedfordshire Archaeological Journal
BHRS   Bedfordshire Historical Record Society

ACKNOWLEDGEMENTS

Stephen Coleman for information from the HER; John Hutchings and members of the Bedfordshire Archaeological Council for many years of productive association; Anna Slowikowski of Albion Archaeology; Claude Ibbett and Richard Parrish for permission to survey their estates.

2. BERKSHIRE

Berkshire lies south of the Thames, with low-lying land in the north next to the river; the Berkshire Downs chalk ridge crosses the centre. London Clay and sandy Bagshot Beds outcrop at the south and south-east. The Downs and the south-east have dispersed settlement associated with irregular fields; the northern part of the county had regular two- and three-field arrangements.

Several open-field maps survive for places in the Clay Vale. In the south, Hull (1949, *BerksAJ* 51: 12) published a map of Benham Valence showing furlongs and three great fields. Walne (1955 *BerksAJ* 54: 13–38) lists eight draft enclosure maps with strips marked. Colvin (1950 *Oxoniensia*, 15: 92) described four maps held by St John's College with open-field information. Tithe maps show some open-field lands in twenty-five parishes (Kain and Oliver 1995).

Gray (1915, 452–4) lists fourteen places that had two-field systems and only two (plus two places not identified) with three fields, which finding has resulted in the generalization that two-field systems were common in the county. The Table lists three more three-field systems and three dispersed demesnes. Glebe terriers of 1634 show there was two-field at Idstone in Ashbury, Aston Tirold, Chaddleworth, Childrey, Chilton, Moulsford in Cholsey, Hampstead Norreys, Harwell, East Ilsey, West Ilsey, East Shefford, and Uffington. Three fields described by glebe terriers occurred at Appleton, Clewer, and Shellingford (Mortimer 1995, xviii) and the Table records three others.

Irregular fields

The structure of the fields of Winnersh and Sonning in *c*.1600 is illustrated by detailed surveys published by Gray (1915, 553–4). Both places have yardlands that include enclosed ground, usually arable. The Winnersh (SU 78 70) estate was 373 acres, split between 12 holdings, in all 14 yardlands, that contained 139 acres of enclosed ground (of which 109 were arable) and 233 acres of open-field land (of which 60 acres were meadow and 173 arable). The average size for a yardland is twenty-six acres. The land distribution shows that Winnersh had two separate field systems, with four holdings having arable in three fields and another four lying in five different fields. There were additionally nine more named areas, occasionally and erratically referred to. Three holdings had land entirely enclosed, but two of them had meadow. An underlying unity in the yardland size was that the enclosed holdings total nearly twenty acres, excluding meadow, and five of the holdings with open field land have nearly twenty acres in the open fields, excluding both meadow and enclosed land. Sonning tithing (SU 76 76) had 231 acres surveyed lying in 12 holdings totalling 11 yardlands, of which 61 acres were enclosed, 141 acres were dispersed in open fields, and 29 acres were meadow. These examples illustrate the complexity of holdings with both open and enclosed land making up yardlands that lay irregularly dispersed.

A survey of lands at Sandhurst (SU 80 76), formerly belonging to Chertsey monastery, was made in 1549. It was 71 per cent enclosed (in small closes of three acres average size) and had three fields, but the tenants' lands lay in only two of them (Kempthorne 1935, *BerksAJ* 39: 192–7). It is an example of the dangers of assuming that three named fields represent a three-course tilth in townships that have a high proportion of enclosure.

This mixture of enclosed and open holding was mapped in 1606 for the estates of Corpus Christi College, Oxford, at Streatley (SU 59 80) and Arborfield (SU 75 67). The

Table App.2. Berkshire field systems.

Place	Demesne type	Field no.	Date	Source
Harwell		2	1210–1322 and 1620	Magd. Berks Charters Harwell 9a and 19a, etc.; G. E. Fussell 1936, Camden 3rd Ser., 53, 96–7
Brightwell	dispersed	2	*c*.1350	The demesne lay in two fields: A. Ballard 1916, 207
Benham Valence	dispersed	3	1468 and 1775	Magd. Berks Charters Benham and Wallington 2; Benham 11 and 166; F. Hull 1948–9, *BerksAJ* 51: 12 [Map]
Denchworth		3	1634	Oxfordshire CRO, MS Archd. Papers, Berks., c185. fol. 70
Sandhurst	dispersed	3	1549	G. A. Kempthorne 1935, *BerksAJ* 39: 192–7, from TNA, LR 2 187, fols. 50–3. Irregular, 71 per cent enclosed
Earley		6	1669	E. W. Dormer 1927, *Berks., Bucks. and Oxon. Archaeological Journal*, 31: 193–8

plans can be related to terriers made in 1516 and traced further back in the deeds of the estates (J. G. Milne 1942, *BerksAJ* 46: 32–44, 78–87). The Streatley property was mainly an assart consisting of closes and had small and irregular-shaped blocks of open-field land with few neighbours. The land was not uniformly distributed and all lay within a small area. The arrangement was the same in 1307 and probably 1180. It is an 'ancient' landscape that was always partly open and partly enclosed, and the view of it mapped in 1606 does not represent a half-enclosed dynamic condition changing from more extensive open field to a completely enclosed state.

Farming

Tenants of Ashbury had to wash and shear sheep in the mid thirteenth century, and those who provided oxen for ploughing had pasturage with the lord's oxen. As well as the agricultural round they carried wool and cheese (C. J. Elton 1891, Somerset Record Society, 5: 51–2). Ashbury ploughing-service and pasturage of oxen with the lord's oxen was recorded as early as 1189 (Jackson 1882, 116). The sheep-folding system of agriculture was practised at Wantage and Hungerford in the mid thirteenth century, with work-service of sheep washing and shearing, providing the wattles of the fold, and carrying it from place to place (Chibnall 1951, 50, 67). Sheep washing and shearing work-service was also recorded at Hungerford by St Frideswide's monastery, Oxford (S. R. Wigram 1896, Oxford Historical Society, 31: 356–8). Brightwalton yardlanders were obliged to carry the lord's fold, and wash and shear sheep; the person chosen to be shepherd had the use of the fold for twelve days and took one fleece called *belwertheresfles*, in *c.*1300 (Scargill-Bird 1887, 58–67).

Robert Loder described his farming at Harwell in remarkable detail from 1610–20 (Fussell 1936). There was a two-field system, and peas, beans, or vetches were grown as a hitch on the fallow (pp. 96–7) or sometimes in the cornfield. Horses were tethered on green vetches (p. 97). Grassed headlands were used for hay (p. 11); eighteen cocks made a load (p. 6). Lands were dunged from a 'shephouse' (p. 108); some meadows were watered (p. xix). Robert Loder was possibly descended from a John Leder who received farm stock in 1484 (Magd. Berks Charters, Harwell 121). Harwell is recorded as two-field from *c.*1210 to 1322 (Magd. Harwell 9A and 19A). Hitching appears to have been done as early as 1360, since no part of the 'falwfelde' was to be sown unless dunged with a cart (Magd. Harwell 21b).

Ridge and furrow

Oxfordshire HER has 1:10,560 scale maps with sketched ridge and furrow for the northern part of Berkshire (probably taken from the RAF 1940s verticals). Sutton (1966, *Oxoniensia*, 29–30: 99–115) plotted ridge and furrow visible on vertical aerial photographs taken by the Ordnance Survey in 1960. His map (fig. 37) shows that almost all examples lay north of the chalk. A plan of ridge and furrow survival at the north of the county in 1990 has been published along with a photograph of the excellent remains at Denchworth (SU 38 92), where 321 acres, 32 per cent of the township area, survived (Hall 2001, 7, 36). The absence of broad ridge on the chalk and on other light soils in the south is probably due in part to the practice of ploughing lands as flat as possible in the early nineteenth century (Mavor 1809, 159–60), as well as to subsequent post-enclosure flat ploughing. Curved strips in arable cultivation in the 1930s are illustrated by the Orwins (1938, pl. 9).

Hardwell and its Saxon charter boundaries are described in the main text (p. 168) (Gelling 1968, 1976 and Hooke 1987). The newly surveyed furlong plan of Hardwell has been checked using RAF vertical aerial photographs, which show ridge and furrow on the claylands but none on the downs (English Heritage, Swindon, RAF 106G/UK/1561 (7 June 1946), frames 3020–21; 106G/UK/1408 (12 April 1956), frames 3275–6, 4272–3, and 4276). Some furlong boundaries are visible on later vertical photographs (Hooke 1988, 132). The remarkably detailed AD 903 bounds of Hardwell (S 369) related to the 2004 furlong map of the surrounding area (Figure 5.4) are as follows. The numbers refer to the text footnotes given by Gelling (1976, 684–6). Items 8–13 are better accounted for than previous interpretations of the boundary in this vicinity, showing that boundary did not exactly follow lines marked on later maps.

1 *spinbroc*, for swinebrook.
2 *riscslæd*, rush slade.
3 *riscsledes byge*, bend in rush slade.
4 *hordwylles weg*, Hardwell Way.
5 *Icenhilde wege*, Icknield way.
6 *ealdan wudu weg*, old wood way; probably wood once grew on the unploughed scarp.
7 *be eastan telles byrg*, by the east of *tellesbyrg* (Hardwell Camp).
8 and 9 *on ænne garan*; *on ænne gar æcer*, probably meaning taking in the triangular shaped furlong there.
10 *andlanges þere furh*, along the furrow.
11 *to anum andheafdum*, to a 'headland', which here would mean the furrow going by the beginning of the furlong boundary.
12 *to anre forierðæ*, the type of furlong boundary much later called a 'joint'.
13 *in to þam lande*. Gelling suggests this means that the boundary belongs to the estate being described.
14 *to þam stane on hrig weg*, to a stone on Ridge Way.
15 *west on anne garan*, 'west on a gore', meaning not clear in terms of boundary markers.
16 *andlanges þære furh*, along a furrow, which in this case is the furrow of a 'headland'.
17 *to anum anheafodum*, to a 'headland'.
18 *dune on fearnhilles sled*, the hill above fern hill slade.
19 *on ane furh an æcer near þæm hlince*, a furrow (of a headland) near the lynchet.
20 *oðærne hlinc æt fearnhylles slæde suðæweardre*, to another lynchet south of fern slade (the low valley there).
21 *on anum heafde*, the head (of a headland).
22 *on ane furh*, a (short length of) furrow.
23 *on ane stanræwe*, on a stone row that presumably marked the furlong boundary (joint).
24 *on hricg weg*, to Ridge Way,

25 *on anne gar æcer*, gore acre. To be consistent with the remainder of the bounds this should have been called a 'furh', unless it refers to a small gore by the Ridge Way, no longer visible.

26 *on hæfde*, (head of) a headland.

27 *andlanges anre fyrh*, along a furrow (of a headland).

28 *to anum byge*, a bend (the headland may have curved slightly to meet the next one).

29 *on ane fyrh*, on a furrow (of a headland).

30 *to anre forierðe*, to a 'forearth', the boundary type later called a 'joint', as 12.

31 *on Icenhilde weg be tellesburh westan*, to Icknield Way by the west of Hardwell Camp.

The remainder of the boundary has not been worked out in detail.

Charlton in Wantage preserves evidence of a large-scale layout of both meadows and arable land. The Tithe map of 1844 (TNA, IR 29 and 30, 2/140) shows meadows at the north with five adjacent owners whose curved parallel strips cross almost the whole width of the township, some 950 yards. The same strips were recorded on a map of 1754. An aerial photograph of the ridge and furrow nearer the village (RAF VAP, CPE/UK/1936 frame 2404 (18 January 1947)) shows four furlongs with strips lying on the same smooth curve as the meadow strips and another two on a very similar alignment to the west of them. It is evident that large parts of the township were laid out with strips across the whole width, about 1,150 yards in length.

BIBLIOGRAPHY

BerksAJ *Berkshire Archaeological Journal*
BRS   Berkshire Record Society

Fussell, G. E. (1936), *Robert Loder's Farm Accounts, 1610–1620*, Camden 3rd Ser., 53.

Gelling, M. (1968), 'The Charter Bounds of *Aescebyrig* and Ashbury', *BerksAJ* 54: 13–38.

Mavor, W. (1809), *A General View of the Agriculture of Berkshire* (London: Board of Agriculture).

Mortimer, I. (1995), *Berkshire Glebe Terriers, 1634*, BRS 2.

Scargill-Bird, S. R. (1887), *Custumals of Battle Abbey* ([Westminster]: Camden Society).

ACKNOWLEDGEMENTS

E. A. Garnish Esq., for information and transcripts of the Charlton (Wantage) Tithe Map. Margaret Gelling for pointing out to me the Saxon charter of Hardwell. R. Salmon, Esq., Compton Beauchamp Estates, for permission to survey Hardwell. Paul Smith and Susan Lisk, Oxfordshire HER.

3. BUCKINGHAMSHIRE

Buckinghamshire has undulating boulder clay at the north, relieved by the River Ouse, which exposed limestone soils. The Clay Vale lies in the centre, with chalk of the Chiltern Hills at the south. There is dispersed settlement on the Chilterns, illustrated by plans of St Leonards (1581) and Hawridge (1550) (Hay 1971, maps 3 and 4). Hanslope and Shenley Brook End represent dispersed settlement on boulder clay. Most of the northern part of the county had regular two- and three-field arrangements of Midland type; on the Chilterns dispersed settlement was associated with irregular field systems.

The field systems of Buckinghamshire were described by Gray (1915, 76–80, 454–6, 552), Beresford (1947–52 and 1953–60), and Roden (1973). Sheringham and the surrounding villages were studied in detail by Chibnall (1965). Information on the Milton Keynes villages is provided by Croft and Mynard (1993), and Reed (1997) has published the earliest glebe terriers from the Buckingham Archdeaconry series.

Many townships have open-field maps such as Great Linford, 1641 (BAS 623/43; Croft and Mynard 1993), Wootton Underwood, and Soulbury (Beresford 1952, 294; at Aylesbury Museum). The Loughton draft enclosure map marks furlongs (Buckinghamshire CRO, IR/143R; Croft and Mynard 1993). Open field was recorded on twenty-seven tithe maps of 1838–49 (Kain and Oliver 1995). Salden has a tithe survey of 1252 that is in effect a field book (Elvey 1975, 214–20).

Regular field numbers

Gray (1915, 454–7) provided lists of two- and three-field systems. Most examples lay north of the Chilterns. Of the ten two-field examples, one was demesne; seven out of the seventeen three-field ones were also demesne. Beresford's study (1967–52) of seventeenth-century glebe terriers showed that 80 of the 104 parishes had two or three fields. The Table gives fourteen examples of two-field, eight of them additional to Gray, and three three-field. Changes from two to three fields at Dunton and Mursley have been given (pp. 40–1). Sherington had two fields in 1312 and before. They were changed to a three-field arrangement between 1514 and 1577 (Chibnall 1965, 108, 221, 273).

Irregular fields

Early grants of land in the Chilterns are often full of references to woods or cleared land, for example at Amersham, a piece in furlong called *Sampsonesbreche* (*c*.1272). There is land, meadow, and wood in Wooburn (1273) and a four-acre piece in Langley in a field called *Wodefeld* lying next to a wood (1200–29). In Beaconsfield in 1190–1203, a hundred-acre heath had a surrounding ditch. A heath of fifty-two acres at *Alkesehulle* was stated to be where a wood once grew in *c*.1222 (Jenkins 1945, BRS 10: 31, 40, 79, 93–4). Chalfont St Peter's probably had a single field. For assarts at Chesham and Crawley see the main text (p. 116).

Demesnes have been mentioned (p. 97); another possible example of the dispersed type is Ilmer in 1338 (Gray 1915, 456; W. Page 1908, VCH, *Buckinghamshire*, ii: 48). Yardlands at Shalstone were eighteen to nineteen acres in *c*.1245 (Elvey 1975, 199, 208). Leadam (1897, 638) records yardland sizes from the 1517 enclosure returns: Dodershall, forty acres; Donnington, Steeple Claydon, and Upton, thirty; Ludgershall and Cranwell, twenty; and Preston, fifteen acres. Others are reported in glebe terriers (Beresford 1953–60, 16: 14–23). The regular ordering of strips at Winslow has been described (p. 119). Other examples are Shalstone (*c*.1265) (Elvey 1975, 163–4) and Broughton (1605), Maids Moreton (1607), Pitchcott (1635), and Bow Brickhill (1639) (Reid 1997).

Table App.3. Buckinghamshire field systems.

Place	Demesne type	Field no.	Date	Source
Beachampton		2	c.1200 and c.1260	J. G. Jenkins 1945/52, BRS 9: 40; Elvey 1975, 372
Drayton		2	1222	M. W. Hughes 1940, BRS 4: 46
Evershaw		2	1289	Elvey 1975, 418–19
Hartwell		2	1270	J. G. Jenkins 1938, BRS 3: 37–8
Kemble [Great]		2	1246	J. G. Jenkins 1945, BRS 10: 147
Loughton		2	c.1220	J. G. Jenkins 1945, BRS 9: 26–7
Shalstone		2	c.1265 and 1289	Elvey 1975, 162–3, 415–16, etc.
Sherington		2	1312	Chibnall 1965, 108
Stokes		2	1198	M. W. Hughes 1940, BRS 4: 13
Thornborough		2	c.1250	Elvey 1975, 244–8
Westbury		2	c.1298	Elvey 1975, 99
Wolverton		2	c.1250	J. G. Jenkins 1945, BRS 9: 47–8; Elvey 1975, 54–5
Chichele	block	6	1557	A. H. Baines 1999, *ROB* 39: 7, 11–17; fields grouped as three-tilth, demesne (Bury Field) enclosed 1526
Kingshill in Little Missenden		3	c.1230	J. G. Jenkins 1938, BRS 2: 68, 113
Lillingstone Dayrell		3	1289	Elvey 1975, 421
Sherington	block	3	1577	Chibnall 1965, 221

Field orders

For Newton Longville in 1291 there was reference to a series of court ordinances (Ault 1972, 84). Fenny Stratford court rolls (1373–82) record fines for various trespasses, with animals in corn, geese in the lord's barley, pigs eating peas, etc. Many villeins refused to do their work-service (W. Bradbrook 1920–6, *ROB* 11: 306–7). Stinting at Fenny Stratford (1625) was ten beasts, two bullocks, four horses, two colts, and forty sheep per yardland (Beresford 1952, 297). Great Linford Court orders made for 1630 state that the yardland stint was two horses, four great cattle, and thirty-six sheep. Furze was collected from a pasture called the ley field (E. Blackmore, in Mynard and Zeepvat 1992, 33–42). Open field orders for 1678 at Grendon Underwood detail the agricultural round. There were for each yardland six cows, three horses, and thirty-three sheep. Horses were not to be tied in the wheat or bean fields after the first land was reaped. In the fallow field every land should have balks one or two feet wide made after June and they should be continued (Eland 1940, *ROB* 13: 285–6). Manorial court orders for 1682 at Sherington covered the usual items (Chibnall 1965, 283–5).

Farming

Gurney has shown that Walter of Henley very probably came from Hanley in Worcestershire and lived most of his life in Buckinghamshire (F. G. Gurney 1946, *ROB* 14: 256–60; A. V. Woodman, 1953–60, *ROB* 16: 216–17). Taxations of 1332 and 1327 give some detail of animals and grain. Oxen still accounted for more than 30 per cent of ploughing animals in parts of the north. There were large numbers of sheep in the Chilterns compared to the northern champagne lands suggesting some sheepfolding (J. C. K. Cornwall 1978, *ROB* 20: 57–75). The Chiltern woods were also used for faggots to supply London (Jenkins 1935, 32–41). Details of demesne farming at Water Eaton in the late fourteenth century show that the yield of grain was between two and four (E. Hollis 1933, *ROB* 12: 165–71). Hitching at Bletchley and Newton Longville has been mentioned (p. 43). A report in 1378 of cultivation in three seasons at Ilmer specified that the crops were corn, peas, and beans, and the third season was fallow (Gray 1915, 456; W. Page 1908, VCH, *Buckinghamshire*, ii: 48).

Several villages had lot-meadows, for example Beachampton in 1707 and Maids Morton in 1607 (Reid 1997, 136; see the main text for Haddenham, p. 18). The introduction of balks, grass ends, etc. has been given for Sherington and elsewhere (p. 26).

Ridge and furrow

North of the Chilterns, especially the Aylesbury Vale, had abundant ridge and furrow. The whole county has been sketch-mapped at the 1:10,560 scale from 1947 RAF vertical photographs (held in the HER). The survival in 1995 has been mapped (Hall 2001, 36). Several glebe terriers refer to ridged lands (Beresford 1952, 295), such as Middle Claydon and Turweston (Reid 1997, 146, 210). Lands at Bradwell and Tattenho, in *c.*1250, had heads at each end (J. G. Jenkins 1939, BRS 3: 5; J. G. Jenkins 1942, BRS 9: 38.

Sherington has reconstructed maps for 1580 and 1300 prepared by using the detailed abuttals of each furlong, the first time such a map had been made (Chibnall 1965, 259–63). Furlong maps surveyed and reconstructed have been made for the Milton Keynes area (Croft and Mynard 1993). All these plans show groups of furlongs with strips in a similar alignment suggestive of a planned layout. A few examples of narrow early modern ridges are known at Dorney Common, west of Eton. The HER had thirty-five records of 'lynchets' in 2003, most of them in the Chilterns. Lynchets at White Hawkridge, Bellingdon, were described in a terrier of 1478 as arable land, and survived as narrow enclosed fields on the tithe map of 1843 (P. Casselden 1987, *ROB* 29: 138).

BIBLIOGRAPHY

BAS   Buckinghamshire Archaeological Society (Aylesbury Museum)
BRS   Buckinghamshire Record Society
ROB   *Records of Buckinghamshire*

Croft, R. A., and Mynard, D. C. (1993), *The Changing Landscape of Milton Keynes*, BAS Monograph Ser., 5.
Elvey, G. R. (1975), *Luffield Priory Charters, Part Two*, Northamptonshire Record Society, 26.

Hay, D. and J. (1971), *Hilltop Villages of the Chilterns* (Chichester: Phillimore).

Jenkins, J. G. (1935), *A History of the Parish of Penn in the County of Buckingham* (London: Saint Catherine Press).

Mynard, D. C., and Zeepvat, R. J. (1992), *Great Linford,* BAS Monograph Ser., 3.

Reed, M. (1997), *Buckinghamshire Glebe Terriers, 1578–1640,* BRS 30.

ACKNOWLEDGEMENTS

Roger T. Bettridge, County Archivist; Mike Farley, Sandy Kidd, David Green, and Julia Wise, HER; and Denis Mynard and Barbara Hurman, with whom I walked the once rural Milton Keynes to survey its furlong boundaries and many fields of ridge and furrow.

4. CAMBRIDGESHIRE

Cambridgeshire has three topographical zones: clayland and chalkland at the south and fen in the north. Some parishes at the south-east have dispersed settlement with ends, greens, and woodland. In spite of the late enclosure, and widespread cultivation on a three-course crop rotation, the county does not fall into the Midland field-system type proper, except in the west. Elsewhere in the county, many townships have a lower percentage of arable than is found in the Midlands and much common ground, although they often ran a three-season cropping course, even on the fen islands. The Silt Fenland had fields with long ditched strips held in severalty without communal regulation.

Gray (1915, 457–60) identified two- and three-field systems and Postgate (1973) discussed the eastern part of the county with East Anglia. Spufford (2000) gives a good overview of open-field economy. Susan Oosthuizen (2006) has provided an account of the county's fields and the VCH volumes supply much information about Cambridgeshire fields. The distinctive field systems of the Silt Fen have been discussed (see p. 73).

Cambridge Colleges, the University Library, and the County Record Office have copies of many open-field maps, and tithe maps of 1839–49 record open field (Kain and Oliver 1995, 55–62). Hildersham remained open until 1883–9. Litlington fields were studied by Crawford in 1937 in his pioneering identification of furlong boundaries, already referred to (p. 163). Cambridge had two open-field systems detailed in a fourteenth-century field book studied by Maitland (1898, 54–6, 125–7). The East Fields were reconstructed by Stokes in 1915 and Mary Hesse in 2000, and the West Fields by Catherine Hall and Jack Ravensdale in 1976.

Field numbers

Gray (1915, 457–60) listed four places with two fields in the thirteenth and fourteenth centuries and twenty-two places with three fields at dates from the thirteenth to fifteenth centuries. The demesnes of Haddenham lay in three fields in 1222, as did that of Wilburton in 1251 (O. C. Pell 1891, *PCAS* 6: 24).

VCH, *Cambridgeshire,* v (2000), describing twenty-two parishes in Longstowe and Wetherley Hundreds in the west of the county, identified nine with three-fields in the

Middle Ages and twelve in *c.*1800 with three or more fields, but grouped as three seasons. Swaffham Bulbec probably had a bare fallow in the fifteenth century since wheat and pease fields are referred to then, as in the Midlands. Bare fallow was used at Little Wilbraham in 1500, when it was stated to be used as common. Burwell still operated a bare fallow in the 1790s.

Irregular fields

Many Cambridgeshire townships had multiple field names. Postgate (1973, 295–300) noted this for the south-east of the county. The names were primarily locational, with fields being grouped into three-season cultivation at an early date. A Waterbeach grant of nine acres in 1176 assigned three acres in each of three fields, yet there were multiple field names in later centuries, proved to be grouped into three shifts. On the eve of enclosure in 1813 there were seven named areas in three shifts (Ravensdale 1974, 86–8).

Both sets of Cambridge town fields had four grouped as three (East Fields in 1155 and 1632; West Fields, 1474). Teversham with four named fields was run as three-tilth in 1636 (Postgate 1973, 298). Spufford (1965, 17–19) showed that Chippenham had eight fields in 1544 and they were even more complicated in the thirteenth century, with holdings not distributed throughout all of them. The glebe of Burrough Green, in 1663, had an arable enclosed parcel of five acres, and ten to thirteen acres in each of four other named fields. There was a further thirty-five acres in the outfield (Palmer 1939, 139). Stetchworth, in 1770, had about 60 per cent old enclosure and three named shifts in which the parcels of a sample farm were unequally distributed. At Weston Colville, a thirty-five-acre holding had strips concentrated in two of six open fields (Postgate 1973, 310–12, plan).

A few places developed more complex field systems in later centuries. Landbeach had three fields in the twelfth and thirteenth centuries, but by *c.*1650 and until 1813 there were four. Cottenham had three fields in the Middle Ages but there were five named fields by *c.*1549, which continued until enclosure in 1847. It was recorded in 1905 by people who had worked in the fields that there was a five-year course with a bare fallow (Ravensdale 1974, 95–109). Cheveley, although three-course in the eighteenth century, adopted a four-course system by 1838 with clover and turnips planted on most of the fallow (VCH, *Cambridgeshire*, v (2000): 51). Both dispersed and block demesnes occurred and have been listed (see pp. 97, 101).

Field orders

A Stretham order for 1607 stated that only cottagers were to put sheep in the lord's fold; the lord gave up commoning rights on 1,600 acres of fen; the commoners could make by-laws regulating fen usage; and holders of thirty acres could common two working horses or mares with foals (Cunningham 1910b, 253–62). Orders for 1614 restricted the times of gleaning, and tenants were to maintain the cow-pasture bank. No cattle were to go in the field until six days after harvest was finished (Cunningham 1910b, 267–74). Orders for Cottenham, made in 1640, concerned the keeping of swine and cattle in the fen, the repairing of fen banks, and restrictions on the amount of (peat) turves to be dug. Dead cattle were to be buried immediately the hide had been removed (Cunningham 1910b, 230–45).

Cropping and farming management

The arable fields of Cambridge City occupied most of the available area, and hay and pasture were in short supply. Sales of corn were used to buy hay from the fen. By the sixteenth century kitchen waste and night soil from the town were used as manure. Crop flexibility in the county is illustrated at Linton, where in 1695 there were twenty-four acres of carrots, peas, turnips, and parsnips on forty strips scattered over fourteen furlongs distributed in seven of the nine fields. At Teversham adjacent strips had wheat, barley, and clover; others had barley, oats, peas, lentils, and tares. No evidence was found that leys were used as a convertible husbandry system (Postgate 1973, 305).

The early fourteenth-century court rolls of Littleport describe the usual offences of breaking into the lord's fold, not cultivating his land, causing damage to the vineyard, and hunting hares in the lord's field. Offences specific to the fens were carrying away sedge and selling it out of the manor, collecting bitterns eggs, and removing oars from a boat. There was an officer to see that court by-laws were kept (Marshall 1881, 79–108).

At Waterbeach in 1344 the freeholders agreed not to have common for more cattle than they could keep in the winter. They were allowed to mow some parts of common for flags to use as fodder when there was a hay shortage. Commoners could dig in Joist Fen but not sell turves outside the parish. No cranes, bitterns, bustards, or herons were to be taken and sold from the commons unless first offered to the lord (1522; Clay 1859, 4–21).

Sheepfolding was practised where there were sizeable heaths and pastures. Chippenham Farm had 1,140 sheep on heathland in 1544. At Willingham and Cottenham the pastures were fen-edge meadows which were suitable for dairy cattle and the lord's sheep flocks, and folds were limited in size (Spufford 2000, 72–3). Orwell copyholders had rights of folding fully described in 1607. Sheep were stinted at one to the acre (of arable), but farm inventories show that in many cases few sheep were kept. Willingham similarly had a stint of thirty sheep to the yardland, and they had to be folded in the lord's fold. Again they were not always taken up since thirty out of forty-four inventories list very few sheep (Spufford 1974, 99, 129). The rights and obligations of folding in cases like these are to be distinguished from an organized folding system operating to manure the common fallow systematically.

Land-use differed between the western clays and eastern pastures. Grass in the form of leys was introduced amongst the arable in the west. The glebe of Toft and Comberton in 1638 contained 2–3 per cent pasture. Leys were used as permanent pasture and were not incorporated into a convertible husbandry system. On the Suffolk border, the glebe of Carlton cum Willingham and West Colville in 1615 had half of its fifty acres in the form of pasture closes.

The survival of open field until a late date may be related to corn production, especially barley, which was supplied to London via King's Lynn or Royston. Some townships had open-field land lying in large pieces (in single ownership) as at Boxworth and Madingley. Vancouver (1794, 104) noted that 'inclosure was not desired'. Lands in

Snailwell were being amalgamated: whereas a 1544 survey at Chippenham described 2,600 strips, in 1712 the same area had only 820 (Spufford 2000, 49–50). If land was occupied in larger pieces it may have partly removed the desire to enclose, and underdraining was possible in some cases, but not always without opposition. A farmer at Barrington hollow-drained his open-field land, but others blocked the outlet drains so that they 'blew up' and they told him not to cultivate turnips on the open field (W. E. Tate 1944, *PCAS* 40: 68).

Woods and assarts

At Kingston a block of eighty acres, the house of Lefeson, and land under the wood, referred to in *c*.1150, are likely to be assarts. Other woods are described at the same date (Hassell 1949, 35, 37). An assart in 'stocking' of four acres lay in a single piece at Eversden in *c*.1200. In the thirteenth century, grants of arable at Silverley were in blocks of seven to twelve acres, the largest being 'old assart'; another was in a field called Woodcroft. West Wratting woods were grubbed up in the early seventeenth century. In the west in 1608, Little Gransden demesne wood comprised 168 acres out of 500. Later, timber was in short supply and sixteenth- to seventeenth-century manorial surveys comment on the shortage, trees being obtained from the hedges at Trumpington and Longstanton (Spufford 2000, 71–2), and probably from the shaws at Abington Piggots.

Cambridgeshire has a large number of open-field maps, field books, terriers, and surveys. Through the use of the early and copious records of the college estates it would be possible to make a comprehensive study of its field systems, especially as the relevant records have been outlined in the relevant VCH volumes.

Ridge and furrow

Ridge and furrow visible on aerial photographs of 1947–63 was plotted by Kain and Mead (1977). It occurred widely on the claylands of the west and the fen islands. Examples at Croydon, Croxton, Clopton, and Bourn have been published (Oosthuizen 1996). On the southern chalklands there was very slight ridging only, as can still be seen at Guilden Morden. Snailwell heath had furrows in 1560 that were formerly ploughlands. Ridges were equated with former arable at Little Gransden in 1608 and Milton in 1683 (Spufford 2000, 43, 72, 76). An engraving of Cambridge made in 1690 shows flat strips separated by furrows (Stokes 1915, plate). Many lands elsewhere were ploughed flat, but on the heavy clays of Cottenham they were still ridged in the 1840s (Ravensdale 1974, 110).

Chalk marls in the south, from Litlington to Isleham, have large furlong boundaries that are visible from the air and were fully recorded on photographs in March 1981 (CUCAP, RC8 series). Long before (in 1937), Crawford had been the first to identify and date them at Litlington. Furlong boundaries survive well on the clayland, and a furlong plan has been made for Stretham (TL 51 75) from fieldwork (by D. Hall and M. Young in 2005). The mapped plan agrees closely with that of the 1836 draft enclosure map (Cambridgeshire RO, 152/P16). The furlongs form a mainly rectangular system with long alignments of strips in as many as ten furlongs, one mile in length.

A study has been made of six adjacent parishes in the south-east of the county (from Burrough Green to West Wrattling (Harrison 2002)). The townships have long linear, approximately parallel, boundaries crossing different soil types. Pre-enclosure map evidence has been used to demonstrate the presence of a system of roads and trackways that ran approximately parallel to the township boundaries. Harrison suggested that the routeways, whether at the parish edges or as central droveways, pre-date the medieval vills and fields, which therefore fitted in with an earlier landscape structure. If the droves and tracks are Roman or prehistoric, it implies a great density of settlement to require so many adjacent parallel tracks to presumed allotted areas of pastures. Such a density is not apparent. On the western side of the chalk ridge the fen-edge parishes have much Roman and earlier settlement, at for instance Chippenham (Hall 1996, 98–100), where there is no particularly linear system of furlongs with tracks making for the heathlands (open-field map, 1712 (Cambridgeshire RO, R58/16/1)). It is likely that the Borough Green area tracks are instead of medieval date.

BIBLIOGRAPHY

CAS	Cambridge Antiquarian Society
Cambridgeshire RO	Cambridgeshire Record Office
CRS	Cambridge Records Society
PCAS	*Proceedings of the Cambridge Antiquarian Society*

Clay, W. K. (1859), *A History of the Parish of Waterbeach in the County of Cambridge*, CAS, Octavo Publications, 4 (Cambridge: Deighton).

Hall, C. P., and Ravensdale, J. R. (1976), *West Fields of Cambridge* (Cambridge: CRS).

Harrison, S. (2002), 'Open fields and earlier landscapes: six parishes in south-east Cambridgeshire', *Landscapes*, 3: 35–54.

Hassell, W. O, (1949), *The Cartulary of St. Mary Clerkenwell*, Camden 3rd ser., 71 (London: Royal Historical Society).

Hesse, M. (2000), 'Field systems in southwest Cambridgeshire', *PCAS* 89: 58–79.

Kain, R., and Mead, W. R. (1977), 'Ridge and furrow in Cambridgeshire', *PCAS* 68: 131–7.

Marshall, W. (1881), 'The court rolls of Littleport', *PCAS* 4: 97–108.

Oosthuizen, S. (1996), *Cambridgeshire from the Air* (Stroud: Sutton).

Oosthuizen, S. (2006), *Landscapes Decoded: The Origins and Development of Cambridgeshire's Medieval Fields* (Hatfield: University of Hertfordshire).

Palmer, W. M. (1939), *History of the Parish of Borough Green, Cambridgeshire* (Cambridge: CAS).

Ravensdale, J. R. (1974), *Liable to Floods* (Cambridge: Cambridge University Press).

Stokes, H. P. (1915), *Outside the Barnwell Gate*, CAS, Octavo Publication, 47 (Cambridge).

Vancouver, C. (1794), *General View of the Agriculture in the County of Cambridge* (London: Board of Agriculture).

ACKNOWLEDGEMENTS

I am grateful to Christopher Evans and the staff of the Cambridge Archaeological Unit for permitting my involvement in their recent research; to Jack Ravensdale, Mary

Hesse, Susan Oosthuizen, Robin Glasscock, David Wilson, Mike Young, and Catherine Hills for access to departmental libraries and discussion, and to Ralph Snudden for pointing out the Fen Ditton map (discussed in the main text, p. 151).

5. CHESHIRE

Cheshire was characterized by irregular and 'one-field' systems with convertible husbandry. The terms 'field' and 'furlong' were used interchangeably. Much of the low ground was ill drained, with marsh and moss interrupted by sandy knolls. Settlement was dispersed and some parishes were large, containing many townships. Wybunbury parish with 17,854 acres had 18 townships (Sylvester 1949) and Prestbury 32. Medieval open-field arable in the county shows a concentration in the west and on the Wirral. It occurred throughout the county except in the east beyond Hyde in the Pennines, and was sparse in the moss lands adjacent to the Mersey. The distribution of arable closely follows the Domesday record of ploughs in 1086 (Dodd 1988), similar to the ridge and furrow.

Cheshire field systems were discussed by Gray (1915, 249–58), Chapman (1953), Sylvester (1957), and Elliot (1973, 41–76, 88–92). A Waverton map of 1737 marks open-field strips that correspond to ridge and furrow surviving in 1988 (White 1988, 8). A 1798 map of Aldford shows curved strips (HER). Ten places had some open field marked on tithe maps (1838–44). The Stockport survey of 1576 describes unenclosed flats, shots, loonts (a Cheshire name for a 'land'), and some closes; there were still a few strips surviving in 1796 (Davies 1960, 54).

Irregular fields

Gray (1915, 249–58) studied fields in the Chester area; the holdings were small and the parcels not grouped into large 'fields'. Terriers of the thirteenth century for Claverton and Newton-by-Chester described dispersed holdings of eleven and fifteen selions. There was no grouping into 'great fields', and 'field' and 'furlong' names were used interchangeably. Elton had lands lying in areas specified as 'fields': *Morefeld, Brom, Morhul, Bothum, Crowegrave, Assefeld,* and *Egmundesheved*. Some selions were next to a heath, others next to a marsh (Tait 1920–3, 261–6). The same mixed usage is found at Manley in *c.*1280 (see the main text, p. 79). Sylvester (1969, 220–1) discussed field numbers found in the county. Most were single-field systems and some were multiple (see main text, p. 80). It is unlikely that the number of names reflected cultivation courses.

A grant of lands at Knutsford in 1430 mentions a townfield and a selion that abuts 'John de Oulegreave's field' and 'William de Are's croft,' suggesting there were some holdings lying in small severalty pieces and not dispersed strips. The Cheshire rod was 8 yards (and so the acre was 10,240 square yards or 2.116 modern acres (Sylvester 1957, 13, 28)). Lands were called 'butts' and furrows 'reins'. Holdings were measured in oxgangs, for example at Gayton in 1291, and Middleton Grange in Aston and Over in 1334 (Brownbill 1914, 63, 85, 95). Tait (1920–3, 237) refers to an oxgang at Church Lawton in *c.*1290. Frodsham manor had oxgangs estimated to be about sixteen acres in *c*.1350 (J. P. Dodd 1982, *THSLC* 131: 24–6). The oxgang at Shotwick was three Cheshire acres (R. Stewart-Brown 1913, *THSLC* 64: 94).

Farming

Tilston 'loondes' (three to an acre) in the town fields could have one beast grazing after harvest had been gathered (White 1995, 23). Detailed regulations for the management of the fields of Alvanley occur in 1710 (White 1988, 6). Convertible husbandry used on the demesnes of Frodsham, Wybunbury, and Tarvin have been described (p. 91).

Ridge and furrow

Chapman (1953, 58) noted the problem of identifying and classifying ridge and furrow: 'open-field type ridge and furrow was not simple to identify since there was continuous gradation of the width of ridges and their straightness or curvature'. Ridge and furrow occurred widely in the county (especially in the south-west), except for the forests, the Pennines, and sandy soils in the east. Its identification can be facilitated by the examination of RAF vertical photographs taken between 1946 and 1948 that are currently held for the south and south-west of the county in the HER. There was then extensive ridge and furrow between Hatton Heath and Hargrave, south-west of Chester (SJ 475 620) (RAF CPE/UK/935 fr. 2198 (17 January 1947)), little of which now survives. Kenyon (1979) studied and mapped ridge and furrow in the townships of Eaton, Eccleston, and Huntington from such aerial photographic evidence. Thompson et al. (1982) similarly studied Bunbury, Haughton, Peckforton, and Huxley. Ridge and furrow was mapped from 1947 RAF vertical photographs and by ground measurement of ridge widths and furrow depths. Several types of early modern ridges are known in the county, both straight and narrow curved strips, and broad ridges that have been modified by splitting (R. Williams 1984, *THSLC* 133: 9).

Chapman (1953, 39–50) found evidence of former open fields at Burton in Tarvin, Farndon, and Clotton Hoofield from the field patterns and names recorded on tithe maps. Sylvester (1957, 17) showed a similar pattern at Barnston in Great Budworth. Curved hedges at Aldersley Green were illustrated by White (1988, 14, 10).

BIBLIOGRAPHY

THSLC   Transactions of the Historical Society of Lancashire and Cheshire

Brownbill, J. (1914), *The Ledger-Book of Vale Royal Abbey*, Record Society of Lancashire and Cheshire, 68.

Davies, C. S. (1960), *The Agricultural History of Cheshire 1750–1850* (Manchester: Manchester University Press for the Chetham Society, 1960).

Dodd, J. P. (1988), 'Landscape history: open field cultivation', *Cheshire History*, 21: 25–8.

Kenyon, D. (1979), 'Aerial archaeology and the open fields—open field agriculture in medieval Cheshire', in N. J. Higham (ed.), *The Changing Past: Some Recent Work in the Archaeology of Northern England* (Manchester: Manchester University Press), 59–65.

Sylvester, D. (1949), 'Rural settlement in Cheshire, some problems of origin and classification', *THSLC*, 101: 1–34.

Thompson, P., McKenna, L., and Mackillop, J. (1982), *Ploughlands and Pastures*, Cheshire County Council Libraries and Museums Monograph, 4 (Chester).

White, G. J. (1988), 'Glimpses of the open field landscape', *Cheshire History* 21: 6–20.

White, G. J. (1995), 'The open fields and rural settlement in medieval west Cheshire', in T. Scott and P. Starkey (eds.), *The Middle Ages in the North-West* (Oxford: Leopard's Head Press), 15–35.

ACKNOWLEDGEMENTS

Adrian Tindall, Joan Collens, and Mark Leah for providing HER data and advising on literature sources.

6. CORNWALL

Cornwall topography ranges from undulating lowland to the high ground of Bodmin Moor. Large parishes contain many dispersed farms and hamlets. A sample of 13 medieval manors had tenants in 233 settlements, some being hamlets of 2–6 messuages but most were single farms. After the fourteenth century there was shrinkage and desertion (Fox 1989, 41–74). The field systems were small and irregular, with outfield cultivation.

Gray (1915, 263–66) briefly discussed Cornish field systems and Rowse (1941, 32–46) showed that open fields once existed in the county. Maps of *c*.1695 indicate scattered strip holdings at Garah and Bollowall (Pounds 1945, 23; Holden et al. 2010, 84, 194–5). Seven parishes had some open field marked in tithe maps (Kain and Oliver 1995). A detailed account of open field is given by Fox and Padel (2000) in their introduction to the Arundell Surveys. The Bodmin Moor report describes monuments and fields of various dates (Johnson and Rose 1994) and good illustrations of different types of enclosed fields are provided in the county landscape characterization study (Herring 1998).

Surviving open fields

The open strip fields next to the church at Forrabury, Boscastle (SX 095 912), are a remarkable survival (Plate 15), lying in blocks of three 'furlongs'. Some uncultivated strips on the east become lynchets as the slope increases, with five-foot risers held by walls. About fifty acres are divided into flat-ploughed, curved strips called 'stitches', divided by uncultivated balks or banks approximately four feet wide and two feet high, on which grow grass, weeds, and brambles. The strips vary in width, the narrowest being twelve to fifteen yards. A few of them were mapped in *c*.1695 (Holden et al. 2010, 282).

P. D. Wood (1963, *CA* 2: 29–33) described the site, publishing a plan and profile. The fifty-acre strips recorded in 1839 were reduced to forty by 1962. Both ownership and tenancies were scattered and intermixed. Cropping was not controlled. The thirty acres of pasture on the slopes (Willapark) were stocked in common all year. Boys were employed in 1940 to keep livestock on the pasture away from cultivated lands. In winter the stitches were stocked in common at the rate of five sheep or two cattle to each acre from November until February. In 1873 the stitches were held similarly from Lady Day to Michaelmas in severalty by the proprietors, who for the remainder of the year stocked in common according to their proportionate rights (Maclean 1873–6, i, 579).

Early fields

There is no evidence of Cornish field systems being communally organized into cropping blocks. 'Fields' of various sizes are mentioned in hamlets (townships) of Bejowan manor in 1575. Trebelzue in St Columb Minor had holdings with 38 parcels (22¾ acres) in the *Easte Felde* common field and 7 parcels (2¾ acres) in the *West Felde*. These are locational fields. Similarly locational in the same parish, at Porth Veor, were holdings with dispersed arable parcels in *Higher Felde* (nine acres), *Middle Feild* (nine acres), and *Lower Feilde* (thirteen acres). Holdings in Trebelzue had small several closes of pasture and some pasture strips intermingled in closes ('parks'), as well as 'rough and furzy pasture in the common heathy down with no division'. At Trebelzue it is specified that holdings had one rood and twenty-two poles in the town place (the green or square) called Trevlesewe (Fox and Padel 2000, 214–19).

Farms in these small townships held exactly equal shares of all types of land and of rent. The hamlet of Porth Veor had rents that were one-eighth of 56s 8d and its multiples (7s 1d), being the total rent stated in 1549, 1575, and in the seventeenth century. Possibly there were once eight farms. Terriers of two of the holdings describe exactly equal lands in all the open 'fields' and closes (Fox and Padel 2000, lxxxiii, 216–17). Similar equal division of acreages and rentals occur in Duchy of Cornwall data for 1337. Tenants of the hamlets of Trewen and Trecaine had properties that were divided exactly in terms of both acreage and rent into, respectively, 8 shares each of 9½ acres, and 9 shares each of 8½ acres. At Trewinnow Vean, the manor of Tybesta in Creed had three tenants holding 10⅓ acres each in 1356–7, but in 1337 there had been a single farm of 31 acres (Fox and Padel 2000, lxxvi, lxxxviii).

The rents of the Arundell estate in Trembleath added up to eighty shillings in rent in 1459, 1499, and c.1586, even though the number of farms varied at each date (Fox and Padel 2000, lxxxvii). This shows that holdings were divisions of whole estates and not irregularly aggregated by, say, one tenant taking on two holdings. More evidence of this share system is found in the glebe of Zennor Churchtown, which in 1727 had one-fifth of the total and one-fifth part throughout the commons on the down and cliff for pasture and fuel (Potts 1974, 177).

Infield and outfield; farming

The infield–outfield at Downinney and elsewhere has been discussed (p. 90). Pounds (1945) described infield–outfield in enclosed field systems as mapped in the Lanhydrock Atlas c.1695–6. Isolated farmsteads with a few dwellings were surrounded by arable and meadow, around which lay a belt of pasture; beyond that were closes of furze and unreclaimed moorland. Redrawn maps were published for Tregenhorne in St Erth, Bosorne in St Just and Creney, and Luxullian (Luxulyan) (see also Holden et al. 2010, 52, 81, 213).

The Atlas gives details of land-use but says nothing about cultivation. Pounds (1945, 22) assumed that the core of arable closes was used as infield, permanently in arable and manured, and the pasture closes were broken up occasionally on a convertible husbandry system. Most of the Lanhydrock maps show enclosed lands, often in single occupation, so they cannot be assumed to represent exactly an earlier open system, except perhaps for the general land-use with an arable inner core. Trevillis was fully enclosed with two town-places and eight holdings (Holden et al. 2010, 308). The closes belonging to each farm were intermixed, which presumably reflects an earlier open arrangement. The mainly open-field map of Predannack Wartha in Mullion shows that about half the main block of land was arable with a detached block of pastures. It was split into eight farms and had three arable furlongs with small groups of stitches in intermixed holdings. The other lands belonging to the farms were also intermixed, but with larger furlong pieces (Holden et al. 2010, 197). One of the seven named commons was arable—this was presumably a convertible intake that would revert to pasture in another year.

An earlier situation, in 1575, is described for Trebelzue in St Columb Minor (Fox and Padel 2000, 215–16), where more than half the pasture of a holding consisted of dispersed strips lying in closes, and so were 'known land' (see p. 183). It seems likely that these were the pasture pieces that periodically became arable.

The customary holding, the ferling, has been discussed (p. 109). Work-service at the manor of Kilcoed in St Mabyn in 1308 for each of nineteen customary tenants included ploughing, mowing, and corn carriage. By 1569, Nether Hellond customary holders owed carriage service only (Maclean 1873–6, ii, 19–22, 287–88).

A few glebe terriers of 1696 and 1727 describe small amounts of remnant open field, usually referred to as 'stitches' or 'quillets'. At Trigg in Tintagel, Trecarne common (field) lands lay next to the church and in 1696 the glebe consisted of four quillets, an acre in all, lying in four separate locations, each with Joseph Fuge as a neighbour (Potts 1974, 160). This implies a regular order, but presumably arises out of the system of re-allotment of tenants' strips.

Carew (in 1602) described the process of burn-baking (Halliday 1953, 101–2), the method of bringing into cultivation wild ground. About May the pasture to be newly broken was cut into turves (a process called 'beating'), which were raised up to dry, after which they were piled into little heaps and burned. Sea sand was brought in, and both sand and turf ash-heaps were spread out before ploughing. Two crops of wheat and two of oats were taken and the land left ley for at least seven to eight years, the farmers moving elsewhere to make another breach. Apart from its use with burn-baking, sand as well as seaweed, both raw and burnt, was used in the sixteenth century to fertilize arable land near the coast (Halliday 1953, 111). The procedure was of long standing because the tenants of Trevelgue blocked the 'sand way' of Tregustick in 1455 (Fox and Padel 2000, 5 n. 5).

Ridge and furrow

Remains of ridge and furrow cultivation are now not common in Cornwall, partly because early enclosure has encouraged its erasure by subsequent severalty farming, and probably also because the system of reallocation of land among varying numbers of tenants meant that strips did not have a long-term stability. Surviving ridge and furrow lies mostly on high ground, much of it early modern in date. A system of curved ridges three to four yards wide and *c.*200 yards long at Lanlavery, Davidstow Moor (SX 159 834), is similar in appearance to medieval broad ridge (Johnson and Rose 1994, 111–12); some lands are seven to eleven yards wide. There are also narrow ridges of three yards width that are likely to be an early modern modification of older ridges (D. Hall, field visit 1999).

Cultivation ridges on Bodmin Moor are difficult to interpret. (For a general discussion with plans and photographs see Johnson and Rose 1994, 103–44; and Maps ii, iii, and fig. 5, with a contribution by Peter Herring.) Late straight furrows, both wide and narrow, are easily identified, but curved narrow ones are more difficult to classify. If they are medieval, then they form lands smaller than elsewhere in the county. According to sixteenth-century surveys strips were normally a rood or half an acre (Fox and Padel 2000, lxxxiii). Rood stitches are recorded for Phillack and Tintagel (Potts 1974, 126, 160). Perhaps the ridges are the result of cultivation in severalty, and have no tenurial significance, the walled or banked enclosures being the smallest unit of the holdings.

Rough Tor site (SX 147 804) is divided into walled enclosures grouped into three units of three. These are the oldest features and seem to be the tenurial units. Within them are irregular, sinuous ridges, 2–2½ yards wide. They do not occupy the whole of the enclosures, unploughable stony parts being left as rough pasture. Since the ridges are not parallel to the walls, the enclosures are unlike closes formed by division of an arable furlong. At the lower, southern, ends of some of the enclosures are banks of soil or 'headlands', indicating that there had been a considerable period of ploughing (a section is illustrated in Kirk, 1899, fig. 68). However, since the furrows continue over the headland they probably represent a later stage of ploughing that has obliterated any earlier form.

At Brown Willy, Herring describes schemes for the development of the walled fields, some of them possibly being outfields. The walled enclosures are to be associated with three small settlements of the thirteenth and fourteenth centuries (Herring, in Johnson and Rose 1994, 107). The date of the ridges is uncertain, but certainly before *c.*1840, when closes in arable cultivation at the time of the tithe survey were now flat and those that were then pasture lay in narrow ridges (P. Herring, pers. comm.).

Remnant strips

Strips defined by low stone banks 6–20 yards apart and 220 yards long are known on much of Cornwall's moorland. Some contain cultivation ridges two to three yards wide and six inches high. Examples are Belowda Beacon, Roche (SW 971 623), Rosenannon Downs, St Wenn (SW 955 672), and on Kit Hill, Stoke Climsland (SX 374 708) (Preston-Jones and Rose 1986, 152–3). The strips divided by stones are likely to be medieval and the cultivation ridges within them are probably early modern.

The same type of narrow closes containing ridges occur at Garrow (SX 147 783) (Beresford and St Joseph 1979, 97–8; Johnson and Rose 1994, figs. 5, 71; Herring 1998, fig. 15) and Fernacre (SX 149 794) (Johnson and Rose 1994, fig. 72). Excavations at Garrow produced thirteenth-century pottery from the buildings and the ploughsoil (D. Dudley and M. Minter 1962–3, *Medieval Archaeology*, 6–7: 272–9). There was also pottery of the seventeenth to twentieth centuries that may account for the post-medieval ridges.

Irregular ridges at Gwithian (ninety by three yards mean) have been assigned a Dark-Age date. The plan does not suggest that they were formed by ploughing and the strips are more akin to lazybeds, an interpretation not accepted by the authors (J. P. Fowler and A. C. Thomas, 1962, *CA* 1: 61–84; Fowler 2002, 130 and 150–2). Field patterns of prehistoric type occur at West Penwith, Chysauster, surrounded by a complex of ovoid and squarish fields defined by earthwork scarps and lynchets. Blocks of these form an irregular pattern quite unlike relict open fields (illustrated in Herring 1998, 26, fig.16).

Furlong-type field boundary patterns are recorded on many tithe maps. At Trevarrian and Tregurian in St Mawgan in Pyndar, the tithe map of 1841 shows a small area on the west that was still open. Preston-Jones and Rose (1986, 152) published the tithe field pattern of Treskilling (1840). Similar field patterns occur at Trevalga near Boscastle (CUCAP, RC8 H28 (1969); oblique aerial photograph in Herring 1998, 28). Altarnun is also well known for its furlong-type field patterns (Herring 1998, fig. 48; CUCAP, HS 12 and 14).

BIBLIOGRAPHY

CA   *Cornish Archaeology*
DCRS   Devon and Cornwall Record Society

Fox, H. S. A. (1989), in R. Higham (ed.), *Landscape and Townscape in the South West* (Exeter: University of Exeter), 41–74.

Halliday, F. E. (1953), *The Survey of Cornwall*, (London: A. Melrose); for a facsimile see *The Survey of Cornwall by Richard Carew*, ed. J. Chynoweth, N. Orme, and A. Walsham (2004), DCRS, NS, 47 (Exeter).

Herring, P. (1998), *Cornwall's Historic Landscape* (Cornwall County Council and English Heritage).

Johnson, N., and Rose, P. (1994), *Bodmin Moor: An Archaeological Survey*, i (London, RCHM).

Potts, R. (1974), *A Calendar of Cornish Glebe Terriers, 1673–1735*, DCRS, NS, 19.

Rowse, A. L. (1941), *Tudor Cornwall* (London: Cape).

ACKNOWLEDGEMENTS

Peter Herring and Peter Rose for information from the HER and a field trip to Forrabury.

7. CUMBERLAND

Cumberland has extensive coastal plains in the west rising to the high ground of the Cumbrian Mountains and Pennines, where there is dispersed settlement. An appreciable area lies above 1,500 feet. The rivers Eden, Lyne, and Esk form long valleys that drain to Carlisle and the Solway Firth. High ground in the fells throughout the county was used for cattle and sheep rearing, often by transhumance. Herders moved their cattle for the summer and lived in cottages called 'shielings' or 'scales'. There were also shielings among sand dunes along the coast. By the seventeenth century many of them developed into permanently settled farms (Ramm et al. 1970, 1–12, 61–70, 79–94; Winchester 1987, 81–99). Cumberland fields were characterized by the occurrence of single-field systems and irregular fields with the use of convertible husbandry and outfield.

Part of the map of Hayton, 1604, shows furlongs with lands up to 700 yards long (Elliot 1973, 44; T. H. B. Graham 1908, *TCWAAS*, NS, 8: 341–51). A map of Renwick, 1815, shows a chequerboard plan of furlongs with three in an alignment (Winchester 1987, 73). Tithe maps of 1839–51 have some open field marked in twenty-two parishes (Kain and Oliver 1995). Gray (1915, 227–42) discusses Cumberland field systems and Elliot (1973, 42–59, 76–81) gives an account of fields and enclosures. Winchester (1987, 59–80) describes many aspects of Cumberland fields.

Graham (1910), quoting from Eden (1797), noted that at Croglin 'a great part of arable land still remains in narrow crooked dales, or raines as they are called'. Castle Carrock had 'dales' or 'doles', 'which are strips of cultivated land belonging to different proprietors separated from each other by ridges of grassland'. Cultivated land at Warwick, 1,126 acres, was similar. Gray (1915, 227–8) found evidence of some open field in the late eighteenth century and he quoted examples of dispersed glebe in 1704 (from Ferguson 1877). A survey of one of the quarters of Holme Cultram, made in 1604, showed that land was dispersed among seven small hamlets that were separate townships. A 1608 survey of the Carlisle region showed similar hamlets with small areas of arable.

Elliot (1959, 85–104) collected evidence for open field in the whole county in the sixteenth century and earlier (plan p. 86). Most places had some open field, except those with very high ground and the belt along the Scottish border. There were pockets of arable in the dales: Wasdale Head, for example, had 345 acres of open field in 1567. Most open field occurred in the lowland, but even here it was often still a low percentage of the township area because of poor, waterlogged soils. The open fields were part of a system of pastoralism; animals grazed on the extensive commons during the summer and were brought to graze on the fallow in the autumn to fertilize the arable fields and because there was a limited amount of hay (Elliot 1959, 86–91). Ten tenants at Wasdale Head each had a small garth of one to two poles and three to ten acres of 'arable and meadow' in the field (Bouch and Jones 1961, 61). This probably means that the arable was convertible. (A photograph of enclosed Wasdale Head can be seen in Hoskins 1955, 84.)

Some aspects of Cumberland fields, with the use of convertible husbandry and outfield (Elliot 1959, 92–3, 97–8) and the occurrence of dispersed demesne, have already been described (p. 98). Strips were usually called 'dales' or 'riggs' and the grass balks between them 'ranes' or 'reans' (Graham 1908, *TCWAAS*, NS, 8: 42–3). Furlongs were named 'flatts' or 'rivings' (Elliot 1959, 87) and sometimes 'brakes', 'fields', 'ings', 'crofts', etc. (see Ferguson 1877).

Field orders

During the sixteenth and seventeenth centuries, Holm Cultram was regulated with by-laws that concerned breaches of manorial custom, rights of way, pasture stinting, and the determination of the boundaries of pastures and townships; each township was to keep a common herd; digging turves for roofing was controlled; and oaks were to be used to repair the sea dyke (Grainger and Collingwood 1929, 204–15). Protective fences around arable fields and field closes opened up for commoning in 1518 at Aspatria and Woodside (Elliot 1959, 101; Winchester 1987, 59–60). The tenants of the Ure-hend (Upper End) of the town of Urswick, in 1559, were not to drive their beasts to the Nederhend, nor the Nederhend to Urehend, till all the corn had been gathered (F. Barns and J. L. Hobbs 1957, *TCWAAS*, NS, 57: 61). Court proceedings at Troutbeck in 1570 referred to disturbance of neighbours' cattle, fishing in the lord's water, and having unringed pigs (Curwen 1923, ii, 50).

Court records representing one third of the county showed that the main business concerned encroachments, grazing, hedges, land and crops, and turbary. Turves (called 'flacks') were used for roofing. At Aspatria it was ordered in 1741 that flacks should be dug in the outfields. In 1679 Cottagers at Bolton having fewer than four acres of land could not cut peat for more than two days. Some lowland mosses had an allotted share for each tenant and they were not allowed to encroach on a neighbouring 'room' (R. S. Dilley 1967, *TCWAAS*, NS, 67: 125–51). Greystoke parsonage had a peat house in 1704 and Orton glebe had two great parcels of moss and the right to cut turf for four days on the moors (Ferguson 1877, 219, 167).

Farming

Oxgangs at Wetheral, in *c.*1210, had work-service to carry the abbot's corn, send one man to mow in the autumn, plough one day, and carry timber to the pond and mill. Villeins were to have all their corn ground at the mill (Prescott 1897, 172). Many Pennine forests had vaccaries; Cockermouth estate when surveyed in *c.*1270 had several. The remote daleheads were used for sheep. Rent at Keskadale was received for a scale (A. J. L. Winchester 2003, *TCWAAS*, 3rd ser., 3: 109–118).

The northern border with Scotland suffered raids and the herders lived in fortified farmhouses called 'bastiles' to protect them from intruders. Animals were kept on the ground floor and the herders lived above. Bastiles ceased to be built in the seventeenth century and many were converted to farm houses with the arrival of less hostile times (Ramm et al. 1970, 61–70).

Ridge and furrow

No broad ridge is recorded in the CUCAP collection nor was observed during a field visit in made in 2003. Riggs are frequently referred to in eighteenth-century terriers and some probably survive. Graham (1913, 1–31) recorded small areas of strips, or groups of a few strips, divided by three-foot grass balks or ranes at Skelton, Ellonby, Hutton in the Forest, Burgh by Sands, and elsewhere. Gray (1915, 227), quoting Eden (1797), said that at Cumrew grass ridges in the fields were 20–40 feet wide and 330 yards long.

Numerous examples of early modern narrow ridges survive. Wiza Beck, Westward (NY 267 450) has straight ridges three yards wide on a steep slope. Similar ones were evident in 2003 at Brocklebank (NY 305 437), Ratten Row, and Caldbeck (NY 322 406) (see HER 3794). The HER records ridge and furrow, some of it specified as post-medieval in date. The stone circle of Long Meg near Hunsonby has narrow rig around and within it (CUCAP, CLW 51). Strip lynchets are known on some steep slopes, being recorded by Graham (1913, 11–12) at Greenhow and Kirkland. Reverse-S curved hedges occur in some lowland townships, for example Brayshaw (NY 051 090). West of Caldbeck reverse-S shaped walls lie on a slope (NY 296 393).

BIBLIOGRAPHY

TCWAAS   *Transactions of the Cumberland and Westmorland Antiquarian and Archaeological Society,* in three series

Ferguson, R. S. (ed.) (1877), William Nicolson, *Miscellany Accounts of the Diocese of Carlile* [sic], *With the Terriers Delivered in to Me at my Primary Visitation* (London: G. Bell).

ACKNOWLEDGEMENTS

Caroline Hardie for providing HER information.

8. DERBYSHIRE

The county has contrasting topography, from the Peak District in the north to the Trent Valley in the south, joined by the central River Derwent. Some northern townships were two-thirds moorland waste. Many parishes contained several townships; Glossop, for instance, had ten (D. Brumhead and R. Weston 2003, *DAJ* 121: 245). Regular two- and three-field systems occurred in the south, while the higher ground of the north contained irregular fields with convertible husbandry utilizing the pastures.

Parts of Over Haddon open fields were mapped in *c.*1528 (E. M. Yates 1964, *Agricultural History Review*, 12: 121–4). William Senior made surveys and maps of the extensive Chatsworth estates in the early seventeenth century (see p. 80). The Senior map of Little Longstone published by Wright (1906, 290) can be compared with the written survey (Fowkes and Potter 1988, 33–6). Buxton and Sheldon were mapped by Senior in 1631 (Jackson 1962, 62–7l). Kain and Oliver (1995, 109–21) record six places that had some open field left in 1841–9.

Field numbers

Gray (1915, 460–1) only refers to three Derbyshire townships, all of which he classed as being three-field, although one of them (Osmaston) is dubious. Jackson (1962) described the extent of open field in the county. Table Add.4 provides more information, with three two-field and eight three-field systems that lie in the valleys of the Derwent, Trent, and headwaters of the Don. Beighton on the Don is probably a satisfactory example. The three seventeenth-century examples in the Peak (Abney, Bradwell, and Wardlow) are unlikely to be regular three-field. Two demesnes were probably in block form. Both 'oxgangs' and 'yardlands' were used; oxgangs are most commonly found in charters (such as those in the Darley Abbey Cartulary (Darlington 1945)). Sizes varied (see p. 111). Some assarts were called 'oxgangs' and were undispersed; at South

Wingfield an oxgang consisted of ten acres in an assart and two acres nearest to the same assart in 1198–1225 (Darlington 1945, ii, 349–50).

Irregular fields

Many published medieval charters offer little open-field information and almost none describes regular large fields. This is striking in the Darley Abbey Cartulary, which has no obvious 'great fields' in the Derbyshire section, but in contrast contains at the end detailed fourteenth-century terriers for three regular fields at Keyworth in Nottinghamshire (Darlington 1945, 530–2). The furlongs of Little Ogston were described in detail in *c.*1270 (probably the whole township except the manor-house and *Elonds* ('inlands') furlong). Among the names were a 'field called the Lower Furlong', a 'furlong called Roger's croft', and another called 'Hengendecroft', showing interchangeable use of 'field' and 'furlong' where extensive fields signifying cropping arrangements did not apply. There were some leys and meadows, and also woods and pastures included (Darlington 1945, 459–61). Catton (1327) had eleven acres described in eleven furlongs, but no great fields are mentioned (Jeayes 1906, 75). A 1413 terrier of Birchover described lands dispersed in many furlongs (a few called 'fields') but not

Table App.4. Derbyshire field systems.

Place	Demesne	Field No.	Date	Source
Abney		2	1658	J. P. Carr 1963, *DAJ* 83: 71
Bradwell		2	1658	J. P. Carr 1963, *DAJ* 83: 71
Staveley		2	1331	R. R. Darlington 1957, *DAJ* 77: 71; terrier of three acres
Allestree		3	1737	Jackson 1962, 60
Ashton on Trent		3	1219	W. H. Hart 1885, *JDANHS* 7: 213. Oxgang 10½ acres
Barlborough		3	1795	Jackson 1962, 61; large areas of commons and closes
Chaddeston		3	1258	W. H. Hart 1888, *JDANHS* 10: 153; three-acre terrier
Daslow	block?		1378	C. Kerry 1901, *DAJ* 23: 3–5; demesne lay in a few furlongs
Glapwell		3	1230	R. R. Darlington 1956, *DAJ* 76, supplement, charters 5 and 134 (dated 1311)
Ockbrook		3	1773	Jackson 1962, 68
Repton		3	1550	F. Williamson 1933, *DAJ* 73: 84–3; small grants
Wardlow		3	seventeenth century	Senior; and J. P. Carr 1963, *DAJ* 83: 71
Wessington	block?		1254	Darlington 1946, 461; six oxgangs in a block; not stated to be demesne

grouped into rubricated great fields (Anon. 1900, *DAJ* 22: 49–51). Similarly, Senior's twenty-three Derbyshire surveys name some pieces called 'fields' that are unlikely to be great fields; nearly all the townships have irregular fields and most have a low percentage of open-field.

Farm management

Convertible husbandry at Great Longstone has been described (p. 88). A glebe terrier of Carsington for 1698 refers to ground set out for ploughing in part of the common pasture called the 'Breck' (Jackson 1962, 65). Convertible husbandry continued to be used in enclosed townships in the nineteenth century. Foolow, in the High Peak, had a rotation of grass for several years, followed by two or three successive crops of oats, then fallow reverting back to grass (J. Beckett and J. E. Heath 1995, *Derbyshire Tithe Files*, DRS 22: 60).

Court rolls of Baslow in the fourteenth century list offences; for instance, owners of cattle trespassing in corn in 1356 had to pay compensation of ten sheaves. Tenants of Middleton and Eyam were fined for pasturing cattle on Baslow moor (C. Kerry 1900, *DAJ* 22: 68–80). Later courts refer to animals trespassing, the cutting of timber in the lord's wood, a ruinous sheepfold in need of repair, and the trespass of a sow and pigs, for which wheat was to be paid in compensation. In 1484 tenants were to make a 'pinfalder' (a penfold for stray animals). Common pastures were kept severally in 1502. Peat was dug on East Moor in 1503 (Kerry 1901, *DAJ* 23: 1–29).

At Temple Normanton in 1488 each oxgang was to repair the town lane with two cartloads of stones or cinders. All tenants were to repair the pinfold (1507). No one should keep animals in the fields except on his own land nor oppress the common (1511). Hedges were to be made around the commons, and horses were to be tied only on their own land (1511; R. H. Oakley 1958, *DAJ* 78: 40–88). At Abney the court of 1654 determined regulations about tethering beasts in the cropped field until harvest had been taken, and controlled the stinting and the times of grazing the stubble. In 1664 the orders were continued and a detailed list was produced in 1683 (C. E. B. Bowles 1907, *DAJ* 29: 136–40). Oxgangholders at Eckington in 1650 could sell coal dug on their own land (Hall 1924, 79).

Ridge and furrow

Ridge and furrow occurs over much of the county; that at Brassington lying within curved hedges is well known (Beresford and St Joseph 1979, 133) and has been discussed by Wightman (1961, *DAJ* 81: 112–17) and Dalton (1991, 89).

Strip lynchets were recorded by Young and Jackson (W. H. Young 1930, *DAJ* 50: 98–100; J. W. Jackson 1957, *DAJ* 77: 62–3, and 1962, *DAJ* 82: 98–103). A fine photograph of lynchets at Wensley, Matlock, has been published (D. N. Riley, in Hart 1981, pl. 8). Open-field patterns are preserved in wall lines as noted by Young, Wightman, and Eyre (1955, 81–4) and Carr (1963 *DAJ* 83: 70–5). Beresford and St Joseph (1979, 99) published a photograph of the Chelmorton narrow walled enclosures. Curved field patterns are evident on modern Ordnance Survey 1:25,000 maps and in online vertical images of Brassington, Taddington, Tideswell, and Wardlow. Very fine photographs of Wardlow curved walls were taken in 1975 (CUCAP, BWH 20, SK 17 75).

BIBLIOGRAPHY

DAJ   *Derbyshire Archaeological Journal*
DRS   Derbyshire Record Society
JDANHS   *Journal of the Derbyshire Archaeological and Natural History Society*

Dalton, R. (1991), 'Maps of the Eggington enclosure award', *DAJ* 111: 85–92.

Eyre, S. R. (1957), 'The upward limit of enclosure on the East Moor of north Derbyshire', *Journal of British Geographers*, 23: 61–74.

Hall, T. W. (1924), *Descriptive Catalogue of the Edmunds Collection* (Sheffield: J. W. Northend).

Hart, C. R. (1981), *The North Derbyshire Archaeological Survey* (Leeds: A. Wigley & Sons).

Jackson, J. C. (1962), 'Open field cultivation in Derbyshire', *DAJ* 82: 54–72.

ACKNOWLEDGEMENTS

Dave Barrett and Andy Myers for HER information.

9. DEVON

Devon has varied topography, much of it undulating and hilly, with high ground towards Exmoor in the north. The granite intrusion of Dartmoor lies at the south-west, reaching 1,800 feet in places. Many parishes comprise several townships with dispersed hamlets and isolated farms of medieval date. Examples are Widecombe (Gawne 1970), Chittlehampton (Figure 1.1), Sampford Courtenay (Fox 1999, 277), and Ottery and Sidbury (Fox 1972, 90). Cullompton has 13 medieval hamlets and farms (M. J. Foster 1910, *TDA* 42: 159–62), and Axminster contained 11 hamlets in its 6,878 acres (W. H. Wilkin 1936, *TDA* 68: 359–73). The dispersed settlement pattern with pastures was associated with small irregular fields and convertible husbandry. More extensive cultivation occurred in East Devon; the fields were enclosed early and therefore have limited records of their open-field state.

Devon field systems have been discussed by Gray (1915, 258–66), Finberg (1952), and Fox (1972, 1975). Some places with open fields were mapped: Braunton (Tithe map of 1840, published by J. B. P. Pheare 1889, *TDA* 21: 200–4), Challacome, Dartmoor, in 1787 (published in Pattison et al. 1999), and Stoke-in-Teignmouth in 1741 (Devon CRO, 5632/81). These places and Ilfracombe were recorded on tithe maps of 1838–52 (Kain and Oliver 1995). The open-field strips, still cultivated, at Braunton Great Field near Barnstable, are well known (Beresford and St Joseph 1979, 43–4; Kain and Oliver 1995, 129). They are flat ploughed with grass balks called 'landsherds' lying between them. The overall pattern is similar to Midland furlongs. Beresford and St Joseph show seven furlongs with aligned strips forming a long curve.

Gray (1915, 412) found little evidence of medieval open fields. Surveys described small closes called 'parks'. He noted the 1566 Dynham survey published by R. P. Chope (1902, *TDA* 34: 418–54) had evidence of intermixed fields at Ilsington and Woodhuish; he also referred to Braunton (ibid. 258–66). Finberg (1949 *Antiquity*, 23: 180–7; 1952,

265–88;) showed that there had been open fields in many parishes. Medieval examples of common grazing on arable were discovered. Fox (1975, 181–6) mapped places that had subdivided arable fields in the thirteenth and fourteenth centuries. East and south Devon had by far the most. Demesne land-use for many manors in these two regions showed that arable predominated. In a sample of *c*.11,000 acres, arable accounted for 70 per cent in east Devon and 79 per cent in the south. Terriers for these regions show that strips were located in 'fields' and 'furlongs', using the terms interchangeably, with no indication that there was any system of 'great fields'.

There is no firm evidence of Devon fields being organized for regulated cropping. The examples below relate to demesnes and are likely to mean 'tilth systems' rather than 'great fields.' Finberg (1952, 287 and 273) concluded that with an abundance of pasture there was no need to leave half or a third of the arable as fallow. He lists several places where small demesnes up to one hundred acres were tilled in two tilths in 1377. Lydford demesne, in 1448, lay mainly in two fields that were sown in alternate years (Yates 1964–8, 147). At Woodhuish in Brixham, the lands of the manor 'for the most part lyeth by londes score in twoe common feldes'. The average for a ferling holding of arable land in the fields and breches was twenty-seven acres (R. P. Chope 1911, *TDA* 43: 281–2). Netherton in Farway demesne was dispersed in 1323 equally among the *Estfeld, Myddlefeld cum Chelshamcrofte*, and in *Westfeld*. Northleigh had a small demesne fairly equally distributed in three fields (Finberg 1952, 272–3).

Previously discussed are the customary holding ('ferling'), convertible husbandry, and demesnes, both dispersed and compact (p. 98). The manor of Stoke Fleming still recognized ferling holdings in 1655; a copyhold entry for one farthing of land had a heriot of one cow (E. Windeat 1884, *TDA* 16: 177).

Fox identified distinctive aspects of Devon medieval agriculture, which included transhumance of animals over considerable distances. Cattle were taken for summer grazing to Dartmoor from all the parishes round it and as far away as North Devon. Some went to grounds adjacent to Exmoor. Cattle were also fattened in the Exe Valley (Fox 1999, 274).

Some work-service is recorded for open-field lands. A ferling of land at Buckland Filleigh in 1266 had the services of ploughing, sowing, and reaping. Another ferling (1268) had the same services and the tenant was allowed to cut turves, but no more than he needed to burn or use for manuring (O. J. Reichel 1909, *TDA* 41: 241–3). The manorial economy of Plympton, Tiverton, Exminster, Topsham, and Barton in the thirteenth century had light work-service. Some freeholders with a ferling of land ploughed 1½ acres. Fertility by 'sanding' was used at Plympton, and beat-burning (burn-bake) at Plympton and Topsham in 1225–6 (K. Ugawa 1962, *TDA* 94: 630–83).

Ridge and furrow

No examples of broad ridge and furrow are known. In a county early enclosed, old ridges will not often survive, because closes under grass are sometimes ploughed and reseeded. Many of them were arable by the 1840s, as shown by tithe map land-use schedules. Convertible husbandry, described in 1808, accounted for seven-eighths of the land being ploughed. The Hartland Survey shows that many closes in 1566 were arable. Dartmoor is a likely location for survival of medieval ridge and furrow, but interpretation is complicated because of widespread early modern ridging.

A probable example of medieval ridging occurs on a steep slope below Upcott Barton, Thorverton, near Exeter (SS 923 042). The ridges are very low-profile, six yards wide and cut by a leet (gutter) that carried spring water along the hillside to feed pasture (Plate 16). The leet is one of several in the same valley and others can be seen in many parts of the county (cf. Griffith 1988, pl. 90, for examples at Tiverton). The Upcott leet is cut by the present field banks and hedges. Assuming a fourteenth–fifteenth century date for the hedges, then the strips are plausibly medieval. Ridges of this type would not survive a single flat-ploughing and would leave no soil marks. Flat strips separated by balks approximately twenty yards apart occur at Holne Moor, Dartmoor. They appear to pre-date the enclosures in which they lie. Unridged strips of around ten to fifteen yards width defined by stone (clearance) balks occur at Black-slade Down (SX 736 757) and elsewhere (A. Fleming 1994, *DASP* 52: 109–17).

Narrow ridges of post-medieval date, 2–3½ yards wide occur at many places on the edge of Dartmoor (A. Fleming 1994, *DASP* 52: 101–9). They often lay in abandoned fields and were later than medieval ridges and lynchets. They may be the result of out-field cultivation of the seventeenth–eighteenth centuries. Butler maps 'ridge and furrow', most of which is early modern (Butler 1991, Maps 16 and 17, etc.). Good examples of late cultivation ridges lie near a side track to Drury Farm (SX 665 776), 2¼ yards wide. They are cut by a roadside quarry and leave a fifteen-yard gap beside the farm track (marked on Butler 1991, Map 18 and p. 139).

Lynchets of prehistoric date can be found on Dartmoor and the remarkable parallel divisions defined by low 'walls' called 'reaves' are extensive. These were thought to be prehistoric in 1848 and Curwen (1927, 277) suggested a late Bronze Age date. Elizabeth Gawne and J. Somers Cocks (1968, *TDA* 100: 277–91) made a detailed study confirming the date. A plan of Sherril (Sherwell) (SX 68 75) shows reaves cut by the later, now deserted, fields, some containing ridge and furrow (Gawne 1970, 64; fig. 6). There have been further detailed studies by Andrew Fleming (1988), and additional areas of reaves were mapped by Butler (1991).

Strip lynchets of medieval date are known on steep scarps in the county, for example above Saunton sands (SS 446 379), where five terraces have risers of fifteen to eighteen feet (*Devon and Cornwall Notes and Queries*, 1940, 23: 131). Massive lynchets occur at Hexworthy (SX 668 729) and Eadon Down, Manaton (SX 736 826) (Gawne 1970, 61–2). Some of the most spectacular strip lynchets occur at Challacombe, in Manaton, Dartmoor (SX 69 79). They lie on the hillsides adjacent to the medieval vill, now reduced to a single farm. Challacombe was first described by Shorter (1938), who published a plan with excellent photographs (see also Hoskins 1954, pl. 47). The terraces lie between 1,100 and 1,350 feet and follow the contours lying in blocks equivalent to 'furlongs' divided by walls or gulleys. They vary from 5 to 30 yards in width and from 110 to 320 yards in length. The risers, from 1½ to 7 feet, are made up of clearance stones. In two areas at the south the strips run up and down the contours and are defined by 'balks' of stones. The whole system of lynchets and closes extends to 800 acres.

Bonney (1971) described Challacombe and noted that many lynchets had narrow cultivation ridges attributed to the eighteenth or nineteenth centuries. He thought the up-and-down strips were later modifications of an earlier lynchet system. In 1613 there were five tenements, which can be traced up to 1880. Other eighteenth-century sources have more detail, showing that the lynchets were called 'landscores' and the 'furlongs' were called 'wares', as shown by Gawne (1970, 57–8). Holdings were scattered.

Pattison et al. (1999, 61–70) describe the linear medieval settlement (fig. 24); it was probably planned. A map of the lynchets made in 1787 (figs. 25–6) shows enclosures near to the village and at the south that were marked on the tithe map. A field survey showed that at the south-east next to Hamel Down, lynchets, not mapped in 1787, existed in enclosed walled fields; east of them are abandoned strips defined by low banks or lines of stones, where there are traces of ridge and furrow and groups of clearance heaps. The lands of three farms mapped in 1787 were completely dispersed amongst the lynchets and closes (fig. 26). The abandoned fields were not mapped. Challacombe was mentioned in the thirteenth century.

A site visit made in 2004 found that the outer limit of the area cultivated in the eighteenth century is defined by a substantial wall with outer ditch. It is all pasture. Bracken, but not heather, has encroached and much of it has narrow nineteenth-century type ridges (3¼ ft). These occur on some of the terraces and also on the south-west of the very steep slopes immediately west. The farm that has no lynchets and was likely to have been permanent 'medieval' pasture. The abandoned closes at the southern end of Challacombe Down are thick with bracken and heather. On the west, the steep surface has rocks and hollows and the closes are not likely to have been ploughed. On the eastern slope are cross-contour ridges three yards wide. The date is uncertain—they may have been medieval outfield. The presence of heather seems to indicate antiquity.

Elsewhere, on steep slopes at Hound Tor (SX 746 788) (340 m, 1,133 ft) outside of the enclosures are terrace-type divisions marked by stones and having rough uneven land between them, similar in principle to the Challacombe cleared strips (mapped in Beresford 1979, 151). A few strip fields separated by grass balks survived at Brixham in *c*.1950 (A. H. Shorter 1950, *TDA* 82: 275, pls. 6–7). The field system of Parracombe, near Exmoor, is remarkably preserved by slight lynchet-type soil banks on the steep slopes of surrounding spurs (Figure App.1). The 'furlongs' had, in most cases, been individually enclosed, causing the modern field boundaries to reflect them (compare Devon CRO, Map 1262/M/E22, nos. 20–2, 26).

Enclosed field patterns

Devon had much pioneering work done on its field patterns. Worth pointed out in 1896 that the tithe map of Compton Gifford (1840) had some very small fields of one to two acres, which were individual or double lands with many curved hedges. Field names preserved sixteenth- and seventeenth-century owners' names (G. N. Worth 1896, *TDA* 128: 714–70). Field-names and hedge patterns at Brixham were discussed by Shorter (1950, *TDA* 82: 271–9), who noted the difference between those adjacent to the town and those farther out associated with farms and hamlets. Next to the town were many small narrow 'rectangular' enclosures with curved hedges, groups of which had the same names (figs. 3–5) that include 'landscore' and 'clinage' (showing the fields are enclosed furlongs). The narrow strip fields included divisions with turf balks two feet high (illustrated by Shorter, pl. 6) and were further characterized by the absence of old

Fig. App.1. Strip fields at Parracombe near Exmoor, Devon, in 1989 (copyright Devon County Council, photograph by Frances Griffith, DAP/LC3/10/01/1989).

farms. Holdings were dispersed in 1840. Farther out, fields were larger and more irregular (fig. 2).

Finberg and Hoskins (1952, 278–81) noted strip patterns fossilized as hedges in several places. They studied twenty-four tithe maps and found that eighteen of them had evidence for strip fields. Those with 'church towns' were often nucleated and surrounded with (hedged) strips, and outlying hamlets had strips of their own. Kenton tithe-map fields were published (fig. 2, p. 280). Blocks of small, strip-like fields were noted in the Ottery St Mary region (R. R. Rawson 1953, *Economic History Review*, 6: 52–4, with plan). The English Heritage analysis of the modern landscape has shown that there are extensive remains of such hedge patterns in the county, except on Dartmoor (Turner 2007, 55; for Devon fields of all types see pp. 27–79).

BIBLIOGRAPHY

DASP   *Devon Archaeological Society Proceedings*
DCRS   Devon and Cornwall Record Society
TDA   *Reports and Transactions of the Devonshire Association*

Beresford, G. (1979), 'Three deserted settlements on Dartmoor', *Medieval Archaeology*, 23: 98–158.

Bonney, D. J. (1971), 'Former farms and fields at Challacombe, Manaton, Dartmoor', in K. I. Gregory and W. Ravenhall (eds.), *Exeter Essays in Geography* (Exeter: University of Exeter), 83–91.

Butler, J. (1991), *Dartmoor Atlas of Antiquities*, i (Exeter: Devon Books).

Fleming, A. (1988), *The Dartmoor Reaves* (London: Batsford).

Fox, H. S. A. (1999), 'Medieval agriculture', in Kain and Ravenhill 273–80.

Gawne, E., and Somers Cox, J. V. (1968), 'Parallel reaves on Dartmoor', *TDA* 100: 277–91.

Griffith, F. (1988), *Devon's Past: An Aerial View* (Exeter: Devon Books).

Hoskins, W. G. (1954), *Devon* (Exeter: Devon Books).

Pattison, P., Field, D., and Ainsworth, S. (1999), *Patterns of the Past* (Oxford: Oxbow), 61–70.

Shorter, A. H. (1938), 'Ancient fields in Manaton parish, Dartmoor', *Antiquity*, 12: 183–9.

Stanes, R. (1994), 'Braunton Great Field management study.' Report for North Devon District Council; copy in HER.

Turner, S. (2007), *Ancient County: The Historic Character of Rural Devon* (Exeter: Devon Archaeological Society).

Yates, E. M. (1964–8), 'Dark Age and medieval settlement on the edge of wastes and forests', *Field Studies*, 2: 133–53.

ACKNOWLEDGEMENTS

Bryony Coles for the discovery and photography of Upcott Barton ridge and furrow. Frances Griffith, Devon County Archaeologist, and John Allan for discussion, HER data, and permission to publish the excellent photograph of Parracombe. Sam Turner for information from the HLC.

10. DORSET

Dorset consists mainly of chalk downland with limestone in the north-west and sand in the south-east, where settlements are dispersed. Most of the central part of the county had townships with two- and three-field systems. There were differences between the limestone region that had extensive fields of Midland type, as at Hinton St Mary (Figure App.2), and the chalklands with smaller proportions of arable linked to ample downland pastures used for sheepfolding (Chilfrome, Figure 1.2).

Dorset had several examples of open-field systems that survived to a late date. Fordington remained open until 1874 and details were recorded of its structure and operation (Moule 1892). Holdings of 40–60 acres were called 'places' and could run a flock of 120 sheep. A redrawn plan of 1779 shows the extensive furlong system (Woodward, in Sharples 1991, 18). Stratton and Grimstone fields, open in 1895 (Pope 1909), were similar to extensive systems of the Central Region. Portland fields still had a few strips cultivated until 1970 called 'lawns' and separated by grass balks called 'lawnsherds'. They have been described by Drew (1947, with photographs), by Bettey (1970, and Bettey 1971, *PDNHAS* 92: 224–9), and by Beresford and St Joseph (1979, 42–3). Tithe maps of twelve townships, dated 1837–49, have some open-field (Kain and Oliver 1995), as at Stourpaine (Figure App.3).

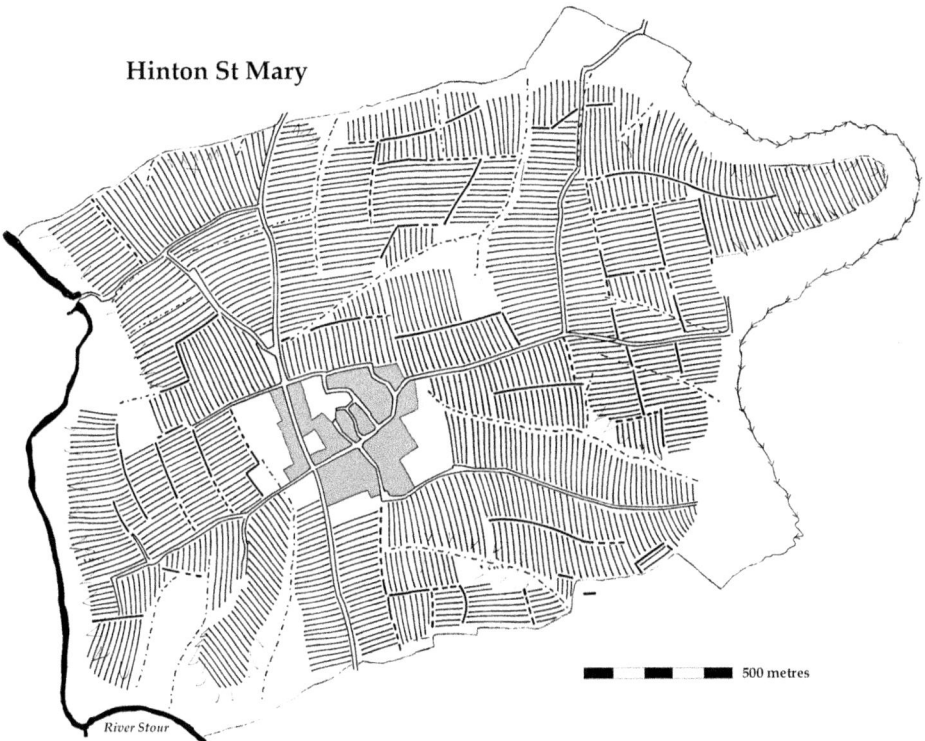

Fig. App.2. Hinton St Mary, Dorset, showing a Midland-type furlong pattern (field survey by D. Hall and P. Martin, 2003).

Field numbers

Examples of townships with two and three fields are listed below. Gray (1915, 461–2) recorded twelve two-field townships before 1400 and fifteen with fourteenth-century monastic demesnes cultivated in three-tilth arrangements. The demesne of Lillington was dispersed in 1215 (London 1979, Wiltshire Record Society, 35: 117–18).

West Lulworth was complex in the eighteenth century (Dorset CRO, D/WLC/P1/15), with five open fields and four old enclosed farms: St Andrews, 771 acres (1284); Bury-gate, 614 acres (1233); Bellhuse, 289 acres (1303); and Bindo Farm, 233 acres (1244). These enclosed farms of medieval date (Mills 1977, 129–31) are likely to be intakes or separate small townships. Similar farms (or small estates) belonged to Forde Abbey at Broadwindsor and nearby places in the twelfth century (Hobbs 1998, 31–48, 78, 91; see p. 118).

Field management

A 1515–16 court roll concerning twelve manors belonging to the abbot of Sherborne refers to by-laws and ordinances for regulating the open fields (J. Fowler 1955, *PDN-HAS* 77: 157–61). Account rolls for Portland and Wyke (Regis) in 1249 provide details

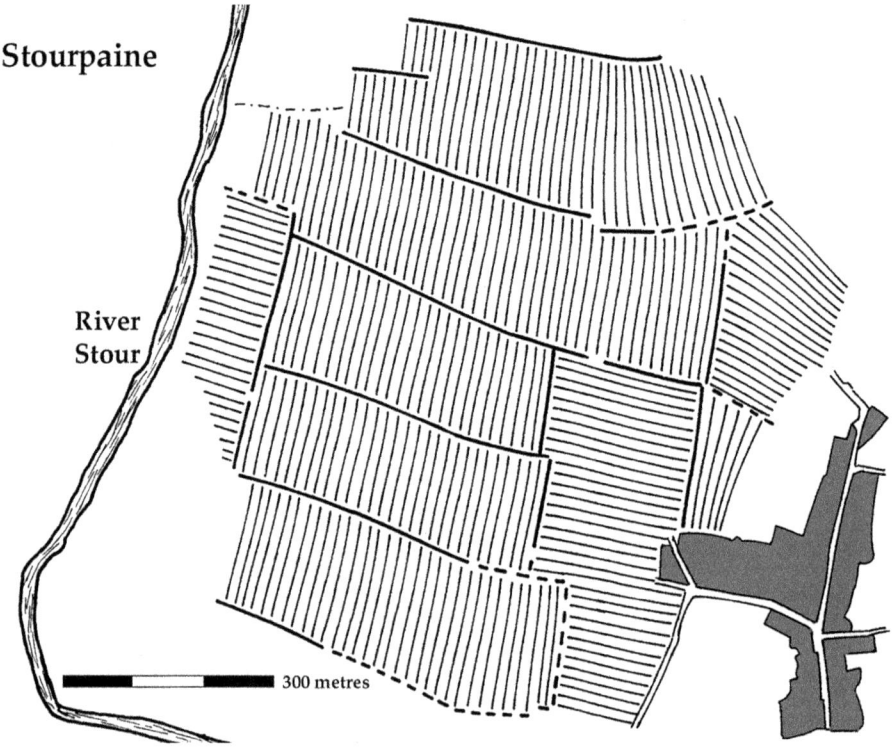

Fig. App.3. Stourpaine furlongs, Dorset. Mapped in 2003; *opposite:* an extract of the 1841 Tithe Map (Tithe Map T/SPN, courtesy of Dorset History Centre).

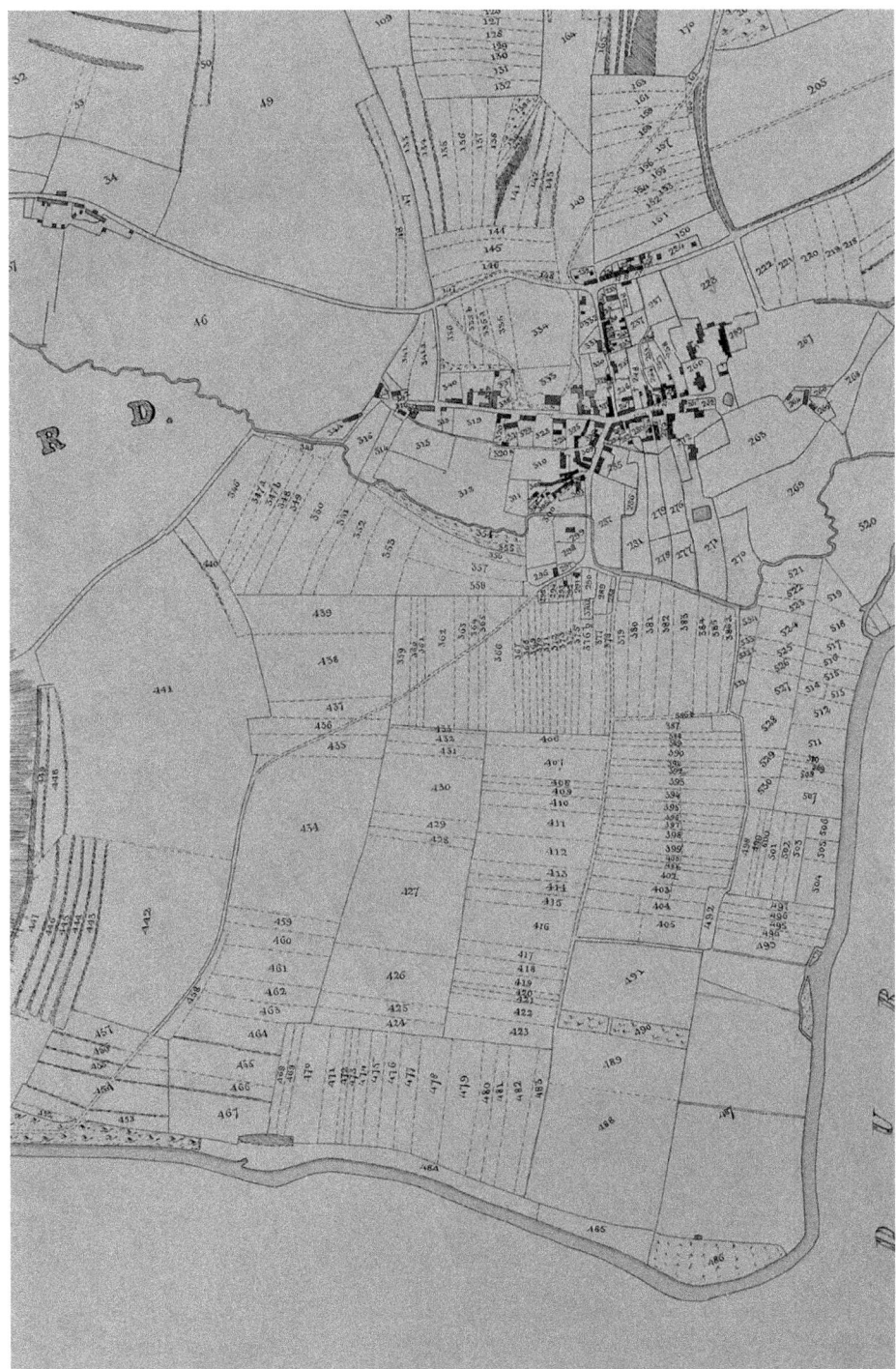

Fig. App.3. Stourpaine Tithe Map, 1841.

Table App.5. Dorset field systems.

Place	Field No.	Date	Sources
Bere Regis	2	1476	Squibb 1943, *PDNHAS* 65: 93; three fields 1818, Dorset CRO, D1/KL6 [only 16 per cent open]
Bloxworth	2	*c.*1190	Historical Manuscripts Commission 1911, 31
Blandford St Mary	2	*c.*1225	W. O. Hassall 1946, *PDNHAS* 68: 46–7
Buckland Newton	2	1313	I. Keil 1965 *PDNHAS* 87: 34–50
Fossell in Winfrith Newburgh	2	1768	Dorset CRO, Map D/WLC/E 18 and 19
Gillingham	2	1608	Also enclosures; Gray 1915, 30, 439
Lillington	2	1215	London 1979, Wiltshire Record Society, 35: 117–18; dispersed demesne
Sturminster Newton	2	1189	Jackson 1882, 137 [and in 1313]
Winterborne Monkton	2	1722	Barnes 1906, 59
Ashmore	3	1653	E. W. Watson 1890, *Ashmore, Co. Dorset* (Gloucester: J. Bellows), 42
Chilfrome	3	*c.*1823	Dorset CRO, D11/1; Figure 1.4
East Burton in Winfrith Newburgh	3	1768	Dorset CRO, D/WLC/E 18 and 19
East Lulworth	3	1770	Dorset CRO, DWLC/P1/2), open field 16 per cent
Hinton St Mary	3	sixteenth century	Gray 1915, 32, 442; nearly half was enclosed
Staplebridge	3	1515	J. Fowler 1955, *PDNHAS* 77: 157–61
Stratton	3	1895	Pope 1909
Wootton Abbas	3	1401	B. Kerr 1968 *PDNHAS* 89: 245

of crops and list fines for open-field infringements such as the ploughing of pasture (J. S. Drew 1944–5, *PDNHAS* 66: 46 and 67: 46). Marling was practised at Chilfrome, which had a 'marlepytt' in 1549 (E. A. Fry 1909, *PDNHAS* 30: 23). Work-service for Povington and Milburne described in the mid thirteenth century shows that sheepfolding was included in the usual agricultural round (Chibnall 1951, 62, 90). Court rolls for manor of Winterborne Monkton in 1673–1788 provide many open-field orders. Dates were decided for the breaking and laying up of the common fields and when animals could go on various leys and meadows. No pigs were to go unringed from the end of harvest until Michaelmas; there were to be no geese or ducks on commons; furrows were to be made and boundary stones placed in the common field (Barnes 1906, 48–56).

Flocks of sheep were managed according to the seasons. Thus, lambing flocks were not folded in the early spring but put on the lower pastures and water meadows. In bad winter weather, they were allowed to take what shelter they could in woodland, if available, or in thorns and furze on the high windswept downs. Other details of Dorset sheepfolding have been given. Animal stints were generous. At Stratton a ten-acre holding could graze two horses, two cows, and forty sheep (Pope 1909, and see Fordington) and a thirty-eight-acre holding at Winterborne Monkton had one hundred sheep, five horses, and five cows (Barnes 1906, 59).

Ridge and furrow

It is likely that most of the arable chalkland had very low profile or grooved ridges. A few can be seen with reverse-S ends at Fordington near the Maiden Castle car park (SY 669 889). They are visible on the Allen aerial photograph (*c*.1935) published in Wheeler's 1943 Maiden Castle Report (Society of Antiquaries, pl. cxviii). The RCHM *Inventories* often refer to ridge and furrow but no illustrations are given. Broad ridge survives near White Horse Farm, Osmington (SY 713 84) (pers. comm. P. W. Martin 2002).

More prominent are strip lynchets that developed on steep slopes, discussed in the RCHM *Inventories* (ii, p. xix; iii, p. xlcvii) and by Taylor (1966, *Antiquity*, 40: 277–84). Many lynchets are marked on Ordnance Survey 1:50,000 maps; visible from the A35 road are Winterbourne Abbas (SY 62 90) and Litton Cheney (SY 55 90). C. D. Drew (1947, *PDNHAS* 69: 53) studied open-field maps of Godmanstone and Sutton Waldron and related them to lynchets on the ground. He found that the evidence of the maps and lynchets 'fit together neatly', showing that the lynchets had an open-field origin. Striking strip lynchets occur at Loders and Bothenhampton (CUCAP, AAT 70 and BKS 73).

Hinton St Mary open fields were reconstructed from fieldwork surveys of furlong boundaries and headlands in 2001 (Figure App.2). The linear earthworks are low and diffuse, difficult to distinguish from ploughed-out hedge banks. Existing hedges are placed on high banks, as in Devon, including those made later than an estate map of 1782. No ridge and furrow or grass strip balks were discovered in any of the coverts. The plan is a standard Midland type, being substantially rectangular within the constraints of roads and the topography. The high percentage of arable would necessitate grazing on the fallow, as recorded in the sixteenth century (Gray 1915, 32, 442). A boundary charter of 944 has several references to headlands suggesting a considerable extent of arable (S 502; B. Grundy 1936, *PDNHAS* 58: 119). Existing curved hedges are likely to follow strips. There were many more hedges mapped on an estate map of 1782, and these were used to obtain accurate alignments for the reconstructed plan.

BIBLIOGRAPHY

PDNHAS   Proceedings of the Dorset Natural History and Archaeological Society

Barnes, W. M. (1906), 'Rolls of court baron of the manor of Winterborne Waste alias Monkton', *PDNHAS* 27: 44–71.

Bettey, J. H. (1970), *The Island and Royal Manor of Portland: Some Aspects of its History, 1750–1851* (Bristol: University of Bristol).

Drew, C. D. (1947), 'Open arable fields at Portland', *PDNHAS* 69: 51–3 + pls. 1–3.

Mills, A. D. (1977), *The Place-Names of Dorset, Part 1*, EPNS 52 (Nottingham: English Place-Name Society).

Moule, J. J. (1892), 'Notes on the manor of Fordington,' *PDNHAS* 13: 152–62.
Pope, A. (1909), 'Some ancient customs of the manors of Stratton and Grimston,' *PDN-HAS* 30: 83–93.

ACKNOWLEDGEMENTS

G. Anthony L-F Pitt-Rivers, OBE DL, for permission to survey Hinton St Mary and examine the 1782 Hinton Estate map. Mr S. Johnston of the Dorset History Centre for permission to publish part of the Stourpaine Tithe Map of 1841. Lt. Colonel J. F. Warren, Lazerton Farm, and R. Moger, Esq., Manor Farm, Stourpaine, for permission to map their fields. Paul Martin for arranging access to Dorset parishes.

11. DURHAM

Durham has lowland in the south along Teesdale and in the valley of the Wear, with a coastal plateau on the east. On the west lie the Pennine spurs and the high Pennine moorlands against Cumberland. Large parishes were common—Sedgefield with 17,471 acres had seven townships (Clifford-Brown 1995, 1). There were open fields on the coastal lowland and in valleys, but arable land often occupied a low proportion of a township area. Convertible husbandry was practised at, for example, Sherburn in 1604. Some of the three-season cultivation systems recorded therefore differ from a 'normal' three-field arrangement.

Open-field maps are known for several places: Ravensworth, 1712, published by Hughes (1952, 133) and Crawcrook, 1794, was redrawn by Butlin with an outline of Bolam, 1786 (1973, 101, 104–5). Only two places had open-field marked on tithe maps (1830 and 1844): Cornsay and Counden (Kain and Oliver 1995). Gray (1915) discussed Durham field systems and Butlin (1973, 93–144) gives an account of fields and enclosures of the county. Various surveys (dated 1183, 1380, and 1418) made for the bishop of Durham's estates show the highest density of arable lay in the south-east and north-east (Dickinson and Fisher 1959, 3–15).

Field numbers

Gray (1915, 462–3; 446) found no examples of two fields, but there were fourteen three-field townships, all of them dating from the first decade of the seventeenth century, except Wolviston (1325). The details of Ingleton in 1608 show that the freeholders' lands were not exactly equally distributed in all three fields in many cases. Butlin (1973, 100) shows a map of the distribution of places known to have had open fields from *c.*1550 to 1800. Most are located on the coastal region or in major river valleys in the east and south of the county.

Irregular fields

Irregular fields occurred at (East) Brandon and Eggleston and places nearby. Gray (1915, 105–7, 534–5) provides some details. East Brandon had five main arable areas and five other places in which tenants' lands were irregularly dispersed; there were variable amounts of meadow and pasture. Eggleston had similarly irregular holdings with greater quantities of meadow than arable in the three open fields. It is possible that convertible husbandry was practised. Single arable fields occurred in a few places in the seventeenth and eighteenth centuries (Butlin 1973, 143).

Field orders and farming management

Oxgang sizes have been given (p. 112). Arable fields were used for common grazing after crops were removed in the sixteenth century. Three-field systems had one field left fallow each year (Butlin 1973, 132–3). A few orders are recorded in court rolls of 1379–81 for Billingham: horses and ponies were not to be placed in the cornfield and cows in the pasture were to be tended by a herdsman; watercourses were to be cleaned, but boys were not to be sent to clean the millpond (Greenwell 1872, 169–71).

The 1183 bishopric survey refers to some work-service. Warden oxgang tenants worked twenty days in the autumn with one man, harrowed thirty-two days with a horse, and carted corn for two days and hay one day. Some holdings in West Aukland had to cart and go on the bishop's errands between the Tyne and Tees, as well as find eight oxen to cart wine (Greenwell 1852, 48; 61–2). Boldon's pinder (penfold keeper) held twelve acres (rent free) for his office, but he was also poultryman; he received one thrave of corn from each plough and provided the bishop with 24 hens and 500 eggs. The smith of Sedgefield had to provide coals; a collier at Escomb had a toft, croft, and four acres for which he found coals for making the ironwork of ploughs at Coundon (Greenwell 1852, 45–50, 61). The Hatfield Survey of 1382 describes agricultural work-service, and records that villeins provided carriage when the bishop visited (Greenwell 1857). Coal mines were a source of revenue at Rainton in 1409–19 (Fowler 1901, 708–13).

Ridge and furrow

Broad ridge, five to six yards wide, occurs over most of the lowland. Examples survive at Egglestone Abbey (NZ 062 151), where there is a block with headlands. The CUCAP

Table App.6. Durham field systems.

Place	Field No.	Date	Sources
Bolam	3	1786	Butlin 1973, 101: fields 67 per cent; for 1604 see Gray 1915, 463
Crawcrook	3	1794	Butlin 1973, 102–5 plan; arable 50 per cent of township
Nether Throston	3	1583	D. Austin 1976, *AA*, 5th ser., 4: 78
Sherburn	3	1604	E. Heppell and P.A.G. Clack 1991, *DAJ* 7: 136–8; arable *c*.33 per cent, evidence of convertible husbandry on plan
Barnard Castle	6	1783	Three largest fields run on thee-tilth system; arable 13 per cent; Butlin 1973, 102

collection has photographs for Denton (NZ 217 187) (AWU9, 1969) and Houghton (NZ 222 217) (CIF82, 1969). Some broad ridge is narrower than that found in the East Midlands. This was commented upon by Bailey (1810, 246), who attributed it to being the best way of dealing with a clay subsoil. A study of ridge and furrow at Sedgefield showed that there was 20–25 per cent surviving in 1947, which had been reduced to as low as 2–3 per cent in 1995. Widths were most commonly four to seven yards (Clifford-Brown 1995, 8–12). Many strip lynchets lie west of Eggleston (NY 99 23–4) and around the Roman fort of Lanchester (NZ 159 470) (CUCAP, CIB72 (1979)). West of Wolsingham (NZ 05 37), enclosed fields have parallel curved hedges up to 500 yards in length.

Numerous examples of early-modern narrow straight ridges are known, for example Rose Hill near Stanhope, Weardale (NY 937 386) (R. Young 1993, *DAJ* 9: 13). A field of ridges next to Ulnaby village site (NZ 228 172) contrasts with wide, curved broad-ridge adjacent to it (CUCAP, AWU14 (1969)). At Park End, Holwick, narrow ridges of eighteenth- to early nineteenth-century type fit within walled fields, but are older than a quarry marked on the First Edition Ordnance Survey 1:10,560 map (1856). Some farms were converting their broad, crooked high-back ridges to straight, narrow ones (Bailey 1810, 246).

BIBLIOGRAPHY

AA   *Archaeologia Aeliana*
DAJ   *Durham Archaeological Journal*

Bailey, J. (1810), *A General View of the Agriculture of the County of Durham* (London: R. Phillips).

Clifford-Brown, H. J. (1995), *Ridge and Furrow in Sedgefield Parish, County Durham* (Durham: Dept. of Adult and Continuing Education, University of Durham).

Dickinson, P., and Fisher, W. B. (1959), *The Medieval Land Surveys of County Durham*, University of Durham Research Paper, no. 2.

Fowler, J. T. (1901), *Durham Account Rolls*, Surtees Society, 103.

Greenwell, W. (1857), *Bishop Hatfield's Survey*, Surtees Society, 32.

Hughes, E. (1952), *North Country Life in the Eighteenth Century* (London: Oxford University Press).

ACKNOWLEDGEMENTS

Niall Hammond, Durham County Archaeological Officer, for discussion of HER data and a field excursion. Robin Daniels, Cleveland Archaeological Service.

12. ESSEX

Saffron Walden in the north-west lies at about 350 feet; the county falls gently to coastal plains and marshes in the south-east. Both relief and geology are published in coloured form by Hunter (1999, 3). Most parishes contained dispersed settlement associated with greens often called 'tyes' (Hunter 1999, 99–104). Writtle has been described (pp. 15–16), and a study of Brightlingsea in *c*.1300 shows that tenements were dispersed over most of the parish that was not marshland (Dickin 1939). A similar dispersion was found at Great Canfield, where nine medieval 'ends' have been identified. In the fourteenth century, thirteen open fields were named (Eland 1949, 100–56). The dispersed settlement was associated with small irregular fields and some enclosed severalty holdings and crofts. There were no recorded regular field systems with communal organization. Enclosure occurred at an early date.

The fields of Essex have been described by Gray (1915, 387–94), Roden (1973), and Britnell (1983). There are references in the economic sections of VCH, *Essex*, v (1966) and volumes published subsequently. Some parishes at the north-west have open-field maps. The Saffron Walden and Little Chesterford map of 1758 has been partly published by Cromerty (1966, cover illustration). Arksden has an earlier map (1733), and several draft enclosure maps mark open fields: Ashdon and Bartlow (1850), Great and Little Chesterford (1804), and Hadstock (1801). References for maps can readily be found on the Essex County Council website at <http://seax.essexcc.gov.uk>. Some evidence of aratral curved hedges has been found in the Southend-on-Sea region by Rippon (pp. 33–4 of his online report at <https://ore.exeter.ac.uk/repository/handle/10036/30031>).

Irregular fields

Gray (1915, 389–90) noted that some places had evidence of dispersed strips in the north-west of the county. However, many yardlands were not dispersed, but lay in a few places only, as explained for Dunmow, Laver, and Havering in the early thirteenth century, and for the Witham area (Britnell 1983). Bumpstead, in 1200, is another example where land lay in small pieces: two acres were situated 'in a corner' and ten acres abutted a bridge called *Deresbrige*. Part of two virgates in Clavering consisted of nine acres below the house of Ralph Pipesinke and nine acres beyond *Fulebrok*. At Stambourne sixty-six acres lay in seven pieces, twenty-two acres being next to a messuage towards the wood, one acre in a croft where the granges were, and one acre in a croft enclosed (Kirk 1899, 20, 24, 62). Kelvedon glebe, described in 1356, had fifty-two acres together next to the vicarage called *Church Field*, and nine in *Lyndeland Field* were enclosed with hedges and ditches (E. F. Hay 1911, *TEAS* 11: 1–9).

A plan of Thaxted open fields and parks in the fourteenth century was reconstructed from a survey of 1393 and other evidence by Newton (1960). As well as Thaxted, there were small settlements at Woodham, Cutlers Green, Boyton End, Bardfield End Green, and Richmonds Green. Thaxted manor lands occupied the greater part of the parish, and there were smaller manors of Priors Hall and Norham, the lands of the latter lying in several detached parts. Thaxted manorial demesne was divided into three areas for cropping, each made up of smaller named pieces. Parks and woods totalled 762 acres, and the sizes of the fields were 258, 331, and 264 acres.

Fisher reconstructed Harlow fields using detailed extents and rentals from Bury St Edmund's Harlow Cartulary. The demesne consisted of 586 acres lying in 7 fields of 15½ to 110 acres. Six of them were styled as 'feld'. They all lay in the north part of the parish, and elsewhere were lands of the manors of Welds, Brendhall, and Kitchen Hall. Some villein messuages were also identified (J. L. Fisher 1940, *TEAS* 22: 239–71).

Other medieval studies have been made for Colchester (J. L. Fisher 1949, *TEAS* 24: 77–127), Lawling in 1310 (J. F. Nichols 1933, *TEAS* 20: 173–98), and Finchingfield (R. H. Britnell 1977, *Essex Archaeology and History*, 9: 107–12). Havering's thirteenth-century charters describe a large number of irregular 'fields' and lands dispersed in a few small blocks only (Westlake 1923, 23–87). Multiple field names, some called 'hides', are recorded in 1527 at West Ham. Many of them were referred to in the Middle Ages (VCH, *Essex*, vi (1973): 74–6).

Some Essex parishes had customary and freehold lands that were undispersed from an early date and probably were never part of any open-field system, for example Felsted and Ingatestone, which have detailed studies of their land-use and field structure. They had block demesnes and large areas of tenement block holdings. At Ingatestone several holdings had 'hide' names that are likely to be early. Common arable was a low percentage (Martin and Satchell 2008, 192). See pp. 184–5 in the main text for further discussion of undispersed 'hides' and 'hydes'.

The demesnes of Writtle and Saffron Waldon have been discussed (p. 115). These and many other manors ran their demesne fields on a three-season system, but with crops planted in non-adjacent furlongs without communal control (Cromarty 1966, Map 3). Further block demesnes at Cressing, Great Easton, and Tilty are illustrated by Hunter (2003, 13, 15, 20). Commoning on the fields was limited, but still occurred at Colchester in the fifteenth century (I. H. Jeayes 1918, *TEAS* 14: 81–9). Work-service for Netteswell in 1220 included the agricultural round, but also referred to hoeing dunged land, undertaking carriage as far as Upminster, gathering nuts, and carrying wood (J. L. Fisher 1930, *TEAS* 19: 111–16). Kirby Hide tenants in 1222 owed service of carriage from the lord's wood and made four hurdles for the fold; they also had a full round of agricultural services as well as repairing six perches of fence around the courtyard (Hale 1858, 43, 47–8).

Ridge and furrow

No surviving earthwork ridge and furrow is known in the county. In 2002 the HER had fourteen entries for medieval-type ridge and furrow. At Little Oakley five furrows 2.5–3 m apart were aligned parallel to a Roman ditch. A late Saxon or early medieval date was suggested (P. M. Barford 2002, East Anglian Archaeology, 98: 69, 82, 198). Furrows were also discovered at Chignall St James (TL 6672 1012), 2.9 to 6.5 m apart (average 4.76 m). They were thought to be of medieval date, and were proved to be later than AD 360 (H. Brooks 1992, *TEAS* 23: 38–50). Since ridge and furrow is so rare in Essex, it is possible that the excavated 'furrows' are trenches of Roman vineyards. A field survey of the north-east part of Orsett and adjacent Horndon (2103) showed there were the low linear banks of furlongs underlying former hedges planted on them (marked on the tithe map). It is therefore possible to map and reconstruct the former open field systems of this part of Essex.

J. H. Round (1918, *TEAS*, 14: 191–208) pointed out that an 1845 map of Colchester distinguishing intermixed parishes showed evidence of open fields from the hedge curves and that one piece was divided by balks and not hedges. Woodgate End in Broxted had enclosed fields in 1594 that look to be taken from open-field furlongs (Hunter 1999, fig. 2, p. 192). Rectangular field patterns at Orsett and elsewhere have been discussed by Rippon (1991, 57).

BIBLIOGRAPHY

EAH   *Essex Archaeology and History,* the renamed *Transactions* from the 3rd ser. (1971–)

TEAS *Transactions of the Essex Archaeological Society*

Cromarty, D. (1966), *The Fields of Saffron Walden in 1400* (Chelmsford: Essex County Council).

Dickin, E. P. (1939), *A History of Brightlingsea* (Brightlingsea: D. H. James).

Eland, G. (1949), *At the Courts of Great Canfield, Essex* (London: Oxford University Press).

Hunter, J. (2003), *Field Systems in Essex*, EAH Occasional Paper, 1.

Newton, K. C. (1960), *Thaxted in the Fourteenth Century* (Chelmsford: Essex County Council).

Westlake, H. F. (1923), *Hornchurch Priory: A Kalendar of Documents in the Possession of the Warden and Fellows of New College, Oxford* (London: Philip Allan & Co).

ACKNOWLEDGEMENTS

David Buckley, Helen Walker, Alison Bennett, Lynn Dyson-Bruce, and Sue Tyler for HER data and discussion.

13. GLOUCESTERSHIRE

The Cotswolds form a high belt of ground rising to 700 feet in the north-east of the county and running south-east towards Bristol. The Forest of Dean lies west of the River Severn. There was dispersed settlement in the wooded country of the Severn Vale and the Vale of Berkeley (Finberg 1957, 44). Frocester had dispersed settlement lying along several roads (E. G. Price 1998, *TBGAS* 116: 10–15, 18) and Dyer provides data on dispersed and 'non-village' settlement in the Cotswolds (2002, *TBGAS* 120: 7–35). Gloucesteshire is primarily a county of regular two- and three-field systems, with limited areas of irregular fields.

Gray (1915, 463–88, 515–20) discussed Gloucestershire fields, and much information is available from the volumes of VCH, *Gloucestershire* (e.g. vol. xi (2001)), of which only a sample has been included in this account. There were late surviving open fields: Upton St Leonards, for instance, had 534 acres until 1897, divided between 80 proprietors. There were then 1,120 strips with meres and balks occupying 14 acres (Slater 1907, 63). Elmstone Hardwicke near Cheltenham had intermixed holdings until 1914 (Finberg 1957, 98); the lands did not have dividing balks (Orwin and Orwin 1938, 45). Each owner farmed as he pleased but there were still common rights after harvest (Slater 1907, 47–50). Westcote was still open in 1956 (Beecham 1956, 41–2).

Finberg published a 1748 map of Hawling (1957, 99; 1972, 486–7). It shows three 'planned' fields with two pastures on hills. The Aston Blank map of 1752 has been redrawn (C. Dyer 1986, *TBGAS* 105: 169). Tate and Turner (1978, 119) noted two other maps and there are also maps for Boxwell and Leighterton from 1794 (1964, *TBGAS* 82: 216). Tithe maps dated 1838–47 have some open field for twenty-seven places (Kain and Oliver 1995).

Regular field numbers

Gray (1915, 30, 438, 463–8) listed twenty-five examples of two-field townships. The Table lists fourteen more and others are given by the VCH. In 1299, two-field systems operated on demesnes at Blockley, Paxford, Bishop's Cleve, and Withington (M. Hollings 1934–50). The Table gives three three-field systems. Gray's eight examples of three-fields were all demesnes. Gray (1915, 88, 516) also drew attention to four-field arrangements in the county, at Welford and elsewhere already referred to (see p. 44).

Irregular fields

Irregular arrangements, characterized by a multiplicity of fields ranging from six to twelve, with holdings irregularly distributed among them, occurred in low-lying townships at Clapton, Frocester, Oxlynch, Horton, Yate, and Frampton Cotterell during the sixteenth century (Gray 1915, 88–93, 517–20). Some of these were partly enclosed. At

Table App.7. Gloucestershire field systems.

Place	Demesne type	Field no.	Date	Sources
Aston Blank		2	thirteenth–seventeenth century	Four fields by 1752, Dyer 1987, 173
Bibury	block	2	1299	Dyer 2007 in Bettey 2007, 69–93
Cirencester		2	thirteenth century	Elrington 2003, 10; Ross/Devine 1977, iii, 774; five acres in each of two fields 1220–50
Cutsdean		2	thirteenth century	Elrington 2003, 16
Farley		2	1263–84	Hart 1863–7, i, 304–5
Frampton		2	1250–65	C. E. Watson 1940, *TBGAS* 61: 87
Frome		2	1194	Ross, 1961, 177–8; one virgate eleven acres in Estfeld and twelve acres in Westfeld
Fyfield		2	twelfth century	Hart 1863–7, i, 271–2
Hawling		2	thirteenth century	Aldred and Dyer 1991, 153
Leighterton		2	thirteenth century	Hart 1863–7, i, 369
Old Sodbury		2	1649	Cave and Wilson 1924, 232
Rodmarton		2	*c.*1200	Ross 1964, 344–5
Saintbury		2	thirteenth century	Elrington 2003, 2
Througham		2	thirteenth century	Elrington 2003,113; Ross 1964, 945; three acres in each of two fields, mid thirteenth century
Deerhurst		3	1248	Elrington 2003, 91–2
Minchinhampton		3	1334	C. E. Watson 1932, *TBGAS* 54: 243
Standish		3	1801	W. E. Minchinton 1951, *TBGAS* 68: 171

Frocester there was were ten unequal open fields cultivated on a three-year tilth by the sixteenth century (E. G. Price 1998, *TBGAS* 116: 10–15, 18).

Demesnes have been discussed as well as yardland sizes and tenurial order (pp. 97, 125). Hampnet yardland was sixteen acres in the seventeenth century, Hazelton forty, Notgrove thirty-eight, and Salperton twenty acres (VCH, *Gloucestershire*, ix (2001): 86, 97, 150, 161). Farmington had 45¼ yardlands each of 48 acres in 1432. With the glebe of 2 yardlands (in 1535) the total of 47¼ equates closely to Domesday's 12 hides at near four to the hide. Commoning for a yardland was two horses, one ox, one cow, and forty sheep (Stevenson 1987, 164). At Frocester a yardland of forty-eight acres had a quarter called a 'fardel' of twelve acres and at Brookthorpe the values were sixty-four and sixteen acres (Hart 1863–7, iii, 24, 142–4). An assart belonging to two yardlands at Edgeworth in *c.*1235 was reckoned separately from them (Ross 1964, 358). Half a yardland at Bromfeld (1263–84) had one acre of assart added to it (Hart 1863–7, i, 221).

Field orders and farming management

Wortley orders for the mid seventeenth century referred to sheep stints (twenty-four to the yardland); gates were to be made between the fallow field and cornfield; there were encroachments; merestones were to be set up; pigs were not to run loose; cattle being driven through the cornfield were to be yoked (E. S. Lindley 1952, *TBGAS* 69: 178–9). Finberg (1972, 486–7) quoted Marshall (1789) that fields near Gloucester were cropped year after year without any whole year of fallow intervening and were called 'every year's land'. On these lands no regular succession of crops was observed. Oxen were still used for ploughing on the Cirencester Bathurst Estate in 1943, as they had been for the previous 200 years (E. M. Clifford 1943, *TBGAS* 63: pl. facing 168).

A complex management of *Lutletone*'s[1] demesne operated in 1265–6. There were two arable areas called 'fields'. One, of eighty-four acres, lay in three pieces called 'forlongs', the other field, ninety-seven acres, lay in seven pieces. The whole was run in three seasons with one third of the area fallow. Additionally forty acres [of fallow] could be 'inhoked' every second year, and the furlongs used were listed (presumably varied in a sequence; Hart 1863–7, ii, 35–6). An area called 'inheching' at Ham, 1243–54, may refer to the practice of inhoking (Walker 1988, 111).

Leys were recorded at Salperton in the early fifteenth century (VCH, *Gloucestershire*, ix (2001): 161). The 1840 tithe map of Westcote marks balks two to three feet wide, called 'meres', lying between every land and belonging to the strip on the north side. Wescote was still open in 1956, when the weeds on balks were kept down by burning (Beecham 1956, 22–44, at 41–2).

Work-service

The manors of St Peter's Abbey record much detail of work-service. At Elmbridge near Gloucester lepers were used to help sow land in the Middle Ages (L. E. W. O. Fullbrook-Leggatt 1946–8, *TBGAS* 67: 281).

---

[1] Often stated to be Littleton, Hampshire, but most of that vill belonged to Winchester Prior and Convent, and then to the Dean and Chapter (VCH, *Hampshire*, iii (1973), 422–3); perhaps the St Peter's manor refers to West Littleton, Gloucestershire.

Ridge and furrow

RAF aerial photographs of the 1940s have been published for parts of Slimbridge (J. R. L. Allen 1986, *TBGAS* 104: 143), Awre, Berkeley, and Oldbury (J. R. L. Allen 1993 and 1998, *Archaeological Journal*, 149: 87, and 154: 19, 29). Many other places have good examples such as Little Farmcote (SP 065 312) (CUCAP, AIJ 60 and 63 (February 1964)) and Stanway (SP 065 302) (AIJ 73–4 (February 1964)). The deserted vill site of Upper Ditchford is surrounded by broad ridge and furrow (M. Aston and L. Viner, in Saville 1984, 285). Ridge and furrow (1993) in the Stow on the Wold region have been mapped; large areas were found at Todenham and Aston Subedge (Hall 2001, 36).

Finberg (1957, 40–43) noted ridging occurred at Eastleach in a furlong called *Hey-ruggede Londe* in early thirteenth century. Lands at Deerhurst were described as ridges in 1248 (Elrington 2003, 92). In the Severn Vale, Marshall recorded that ridges were 2½ feet high but 4 feet was not unusual. Straight early modern ridges occur near Berkeley (SY 634 945) (1992, *TBGAS* 112: 92) and in the Severn Estuary at ST 438 859 and ST 439 849 (J. R. J. Allen and S. J. Rippon 1995, *TBGAS* 113: 80–1).

The county has many remarkable strip lynchet systems, three exceptional ones being Marshfield (CUCAP, AIP 46 and 49), Horton Hill (ST 742 767) (CUCAP, RMC 77 (1973)), and Hawkesbury (ST 767 837) (CUCAP, RIO 93 (1964)). A photograph of lynchets at Hutton Hill near Dyrham was published by Finberg (1975, pl. 15). The distribution of strip lynchets in the Cotswolds had been plotted and discussed in terms of their relation to geology (G. Whittington 1961, *TBGAS* 79: 212–20).

BIBLIOGRAPHY

TBGAS   *Transactions of the Bristol and Gloucestershire Archaeological Society*

Finberg, H. P. R. (1957), *Gloucestershire Studies* (Leicester: Leicester University Press).
Grundy, G. B. (1935–6), *Saxon Charters and Field Names of Gloucestershire* (Gloucester: Bristol and Gloucestershire Archaeological Society).
Saville, A. (ed.) (1984), *Archaeology in Gloucestershire* (Cheltenham Art Gallery and Museum, and Bristol and Gloucestershire Archaeological Society).
Stevenson, J. H. (1987), *The Edington Cartulary*, Wiltshire Record Society, 42.
Tate, E. (1944), 'Gloucestershire enclosure acts and awards', *TBGAS* 64: 1–70.

ACKNOWLEDGEMENT

Jan Wills, County Archaeologist, for access to HER data and air photographs.

14. HAMPSHIRE

Hampshire consists mostly of chalk with gravels and sands in the Southampton region at the south. Some parishes at the north have dispersed settlement and medieval sites or farmsteads taken out of wooded downland before 1397, such as Woodcroft in Halton (Cunliffe 1972, 8–9). Selborne similarly has dispersed settlement on the edge of Woolmer Forest (G. I. Merion-Jones, 1972, *PHFC* 29: fig. 1, opp. p. 7). Many parishes contained multiple townships, for example Longparish, which had five (Chapman and Seeliger 1997a, xxiii). The fields of Hampshire are complicated. There is limited evidence for some two-and three-field arrangements over much of the county, and equally there is much dispersed settlement associated with small irregular fields and crofts.

Fields

Several parishes have maps with detail of open fields (Chapman and Seeliger 1997a). Tithe maps showing open field are Kingsclere (1843) with nearly all the system intact (Chapman and Seeliger 1997b, fig. 9), East Worldham *c.*1841 (Chapman and Seeliger 1997a, pl. 4), and Selborne (redrawn by Merion-Jones, 1972, *PHFC* 29: fig. 1, opp. p. 7). Kain and Oliver (1995) list sixteen places with some open-field recorded on tithe maps (1839–45). A map of Portchester fields in 1405 has been reconstructed from a field book (Munby 1985).

Gray (1915, 466–8) found seven townships with medieval two-field systems, and two demesnes on the Isle of Wight, each with two fields. He recorded six townships with satisfactory evidence of three fields, and four that had demesnes in three seasons. A demesne at Wroxhall, Isle of Wight, had three fields, and three seasons, probably representing fields, operated in 1343 at Timsbury, Nether Wallop, Longstock, Lower Eldon, and East Dean (N. S. Rushton and C. K. Currie 2001, *PHFC* 56: 212). Mapledurwell was complex, with a double field system of three fields at the north and five at the south in 1616 (S. Waight 1996, *PHFC* 51: 170).

The Table gives a few more examples of field systems and all the information suggests the occurrence of two- and three-field systems over much of the county apart from the New Forest. However, the charters of Magdalen College, Oxford (largely those of Selborne Priory), show that there were many irregular fields in the thirteenth century and townships where 'field' meant furlong: Selborne, Worldham (Macray 1891–4, i, 29–30, 38, 52, 81), and Braishfield, whose land lay in crofts enclosed with a quick hedge in 1327 and 1330 (Magd. Hants charters, Somborne 48 and A66). Enham in Andover was similar (Enham charters, 115, 138). Although both Enham and Andover had fields with cardinal names, they were for locational purposes, not for cultivation regimes. At Skyres in Wootton St Laurence (1260–70) there were many small pieces with 'trees around', 5½ acres of pasture where there was a wood, and crofts with ditches and trees (Magd. Skyres charters, 18, 55, 71).

Demesne

A Bramdean grant (1234) of 140 acres lay in three pieces (Magd. charter Bramdean 24). The demesnes of Linkenholt (1266) were cultivated with one third fallow (Hart 1863–7, iii, 41), lying in several unequal large pieces. Mottisfont's three-season demesne in 1342 probably lay in a block (Rushton and Currie 2001, *PHFC* 56: 206). Demesnes at Hambledon, Fareham, Alresford, and Cheriton lay in named pieces in 1410, probably being dispersed; some are called 'furlongs' (Page 1999, HRS 16: 288, 293, 334, 346). Lideshete in Bramshott had had a block called the *Great Burifeld* in 1240, stated to be demesne, and probable block demesnes occurred at West Tisted

Table App.8. Hampshire field systems.

Place	Field no.	Date	Sources
Somborne	2	1340	Rushton and Currie 2001, *PHFC* 56: 212
Waltham, North	2	1410	Page 1999, HRS 16: 268–70
Basingstoke	3	1444	Baigent and Millard 1889, 617–22
Monxton	3	1807	Tenants down and farm down (Chapman and Seeliger 1997a, 288)
Portchester	3	1405	Munby 1985
Worthy, Headbourne	3	1632	Finn and Johnson 1999
Worthy, Kings	3	1750	Finn and Johnson 1999

and Hartley Mauditt in Alton in the fifteenth century (Macray 1891–4, 25, 108, 117). Foxcotte's block demesne of 591 acres was mapped in 1616 (R. A. Jones 1985, *PHFC* 41: 195).

Farming

Crop and stock details are recorded in the Bishopric of Winchester's records. In 1209 yields were low and ewes only produced one lamb each. For some manors (Ashmansworth, West Meon, East Woodhay, High Clere, Bishops Sutton) tenants paid a fine for not putting their sheep in the lord's fold (Hall, 1903, xxi, xliv–xlvi). In 1302 the yields on the bishopric estate for wheat lay in the range of two- to five-fold (Page 1996, detail given under each manor).

Sheepfolding was widespread in the county. On the Bec Abbey manor of Combe in *c.*1250, the half-yardlanders washed and sheared sheep, provided stakes and hurdles for the fold, and carried it from place to place. Cheese and wool were carted to Southampton (Chibnall 1951, 43). Linkenholt villeins washed and sheared sheep for two days in 1266; at Littleton[2] the 'doun' pasture was let (Hart 1863–7, iii, 36, 61). The manor of Old Alresford was manured by folding in 1492 (Greatrex 1978, 166).

Work-service for Priors Dean, Colemore, and Candover was listed in detail in *c.*1230 (Hanna 1988, HRS 9: 111–14), and for Hilsea and Wymering (near Portsmouth) in the early fifteenth century (R. F. Bigg-Wither (1916, *PHFC* 7: 1–19). At Candover the work included carrying sheep to be washed and sheared, making hurdles, and moving the fold. Vetches were to be made into cocks. Littleton agricultural work-service specified planting beans in 1266 (Hart 1863–7, iii, 37–8). The court rolls of Basingstoke note various field regulations. A tenant had ploughed and encroached on the heads of all his neighbours' lands in 1457; no pigs were to enter the field unringed (1571). The hayward's salary was to be paid at the rate of fourpence for each yardland (Baigent and Millard 1889, 285, 347–50).

Martin, Rockbourne, and Whitsbury had fields called *Burnbake* in 1788 and 1798 (Chapman and Seeliger 1997a, 122, 51). This probably refers to the technique of soil

---

[2] See Gloucestershire (p. 263 n. 1) for the possible identity of 'Littleton'; West Littleton, Gloucestershire, has a 'down'.

preparation by 'denshering' (see Wiltshire, p. 337). Kings Worthy had pasture called 'old baked land' (Finn and Johnson 1999) that may once have been cultivated in the same manner.

Ridge and furrow

Ridge and furrow currently survives at East Woodhay in the north-west at SU 409 624. Not much broad ridge and furrow is known in the county, there being only thirty-eight records of ridge and furrow of all types on the SMR (2003). The distribution is fairly uniform across the county excepting the south-west. Curved hedge patterns likely to reflect early enclosure around open-field furlongs have been identified (Lambrick and Bramhill 1999); they have the same distribution as the ridge and furrow.

East Worldham tithe map, *c*.1841, shows hedges set around furlongs containing strips in mixed ownership (Chapman and Seeliger 1997a, pl. 4). Early clearance of wood left irregular enclosures with shaws, as shown on a 1740 map of Cranbury (Hampshire CRO, 76M83/1).

The HER (2003) has many records of strip lynchets that occur over most of the county except in the south. Most are likely to be of medieval date, such as those at Overton (SU 503 488), visible on an RAF vertical photographs of 1947 (CPE/UK/1973, frame 4140). A fine set was photographed near Winchester at SU 485 278 in 1956 (CUCAP, SF 39).

BIBLIOGRAPHY

HRS   Hampshire Record Series
PHFC   Proceedings of the Hampshire Field Club

Baigent, F. J., and Millard, J. E. (1889), *The History of the Ancient Town and Manor of Basingstoke* (London: Simpkin, Marshall, and Co).

Finn, P., and Johnson, P. (eds.) (1999), *A History of Worthy Villages* (Headbourne Worthy: Worthys Local History Group).

Macray, W. D. (1891–4), *Calendar of Charters and Documents relating to Selborne and its Priory*, 2 vols. (London: Simpkin Warren).

Greatrex, J. (1978), *The Register of the Common Seal of the Priory of St Swithun, Winchester, 1345–1497*, HRS 41.

Lambrick, G., and Bramhill, P. (1999), *Hampshire Historic Landscape Characterisation* (English Heritage and Hampshire County Council; copy at the HER).

Munby, J. (1985), in B. Cunliffe and J. Munby, *Excavations at Portchester Castle* (London: Society of Antiquaries of London), 281.

Page, M. (1996), *The Pipe Roll of the Bishopric of Winchester, 1301–2*, HRS 14.

Page, M. (1999), *Pipe Roll of the Bishopric of Winchester, 1409–10*, HRS 16: 288, 293, 334–6.

ACKNOWLEDGEMENTS

David Hopkins and Bruce Howard for providing much information from the HER.

15. HEREFORDSHIRE

Herefordshire is drained centrally by the rivers Lugg and Wye, with high ground on the west and south. The River Teme forms part of the eastern boundary. Many parishes have complex dispersed settlement, as described for Acton Beauchamp (Pratt 2000–2) and Marden (Sheppard 1979).

The fields of Herefordshire are nearly all irregular with multiple parts, associated with dispersed settlement. Use of the terms 'field' and 'furlong' is interchangeable. There is evidence for a few examples of two- and three-field arrangements in the main river valleys.

Gray discussed Hereford field systems and published details of several surveys (1915, 93–7, 139–53, and App.). Roderick (1949) studied late medieval field systems from the evidence of charters and surveys. The fields marked on a county map produced by Sylvester (1969, 220–1) were probably 'one-field' systems with irregular holdings. Stretton Grandison and Ashperton have a late seventeenth-century map that shows open land and many enclosures (O'Donnell 1999, *TWNFC* 49: 21). Tithe maps of thirty-one places record some open-field land (Kain and Oliver 1995). Gray (1915, 144–9) redrew plans of Holmer and Risbury in Stoke Prior, both dated 1855, showing small areas of open field and referred to late examples of multiple fields at Much Cowarne, Madley, and elsewhere.

Field numbers

Gray (1915, 468–9) considered that the county lay within the Midland system of two and three fields. However, he was not able to discover any two-field systems and listed four townships that had their fourteenth-century demesnes in a three-tilth arrangement, which is not necessarily proof that there were three fields. Hennor (1608), is a satisfactory example (Gray 1915, 37, 447). Multiple fields occurred in 1608 at Stoke Prior (seven), Risbury (four), Stockton (nine), Hamnish (four), and Kimbolton (six) (Gray 1915, detailed surveys 447–9).

Kingstone was noted by Gray (1915, 143) to have three small fields of unequal size in 1812. Additional evidence shows that there had been three fields cultivated on a three-year tilth in the eighteenth century, and that some closes were farmed in the same way. The fields lay in the south of the parish, but there had formerly been others. One was demesne separated from the adjacent field by a hedge that had to be made good when one was sown and the other left for fallow grazing (D. J. Coleman 1990, *TWNFC* 46: 407–21). A view of the whole area in its open state is not reconstructable.

Some evidence for two-tilth cultivation was found for Much Cowarne, Fromington, and Stretton in the fourteenth and fifteenth centuries. Eyton had twelve acres in three fields and forty-eight acres at Withington in 1452 were said to lie in three fields. The Aconbury (sixteenth-century) and Ivington (1327) demesnes lay in three fields, but there were others (Roderick 1949, 56–7). The Table lists four other three-field systems, some with considerable amounts of old enclosure. They mostly lie in major river valleys (Wye, Teme, and near the Lugg headwaters) where extensive field systems might be expected to lie on fertile soils.

Three block demesnes have been identified. A grant of 123 acres of land made to Tillington chapel in 1394 (Capes 1914, 30–6) seems to be a block demesne. From

Table App.9. Herefordshire field systems.

Place	Demesne type	Field no.	Date	Sources
Bunsill		3	1326–44	A. J. Roderick 1949, *TWNFC* 33: 57
Burghill and Tillington	block		1394	Capes 1914, 30–6
Hampton Bishop		3	1516	A. J. Roderick 1949, *TWNFC* 33: 61
Hennor		3	1608	Gray 1915, 37 and 447; one holding, 77 per cent open
Mawley and Prysley		3	1578	Gray 1915, 449; only 27 per cent open
Overton		3	1586	P. Cross 1997, *TWNFC* 49: 48–9
Richards Castle	block		1680	P. Cross 1997 *TWNFC* 49: 56–7; demesne called Bury Field; other fields complex
Whitbourne	block		1577	P. Williams 1972, *TWNFC* 40: 333–47; yardland of sixty acres
Woofferton		3	1713	Cross 1997, *TWNFC* 49: 49–51; about 60 per cent open

internal evidence it lay on the west and north of Tillington and Burghill towards Stretton Sugwas, Credenhill, and Badbage Wood. Neighbours were mainly the lord or institutions, and a separate field of the villeins is referred to. The grant lay in seven fields of relative sizes of six to twenty-nine acres, but in parcels mainly a half or one acre.

Irregular fields

Ballingham settlement was dispersed in the thirteenth century, with three centres as well as Ballingham village and church (Kilforge, Seisals Farm, and Duns Farm). There were at least four fields; most of the land seems open (G. H. Martin 1952, *TWNFC* 34: 70–5). Much Marcle charters refer to about thirty fields, Brinsop fifteen, and Breinton six. Some of the fields were grouped together as at Stockton and Ivington in the fourteenth century. The complexity of Ivington fields is shown by land lying 'in le Westfield in a field which is called Waltonfeld'. Newton field at Wigmore was made up of four other fields (Roderick 1949, 58–60). Very possibly the charters are using the word 'field' for what elsewhere would be called a furlong. A study of Whitbourne (Williams 1972, *TWNFC*, 40: 333–55) showed that common arable (nine areas, of which five were called 'fields') lay in the north of the parish and that at the south were severalty farms with medieval names.

Details of some of the bishop of Hereford's manors were made by Swithin Butter-field in 1580: Bosbury, Colwall, Coddington, Eastnor, and Whitbourne (HCRO, A59/A/2). Analysis of the fields shows that they were all irregular. Coddington had thirteen large areas of open arable, seven of them called 'fields,' but holdings were irregularly dispersed. One of ninety-seven acres lay in four fields, but a holding of forty-four acres lay in six, only one of them the same as the ninety-seven acres.

Colwall and Eastnor yardlands were not described in terms of great fields, but in a complex array of pieces, crofts, and closes. Bosbury was also complicated. The first four customary tenants had dispersed lands in various combinations of four and five out of six fields, as well as blocks of closes of twelve to forty-one acres. The next three had land in only one of these fields and in other different ones. The explanation is possibly the existence of different townships in the 4,767 acres of the parish. Glebe terriers show that multiple irregular fields were the norm, for example in Ross Deanery, Fownhope (1589), Hope Mansell (1589), Upton Bishop (1607), and Walford (1614) (HCRO, HD2/4).

A survey of Acton Beauchamp made in 1594 shows it was mostly enclosed with four remnant arable open fields (270 acres or one-sixth) lying in various parts of the township (Pratt 2000–2). Tenants' holdings were dispersed irregularly within them and farms had enclosures, some of which were arable. There was a large area called 'customary land' that was probably former arable converted to pasture, since it was rent free and let out by the 'yardland'. It is very probable that Herefordshire irregular fields were partly caused by assarting. Roderick (1949, 62–5) gave medieval evidence for assarts and 'stokking' names. The Domesday Survey, in 1086, refers to assarts at Much Marcle, Leominster, Weobley, and Fernhill.

Work-service

The customary tenants of Bromyard owed services of mowing and making hay, and looking after one perch of the mill pond [bank] in 1285 (HCRO, AA 59/A/1, p. 88). At Kingsland, the work-service of 30½ yardlands in 1390 was similar (E. K. Cole 1958, *TWNFC* 35: 168–76).

Ridge and furrow

Broad, curved ridge and furrow occurs over most of the county, except at the south-west and with little at the north-west. Some is preserved in woodland, as at Whitemans Hill (SO 751 480) and Halesend Woods (SO 739 500) near Storridge (T. Horvard 2004, Herefordshire Archaeology Report, 72, HER). Strips at Stretton Grandison were described as 'ridges' in 1615 (O'Donnell 1999, *TWNFC* 49: 20).

Lands were ridged at Bosbury in 1578 (HCRO, Butterfield Survey, AA59 A2, fol. 37), and at Brampton Abbots and Hampton Bishop, both in 1618 (HCRO, glebe terriers, BA HD2/4/1). Strip lynchets are preserved in woods, such as Frith Wood, Ledbury, where they lie on a steep slope. The longest is 300 yards; others are 200–250 yards, the largest having a 5-yard riser and a platform 8–10 yards wide. They slope down a little at the ends for drainage.

Some enclosed fields have curved hedges indicative of former open fields over most of the county except in the south-west, amounting to about 15 per cent of the total (P. White, HLC in HER). Examples are Brimfield (SO 52 67) and Orleton (SO 48 67). This distribution can be compared to the similar online 'ridge and furrow' plot. It is also the same as the county map of tithe field names searched for 'furlong' (306 SMR entries). Longer hedge lines at Much Marcle may indicate that the open-field strips were laid out on a planned scale.

BIBLIOGRAPHY

HCRO   Herefordshire County and Diocesan Record Office
TWNFC *Transactions of the Woolhope Naturalists' Field Club*

Capes, W. W. (1914), *The Register of John Trefnant, Bishop of Hereford* (Hereford: Cantilupe Society).
Roderick, A. J. (1949), 'Open-field agriculture in Herefordshire in the later Middle Ages', *TWNFC* 33: 55–67

ACKNOWLEDGEMENTS

Keith Ray, Tim Hoverd, and Paul White for HER information and a field trip.

16. HERTFORDSHIRE

Hertfordshire has high ground in the north-western half of the county, of which the north and east is chalk partly covered with boulder clay. The remainder has soils derived from London Clay. There is much dispersed settlement often associated with greens. The fields of Hertfordshire were predominantly irregular with limited evidence of regular field systems, in spite of Seebohm's study of Hitchin.

Seebohm (1883) described Hitchin in his famous exposition of what became known as the 'Midland system'. A survey of the fields and town made in 1676 has been published (Howlett 2000). The fields of Hertfordshire have been discussed by Gray (1915, 369–81), Levett (1938, 182–7), and Roden (1973).

A 1704 map of Cole Green, Panshanger, shows a block of enclosure north of the vill, surrounded by six named fields, each consisting of a few furlongs (H. Prince 1959, *TEHAS* 14: 49). Layston had open fields extending to 1,032 acres in 1774. Wymondley and Hitchin remained open into the nineteenth century and plans of both were published by Seebohm (1883, 432 and frontispiece); for Wymondley in 1803, see Finberg (1972, 92–3). Open field is recorded on fifty tithe maps (Kain and Oliver 1995, 782).

Field numbers

Gray (1915, 376) recorded that there were two fields at Hexton during the fourteenth century, and the glebe of Aldbury was probably in two fields in 1628 (Brigg 1897, 20–2). Three fields were recorded at Bovington near Hemel Hempstead in 1225 (J. G. Jenkins 1945, Buckinghamshire Record Society, 10: 73). Some demesnes of St Alban's manors were run on a three-course tilth in 1332 (Levett 1938, 184, 338–9); these were not regular three-field systems. In the early seventeenth century Amwell had a small glebe dispersed in three fields (S. G. Doree 1989, HRS 5: 216). Walsworth, next to Hitchin, had three fields described in 1676. One sample holding of ten acres was equally dispersed, but another of twenty acres lay as three, ten, and seven acres in the three fields (from data in Howlett 2000, 35–40). Gray (1915, 370–1) noted the open fields at Kelsall, and Lannock had respectively six and three fields at the end of the eighteenth century. Long Marston had three fields. Elsewhere in the county irregular fields of the Thames Basin type were common.

Hitchin's field system as recorded in 1816–17 was used as a standard example by Seebohm (1883) to explain open-field structure and its operation for the whole country. It is actually typical only of field systems lying in North Hertfordshire and not even the remainder of the county. He explained how the land was divided into strips and shotts or furlongs, going into details of headlands, balks, gores, butts, and lynchets. A detailed example of Hitchin's *Purwell Field* showed how strips of many proprietors were intermingled. Hitchin had six fields in the nineteenth century, and court orders for 1819, published by Seebohm (1883, 443–53), provide details of how they were grouped into three tilths with a bare fallow, and had 'immemorially been kept … in three successive seasons'. The three-tilth arrangement differed from the Midland norm in that two fields making up each season were not physically adjacent. Seebohm showed a plan of a dispersed farm, the yardland. This appears to be a spurious reconstruction using the Hitchin map as a base for what was probably the correct form of a dispersed yardland at Winslow, Buckinghamshire, for which Seebohm had published and discussed a terrier of 1361. Land had ceased to be measured in yardlands at Hitchin before 1676, and there is nowhere in the county a record of such a dispersed yardland known at any date.

The fields of Hitchin were described in a survey made in 1676; there were then ten named areas (Howlett 2000, xv–xvi). Analysis of the distribution of ten sample holdings, where the fields are clearly described, shows that very few (none probably) were dispersed among all of them. One (of 42 acres) lay in 6 fields, three (of *c*.12 acres) were in 4 fields, four holdings (12–20 acres) were in 3 fields, and two (10½ and 16½ acres) lay in 2 fields. Most of the holdings were irregularly distributed among whatever fields they occupied and most of the individual parcels averaged one to two acres with pieces up to nine acres in some cases. It is difficult to reconcile these data with a regular three- or six-field system. It is probable that the three-tilth cycle described in 1819 had been superimposed on an irregular field system, rather than the state of holdings found in 1676 derived from a once regular three-field system by a process of exchange and engrossing.

Irregular fields

Gray (1915, 379–80) gives several examples of terriers describing dispersed lands not grouped into great fields and often consisting of large pieces. Half a yardland at Alswick in the thirteenth century consisted of 17½ acres divided in ten parts. Terriers of Kensworth, in *c*.1490, described irregular open parcels not grouped into any areas. Two half-yardlands of about twelve acres consisted of seven and five parcels only and not lying in the same furlongs. Cheshunt manor had irregular fields in 1621. Holdings were scattered irregularly in four main fields as well as in thirteen other areas, and there were variable amounts of enclosed arable and pasture. No holding had its land in more than five of the fields according to Gray (1915, 550). Further detailed examples of irregular fields are given for Kings Walden (sixteenth century), Weston (mid seventeenth), Ware (sixteenth), and Edmonton. Weston had a yardland called *Bondsland* that lay in a single field (Gray 1915, 372–6).

The demesne of St Albans Abbey at Tyttenhanger had three seasons, but irregular fields. Two seasons each had two pieces called 'fields', but the third had pieces of ninety-three and seventy acres plus crofts of four and six acres. Codicote, in 1331, lay in three unequal seasons (135½, 175, 152 acres) and in twelve pieces, of which seven were called 'fields'. The tenants' lands lay in five other fields. Two acres at Kingsbury, in 1310, were allowed to be enclosed in the common fields with a ditch surrounding with common consent of the tenant's neighbours (Levett 1938, 183–7, 339–40).

Roden discussed irregular fields on the Chilterns. Multiple fields were often grouped into three tilths, using blocks of lands that were not adjacent. A reconstructed plan of Kings Walden in 1600 showed complex fields. It was about one third enclosed and had twenty-six named 'fields' as well as some wood, common, and greens. At Knebworth, strips held by families living in three hamlets at the west of the township shared eight fields, but none had land in all eight (Roden 1973, 331–7). Brigg (1897) gave details of eighteen places in the seventeenth century. There were multiple fields with parcels irregularly dispersed and arable closes.

Farming

Inheritance in the east of the county was by Borough English (L. L. Rickman 1928–33, *TEHAS* 8: 303). Arable in closes is recorded at an early date, for example at Baldock a half acre of arable lay in a close in *c*.1222 (Gibbs 1939, 59 (no. 82)). An order for Welwyn rectory manor in 1287 stated that men and women who were able to reap were not to glean after the fashion of paupers (Ault 1972, 82). In 1271 Ramsey Abbey tenants at Therfield owed carriage service to Ramey, Cambridge, London, or Ware. The agricultural round included thrashing specified numbers of sheaves of barley, *bere*, and beans (Hart and Lyons 1884–93, i, 45–8). Codicote work-service was fully described in 1331 (Levett 1938, 200–1). In the seventeenth century Hitchin had field regulations for stinting and grazing the commons (Howlett, 2000, xv). Seebohm (1883, 443–53) published full court orders that regulated the fields and commons of Hitchin in 1819.

Ridge and furrow

Broad ridge once occurred over most of the county. A county plot of information on RAF photographs of *c*.1947, excluding the north-west, produced eighty-two records (2003), with few in the urbanized south-west and none in the east against Essex. The areas were mostly small. Additionally, the English Heritage National Monuments Programme plotted data from all known aerial photographs except the RAF verticals. Early modern straight ridges are known on Norton Common, Letchworth (TL 219 334) (HER 4531).

Late surviving open fields at Bygrave were ploughed in unfenced strips until destroyed in 1922. In 1912 common arable fields amounted to nearly 1,000 acres, with grass balks dividing the furlongs. Three farmers held most of the land. Occupiers of the arable could graze livestock over the whole field after harvest until winter ploughing (Finberg 1972, 483–5; W. Page 1912, VCH, *Hertfordshire*, iii: 211–12). Part of Clothall was never enclosed and some detail was recorded on the Ordnance Survey First Edition map. In 1912 there were still 600 acres open, most of it shared between three landowners (W. Page 1912, VCH, *Hertfordshire*, iii: 220). Seebohm published an etching of Clothall lynchets in 1883 and a photograph of the strip fields was published in 1910 (Cunningham 1910a, frontispiece). It shows sinuous curves of flat-ploughed lands separated by upstanding grass balks. A plan of the late eighteenth-century field patterns of Codicote reflect the open-field furlongs (Roden 1973, 368). The hedge patterns at Redbourn, Knebworth, Kings Langley, and Bovingdon resemble furlong shapes (Munby 1977, 169–70).

BIBLIOGRAPHY

HA   *Hertfordshire Archaeology*
HRS   Hertfordshire Record Society
TEHAS   *Transactions of East Hertfordshire Archaeological Society*

Brigg, W. (1897), *Hertfordshire Genealogist and Antiquary*, 2 (Harpenden: W. Brigg).
Gibbs, M. (1939), *Early Charters of the Cathedral Church of St. Paul, London*, Camden 3rd Ser., 58.
Howlett, B. (2000), *Survey of the Royal Manor of Hitchin, 1676*, HRS 16.
Munby, L. M. (1977), *The Hertfordshire Landscape* (London: Hodder & Stoughton).

ACKNOWLEDGEMENT

Stewart Bryant, Hertfordshire County Archaeologist.

17. HUNTINGDONSHIRE

The River Ouse traverses the south of the county, affording slopes with good agricultural soils, and the Nene forms the northern boundary. Between them lies higher ground covered with heavy boulder clay and some dispersed settlement. At the north-east the ground falls to the Fenlands of the Wash, lying below sea-level. Huntingdonshire is predominantly a county with irregular fields, but regular two- and three-field arrangements are found at the north-west and south.

Gray (1915, 469–70) listed some Huntingdonshire parishes in his Gazetteer of two- and three-field systems. Ambrose Raftis and his school have studied the records of Ramsey Abbey. Huntingdon Record Office holds copies of open-field maps on open access; a published catalogue (P. G. M. Dickinson 1968, *Maps in the County Record Office, Huntingdon* (St Ives: Huntingdon and Peterborough County Council)) lists many of them. Maps of Leighton Bromswold (1680) and Glatton (1613) are published (VCH, *Huntingdonshire*, iii (1936): 86, 176). Some open-field is recorded on six tithe maps dated 1839–49 (Kain and Oliver 1995, 232–3).

Gray (1915, 469–70) listed four places with two fields in the thirteenth and fourteenth centuries and five with satisfactory evidence of three fields, three of them thirteenth century. Extents of Needingworth and Wistow refer to meadow being thrown into the fallow field every third year (Hart and Lyons 1884–93, i, 295, 353), showing that there were three-field systems in 1251–2. The Table lists three others of fifteenth- and seventeenth-century date, and one irregular system at Diddington in 1631. Five unequal fields were grouped on a three-season cropping with a bare fallow, but of the eleven holdings five lay in five fields, four in four, one in three, and one in one field. Diddington lies in a wooded area at the north-west, where assarting probably accounts for the irregularities.

Table App.10. Hertfordshire field systems.

Place	Demesne type	Field no.	Date	Sources
Diddington		5	1631	Field book; Northamptonshire CRO, BRU ASR 138; irregular
Elton	block	3	1605	Elton, (Hart 1999), p. li;
Glatton	block?	3	1613	VCH, *Huntingdonshire*, iii (1936): 176; three fields plus Bury Field, presumably demesne
Lutton, East End Fields		3	1445 and 1622	1445 Cartulary of Geo. Fraunceys (Northamptonshire CRO, Buccleuch, pp. 403–8); 1622, Northamptonshire CRO, W(R) 311,
Wistow	dispersed	3	thirteenth century	Hart and Lyons 1884–93, i, 353; P. M. Hogan 1988, *Agricultural History Review*, 36: 117–31.

The Ramsey Abbey Cartulary has statements of the number of hides, yardlands, and their acreages referred to in the text. Some yardlands were called 'acremanlands'. At Warboys, in the twelfth century, it is explained that holders of these lands worked the abbot's plough (Hart and Lyons 1884–93, iii, 257). Domesday hides and yardlands are discussed in the text. Saxon charters of the tenth century record the same hidage as Domesday (Hart 1968, *PCAS* 61: 55–66; Roffe 1989, 13). The five hides at Water Newton recorded in AD 937 descended to Thorney Abbey (S 437; Hart 1966, 150–5).

Field orders

Various items are recorded for Elton; in 1279, Reynald Beneyt ploughed three furrows to his headland from all the strips that abutted to it. In 1312 the village meadow was trodden down by beasts against the 'belawe' (by-law), showing that there were court orders regulating the fields at this early date. In 1320 the servant of John Wagge allowed four and five beans to fall in one hole (Ratcliff 1946). The court rolls of Ramsey contain cases of open-field infringement in the fields of Ramsey and Bury. Orders were specified from 1377 onwards (DeWindt 1990, 50).

Cropping and farming management

Court rolls, account rolls, and extents of the Ramsey Abbey manors provide much detail of work-service and the farming round. From 1279 to 1351 the Elton manor employed eight ploughmen, a carter, a cowherd, a swineherd, and up to three shepherds. Ploughing was further assisted by boon works and all harrowing, harvesting, and threshing work was done with boon service. Oxen were used for ploughing, probably in teams of four (Ratcliff 1946, xliii–lix). Wool production varied from 103 to 356 fleeces. Posts and forty hurdles were used for the fold (1287). Barley and wheat were the main corn crops, with smaller amounts of oats, peas, and beans. Rye was not cultivated every year. For 1297 the grain total was 610 quarters of which wheat and barley each comprised 36 per cent, oats 17 per cent, peas and beans 5 per cent, rye 5 per cent, and drage 1 per cent. In 1287, 4,300 turves were supplied as fuel 'from the marsh' (i.e. the fen) and 132 cartloads of stone were taken from Elton by river transport. The carters also helped with transport farther away; large stones intended for an altar and a latrine were taken from Barnack to Ramsey. Faggots were bought in Fotheringhay Park and wood for plough beams from Castor Wood, timber for the mills was bought at Castor and Thornhawe Woods (1314). In 1297, 1000 bundles of rushes were brought from Whittlesey Mere by boat from Yaxley to Elton. They were used to thatch the grange and for kindling. Stone slates for the grange porch came from Southorpe (Northamptonshire) and pigs were pannaged in Cliffe Park (1308). Elton sheep were milked in 1287, and in 1351 it was stated that the milk was used for making cheese (ibid. 24, 367–79).

On the Ramsey manors stock-lists of horses and oxen show that oxen predominated in Wistow, Upwood, Broughton, and elsewhere in the thirteenth to fourteenth centuries, but at Houghton oxen were less important after 1400, with only horses being used for ploughing during 1445–60 (Raftis 1957, 132–6). Grain and animals at Godmanchester were listed for 1309 (Raftis 1982, 106). Several economic studies of the estates have been made (E. Karakacili 2005, *Journal of Economic History*, 64: 24–60).

The most remarkable early farming details come from Wistow, where late fourteenth-century account rolls detail crops sown and yields by furlong and part furlong (M. P. Hogan 1988, *Agricultural History Review*, 36: 117–31). An unusual feature is that legumes, wheat, and barley were grown in equal quantities, with small amounts of oats. The cropping cycle was three year, with a bare fallow. Legumes were concentrated in one field in each year, but not quite exclusively. Since the demesne was dispersed, it follows that the villeins also had a bare fallow and two cropped fields. The system is an early example of the scheme widely adopted in later centuries of a three-year cycle of corn, peas and beans, and bare fallow.

The thirteenth-century custumals of Ramsey Abbey manors describe work-services of the normal agricultural routine of ploughing, sowing, reaping, etc. Villeins had to collect wood and thorns and take them to each manor or to Ramsey. The thorns came from the demesne woods, the marsh of Upwood, and hedges and closes at Houghton. Hay was carried from Houghton (with its ample meadow) to Abbots Ripton (which had little meadow). Broughton villeins were liable for carriage to Elton, Ellsworth (Cambridgeshire), Shillington (Bedfordshire), or London. Wistow villeins had to work in the vineyard.

A view of villagers' stock is obtained by from 1290 taxation details of the animals and grain belonging to each named person. There were relatively few sheep. Common grain crops were wheat, barley, peas, beans, and drage, but no rye. Great Raveley also produced oats. Ramsey's manor at Biggin had 2 cart horses, 6 stots, 18 oxen, 31 cows, 1 bull, 16 yearling calves, 6 three-year bullocks, 6 two-year bullocks, 15 yearling bullocks and yearling mares, 100 sheep, 40 pigs, 3,000 [bundles of] reeds, 2 iron carts, and very many eels (Raftis and Hogan 1976, 31–8).

Ridge and furrow

Ridge and furrow occurred widely through the county, except in the fen. A plot of that surviving in 1988 has been published (Hall, 2001, 36). Little now can be found except at Bythorn (TL 05 75) (285 acres, 24 per cent) and at adjacent Catworth. Furlong boundaries survive well on the clayland and complete furlong maps have been made for Broughton, Eynesbury Hardwicke, Fenton, Holme, Abbots Ripton, and Kings Ripton (Hall unpublished). Haddon furlongs have been reconstructed by field survey and aerial photographic evidence (S. Upex 2002, *Archaeological Journal*, 159: 83–4). A photograph of Leighton Bromswold shows an intact area with part of two furlongs lying adjacent to a recently ploughed field where strips can be seen as light and dark marks (CUCAP, ATY 41 (1968)). Four large furlong boundaries are visible as linear soil banks.

BIBLIOGRAPHY

PCAS   Proceedings of the Cambridge Antiquarian Society

DeWindt, E. B. (1990), *The Court Rolls of Ramsey, Hepmangrove and Bury, 1268–1600.*

Raftis, J. A. (1957), *The Estates of Ramsey Abbey* (Toronto: Pontifical Institute of Mediaeval Studies).

Raftis, J. A. (1974), *Assart Data and Land Values, 1200–1350* (Toronto).

Raftis, J. A. (1982), *A Small Mediaeval Town: Godmanchester* (Toronto).

Raftis, J. A., and Hogan, M. P. (1976), *Early Huntingdonshire Lay Subsidy Rolls* (Toronto).

ACKNOWLEDGEMENTS

Ambrose Raftis and Edwin and Anne DeWindt for discussion of Ramsey Abbey matters over many years.

18. KENT

The county has several east–west bands of differing geology that have a marked effect on the landscape and settlement. At its north is low-lying London Clay, then there are high ridges of chalk on the North Downs, followed by Greensand, the Low and High Weald, and Romney marsh at its south (mapped as regions in Baker 1973, 378). There is much dispersed settlement. Ightham had at the end of the fifteenth century seven hamlets and many dispersed farms (E. Harrison 1940, *AC* 48: 172). Dispersal was encouraged by the custom of gavelkind or partible inheritance: newly divided holdings could have a messuage built upon each part at Wingham in 1285 (Baker 1965, 156–7). Dispersion also occurred because large settlements had detached parts of the Weald called 'denns' to use for fuel, timber, and for swine pastures (Reaney 1961, *AC* 76: 58–74). Many became assarted and had farms built in them. Some estates had detached parts of Romney Marsh; a piece called *Sellinge Farm* belonged to Wilmington in AD 700 (now part of Sellindge parish; S 21; G. Ward 1936, *AC* 48: 20–7).

Kent field systems were described by Gray (1915, 272–304), Du Boulay (1966, 118–42), and Baker (1973). Most of the county was enclosed early and only 41 out of 187 maps of seventeenth-century date showed some parcels of land with unenclosed boundaries (Baker 1965b, 16).

Field numbers

There were no 'great fields' in the county as land was held in severalty. A description of Kent fields has been given, explaining the field structure of yokes—small blocks of land shared among a few families, which had fragmented ownership caused by partible inheritance (see pp. 61–4). Some demesnes were cultivated on a three-year tilth or season. There were also areas of irregular fields, some discussed, to which Deal may be added. Deal remained partially open until after 1900 and was commented upon by Gray (1915, 276–7) and Baker (1963). A plan of an 18.8 acre holding in 1796 had 2 closes and 16 scattered lands limited to some shots only, but lying in no regular arrangement. Maps of 1734 and 1766 together with a rental of *c*.1750 allow a reconstruction of the open-field land of one of the three manors, some 229 acres, which was half the area of the enclosed ground. The pattern of the shots was rectangular and holdings were in dispersed blocks with some small strips. Enclosed lands were similarly dispersed.

Limited charter evidence shows that parcels were small in the late Middle Ages. A grant of 1289 describes seventeen acres in *Skottesteghe* and *Horsesteghe* held by several people called Skot and Hors, which is a similar terminology to Gillingham and Wrotham, where yokes have thirteenth-century personal names and are found shared by presumed descendants. Deal is difficult to interpret. The thirteenth-century evidence makes it similar to other places with subdivided yokes. From the fourteenth century onwards it seems to have had intermixed small strips that slowly become partially consolidated.

Field orders and farm management

Court orders at Ightham (fifteenth–eighteenth centuries) were concerned with trespass, overstinting the common, animal trespass in crops, and failure to clean drains or maintain fences. There are no orders suggesting any organized cropping and all the holdings appear to be enclosed (E. Harrison 1936, *AC* 48: 212–18).

Yalding had two acres of marl land in 1263 (Anon. 1861, *AC* 4: 312). Folkestone demesne, in 1271, had the usual agricultural work-service for arable and meadow. The park had to be fenced every four years (1,980 yards). Work-service on the demesne of Gillingham was described in 1285 (Baker 1964, 13). Each yoke had slightly different services to render on the lord's land for the agricultural round. Whatever tenants sowed they had to reap, bind, and carry to the grange. Some had to repair the wall around the manor-house, others had to provide stakes and hurdles from the lord's wood for the manorial fold. There were also two ploughmen and a carter who were paid officers. Wheat and sheep provided most of the income. Wye manor work-service was based on yokes in the late thirteenth century (Scargill-Bird 1887, 124). Deal demesne in the fourteenth century had paid ploughmen, a shepherd, and a part-time oxherd. Five acres of land were manured by folding and another four with carted manure (Baker 1963, 114–15). At Otford, there were different crops on the same furlong in any one year, changing without any apparent regularity during the fourteenth and fifteenth centuries (F. R. H. Du Boulay 1959, *AC* 73: 120–1). Large areas of some furlongs dropped out of the record and presumably were left as leys.

Surveys of the demesnes belonging to the Archbishop of Canterbury, made in 1283–5, listed woods and also denns, which, as mentioned, were small pieces of detached woodland on the Weald, often many miles distant. Most of them can be identified on maps of the sixteenth century and later. Pigs were driven to them for mast. Some denns were already being colonized in 1086 with tenants who had ploughs and paid rent. The archbishop later accepted rents for denns that had been damaged or ploughed (F. R. H. Du Boulay 1961b, *AC*, 76: 75–87). Several denns are recorded in Anglo-Saxon charters (P. H. Reaney 1961, *AC* 76: 69–74) and Domesday records four denns at Leeds and one at Hardres (Williams and Martin 2003, 19, 22). The many denns of Wye, Wrotham, and Otford lay on droves (Witney 1976, 88).

The common of Ightham was about 560 acres, or 29 per cent of the township. It consisted of wood, underwood, and pasture. The wood belonged to the lord and there was much unauthorized removal of timber and firewood in the early sixteenth century. There were also common grazing rights that sometimes exceeded the stint made by the manorial court (1561). Brakes were mowed. Small amounts of land were allowed to be enclosed between 1739 and 1767 and two people had houses built (E. Harrison 1936, *AC* 48: 204–11).

Ridge and furrow

Broad ridge probably rarely occurred in the county because the two-way wrest plough was used for flat ploughing. Early enclosure gave time to obliterate any medieval ridging of low profile. Boys (1813, 16–17) noted that the rich loams of East Kent were easily worked and needed no ridges or water furrows. The HER has twenty-five records of 'ridge and furrow' (2001), distributed mainly on low ground. Not all sites are differentiated into medieval and post-medieval, but some are catalogued as of medieval type (as at Birling (TQ 685 598), HER, KE14141). The Cambridge Collection records one photograph of a few cropmarks that may be furrows at Eynsford (TQ 526 664) (CUCAP, SR 68–9 (1965)).

Early modern straight ridges occur at Snargate (TQ 960 252) (HER, KE3291) and elsewhere. Ridge and furrow in Kent was studied by Mead and Kain (1976), using vertical photographs taken during 1946–63, and they published a county distribution map. Most examples were straight and were contained within existing field boundaries. They were likely of early modern origin, formed on claylands before the introduction of underdraining. Late eighteenth-century reports confirm that ridges were made for drainage. The HER had records of forty-three sets of lynchets in 2001. Some were once quite large, such as that reported by O. G. S. Crawford at Plaxtol (TQ 600 532), which was ten feet high (HER, KE1340). The HLC has categorized field types 1.6 and 1.16, which are very probably the result of hedges being planted around medieval fields. An example of type 1.6 is at Sutton Valance (TQ 82 49), where the patchwork of curved hedges suggests that at least some of them were set around strips. Brenchley (TQ 68 42) is an example of type 1.16. Both are shown on the 1862 First Edition Ordnance Survey map.

Baker (1965b) studied the field patterns of seventeenth-century Kent. The smallest fields, under four acres, occurred on the Low Weald Plain and the largest, eleven acres, in East Kent and Romney Marsh. Wealden fields were very irregular in shape (Horsmonden 1605, Brenchley 1603), with east Kent having rectangular fields (Hougham

1630, Guston 1640). Thick wooded hedges called shaws were common in the Weald. At Eastwell, a map dated 1732 shows about 1,000 acres with a large number of small closes (over 100) having very irregular shapes, not at all 'furlong like' (Northamptonshire CRO, Map 2189).

BIBLIOGRAPHY

AC   *Archaeologia Cantiana*
KR   Kent Records

Baker, A. R. H. (1962), 'Some early Kentish estate maps and a note on their portrayal of field boundaries', *AC* 77: 177–84.
Bishop, T. A. M. (1938–9), 'The rotation of crops at Westerham, 1297–1350', *Economic History Review*, 8: 38–44.
Boys, J. (1813), *A General View of the Agriculture of Kent* (London: Board of Agriculture).
Mead, W. R., and Kain, R. J. P. (1976), 'Ridge-and-furrow in Kent', *AC* 92: 165–71.
Witney, K. P. (1976), *The Jutish Forest: A Study of the Weald of Kent from 450 to 1380 AD* (London: Athlone Press).

ACKNOWLEDGEMENTS

I am grateful to John Williams and Paul Cummings for discussion and information from the HER.

19. LANCASHIRE

Lancashire has extensive coastal plains in the west rising to the high ground of the Pennines, with Furness being a detached part lying south of Cumberland. Parishes were large and contained many townships, Chidwall being ten miles long with ten townships in 1653 (R. G. Dottie 1986, *THSLC* 135: 15). Dispersed settlement was partly caused in the dales by vaccaries developing into small farms and hamlets: Wyresdale had twenty-one vaccaries in 1324 and 1346 lying along four miles of the dale from Green Bank to Tarnbrook (W. Farrer 1907, Chet. Soc. 54: 177; W. Farrer 1915, LRS 70: 129). Lancashire was a county with irregular fields that were associated with convertible husbandry and large areas of pasture.

Small areas of open field survived in 1623 at Caton (twenty-eight acres) and at Ince Blundell in 1729, plans of which were reproduced by Youd (1962, 5–7). Tithe maps (1843–8) recorded nine places with some open-field (Kain and Oliver 1995). Gray (1915, 242–9) discussed Lancashire field systems, Youd (1962) made a more detailed study, and F. J. Singleton (1962, *THSLC*, 115: 33–40) gave a summary. Elliot (1973, 41–76, 84–9) included Lancashire.

Irregular fields

Single-field systems have been described (p. 79, from Youd 1962, 20–34). Field numbers varied but were not related to cropping regimes and were sometimes a 'field' was a furlong. Holdings were irregularly dispersed, for example at Bolton le Sands and Chidwall in 1653 (R. G. Dottie 1986, *THSLC* 135: 18–19). Three fifteenth-century terriers of lands 'in diverse places' in the fields of Farington described holdings of 7½, 11, and 12 acres lying, respectively, in only 4, 5, and 2 named pieces. None of the holdings lay in the same furlongs, except for a single case (W. A. Hulton 1853, Chet. Soc. 30: 66–8). Salford had locational areas called Little Field and Middle Field, as well as many arable crofts and other areas (furlongs). Most of the owners, except the largest two, had lands limited to one or a few places and they were not distributed throughout the township (H. T. Crofton 1905, *TLCAS* 22: 171–9).

Convertible husbandry

Convertible husbandry at Deadwinclough in Rossendale (1528), Penwortham Priory (1543), and Carnforth in the eighteenth century has been described (p. 92). Demesnes occurred in block and dispersed form. Lands of the tenants lay in oxgangs, but assarts and crofts were common. Colne manor distinguished pieces called 'rodland' in 1322 (Farrer 1897, 482–5). A hedge called the 'ringyard' surrounded many townland arable fields, for example at Chatburn in 1530 (Farrer 1897, 91). Some ring hedges survived as large ovals in later enclosed hedged landscapes (Atkin 1985).

Work-service

In 1258 oxgang tenants at Burnley and Padiham had to plough and reap one day each (W. Farrer 1903, LRS 48: 213–16). At Gorton, in *c.*1320, they ploughed, harrowed, reaped, and carried goods as far Chesterfield. Singleton oxgang tenants in 1322 did agricultural work and carried provisions to Richmond, York, Doncaster, Pomfret, and Newcastle (W. Farrer 1907, LRS 54: 50–2, 132). Ashton-under-Lyne work-service (1422) included carting ten loads of turves from *Doncanmoss* (J. Harland 1868, Chet. Soc. 74: 94).

Field orders

Court rolls of the early fourteenth century at Colne recorded copyhold entry, breaking into the lord's fold, escape of animals, debts, and trespass. Ightenhill tenants broke a 'byelaw' (W. Farrer 1901, LRS 41: 3–48). There were no orders regulating the common fields. The courts of Prescot, 1510–75, presented tenants for digging on the pasture of the town moss and taking wood out of the lord's park. In 1536 they were to clean watercourses and remove trees that 'shadowed' the windmill, and were not to cart dung over Churchfield, except on their own land (F. A. Bailey 1937, LRS 89: 73–189).

Clitheroe courts in the early sixteenth century referred to the common pastures of Chatburn being overstocked with sheep, cows, and horses. Pigs were not ringed and slates were dug. Town fields were mentioned at Worston and Chatburn (1510, 1508); oxgangs were still used and 'rodeland' was distinguished from it (Farrer 1897, 25–91). In Leyland (1612) no one was to plough or feed on any balk or partition of the lands without the consent of the neighbouring farmers. Balks were occasionally called 'reans', one being ploughed up at Lowick in 1646 (Youd 1962, 15, 17). By-laws of

Ashton-under-Lyne made in 1682 included orders not to waste turf rooms in the moss nor let them to a 'foreigner'; there were not to be cattle on the moss from 1 May to 11 August, and moss reeves were to be appointed (E. H. Rideour 1929, *TLCHS* 80: 142).

Ridge and furrow

Only two broad-ridge sites are recorded in the CUCAP collection, both west of Preston; Hutton (CNS 26 and 28–9, at SD 470 270), and Longton Marsh (CNS 25 (1980), at SD 468 258), which show several furlongs lying in fields with curved hedges. Millward (1955, 41) published a photograph of broad ridge at Broughton in Furness.

The Lancashire HER had 276 records of ridge and furrow (in 2000), most of them being early modern in date. Good examples can be seen near the settlement site at Bracewell and Stockton, north of Colne (SD 86 48) (HER, PRN 3298; photo N1263). There are curved broad ridges (some with a central later furrow place on the ridge top) as well late straight narrow ridges. Headland linear earthworks have been identified near Highfield Moss, Lowton (SJ 625 955), and reverse-S cropmarks observed near Bedford Hall (SJ 676 991) (Hall et al. 1995, 104, 28). Many examples of narrow late ridges can be seen in the Manchester area. Some lie on the peat of White Moss near Alkrington (SD 879 036) that was not drained until *c.*1840, and so give a firm chronology (Hall et al. 1995, 112–13). Others can be seen at Denton and Reddish (in 1993). Curved-hedge patterns can be seen on many First Edition Ordnance Survey maps, for example near Morley's Hall, Astley (Hall et al.,1995, 122–3).

BIBLIOGRAPHY

Chet. Soc.   Chetham Society
LRS   Lancashire Record Society
THSLC   *Transactions of the Historic Society of Lancashire and Cheshire*
TLCAS   *Transactions of the Lancashire and Cheshire Antiquarian Society*

Hall, D., Wells, C. E., and Hickory, E. (1995), *The Wetlands of Greater Manchester* (Lancaster: Lancaster University Archaeological Unit).
Millward, R. (1955), *Lancashire: An Illustrated Essay on the History of the Landscape* (London).

ACKNOWLEDGEMENTS

Peter Iles for providing HER information and Philip Mayes for introducing me to Lancashire through the Manchester Wetlands Survey.

20. LEICESTERSHIRE

The north-flowing River Soar and its tributaries form the central drainage system of Leicestershire, with the Welland in the south-east. The highest ground, over 600 feet, lies in Charnwood Forest and towards Rutland on the east. Some parishes contain more than one township: for instance, Bringhurst (3,612 acres) contains Drayton and

Great Easton. Leicestershire was a county with large-scale, regulated field systems, mostly two- and three-field, from an early date.

Gray (1915), Hilton (1954), and Thirsk (1973) discussed the medieval fields. Several open field maps are known: Congerstone, Stathern, Swannington, and Whitwick (Beresford 1948, 96); also Sileby and Ratcliffe on the Wreake (Hartley 1989, 4–5). Stathern (1792) and Barsby (1794) open-field maps have new enclosure hedges superimposed on them (Hoskins 1955, 144). Tithe maps of several places indicate open fields such as Enderby, 1849 (Kain and Oliver 1995, 282).

Gray (1915, 470–1) listed eleven parishes with two fields and eighteen with three. Additional data are given in the Table. Yardland sizes ranged from eighteen to twenty-four acres. Frolesworth yardlands were twenty-one acres in 1463 (G. Farnham 1922, *TLAHS* 12: 194), Baggrave eighteen acres in 1517 (Leadam 1897, 229–32), and Kirby Muxloe yardland twenty acres in 1634 (Beresford 1948, 124). Bottesford was assessed in oxgangs in 1524 (Thompson 1933, 173). Three demesnes lay in a block and one was dispersed.

Management

Peas and beans were grown extensively during the sixteenth century according to the evidence of probate inventories, sometimes being as high as 46 per cent of the recorded crops (W. G. Hoskins 1942, *PLAHS* 22: 80; Hoskins 1950, 160–73). Barley was the main cereal crop (average *c.*38 per cent) with little wheat (*c.*14 per cent) and occasional rye (5 per cent). The high percentage for legumes continues the cropping regime of monastic demesnes of the fifteenth century, where it was high (*c.*30 per cent) with additional oats (*c.*8 per cent) and low wheat and rye (Hoskins 1950, 166).

Grass was introduced into fields at early dates: Denton had a grass headland in 1402, and leys were recorded at Allington in 1468 and Great Bowden in 1507 (Thompson 1933, 212, 107, 185). Arnesby in *c.*1550 had 42 per cent leys and pasture on a 58½-acre farm, dispersed in three fields. Leys amounted to 16 per cent of the arable at Lutterworth in 1607. The average land-use for each yardland was 0.7 acres enclosed, 13 arable, 2.4 leys, and 2.8 acres of meadow. At Wigston Magna sample farms had 17 per cent leys in 1577; in 1659, 21 per cent; and in 1639, 10 per cent (Hoskins 1950, 138–54). Yardland stints are recorded for many places: Billesdon (1609) eight cows and forty sheep; Rolleston (1658) four cows and twenty-two sheep, and a cottager had two cows and ten sheep (VCH, *Leicestershire*, v (1964): 10, 19). A West Langton yardland was stinted at four cows and thirty sheep (Hill 1867, 69).

Field orders

Wymeswold fields regulations of *c.*1425 are listed in the main text (see p. 129). Later examples are available in court rolls. At Dadlington, tenants were to repair the *Fennewey* (1429) and 'pinfold' (1478; Leicestershire CRO, 2D71/34 and 37). There was a series of manorial 'by-laws' by 1484, and tenants were presented to the court for allowing horses to trespass in the corn in 1520 (DE 226/4/2). Detailed lists of open-field orders were made for the years 1592, 1617, and 1632 (6D40/4/1 and DE 226/4/10 and 16). They included the usual items restricting the number of animals allowed for each

Table App.11. Leicestershire field systems.

Place	Demesne	Field No.	Date	Sources
Allington		2	1275	Thompson 1933, 95; six fields in 1468, possible three-season
Chadwell		2	c.1535	Thompson 1933, 541
Foston		2	1295	Thompson 1933, 238–6
Barkby	dispersed	3	thirteenth century	Hilton 1954, 158
Bowden, Great		3	1507	Thompson 1933, 183–5
Leicester, East Field		3	1764	Bilson 1925, *PLAHS* 14: 1–20. thirty-eight yardlands; Fowkes and Potter 1988, 167–74, pl. 8 (1624)
Leicester, South Field		3	1708	Bilson 1925, *PLAHS* 14: 1–20
Leicester, West Field		3	1448	VCH, *Leicestershire*, iv (1958): 383. In 1279 and 1627, thirty-one yardlands
Kirby Bellars	dispersed	3	fourteenth–fifteenth century	Hilton, 1954, 157, 161 DP
Oadby		3	1380	Thompson 1933, 416–17
Shepshed	dispersed	3	thirteenth century	Hilton 1954, 158
Slawston		3	fifteenth century	Northamptonshire CRO, Bru E.x.1
Staunton Wyville	block	3	1635	Northamptonshire CRO, Map 1352, fols. 26–37
Wigston Magna		3	1393	Thompson 1933, 500–8 (and in 1417); Hoskins 1937, 170; ninety-six yardlands in 1765; yardland twenty-two acres
Cranoe		4	1635	Northamptonshire CRO, Map 1352, fols. 26–37; half enclosed
Glooston	block	4	1635	Northamptonshire CRO, Map 1352, fols. 26–37
Langton, West	block		1743	Hill 1867, 68
Norton, Kings	dispersed		1360	Hilton, 1954, 157; yardland twenty-four acres

yardland and ordering the repair of fences. Horses were not to be put in the meadows until hay was carried; peas were only to be gleaned on the tenants' own land. Work-service at Saddington in 1316 supported the usual agricultural round (G. F. Farnham 1930–3, iv, 8).

Ridge and furrow

There are extensive remains of ridge and furrow. Leicestershire had the best survival in the country in 1991. It is distributed over most of the county with little around Charnwood and most in a broad swathe in the centre from the Welland Valley to Melton Mowbray (Hall 2001, 36, 34). R. F. Hartley has sketch-plotted ridge and furrow visible on RAF 1940s vertical photographs for the whole county at the 1:10,560 scale, some of it published (Hartley 1984, 1987, 1989). A few strip lynchets occur at Belvoir and Knipton (SK 822 308) (CUCAP, NR 14–15 (1954)), some with ridge and furrow on top of them. Hilton (VCH, *Leicestershire*, ii (1954): 159) published a photograph of Kilby.

Furlong plans have been reconstructed from an archaeological survey of the Bosworth battlefield area: Dadlington, Shenton, Stoke Golding, Sutton Cheney, and Upton Hall (in Foard and Curry 2013). A Kibworth Harcourt surveyed plan (D. Hall, unpublished) agrees with the draft enclosure plan held at Merton College, Oxford. Post-enclosure ridges near Burrough on the Hill are illustrated by Hoskins (1955, pl. 45 opp. p. 144).

BIBLIOGRAPHY

TLAHS   Transactions of the Leicester Archaeological and Historical Society

Bilson, C. J. (1925), 'The open fields of Leicester', *TLAHS* 14: 1–20.

Foard, G., and Curry, A. (2013), *Bosworth 1485: A Battlefield Rediscovered* (Oxford: Oxbow).

Hartley, R. F. (1984), *The Medieval Earthworks of North-Western Leicestershire*, Leicester Museums and Art Gallery Archaeological Report, 9.

Hartley, R. F. (1987), *The Medieval Earthworks of North-Eastern Leicestershire*, Leicester Museums and Art Gallery Archaeological Report, 13.

Hartley, R. F. (1989), *The Medieval Earthworks of Central Leicestershire*, Leicester: Leicestershire Museums Publication, 103.

Knox, R. (2004), 'The Anglo-Saxons in Leicestershire', in P. Bowman and P. Liddle (eds.), *Leicestershire Landscapes*, Leicester Museums Archaeological Fieldwork Group, Monograph no. 1.

Liddle, P. (1996), 'The archaeology of Anglo-Saxon Leicestershire', in J. Bourne (ed.) *Anglo-Saxon Landscapes in the East Midlands* (Leicester: Leicestershire Museums, Arts and Records Service).

Thirsk, J. (1954), 'Agricultural History 1540–1950', in VCH, *Leicestershire*, ii: 199–220.

ACKNOWLEDGEMENTS

Peter Liddle, Fred Hartley, Richard Knox, and Richard Pollard for information from the HER. Glenn Foard and Mark Page for access to the Dadlington court roll transcripts used in the Bosworth Project.

21. LINCOLNSHIRE

Lincolnshire has a varied topography, with a central belt of undulating limestone running north from Stamford, chalk wolds in the north-east, and the extensive wetlands of Wash Fenland. More wetlands lie in the north next to the Isle of Axholme and along the Humber. Heath grounds developed on some limestone soils. Parishes on the western fen edge have elongated shapes taking in various soils, and in the Silt Fen, Holbeach (20,981 acres) is 15 miles long.

There are two main field types in Lincolnshire, those of the Silt Fenland, with ditched strips, and those of the upland west of the fen, which are similar to the regular two- and three- field systems of the East Midlands.

The field systems have been described by Gray (1915), Hallam (1965), and Thirsk (1973). Open-field maps survive for much of the county, since much of it was enclosed late, for example East Keal (1757) (LAO, Misc Dep. 2/1), and Toynton All Saints with Toynton St Peter (1614) (LAO, 5ANC 4/A/4). Part of Broxholme map of *c.*1600 was published by Beresford and St Joseph (1979, 34–5). Tithe maps of the 1840s show open-field lands in several places. Late field systems were studied by Swales using parliamentary enclosure papers (1934–6). The Russells also studied enclosure papers and published maps showing the extent of the great fields and pastures (1974, 1987). As well as two- and three-field systems there were examples of four fields; some with paired names suggest that older two-fields had been split. Several townships having two fields in *c.* 1800 were also two-field in *c.* 1200. The information was used to compile the Lincolnshire part of Figure 0.1.

Gray (1915, 473–6) provided thirty-nine examples of two-field and eight of three-field systems (thirteen and four of them, respectively, being demesnes). Hallam (1965, 231–6) found that in the twelfth century most upland vills had two fields and a few had three or four fields. He summarized the early twelfth- and thirteenth-century evidence published both by Stenton (1920, 1922) and by the Lincoln Record Society in the seven volumes of the *Registrum Antiquissimum*. There were sixty-seven two-fields (plus two of fourteenth century date) and five three-field arrangements. More information is given in the main text (p. 39). Hence the county can be interpreted as being a predominantly two-field region, even in more recent centuries. There is limited evidence of 'inhocking'.

Other examples of two- and three-field systems have been noticed, but they do not alter the conclusions of Hallam. Further examples of two-field systems are Wyham (1219) (D. M. Stenton 1934, Selden Society, 53: 436), Harlaxton (1234) (Massingberd 1896, 262), and Marston (1630), with an enclosed block demesne (Northamptonshire CRO, Bru. ASR 132, 33). An additional three-field example is Messingham near Gainsborough (1643) (Hall 1935, 56). Field-number changes at Corringham have been described (p. 41).

Fenland fields

The early silt-fen fields with their ditched strips have been discussed (p. 75). Fleet fields are described in a field book made in 1315 (Neilson 1920). They were divided into thirty-nine blocks ('furlongs', most probably flat-ploughed ditched strips) called 'inlikes'.

The holdings of both freeholds and tenancies were called 'freelonds', 'werklonds', 'molelonds', and 'Mondaylonds'. The last two were quarter-oxgangs; the most usual were werklonds, which were half a bovate in extent, about twenty-five to thirty acres. No werklond lay in one place and very few lay together in adjacent inlikes. Most are dispersed without any apparent system in twelve or so inlikes. For example the werklond of Richard son of Nigel was in 15 parcels of 3 roods to 5 acres, being 29¼ acres in all and dispersed in 10 inlikes, which lie over a large part of the township (Neilson 1920, lxvii). The demesne extended to over 1,000 acres and lay in 18 of the 39 inlikes. It consisted of large pieces, some of them enclosed. There is no internal information on how these holdings were farmed and whether there were any common rights on the arable (Neilson 1920, lxxxiii). All the siltland parishes have full descriptions of their lands in field books (also called 'acrebooks') dating from the seventeenth and eighteenth centuries (Hallam 1965, xiv–xv).

Cropping and management

Thirsk (1973, 248–53) discussed the many manorial court records that detail field management and farming in the different regions of the county. Oxgangs and the regular tenurial order implied at Hough on the Hill have been described. Further examples are found in a late twelfth-century charter of Edlington, which granted an acre next to William son of Lamfer and other acres next to the same William in the other part of the vill (K. Major 1950, LRS 41: 152). Seven lands at Fillingham in *c*.1225 had the same neighbour six times, and six pieces at Nettleton, in *c*.1235, all lay between Matilda and Roger (C. W. Foster 1937, LRS 32: 94–5, 145–6).

Both compact and dispersed demesnes are known (see pp. 97, 102). Another dispersed demesne is Castle Bytham, which in 1316 had a demesne lying in blocks in many furlongs grouped into three fields of unequal size (Platts 1985, 86).

Ridge and furrow

Ridge and furrow occurred over most of the county outside of the wetlands (for the survival in the south-western part in 1993–4 see Hall, 2001, 36). Uffington and Deeping St James had some lands 300 yards long in 1947 (RAF VAP CPE/UK/1932/frames 3068 and 1076–80). Striking vertical photographs of West Deeping cropmarks were taken in 1976 (CUCAP, RC8-BO 220–4), with long alignments of strips and chequerboard patterns.

The heavy clays of Humberside still have good examples of ridge and furrow, for example at Goxhill (TA 108 220), where there are steep curved ridges that are older than the hedges set at enclosure in 1773 (Tate and Turner 1978, 162). Many photographs are held at CUCAP, such as Ailby and Alford (TF 438 767 and TF 447 764) (BOR 105–9 (1973)), and Gayton le Wold (TF 228 846) (CID 38–40 (1979)). Examples of narrow straight ridges without heads of post-medieval date lie in the precinct of Louth Park Abbey, dissolved in 1536 (Start 1993, 55).

In the silt fen much of the land in the south was cultivated in long ditched strips that were flat ploughed. A few examples mostly of short length, survive at Quadring. North of Boston, Waynfleet St Mary has both ditched and ridged strips. Saltfleetby St Peter, east of Louth, has a large area with blocks of ridged strips up 1,100 yards in length recorded on 1940s RAF photographs, some still visible on satellite images (e.g. at TF 44 90; see Gardiner 2009, 2). Long-ditched strips have also been identified in the south near the Welland Valley at West Deeping (J. P. Hunn 1998, fig. 12).

Orwin drew attention to the survival of open fields in the Isle of Axholme at Epworth (Figure 0.2), West Butterwick, and Haxey (C. S. Orwin 1937–8, *Economic History Review*, 8: 125), of which a photograph of Epworth had been published by Slater (1907, frontispiece) and Venn (1933, pls. v and vi). It was noted, in 1935, that the region was 'until recent years perhaps the most primitive in the Midlands' (Hall 1935). Townships in the Isle still have many examples of flat-ploughed, curved strips surviving, unenclosed and farmed in dispersed holdings (see CUCAP, RC8-GA 029 (1964)). They were perhaps always ploughed flat. Belton (2004) has a large central area called Belton Field (SE 78 07) with curved strips, many of them very narrow, perhaps consisting of just one or two lands. There is an occasional green balk that forms a boundary or footpath (T. Fulton 2010, *Landscape History*, 31: 71–2). Only furrows mark the strip boundaries of most lands. Furlong boundaries are well preserved at Pointon.

BIBLIOGRAPHY

LAO   Lincolnshire Archives Office
LRS   Lincolnshire Record Society

Boyd, W. K. (1896), *Abstracts of Final Concords*, i (London: Spottiswoode & Co.).

Gardiner, M. (2009), 'Dales, long lands and the medieval division of land in Eastern England', *Agricultural History Review*, 57: 1–14.

Hall, T. W. (1935), *Charters…relating to Land near Sheffield* (Sheffield: J. W. Northend).

Hunn, J. P. (1998), in A. P. Johnson, *Land at West Deeping (King Street), Lincolnshire* (Oxford: Oxford Archaeological Associates Ltd).

Massingberd, W. O. (1896), *Abstracts of Final Concords, Temp. Richard I, John, and Henry III* (London: Spottiswoode & Co.).

Platts, G. (1985), *Land and People in Medieval Lincolnshire* (Lincoln: History of Lincolnshire Committee for the Society for Lincolnshire History and Archaeology).

Russell, R. C. (1974), *The Logic of Open Field Systems: Fifteen Maps of Groups of Common Fields on the Eve of Enclosure* (London: Bedford Square Press for the Standing Conference for Local History).

Russell, R. C., and Russell, E. (1987), *Parliamentary Enclosure and New Landscapes in Lincolnshire* (Lincoln: Lincolnshire County Council).

Start, D. (1993), *Lincolnshire from the Air* (Sleaford: Heritage Lincolnshire).

Swales, T. H. (1934–6), 'The Parliamentary Enclosures of Lindsey', *Lincolnshire Architectural and Archaeological Society, Reports and Papers*, 42: 233–74; and 43: 85–120 [also titled NS 1, pt. 1].

ACKNOWLEDGEMENTS

I am grateful to Tom Lane, Hilary Healy, Martin Redden, Jonathon Hunn, Rog Palmer, Ron Parker, and Mark Bennett for information about various aspects of Lincolnshire fields. Terry Fulton kindly pointed out the fields of Belton and provided a copy of the tithe map.

22. MIDDLESEX

Middlesex has low land alongside the Thames and modest hills in the north towards Hertfordshire. The underlying geology is London Clay with gravelly Bagshot Beds in the west. The county had much dispersed settlement associated with greens, woods, and irregular multiple fields, with no evidence of communal agricultural procedures. Some property descended by partible inheritance and Borough English.

Gray discussed Middlesex fields, noting that some open fields in the west survived into the eighteenth century but that mostly the fields were irregular and enclosed early. A 1604 survey of Edmonton lists four fields in which holdings are dispersed in one, two, or three fields and there are six other areas called 'fields' occasionally referred to, as well as enclosed arable and pasture. Similar irregularities occurred elsewhere in the thirteenth century (Gray 1915, 381–4, 551). The economic sections of the VCH *Middlesex* volumes confirm Gray's findings.

Thus at Stepney, field names from *c.*1200 did not suggest a system of communal management or crop rotation within the parish. Individual holdings were very irregularly distributed between different fields in *c.*1380. Land of freehold estates and customary tenants lay intermixed in small strips in the open fields in *c.*1400, although there were some closes and blocks of strips (Baker 1998, 52–63). The holdings of St Alban's monastery at Barnet in 1360 each lay in ten to twelve pieces, nearly all enclosed with hedges (Levett 1938, 187). The late sixteenth-century surveys and maps held at All Souls' College, Oxford, for Edgware, Kingsbury, Willesden, and elsewhere show the same complicated irregular arrangements (listed in Trice Martin 1877, 287).

The county needs reconstruction of examples of its medieval fields using the many detailed records available from monastic sources, especially those having fields called 'hyde', such as at Edmonton.

BIBLIOGRAPHY

Baker, T. F. T. (1998), 'Stepney', in VCH, *Middlesex*, xi: 52–63.
Martin, C. Trice (1877), *Catalogue of the Archives in the Muniment Rooms of All Souls' College* (London: Spottiswoode & Co.).

23. NORFOLK

Norfolk has underlying chalk, much of it covered with boulder clay at the south. The sandy Brecklands lie at the south-west, with fenland peat and silt in the west. Norfolk had many large greens and linking droves which have been described (see the main text, pp. 28–9; Williamson 1993, 167–9; Williamson 2003, 91–104, 160–179; Macnair and Williamson 2010, 106). Some greens contained a church and an early nucleus of settlement; medieval settlement spread around many other greens. Settlement was further dispersed by assarting; at Horsead, in the twelfth century, 35 assarts were made (170 acres) with houses built on 19 of them. Cawston, in 1291, had 109 acres of assarts (Campbell 1981–3, *NA* 38: 20). The fields of Norfolk were irregular, often with multiple areas called 'precincts' or 'fields' that were locational descriptors and were not related to cropping systems. Large-scale sheepfolding developed in the late Middle Ages, which affected the course of husbandry.

Manors were extremely complicated, as noted by many authors, and evident already in the Domesday Survey. Detail is provided by a 1298 extent of Bradcar manor in Shropham. There were 58 freeholders holding 289 acres, most of the area belonging to three people. However, nearly all of their land, each with a fold course, lay in the nearby vills of Great Breckles, Great and Little Hockham, and Snetterton. The other freeholders had land in Bradcar. Eighteen tenants holding seventy-one acres were called 'molmen' and seven customary tenants held fifty-four acres. The demesne was 301½ acres. Tenants of nearby Larling pastured beasts on Shropham common (W. Hudson 1900, *NA* 14: 23–32).

Peter Wade-Martins (1980b figs. 15, 33, 37) reproduced plans of Longham, Weasenham All Saints, and Weasenham St Peter (late sixteenth century). These and the plans of Tittleshall and Sutton (1596) have rectangular furlongs. Campbell (1983, *NA* 38: 5–32) published several plans, including Horstead with Staninghall (1586 reconstruction (Corbett 1897)) and Lessingham (1587). A 1629 map of Hargham, lying in the Breckland, shows an arable area divided into about seventeen furlongs. Large heaths and moors occupied more than half of the township (Davison and Cushion 1998–2001, *NA* 43: 265). Creake maps of 1610 were used by Mary Hesse to reconstruct a plan of 1475 (1998, *NA* 43: 80–1). A series of ten furlongs lay on the same alignment. Earlier terriers use the same furlong names as in 1610 and 1475, and it is likely that the same pattern was present in the thirteenth century. A plot of the lands belonging to Creake Abbey before 1350, insofar as they can be identified, shows a concentration towards the western part of the township. Considerable open field is indicated on maps of Scole (late sixteenth century), Langley (*c*.1625), Raveningham (1635), Brooke (late sixteenth century), and Morley 1629 (S. Wade-Martins and T. Williamson 1995, NRS 58: 5). A Cockley Clay map of 1722 shows extensive open field with ten furlongs aligned one above. At the south-west were large areas of sheepwalks on the Brecks (Norfolk CRO, BL 47/1).

Fields

Norfolk irregular fields and customary holdings have been described in the main text (pp. 64–6; from Gray 1915, Postgate 1973, Campbell 1981b, and Williamson 2003).

Partible inheritance

The Martham survey of 1292 shows that partible inheritance occurred. It also was found at Coltishall, Aylsham, Belaugh, and Burgh, and probably at Cawston, Lessingham, Thurne, and Worstead. The court rolls of Martham and Coltishall record many instances of holdings passing to more than one heir between 1275 and 1348. Other manors at Hevingham, Horstead, and Marsham had impartible inheritance (Campbell 1983, *NA* 38: 22–3).

Cropping and farming management

Sheepfolding has been explained (pp. 69–72). Townships were divided into a few large areas called 'courses' where the sheep of two or more manors pastured on commons and on the fallow. The courses did not correspond to the precincts nor to the arable ownership and tenancies. The sandy soils of Horsford were used for a fold course in the early modern period. The flock had rights over open-field and half-year lands from October to March, over 600 acres of heathland for six weeks in summer and six weeks in winter, and over commons (T. Barrett-Lennard 1917–19, *NA* 20: 57–73). In 1432 Binham Priory made an agreement with its tenants about folding. The tenants could have pasture for their cattle every year, fenced in with stakes, within the several pasture of the prior from September to March. Sheep could also pasture there from October to June at the rate of four sheep to each [arable] acre of an individual tenant's holding, without hindering the fold course of the prior in his separate field. The prior gave up the right to common pigs on any of the common pastures of Bingham from January to July (E. B. Burstall 1957, *NA* 31: 211–15).

Bradcar work-service (1298) included ploughing, carrying, hoeing, mowing, haymaking, and carrying, as well as turf digging and cleaning the millpool and church close ditch (W. Hudson 1900, *NA* 14: 31). Work-service at Martham in 1291, involved hoeing, carting manure, harrowing, threshing, making malt, and ditching. There was turbary in the South Fen (W. Hudson 1919, *NA* 20: 286–7). Details of agricultural practice are recorded for Thetford Priory from the period 1518–40: brakes were cut for five days; crab apples were stamped [for cider?]; starlings scared; and payments were made for four loads of hurdles and for 'spredyng of mokke' (Dymond 1995, part ii). The Ramsey Abbey Fenland manors of Hilgay and Snoring had rents of 10,000 eels in the twelfth century (Hart and Lyons 1884–93, iii, 287). At Norwich Priory gardens, clay pots were provided for honey and milk, and sixty alders were used to make stakes for hurdles in 1330. Onions and garlic were being grown in 1387 (C. Noble 1997, NRS 61: 32, 35).

Ridge and furrow

Young (1804, 54, 190) recorded that in 1776 there were broad high ridges at Wallington near Marshland, and that the swing plough was used in the south-west of the county. Silvester (1989, *NA* 40: 286–96) showed that the distribution of ridge and furrow surviving in the county in the 1940s was mainly western (map and illustration of Hilgay and Stradsett). He noted that there seemed to be no evidence of ridge and furrow in central Norfolk and that Marshall stated in 1795 that Norfok soils were sufficiently absorbent to require neither ridge nor furrow. Liddiard (1999) proposed that there had once been a wider distribution over the county, much of it having been removed by agricultural practice before 1940. Enclosed grounds were ploughed by the method of 'stitching', that is, raising narrow ridges; but overall the land surface was kept flat by cross-ploughing. Underdraining increased during the eighteenth century (S. Wade-Martins and T. Williamson 1995, NRS, 58: 27) and in conjunction with cross-ploughing would remove ridge and furrow. Brandon Warren had ridges recorded in its shift field in 1603.

Furlong boundaries have been mapped at South Creake, Fransham, and Dickleborough, and observed at Marham and Barton Bendish (D. Hall, unpublished). At South Creake they corresponded exactly to a map of 1610 (referred to in Hesse 1998, 81).

Field patterns

Field boundaries at Brancaster have been mentioned (p. 160). Curved hedges of open-field type are known at Tharston. Coaxial fields have been discussed and the likelihood that they are of medieval origin established (see Suffolk (p. 320), and Martin and Satchell 2008, 214–16).

BIBLIOGRAPHY

EAA   East Anglian Archaeology
NA   Norfolk Archaeology
NRS   Norfolk Record Society

Liddiard, R. (1999), 'The distribution of ridge and furrow in East Anglia', *Agricultural History Review*, 47: 1–6.
Young, A. (1804), *A General View of the Agriculture of Norfolk* (London: Board of Agriculture).

ACKNOWLEDGEMENTS

Alan Davison, Mary Hesse, Robert Liddiard, Peter Wade-Martins, Andrew Rogerson, Robert Sylvester, and Tom Williamson for discussion and field visits.

24. NORTHAMPTONSHIRE

Northamptonshire is a lowland county lying in the East Midlands, the ground falling from a few highs of 700 feet in the west to fenland now below sea-level near Peterborough. Most of the county was subject to medieval arable agriculture. The principal 'waste' was Borough Great Fen (8,900 acres), commonable to all the Nassaburg Hundred and formerly to all the Peterborough Abbey estates. The three royal forests of Rockingham, Salcey, and Whittlewood amounted to about 63,000 acres of wood, woodland-pasture, and deer parks, and were partly commonable to nearby townships. Northamptonshire is a county consisting almost entirely of two- and (primarily) three-field systems. Only a few irregular arrangements are known, but some townships had multiple fields in the early modern period, run as three-tilth. A few five-field systems developed that were parallel two- and three-field (see the main text, p. 44).

The open field history has been described (Hall 1995). Plans of the furlongs for the whole county at 1:25,000 scale have been published (Hall 2009 and 2013). A few items referred to in the main text are further discussed below to provide references.

The county plan of Saxon sites discovered by fieldwork given in Plate 9 has been updated from the previously published versions (Foard et al. 2009, 46–54; Williamson et al. 2013, plate 15a). The updating has included checking soil type on which the sites lay. There are additional sites coinciding with Roman rural ones, many lying on clayland, that have a few Saxon sherds, indicating a widespread initial Saxon presence before abandonment in favour of better soils and locations. Details of the numbers of sherds and other finds will be given elsewhere (Hall and Martin unpublished). Some Middle Saxon fabrics of Ipswich Ware have been found and also Maxey Ware, the latter being largely confined to the Peterborough region. This is probably to be interpreted as indicating that the county lies outside of the main distribution areas of these fabrics rather than all the sites without Middle Saxon fabrics are of Early Saxon date.

Domesday assessments and customary holdings

The interpretation of the relationship between Domesday hides and ploughlands with yardlands has been discussed fully (see pp. 197–8). It remains a difficult theme. For Aynho and Whilton both hides and ploughlands can be correlated with later yardland numbers. In many cases hides and yardlands can be shown to have a limited number of ratios, but ploughlands do not. Williamson et al. (2013, 69–70) have been unable to make much sense of ploughlands by interpreting them as 'farms'. Ploughlands are only found in the Domesday Survey, not before or after, but hides occur in other medieval assessments, as well as in earlier Anglo-Saxon charters. These are therefore the taxation units that seemed to matter in the eleventh and twelfth centuries. As explained, Domesday hides can be compared with the twelfth-century Northamptonshire Survey, which does not mention ploughlands (p. 201). Generally the hidages are similar in both surveys, which confirms that 'hides' were the chief taxation units.

Mears Ashby was assessed as four hides in 1086 and in *c.*1124. The assessment normally represents the taxation levied on the villein land, and excludes the demesne. This is consistent with a 1577 field book which shows there was a separate lordly block demesne and other fields which were divided into a regular tenurial cycle of forty lands throughout, thus suggesting that the assessments were made on villein land only, taking ten yardlands to the hide as often found in the county. The total of 52½ yardlands for the township in the eighteenth century suggests 12½ for the demesne, which equates to the one ploughland in demesne in 1086. The other seven ploughlands recorded in 1086 do not readily relate to the yardlands (sources in Hall 1995).

Not all hidages and ploughlands are explicable in terms of yardlands. There are many reasons for this—the later yardlands may not refer to the same area as 1086, or some Domesday assessments may be composite in that they consist of both the 're-duced' figures normally used (commonly taking ten or twelve to the hide) and unreduced at four to the hide. Rushden is an example (Hall 1995, 86). The Anglo-Saxon charter value for the abbot of Peterborough's holding at Lutton, nine yardlands (S 1110), is recorded in Domesday as 2½ hides (*recte* 2¼ hides), so four yardlands and not ten or twelve were grouped into a hide. Such problems as these will confuse and confound numeric calculations. Some values appear to make no sense, but the significant point is that a clear interpretation *can* be given in many cases. Such correlations cannot be accidental and demonstrate an underlying ancient organization.

ACKNOWLEDGEMENTS

I am grateful to Paul Martin for many years of fieldwork undertaken in this county and elsewhere; to Glenn Foard for discussion and organizing many stimulating projects; to Stuart Storey-Taylor and Tom Williamson for the Rockingham Forest Trust and UEA Midland Projects, respectively, who enabled the GIS mapping of all Northamptonshire fields; and to Tracey Partida for the immense work of incorporating 400 parish furlong maps drawn at the 1:10,560 scale into a single GIS.

25. NORTHUMBERLAND

Northumberland topography rises from extensive coastal plains in the east to high ground of the Pennines in the west and the Cheviots in the north-west. An appreciable area lies above 1,000 feet. Fells throughout the county were used for cattle and sheep rearing, often involving transhumance and shielings and bastiles. As with Cumberland, many shielings developed into permanently settled farms (Ramm et al. 1970, 1–12; Map 2). The border against Scotland, and as far south as Corbridge and Hexham, suffered raids. New Minster Abbey was located on the south side of a valley because the monks were 'always glad to have a river between themselves and the Scotch' (Fowler 1878, xv). The south-east was badly affected by cattle raiders in 1573. Many parishes were large—Warkworth contained eighteen townships in its 17,455 acres (J. C. Hodgson 1899, *HN* v: 14).

The fields of Northumberland in the upland valleys were small and irregular. Larger two- and three-field arrangements occurred on the lowland near the coast. Many fields within a given system were unequal and ran alongside large commons, some of which participated in convertible husbandry, so that the extent of the fields varied from year to year.

Gray (1915, 206–27) discussed Northumberland field systems and Butlin (1973, 93–144) gives an account of the fields and enclosures of the county, developing the data presented in 1964. Much information, including details of surveys and terriers, is given in the volumes of the *History of Northumberland* (1894–1940). Fallowfield deserted village has a plan of *c.*1583 showing schematic indications of strips in the West, Middle, and East Fields (M. W. Beresford 1955, *Medieval Archaeology*, 10: 165–7). Four places had open-field marked on tithe maps of 1839–47 (Kain and Oliver 1995).

Field numbers

Gray (1915, 217–20), drawing upon the records of Hexham Priory (Raine 1865) and other sources (Fowler 1878 and Page 1893), was unable to find any evidence of medieval division into two or three fields. Further inspection shows this to be true, except perhaps for Leighton. Gray noted, too, that 'fields' mapped and described by surveys of the sixteenth and seventeenth centuries did not clearly indicate an equal division (ibid. 208–16). Butlin (1973, 100) produced a map of the distribution of places known to have had open fields from *c.*1550 to 1800. Most are located on the coastal region or in major river valleys. Data from 115 surveys of the sixteenth and seventeenth centuries showed there were two places with two-fields, thirty-six with three (34 per cent), ten had four, one had five, and one had six named fields. 'Common' or 'town fields' without any indication of grouping into fields were recorded in fifty-one townships (Butlin 1964, 101). Two fields are mentioned in a grant of six 'dales' at Leighton in the thirteenth century (Fowler 1878, 85).

The field systems are complicated. Of those with three fields, few fields are equal sized and many have a low amount of arable as a proportion of the total township area, there being much common pasture. This enabled convertible husbandry to be incorporated into the operation of a three-field course of husbandry, as recorded at Acklington, Guysance, and Rennington. Ten places in the Table with three to five fields had holdings distributed throughout all the fields. It is possible that the field names are largely locational, but the arable was organized into three groups of equal area that varied in detailed location from year to year as leys and pasture were temporarily taken into cultivation. That there was considerable manorial organization is proved by the occurrence of detailed field orders recorded in court rolls, including crop regulation.

The scarplands and inner vales of the Tyne had holdings distributed throughout the principal named fields. Inghoe had 4 fields, in *c*.1629, and a large common of 938 acres. Three common fields at similar dates are recorded at Kirkwhelpington (Alnwick) and at Clarewood, Aydon, and Halton Shields (Butlin 1973, 120–3). The Table identifies four block demesnes and one dispersed. Farms were called 'oxgangs' or 'farmlands'.

Field orders

At Warkworth, in 1473, the manorial court noted that salmon had been poached, swine had strayed in the field, and the lord's pinfold needed repair (*HN* v (1899): 144). Hartley orders for 1495–1567 prevented cottagers from keeping pigs or geese outside their houses; tenants were not to plough the green field (i.e. fallow); beasts could not be pastured on balks; gleaning was not to begin until all sheaves had been removed; and swine were to be taken in every night. In 1567 every tenant was ordered to leave one foot of ground between every land (for greensward; H. H. E. Craster 1909, *HN* ix: 119).

Seaton Delaval's sixteenth-century orders include various aspects of management. Cattle were to be brought in each night from the pastures; horses or mares were to be tethered only on the tenants' rygs; nags were to be kept in common herds; and swine were to be kept in a herd and brought back at night. Grass growing upon a balk was cut for hay. Cottagers could not put cattle in any hained field belonging to the tenants until the tenants' oxen had depastured for one month. Every man had to sow the same seeds as his neighbours. Kine were put on stubble but there was a fine for oxen let out before corn was carried. The meadows and arable fields were protected from the commons by a surrounding dyke that was repaired as needed. In 1537 a bank was to be made as high as could be reached with a spade; there were gates in it for stock to traverse (H. H. E. Craster 1909, *HN* ix: 194–6). These orders show there was a controlled communal cropping system on a uniformly cropped sown field. Dilston's late sixteenth-century courts regulated the common stint. Corbridge court orders for 1674 controlled the pasturing of horses; all hedges about cornfields and pastures were to be repaired at the usual times. The herders of the cornfields were ordered to bring overstinted animals to the common fold each day; swine were to be ringed (ibid. 134).

Convertible husbandry at Shoreston, Slaley, and Guysance has been described (p. 92). Before 1805 the tenants of North Middleton in Hartburn (1,128 acres) determined what parts of the fields should be arable, meadow, and pasture. Stints (285½) were apportioned to each of the fourteen farms. There were also small parcels of dispersed cottage lands (in the hands of the major landowners) which were sometimes arable, pasture, or meadow, according as the sequence of land-use determined (Dendy 1894, 138–9). This seems to be a form of convertible husbandry.

Table App.12. Northumberland field systems.

Place	Field no.	% of Township	Demesne	Distribution of holdings	Date	Source	Notes
Acklington	3	39			1616	*HN* v (1899): 372–6	Plan Butlin 1973, 118; fields equal; convertible husbandry practised
Alnmouth	1	16			1624	Butlin 1973, 116	Small area of townfield
Aydon	3				1562	*HN* x (1914): 367	
Bilton	3	39		uniform	1624	Butlin 1973, 116, plan	
Chatton	4	16		uniform	1616	Butlin 1973, 121, plan	Unequal fields
Corbridge	5	37?	1635 dispersed	uniform	1777	Butlin 1973, 105–8; *HN* x (1914): 124–40	Chequered furlong pattern
Denwick	4	29		uniform	c.1624	Butlin 73, 113, plan	Pasture in one field
Fallowfield	3					Beresford 1966, *Medieval Archaeology*, 10: 165–7	Fields near equal
Great Whittington	3				1687	*HN* x (1914): 424	Unequal fields
Guysance	3				1618	*HN* ix (1909): 22, 210, 124; Butlin 1973, 117	Fields similar in size; used convertible husbandry
Hartley	3			uniform	1536	Butlin 1973, 119	Equal-sized fields
High Buston	1	42		uniform	1616	*HN* v (1899): 204–12	
Inhoe	4	36	block		1629	Butlin 73, 123, plan	Unequal fields
Leighton	2		block		thirteenth century	Fowler 1878, 85	Grant of six and four dales in West and East fields
Lesbury	4	63		uniform	1624	Butlin 1973, 116, plan	
Long Houghton	3	39	block		c.1619	Butlin 1973, 114–15	Fields very unequal

Township					Date	Reference	Description
Ovington	3	54			seventeenth century	Dendy 1894, 129	Equal-sized fields
Preston	2			uniform	1649	Adamson 1887, *AA*, ns, 12: 172–90	Slightly unequal fields
Rennington	3	27		uniform	c.1624	Butlin 1973, 113, plan	Fields unequal; large common plus intakes
Rock	3		block		1599	*HN* ii (1895): 128	
Seaton Delavel	3		block		seventeenth century	Butlin 1973, 120	Equal-sized fields
Shilbottle	3	33		uniform	c.1624	*HN* v (1899): 462	Fields unequal; Butlin 1973, 111–14
Tynemouth	2			uniform	1649	Adamson 1887, *AA*, ns, 12: 172–90	Unequal fields with commons; three fields in seventeenth century; Butlin 1973, 119

Upland fields

The highland regions had small settlements, little arable, and relied on cattle rearing and their use of transhumance. Coket Moor, 300 acres of waste and 300 acres of moor, had a messuage with 12 acres in 1325 (E. Miller 1953, *Proceedings of the Society of Antiquaries of Newcastle*, 5th ser., 5: 189–91). A site at Low Cleughs Burn, Corsenside, situated at 210 m (NY 875 868), has been excavated (M. Adams and P. Carne 1995, *DAJ* 11: 85–95). In 1605, Wark had 18 tenants on 111 acres of arable (which became common after cropping), 193 acres of pasture, and 97 acres of meadow. In North Tynedale 67 farms with 468 acres meadow had 841acres arable, 1,140 acres of pasture, and 9,750 acres of common waste; the farms were scattered (Butlin 1973, 125–8). Harbottle manor had eighteen places for its summer grounds; the season normally finished in August. Partible inheritance operated and members of the same family often shared a holding, such as Dunterley in North Tynedale, where 150 acres of arable meadow and pasture were shared between five people surnamed Charlton.

Tenurial cycles and field rearrangements

Several open-field townships were divided into two parts, each having newly allotted scattered strips in the two halves. Long Houghton was divided before 1567. At Chatton, because the soil varied, the land was re-allotted rig by rig, an East End and a West End being defined by a hedge separating them (E. Bateson 1895, *HN* ii: 369). Rock was divided into two in 1599; there had been three fields previously. The division placed five farms on the north side with 214 acres of arable and pasture and 200 acres of moor. On the south side were 7 farms with 301 acres of arable and pasture and 280 acres of moor. Each farm had forty-three acres of arable and forty of waste (ibid. 128–9). Acklington was not recommended to be split in 1567 because of soil inequality, but a division had been made into a north side and south side before 1702 (*HN* v (1899): 369–2). Shilbottle had four farms ('husbandlands') consolidated and taken out of the three main fields before 1567. Charlton was divided in 1685 (E. Bateson 1895, *HN* ii: 308).

Ridge and furrow

Ridge and furrow occurs over most of the lowland and in river valleys. Vertical air photographs taken in 1991 by Northumberland County Council showed much then surviving in the lowlands, at Haggerston Castle (NU 04 44), North Charlton (NU 16 22, NU 18 22), and elsewhere. Furlong patterns are often complex, fitting in with topography and drainage. Tim Gates has kindly provided a distribution map of recently surviving ridge and furrow, gained from his extensive aerial photography over the county. The largest block of survival is around Hexham, to Haltwistle in the west, and Otterburn northwards. An excellent headland occurs near Little Bavington (NY 985 782). The CUCAP collection contains outstanding photographs in Boxes ATV and AWT, taken in 1967 and 1969.

An example of broad ridge with a large headland (NZ 136 787) was observed south of Ogle during a field visit in made in March 2002. The Ingram area has many interesting sites. At NU 006 152, south-west of Ingram, a Roman settlement with stone-built round houses lies on a small rocky hill and is surrounded by broad ridge and furrow.

The steepness of the ridging would suggest a sixteenth-century date if it were found in the East Midlands. West Whelpington broad ridge, 9–12 yards wide and up to 1.8 yards in depth, was plotted in 1960. One of the furlongs has strips 630 yards long (M. G. Jarrett 1962, *AA*, 4th ser., 40: fig. 2).

Numerous examples of early modern narrow ridges survive. Some at Belling Law have been illustrated (G. Jobey 1977, *AA*, 5th ser., 5: pl. 1b opp. p. 104). Narrow straight ridges, three to seven yards wide, at West Welpington fit within new eighteenth-century enclosed fields, first recorded in 1722 but abandoned by 1840 (D. H. Evans, M. G. Jarrett, and S. Wrathmell 1988, *AA*, 5th ser., 16: 160–7). Great Whittington has an example of narrow straight ridges overlying curved broad ridge orientated at a different alignment (NZ 002 757) (CUCAP, AWT 36 (1969)). Strip lynchets were identified at Housesteads below the Roman fort of Borcovicium (W. P. Hedley 1931, *Antiquity*, 5: 351–4 and pl. 1). Some lynchets in the Kirktown region are probably of prehistoric origin (P. Topping 1982, *Northern Archaeology*, 3: 21–31).

BIBLIOGRAPHY

AA   *Archaeologia Aeliana*
DAJ   *Durham Archaeological Journal*
HN   *A History of Northumberland*

Butlin, R. A. (1964), 'Northumberland field systems', *Agricultural History Review*, 12: 99–120.
Fowler, J. T. (1878), *Chartularium Abbathia de Novo Monasterio*, Surtees Society, 66.
Ogle Millennium Group [2000], *Ogle: A Northumbrian Village*.
Raine, J. (1865), *Cartulary of the Priory of Hexham*, ii, Surtees Society, 46.

ACKNOWLEDGEMENTS

Paul Frodsham, Northumberland National Park Archaeologist, for a field trip in the Ingram Valley and discussion. Tim Gates for detailed information on ridge and furrow distribution in the county and discussion. John H. H. Peile for a visit to Ogle. Elizabeth Williams of the County HER for information and photographs.

26. NOTTINGHAMSHIRE

Much of the county is dominated by the valley of the River Trent, entering at the south-west and flowing north-east and north. Away from the river soils are predominantly clays. Sherwood Forest in the west lies on sandy ground derived from Bunter Sandstone; it has large parishes and was well wooded in the twelfth century (D. Crook 1984, *TTSN* 88: 14). The south and east of the county falls into the Midland area, where extensive three-field systems survived until the eighteenth century, and there was still 30 per cent open field in 1800 (Barnes 1997). Irregular fields with brecks operated in the Sherwood Forest region.

Many open-field maps are known for Nottinghamshire. The best known is Laxton, 1635, which has been fully described by the Orwins (1938) and Beckett (1989). East Bridgford has a fine map of 1616 (Ashikaga and Henstock 1996). Copies of the maps made by William Senior for the Earls of Devonshire are held at Nottingham University; the accompanying written surveys have been published (Fowkes and Potter 1988). Tithe maps of 1836–9 show some open fields in eight townships (Kain and Oliver 1995, 389–93). Nichols (1987) lists maps made before 1800.

Lowe (1798) states that three fields were common in the Vale of Belvoir. Gray noted seven three-field townships in the county and found no examples of two-field systems. Surveys of seven manorial demesnes (Ratcliffe on Soar, Preston, Brimesley, Laxton, Wandesley, and Trowell) showed they had a three-course tilth, and possibly three fields, during the second quarter of the fourteenth century (Blagg 1939, 28–92, 166). Several of these townships had land left uncultivated because it was sandy or because of a lack of tenants. In the second half of the fourteenth century three-tilth demesnes were recorded at Normanton on Trent and Tuxford, while the demesne at *Knesale* was two-tilth (Train 1949, 108, 1, 44). Recent work has identified two block and two dispersed demesnes, as well as three two-field systems (of early date) and fourteen three-field examples, one of them irregular. Higher numbers of fields are four (four instances), and five (a single example). At the west in Sherwood Forest were irregular fields with brecks (see p. 93), where Sutton in Ashfield had six fields. The Senior survey of this township shows it to be complex (Fowkes and Potter 1988, 39–43). The Earl of Devonshire owned 42 per cent, of which 35 per cent was open, lying in 6 fields of sizes ranging from 7½ to 55 acres. Of the thirteen tenants, eleven held more than seven acres and the percentage of each holding open-field ranged from 19 to 78 per cent. One tenant had land in all six fields, six had different combinations of five fields and four in different sets of four fields. Hence there was not a regular arrangement; possibly spring crops were grown on the open-field land and winter crops in the some of the ample enclosures.

Yardlands (one of eighteen acres) and oxgangs (mostly twelve acres) were used in the county at dates before 1500. There is no particular geographical distribution, both being used in the north and south of the county.

Management

Sets of by-laws for Orston have been listed by Barnes (1997) and Laxton orders from 1686 were given by the Orwins (1938, 172–81). Clayworth *Rector's Book* provides details of open-field farming on the glebe in the late seventeenth century, including the value of produce and how tithe in kind was collected. Turnips were sown in 1691 (Gill and Guilford 1910). Convertible husbandry around Sherwood Forest using brecks has been described (Fowkes 1977). A half oxgang of demesne at Elkesley, south-east of Worksop, had in 1206 'brecks wherever and how many pertain to a bovate in my fee' (Timson 1973, 222–3).

Ridge and furrow

Ridge and furrow survives on clayland, where it was once widespread. A county map of what was visible on RAF vertical photography in 1946 has been prepared (National Mapping Programme, HER). Good examples have been recorded by oblique photography in 1972 at Headon (SK 748 549) and Owthorpe (SK 673 356) (CUCAP, BHW 6–12 and 61). Reconstructed maps have been published for Nottingham in *c.*1480

(Butler 1950, *TTSN* 54: opp. p. 44), and Newark in the sixteenth century (Barley 1949, *TTSN* 53: 16), Beeston in 1600 (Cossons 1958), Lowdham in 1736 (Hoskins 1974, 69), and Orston in 1793 (Barnes 1997, 27).

Sanderson's 1835 map of the Mansfield area shows many narrow, curved, strip-like hedges, most strikingly at Mansfield Woodhouse (Figure 5.1), where the hedges have a very great curvature with a general alignment of more than a mile in length.

BIBLIOGRAPHY

NAO   Nottinghamshire Archives Office
Thoroton   Thoroton Society, Record Series
TTSN   *Transactions of the Thoroton Society of Nottinghamshire*

Beckett, J. V. (1989), *History of Laxton* (Oxford: Blackwell).

Blagg, T. M. (1939), *Abstracts of the Inquisitions Post Mortem relating to Nottinghamshire 1321–1350*, Thoroton, 6.

Cossons, A. (1958), 'Early enclosure in Beeston', *TTSN* 62: 1–10.

Hoskins, F. N. (1974), 'Marlock House Farm; a post-enclosure farm', *TTSN* 78: 68–74.

Gill, H., and Guilford, E. L. (1910), *The Rector's Book, Clayworth, Nottinghamshire* (Nottingham: Henry B. Saxton).

Foulds, T. (1994), *The Thurgarton Cartulary* (Stamford: Paul Watkins).

Lowe, R. (1798), *A General View of the Agriculture of Nottingham* (London: Board of Agriculture).

Sanderson, G. (1835 [2001]), *Map of the Country Twenty Miles round Mansfield* (repr. Nottingham: Nottinghamshire and Derbyshire County Councils).

Scurfield, G. (1986), 'Early seventeenth century Worksop and its environs', *TTSN* 90: 46–56.

Timson, R. T. (1973), *Blythe Cartulary*, ii, Thoroton, 28.

Train, K. S. S. (1949), *Abstracts of the Inquisitions Post Mortem relating to Nottinghamshire*, Thoroton, 12.

ACKNOWLEDGEMENTS

Mike Bishop and Virginia Baddeley for information from the HER.

27. OXFORDSHIRE

Oxfordshire has a varied topography, running from the Cotswolds in the north, spanning the low-lying land of the Clay Vale, and extending to the Chiltern chalk ridge in the south-east. In the south, field systems were irregular, with mixed usage of 'field' and 'furlong', which terms were used for closes held in severalty as well as for common-field arable. The northern systems were predominantly Midland type two- and three-field, but some large Cotswold parishes had multiple townships and assarts. There are many open-field maps held by Oxford Colleges and twenty-nine tithe maps record some open field. Mowat published fifteen maps of nine Oxfordshire townships in 1888, most of them open field. Field systems have been studied by Gray (1915) and Roden (1973),

Table App.13. Oxfordshire field systems.

Place	Demesne	Virgate (acres)	Field No.	Date	Sources
Barton			2	thirteenth century	Magd. Oxon Cat, i, 2–4
Balscot			2	1200	H. E. Salter 1930, ORS 12: 10
Brize Norton		40	2	1187	H. E. Salter 1947, ORS 25: 68
Chipping Norton			2	1250	Magd. Charters, Chipping Norton 7 and 8
Chalford			2	1300	Lobel 1935, ORS 17: 86
Clifton			2	1240–50	Magd. Charters Clifton and Deddington 10 and 25
Dean			2	1204	Lobel 1935, ORS 17: 85
Dornford			2	thirteenth century	VCH, Oxfordshire, xi (1983): 274
Duns Tew 1436	dispersed	25	2	1436	H. E. Salter 1934, OHRS 97: 239
Great Tew			2	1249	Calendar of Inquisitions Miscellaneous 1219–1307, i, no. 86 (1916)
Little Tew			2	1260	H. E. Salter 1934, OHS 97: 223
Hampton Gay			2	c.1195	H. E. Salter 1936, OHS 101: 41
Lodewell			2	1257	H. E. Salter 1934, OHS 97: 203–4
Oddington			2	1230–50	Magd. Charters, Oddington 2a, 8, 11a, 13a, 14a
Sibford	dispersed	24	2	c.1190–1220	A. M. Leys 1940, ORS 22: 257–61
Tackley			2	1220	Magd. Charters, Tackley 2
Weston on the Green			2	1230	H. E. Salter 1936, OHS 101: 15
Wootton			2	1280–1300	Magd. Charters Wooton and Slape 2; Gray (1915, 510) has it irregular in 1606

Eynsham			3	seventeenth century	E. Chambers 1936, ORS 18: 94
Garsington			3	1230–40	Magd. Charters Garsington 22a
Marston			3	1279	G. N. Clark 1925, ORS 6, suppl., p. 6
Shenington	block		3	1732	VCH, *Oxfordshire*, ix (1969): 143–5; each of the three were divided into four for cropping
Stoke Talmage		c.22	3		A. M. Leys 1938, ORS 19: 149
Warborough			3	1206	H. E. Salter 1930, ORS 12: 25
Leigh North			8	1581	B. Schumer 1975, *Oxoniensia* 40: 314–15
Attington in Tetsworth				mid fifteenth century	H. E. Salter 1948, ORS 26: 174–92; field book
Cuxham	block			twelfth century–1728	Harvey 1965, 20
Deddington	dispersed	c.28		1448	Northamptonshire CRO, C(A) 3309, pp. 4–29
Stoke Talmage	block	c.28		1150–1211	H. E. Salter 1947, ORS 25: 96–9

and much detailed information about fields is also available in VCH, *Oxfordshire*, v–xv (1957–2004).

Gray (1915, 486–93) lists fifty-two places that had two-field systems, many of early date, and twenty-six with three fields. Surveys by him of Shipton under Wychwood (regular two-field) and Hanborough (regular three-field) illustrate the details (ibid. 438, 430). Data from the VCH volumes published up to 1972 for 148 townships with information before 1300 show that 118 were two-field and only 28 had three fields (Roden 1973, 348). In VCH, *Oxfordshire*, xi–xiii (1983–96) there is evidence for eighteen two-field systems and four three-field ones before the fifteenth century. During the early modern period nearly all had four fields or more. Table App.13 lists eighteen early two-field and six three-field arrangements, and records three block demesnes and three dispersed.

The late survival of two-field systems in the north of the county has been described (p. 42); they seem to have developed from the practice of 'hitching'—the partial sowing of the fallow field in a two-field arrangement. Gray (1915, 493) records eight examples of four-field systems, including West Adderbury and Wigginton. Some places developed more complex arrangements; Kingham had six 'quarters' in 1850 and Great Tew in 1861 had eight 'quarters' (Gray 1915, 125–30). There were many townships with two fields in the thirteenth century that developed into four-field arrangements (see the economic sections of VCH, *Oxfordshire*, xi (1983): Wootton Hundred (Northern Part).

In the south, field systems were irregular and field size was variable (Roden 1973, 328). Examples of southern field systems are at Caversham, Ewelme, Watlington, Benson, and Warborough (Gray 1915, Appendix VI). Yardlands varied from twenty-four to thirty-eight acres. Ewelme had four fields plus fifteen other areas. Some holdings were dispersed in all four fields irregularly; five holdings lay in the same three fields, which may be evidence of a separate township. Watlington had eight fields plus five miscellaneous areas and many holdings were dispersed in four to seven fields. The glebe of Mapledurham in 1454 suggests the same type of mixed holding, there being four crofts, thirty-three acres, as well as open-field holdings. In the seventeenth and early eighteenth centuries Chausey Heath was common for five years, being sowed two years in seven (Cooke 1925, 108; Apps. III and VII). The charters of Magdalen College, Oxford, for Chalgrove and Golder describe complex fields and many component townships—Golder had five in all (VCH, *Oxfordshire*, viii (1964): 141).

Grass leys and balks were widespread in the county, and the annual allotment of Yarnton meadows has been described. Yardlands varied in size from fifteen to twenty-eight acres (see main text, p. 114, and Table, App. 13). Ault (1972) listed early court orders for field regulation. At Kirtlington, in 1538 a yardland was stinted at eight oxen, horses, or cows; and in 1590 balks were to be made between strips (M. Griffiths 1980, *Oxoniensia*, 45: 274). The yardland stint of Dunthrop in 1187 was twenty-five sheep, two cattle, and a pig (VCH, *Oxfordshire*, xi (1983): 138).

Ridge and furrow

The Rev. H. E. Salter reconstructed the open fields of Oxford in 1939 as they may have been in the 1380s from historical sources (see the main text, p. 152). Clanfield furlong plan was reconstructed in 1963 by archaeological survey of the boundary soil banks in modern fields (Pocock 1968; and see p. 150), and in 1977 the plans of Shenington and Alkerton were similarly mapped (Hall unpublished). The Shenington plan agrees with the 1732 open-field map (redrawn in VCH, *Oxfordshire*, ix (1969): 144). Fine examples of strip lynchets marked on the Shenington 1732 map (SP 371 423) were still surviving in 1962 (photo CUCAP, AEY 86).

North of the Chiltern Hills there were ridged fields of the Midland type. Sutton (1964–5, fig. 37, p. 101) plotted ridge and furrow visible in 1960 on Ordnance Survey vertical aerial photographs; almost all examples lay north of the chalk. Oblique photographs taken of Merton (SP 59 12) in 1972 reveal extensive remains (CUCAP, BHV 63–5). A plan of ridge and furrow survival in 1991 has been published showing that much was still to be found at Chastleton (SP 24 95) (Hall 2001, 7, 36).

BIBLIOGRAPHY

OHS   Oxfordshire Historical Society
ORS   Oxfordshire Record Society

Cooke, A. H. (1925), *The Early History of Mapledurham*, ORS 7.
Mowat, J. L. G. (1888), *Sixteen Old Maps of Properties in Oxfordshire* (Oxford: Clarendon Press).
Sutton, J. E. G. (1964–5), 'Ridge and furrow in Berkshire and Oxfordshire', *Oxoniensia*, 29–30: 99–115.

ACKNOWLEDGEMENTS

Paul Smith and Susan Lisk for information from the HER, and Paul Martin for help with Alkerton and Shenington field surveys.

28. RUTLAND

Rutland has undulating topography rising to over 500 feet in the west with boulder clay at the north and limestone in the south and east. Gray (1915) recorded some Rutland data and much information is given by Ryder (2006, Appendix I). Rutland fields in the seventeenth century were nearly all three-field of the regular Midland type.

Tithe maps of Barrowden (1,997 acres, 93 per cent open), North Luffenham and South Luffenham (Northamptonshire CRO, T286, T277, T271) show the open fields in great detail. Many Barrowden furlongs have strips lying north–south, and six lie in a curved alignment 1,400 yards long. There were 2,790 strips, some only 12 feet wide, with green balks that were a 'nursery of weeds' (Slater 1907, 64–5).

Gray (1915, 494) listed no parishes with two fields and only two with three, Essendine (1334) and Whissendine (1337), both demesnes. The demesne of Tinwell (1399) lay unequally distributed in four named fields. The state of the fields in the seventeenth century is described by glebe terriers (NRO, Boxes X623–7) and can be supplemented using data from tithe maps (NRO, T243–T288) and Ryder (2006). Information for the fifty listed townships has been summarized (see p. 39), most of them being three-field.

Whissendine's five fields were unequal, and probably formed a double-field system of three and two fields, the two being used for rye in alternate years independently of the other three (see Newnham, Northamptonshire, p. 44). At Ryhall, brakes and the rabbits living among them were both tithable (1709). The management of grass to supplement meadow and pasture at Ayston, Glaston, and Wardley has been discussed (p. 23). Wardley yardland (twenty-two acres) had commons for four horses, six cows, and thirty sheep in 1633. It had more pasture than many places, with common ground on the hill next to Ayston and in Beaumont Chase lying next to it. Bisbrooke yardlands had a stint of sixty sheep, six cows, and four horses in 1795 (NRO, X2830, Bdle 11, pp. 5–6). Other yardland sizes have been given.

Field orders

Rutland has very many records of field regulations. Those for Exton (1757) have been published (Ryder 2006, 69–71). Seaton had some field regulations for the sixteenth century and many later ones. In 1567 fences between neighbours were to be repaired and every cottager was to have two loads of wood before Martinmas. The millers of Seaton and Thorpe were to maintain flood gates so that meadows were not 'overflowen' at any time (NRO, TB 449).

Seaton and Thorpe by Water have a comprehensive set of orders for 1654 (NRO, TB 469):

Sheep and horses over two years old were not to be kept above the stint;
no person should glean barley before both sides [of the land] were shocked;
no one should gather peas 'white or gray' (except on their own land) without leave of the owner;
no one was to glean any corn or grain before sunrise nor after sunset;
every householder should find one person to gather stones four days yearly, and every labourer two days;
no one was to keep byherds nor tie other 'neat' (bovine) beasts [on leys, etc.] after lammas;
no one was to break hedges or cut any quick out of any other person's hedges;
the pinder could take a halfpenny for every beast or hog impounded and one penny for animals from outside of Thorpe;
no one was to have colts or fillies following their teams unless they were sucking foals; dunghills in the street were to be carried away;
horses, colts, kine, or neat beasts were to be out of the fallow field on 25 March;
every master could winter ten sheep for his servant;
'the jury shall meet at the knowlinge of the bell by the foreman by 8 in the forenoon' on Monday after Michaelmas and Monday after May Day to view encroachments.

These are routine orders for which default fines were 1s.–3s 4d. Much higher fines were set for infringements that compromised the effective operation of the fields:

No sheep were to go in the field unfolded from May Day until Michaelmas (except if they lay in the commons), 10 shillings;
no one should keep any geese after 1 November unless on their own several ground, 10 shillings;

no several [great] field was to be broken with sheep without consent of the major part
of jury, 10 shillings;

no person to let any winter common, £5.

Ridge and furrow

A plan of ridge and furrow surviving in 1991 found that most lay in the west of the
county, with Belton and Braunston having thirty-six and 34 per cent survival (Hall 2001,
38, 58–9). Outstanding RAF vertical photographs of Braunston (17 January 1947, CPE
UK 1932 frame 3032) have been published by Hartley (1983, 34–5) along with plans
covering the whole county showing ridge and furrow recorded on early RAF sources
(ibid. 55–67). Glaston had extensive survival in 1947 as did Whissendine (Ryder 2006,
6). Ryder has made studies of ridge and furrow profiles (2006, 74–8).

BIBLIOGRAPHY

NRO   Northamptonshire Record Office

Hartley, R. F. (1983), *The Medieval Earthworks of Rutland*, Leicester: Leicester Museums and Art Gallery Archaeological Report, no 7.

ACKNOWLEDGEMENTS

Peter Liddle, Fred Hartley, Richard Knox, and Richard Pollard for HER data.

29. SHROPSHIRE

Shropshire topography is dominated by the River Severn and its tributaries, with the
high ground of Clun Forest, Long Mind, and Wenlock Edge to the south-west, and
Wales on the west. Some parishes were large with many townships: Myddle, with 6,903
acres, had the chapelry of Hadnell at the far south-east with 2,200 acres and six other
townships (Hey 1974, 14, 19). Settlement became dispersed in the seventeenth century
by squatters occupying commons (J. P. Dodd 1957, *TSAHS* 55: 1–31; VCH, *Shropshire*,
x (1998): 96). There was much 'waste' in areas with high ground that included mosses.
Part of Oswestry lordship (10,679 acres) was surveyed in 1602, in which waste
amounted to 5,595 acres, along with parks 431, forests 1,473, and encroachments 573;
farms occupied 2,446 acres (Slack 1951).

The Cartularies of Haughmond and Shrewsbury Abbeys along with other sources
record several three-field systems. Some of them are irregular or have an appreciable
amount of enclosure, and the extent of regular field systems is not clear. Some irregular
fields are known and the extensive pastures in the county were probably used for convertible husbandry or outfield.

The county field systems were studied by Gray and many references can be found in
the economic sections of recent VCH *Shropshire* volumes. Paul Stamper and others
have reviewed the county's agriculture (VCH, *Shropshire*, iv (1989): 1–142). Church
Aston and Longford were mapped in 1681 (Sylvester 1969, 243). The fields of Market

Table App.14. Shropshire field systems.

Place	Field no.	Date	Sources
Balderton	3	*c.*1225	Rees 1985, 43
Betton	3	1265	Rees 1975, 85
Bridgnorth	3	1739	G. E. Fussell 1953, *TSAHS* 54: 8–9
Broseley	3	1620	VCH, *Shropshire*, x (1998): 259; open land a low percentage of total
Church Aston	3	1681	One field detached; about 60 per cent enclosed; VCH, *Shropshire*, iv (1989): 124, with plan
Downton	3	1298	Rees 1985, 75; irregular
Grinshill	3	1323	Rees 1985, 86; irregular; two terriers, one does not name the third field; also three fields in 1255, ibid. 84
Haughton	3	1327	Rees 1985, 108
Highley	3	1332	Nair 1988, 10; in the sixteenth century, irregular and 38 per cent enclosed (ibid. 13)
Newton	3	1326	Rees 1985, 164–5
Poynton	3	1412	Rees 1975, 265–7
Upington	3	1320	G. Morris 1886, *TSAHS* 9: 332–3; slightly irregular

Drayton were still extensive in 1780 (VCH, *Shropshire*, iv (1989): 170, plan from Shropshire CRO, 1096/1). Four tithe maps (1838–5) have some open-field marked: Munslow, Oldbury, and Langley; Ruyton; Stokesay Newton; and Wettleton (Kain and Oliver 1995, 415–27).

Field numbers

Gray (1915, 37–8, 66–9) found no examples of two-field systems; he listed some three-field systems, but when demesnes of uncertain type and places with significant percentages of enclosure are removed, there is not much satisfactory evidence of Midland-type three-field systems (except Faintree and perhaps Shawbury). Sylvester (1969, 220–1) produced a county map showing the distribution of field-numbers then known; they probably have little relationship to cultivation regimes. VCH, *Shropshire*, viii (1968) shows that most townships in western Shropshire had three fields at some stage and VCH, *Shropshire*, x (1998) also found evidence in the centre-south. These systems need further analysis to discover whether tenants had land uniformly distributed in all the fields, and how the fields relate to the land-use of the whole township.

The Table lists twelve places with records of three fields. Some of them are irregular and have low amounts of arable as a percentage of the township area, where known. Yardlands and demesnes have been described (pp. 103, 114).

Gray (1915, 38) found evidence for irregular fields near Morfe Forest in the early seventeenth century. Other examples are Betton in Hales where a 1411 survey describes what is probably the whole township. There were seven 'field' names, some of which were probably furlongs. Some of the smaller pieces were called 'furlongs' and others 'bruch' (Rees 1975, 263–4). Much Wenlock had sixteen 'fields' in the early eighteenth century, some of which were furlongs lying in larger fields (Sylvester 1969, 237–8).

These may have been grouped together or were cultivated on a convertible husbandry system. Convertible husbandry operated at Linley in 1309, where two pieces of land were brought back into cultivation, and assarts recently allowed to lie fallow (as leys) in Linley wood were to pay tithes when cultivated again (Rees 1985, 148).

Farming

Marl was dug at Preston Boats in 1259 (Rees 1985, 179). Work-service at Childs Ercall in 1280 for an oxgang was two ploughings, mow corn, mow the lord's meadow, and wash sheep. Money payments instead of the service were recorded (W. G. D. Fletcher 1908, *TSAHS*, 3rd ser., 8: 365–6). Some townships had a bare fallow, as at Hardwick, 1359, where assarts were to be common after harvest for two years and all year when fallow. Land at *Simondesley* (near Eaton) was similar (Rees 1985, 59, 232). Field orders for Stretton note that the common fields of Whittingslow township were overburdened in 1567 (C. H. Drinkwater 1904, *TSAHS*, 3rd ser., 4: 120).

Ridge and furrow

Broad ridge occurs in most parts of the county. The SMR records some as specific 'sites' (134 in 1998) and information from RAF verticals and other sources is recorded in manuscript at 1:25,000. Large areas of broad ridge can be seen at Burley in Corvedale, along the River Vyrnwy and elsewhere (Watson and Musson 1993, 76–7). Ercall Magna (SJ 59 18) (CUCAP, K17 U 94 (1970)) and Berrington (ST 51 08) (CUCAP, RC8-EL 122) have several furlongs revealed as cropmarks.

Narrow straight ridges are known on Long Mind and Stapeley Hill (Watson and Musson 1993, 12, 25; 22). The ridges at the latter site pre-date late eighteenth-century mineral extraction pits. They probably represent early-modern cultivation of the waste. At Hodnet (SJ 787 302) broad ridge has been overploughed with narrow ridges (CUCAP, ATX 50–2 (1968)).

BIBLIOGRAPHY

TSAHS   *Transactions of the Shropshire Archaeological and Historical Society*

Hey, D. G. (1974), *An English Rural Community: Myddle under the Tudors and Stuarts* (Leicester: Leicester University Press).

Nair, G. (1988), *Highley: The Development of a Community, 1550–1880* (Oxford: Basil Blackwell).

Rees, U. (1975), *The Cartulary of Shrewsbury Abbey* (Aberystwyth: National Library of Wales).

Slack, W. J. (1951), *The Lordship of Oswestry, 1393–1607* (Shrewsbury: Shropshire Archaeological Society).

Watson, M., and Musson, C. (1993 [1999]), *Shropshire from the Air* (Shrewsbury: Shropshire Books) [quoted page numbers are from the 1999 reprint].

ACKNOWLEDGEMENTS

Penny Ward, HER, and Mike Watson.

30. SOMERSET

Somerset has the high ground of Exmoor and the Brendon Hills in the west, with the Quantocks and Mendips lying either side of the Levels wetlands that drain to Bridgwater. Some large parishes often comprise separate townships. Martock (7,226 acres) had nine townships, all with separate field systems, and there were several scattered commons (Bush 1978, 78–82). Crowcombe on the west of the Quantocks (3,271 acres) has five main settlements. Many of the smaller farmsteads have medieval origins, being recorded from the thirteenth century onwards (Bush 1985, 54–8). The early nature of the dispersed settlement is illustrated by Spaxton near the Quantocks, which had nine estates listed in 1086, most of which descend as hamlets scattered among additional farms (Dunning and Siraut 1992, VCH, *Somerset*, vi: 111–19). Similarly, North Petherton had thirteen estates in 1086 (ibid. 300). Somerset had many examples of two- and three-field arrangements in the south-eastern part of the county. On the west and especially near to Exmoor there were irregular fields with outfield and convertible husbandry, making use of the extensive pastures.

Dispersion occurs on and near Exmoor. Many medieval settlements, now single farms or quite deserted, have been identified using the Lay Subsidy lists of 1327. Thus at Langford Budville (1853 acres) eight small settlements were located (S. J. Swainson 1935, *PSANHS* 80: 11–16). Aston's study of Exmoor, and Brompton Regis, using additional sources, showed that some farms were in existence by the twelfth century (M. Aston 1983, *PSANHS* 127: 71–104; plans). Some of the small settlements on the edge of the Mendips may have developed from shielings that had an early (Roman) origin (P. Pattison 1991, *PSANHS* 135: 103–4).

Gray (1915) discussed the fields of the county and a general account of the medieval landscape is given by Aston (1988), as well as information about the open fields. Much open-field detail is provided by VCH, *Somerset*, iii–ix (1974–2006). Open field maps are known for Alford, Shepton Beauchamp, South Cadbury, and Stoke sub Hamdon (Aston 1988, 91). A redrawn plan of Englishcombe in 1611 shows two great fields, as there had been in the fourteenth century (Aston and Iles 1996, 111–12). Tithe maps of 1833–44 record open fields in sixteen townships, all lying in the east of the county (Kain and Oliver 1995, 434–49). There were still some open strips ploughed flat at Westonzoyland in 1943 (B. H. StJ. O'Neil, *PSANHS* 88: 80–2).

Field numbers

Gray (1915, 439, 441, 494–6) found twenty medieval examples of two-field systems and three three-field plus four demesnes that left one-third waste. At South Stoke, in addition to two fields there were small areas of enclosed arable meadow and pasture. The Table records nine more two-field and two three-field, not including data from the VCH.

Whitfield (1981) studied the south-east of the county and concluded that most townships had open-field systems. He quoted from glebe terriers and published medieval documents giving very clear evidence of two- and three-field arrangements. There was evidence for the intermingling of small parcels and for commoning on the arable and meadows after the crops had been removed. Aston (1988, 90) plotted the numbers of open fields then known. Places with multiple or no known open fields occur on or close to high ground. The VCH *Somerset* volumes provided many examples of three-fields in the

Table App.15. Somerset field systems.

Place	Demesne type	Field no.	Date	Notes
Frome		2	thirteenth century	Ross 1964, 177; one yardland is twenty acres
Glastonbury		2	1260	Elton 1891, 192
Hinton		3	1554	Tawney and Power 1924, i, 60–3
Mere		2	twelfth century	Jackson 1882, 29
Milton Cleveson		2	1222	Hobhouse et al. 1894, SRS 8: 28
Milverton		2	*c*.1187	Weaver 1909 SRS 25: 11
Mudford		3	1554	Tawney and Power 1924, 60–3
Pennard, East		2	thirteenth century	Jackson 1882, 40
Shapwick	dispersed?	2	fourteenth century	Fox 1981, 80, 105 n. 11
Stoke Trister	dispersed?	2	1562	G. R. Straton 1909, 424; 22 acres in West Field, and 32½ acres in East Field
Street		2	1240	Elton 1891, 12

south-east of the county. The volumes of VCH *Somerset* now available (iii–ix) cover the central swathe of the county from Bruton to the edge of Exmoor in the west; they deal with all types of topography from the Levels to the high ground. Extensive two- and three-field systems occur mainly in the south-east, as Gray discovered.

Podimore fields were changed from two to three in 1333 (H. S. A. Fox 1986, *Economic History Review*, 2nd ser., 39: 526–48). There are several other examples of townships that had two fields at an early date and more later. Crewkerne was two-field in the thirteenth century but had three in the sixteenth (Dunning 1978, 19–20). Podimore cropping plans from 1311 to 1315 showed that 'inhoks' were occasionally sown in the fallow field (Fox 1986, 532). Aller had two fields; Charlton Adam had two in the thirteenth century that by the mid sixteenth century had been split into four (Bush 1974, 62, 82). Glastonbury had four fields recorded during the sixteenth century (Dunning 2006, VCH *Somerset*, ix: 50–2) and Stoke sub Hamdon had fifteen common fields in 1615. A whole 'place' (i.e. yardland) had ten 'rother-beasts' (oxen), a horse, and sixty sheep. Hamdon Hill was a common for sheep only, open all year (J. Batten 1894, *PSANHS* 40: 260–1).

Gray (1915, 524–8) gives details of early seventeenth-century surveys as examples of irregular fields. Most except Bruton were in an advanced state of enclosure and the small irregular open holdings probably have little relevance to earlier field structure.

Outfield and convertible husbandry

There is little evidence for open-field arable on the higher ground of the west, for example at the eastern end of the Brendon Hills, where it is likely that infield–outfield occurred (Aston 1988, 91). Other commons had limited and intermittent cultivation. Parts of Landacre on Exmoor were enclosed and used to grow crops in the thirteenth century. Copyholders of Monkham claimed pasture on Exford commons in 1638 when they were not in tillage; at Withypool part of the common was tilled in 1678 and 'there were signs that several other parts of the common…have been anciently tilled' (O. Hallam 1978, *PSANHS* 122: 44). Parts of the common of Crowcombe by the Quantocks that were ploughed in 1405 and also during the 1430s, were still ploughed in the eighteenth century (Bush 1985, 58). At Bicknoller, rye was grown on Bicknoller Hill in the sixteenth century (Dunning 1985, 15). Outfield cultivation was recorded on the Brendon Hills before 1830 (J. Savage, quoted by Aston 1988, 85). The common of Dunkery Hill in Wootton Courtenay was licensed by the lord to sow corn in 1550 and 1551 (Aston 1988, 87).

In the areas of dispersed settlement, some of the outlying hamlets and farms had their own field systems. Mention of the Great Field of Leigh (now Leigh Farm in Crowcombe) in 1352 may refer to a single-field system (Bush 1985, 58). Frequently on the areas of dispersed settlement no historical evidence of open field survives, presumably because there were single farms in assarts or very small townships that had been early enclosed. Thus in the Brandon Hill region no evidence of open fields was found at Huish Champflower, where there are many farms of medieval origin (Dunning 1985, 83, 136), and similarly at Raddington. The wooded area of Chaffcombe did not refer to open fields and a rental of 1444 showed that land lay in closes (Bush 1978, 125).

MacDermot (1911) found many references for outfield cultivation on the edge of Exmoor. In 1279 the forest was encroached by 'men who work waste ground to make enclosures to sow corn'. In 1540–55 several people took in and enclosed from Withypool Common parcels of fifteen acres and ten acres and tilled them with oxen. In 1567 part of Monkham common had been ploughed up, sown with corn, fenced by a wall of earth and turfs, and (surrounded) with thorns and stakes to preserve it from (grazing) cattle and sheep. In 1675 corn was extensively grown in temporary enclosures on the commons. George Culley of Withypool referred to two pounds; he noted that part of the common had recently been tilled (1678) and that there were signs that other parts then laid down had been anciently tilled (ibid. 55, 446–50, 293–4, 343, 362).

Demesnes and yardlands have been discussed (pp. 99, 114). Additional dispersed examples are at Mells (420 acres) lying in the East and West Fields *c.*1260, most of them called 'furlongs'. Marksbury, Pilton, and Winford were similar (Elton 1891, 200–23). Porlock demesne in 1306 lay in large seven-furlong pieces and some separate closes (C. E. H. Chadwyck-Healey 1901, Appendix A).

Cropping and farming management

Work-service on the Glastonbury manors is described in 1189 (Jackson 1882, 159–66), with full details for Lympsham. Glastonbury service included making charcoal for the cook. Further details were listed in *c.*1240 for all the manors (Elton 1891, SRS 5: 12, 237, 242). Most townships are similar to Butleigh, performing the usual agricultural round of haymaking, ploughing, reaping, carriage, and providing a horse for harrowing. Occupants of places near the abbey had to work in the vineyard and maintain a fixed length of Pilton park wall. Some cottars had the work-service of taking water to the reapers for sharpening scythes. At South Brent some of the tenants had to maintain the seawall (called 'wokarii' in *c*.1260 and 'moormen' in 1513; ibid. 39).

Somerset court rolls reveal evidence of local agriculture. Puriton tithes included cider, leeks, and garlic in the thirteenth century (Weaver 1909, SRS 25: 11). Clayhanger had a marl pit in the thirteenth century (Bates 1899, SRS 14: 55). Curry Rivel court rolls of 1348–9 refer to copyhold orders for drainage, offences of trespass by ploughs and sheep in the common, and regulation of strays. There are no items about the regulation of crops (J. F. Chanter 1911, *PSANHS* 56: 99–105). Leys and grass ends were established at Long Load in Martock by the seventeenth century (Bush 1978, 94).

Ridge and furrow

Broad ridge occurs at Alford (ST 61 32) (CUCAP photograph, ANN 32 (1966)) and elsewhere in the Brue Valley. The practice was mostly to be found in the south-east of the county (Aston 1988, 85–6, with a plan of Alford on 88). Good examples are at West Camel, Lovington, and Nether Adber (a deserted site in South Cadbury, CUCAP, ANN 33 (1966); Croft and Aston 1993, 58–9; Adber mapped in Whitfield 1981, 17). Other are at Hinton and West Mudford (Aston 1977, 44) and Crooks Marsh, Avonmouth (Aston and Isles 1996, 112). A computerized plot of records made by the HER (154 in 2000) shows the south-eastern distribution, with some recorded in the north and on the edge of Exmoor, but none around the levels.

A large area of extremely low profile and narrow strips situated on Winsford Hill, Exmoor (SY 87 34), are the remains of outfield cultivation of the commons, operating intermittently from the thirteenth to the seventeenth centuries, as described by McDermot. They were photographed by Aston (1978, *PSANHS* 122: 45) and mapped by English Heritage in 1999 (Swindon, Exmoor Archive).

Strip lynchets are prominent at Yarlington (ST 66 29), South Cadbury, and Westbury sub Mendip (Aston 1988, 87), and are well preserved on Glastonbury Tor (Croft and Aston 1993, 41). Other examples are at Creech Hill (CUCAP, AJJ 49 (1964)) and Sodbury (Aston and Isles 1996, 121). A plot of lynchet records on the SMR (110 in 2000) shows a similar distribution to that of ridge and furrow, but occurring in hilly countryside.

Hardwick (1978) studied the strip lynchets of South Cadbury, noting that much previous research had 'been futile in that it has failed to recognize that lynchets are the results of adapting the strip-farming to slopes.' A map of *c*.1830 showed the distribution of holdings that included lynchets. Lynchets also fitted into the township boundary as part of the field system and suggesting they were not older. For a plan and photograph of the lynchets see H. S. Gray 1914 (*PSANHS* 59: 9, pl. IIIA). Strip lynchets at Dinder have been excavated (G. Whittington 1976, *PSANHS* 120: 39–44). Early 'Celtic fields' are preserved on the eastern Mendip slopes at Bleadon, some partly overploughed and containing ridge and furrow (Aston and Isles 1996, 100–1). Lynchets at Ramspit, Weston sub Mendip, have been mapped (P. Pattison 1991, *PSANHS* 135:102 and fig. 2).

Field patterns

Hedge patterns preserving the curves of furlongs are common in the county. A photograph near Cheddar shows narrow curved fields as well as lynchets (Aston 1988, 83).

On the high ground at Stawley there are irregular fields and dispersed farms (Croft and Aston 1993, 56).

The Historic Landscape Characterisation, a detailed analysis of hedge patterns and other features, indicates that there had once been widespread open field across the county amounting to about 24 per cent of the area. It did not occur on the hilly ground of Exmoor, the Brendon Hills, the Quantocks, and the Blackdown Hills, all over 800 feet. The area most confidently interpreted as former open-field occurred predominantly on the lower ground in the east and south-east of the county (Aldred 2001, 24–7).

BIBLIOGRAPHY

PSANHS   *Proceedings of the Somerset Archaeological and Natural History Society*
SRS   Somerset Record Society

Aldred, O. (2001), *Somerset and Exmoor National Park Historic Landscape Characterisation Project 1999–2000* (Taunton: English Heritage and Somerset County Council).
Aston, M. A. (1977), 'Deserted settlements in Mudford Parish, Yeovil', *PSANHS* 121: 41–53.
Aston, M., and Gerard, C. (1999), 'The Shapwick Project, Somerset', *Antiquaries Journal*, 79: 1–58.
Aston, M., and Iles, R. (1996), *The Archaeology of Avon* (Bristol: County of Avon).
Bush, R. J. E. (1974, 1978, 1985), in VCH, *Somerset*, iii–v.
Croft, R., and Aston, M. (1993), *Somerset from the Air* (Taunton: Somerset County Council).
Chadwyck Healey, C. E. H. (1901), *The History of the Part of West Somerset comprising the Parishes of Luccombe, Selworthy, Stoke Pero, Porlock, Culbone and Oare* (London: Henry Sotheran and Company).
Currie, C. R. J. (1999), VCH, *Somerset*, vii.
Dunning, R. W. (1974, 1978, 1985, 1992, 2004), in VCH, *Somerset*, iii–vi, viii.
Elton, C. J. (1891), *Glastonbury Abbey Rental and Customary*, SRS 5.
Hardwick, J. (1978), 'Strip lynchets: the case study of South Cadbury', *PSANHS* 122: 29–36.
Straton, C. R. (1909), *Survey of the Lands of William, 1st Earl of Pembroke* (Oxford: Printed for the Roxburghe Club).
Tawney, R. H., and Power, E. (1924), *Tudor Economic Documents*, 3 vols. (London: Longmans).

ACKNOWLEDGEMENTS

Robert Croft, Oscar Aldred, Richard Brunning, Michael Aston, Ralph Fyfe, Stephen Rippon, and Robert Wilson-North for information, discussion, and site visits.

31. STAFFORDSHIRE

Staffordshire has low ground in the centre along the Trent Valley, with the heathy ground of Cannock Chase in the south and part of the Peak District in the north. Some parishes are very large: Eccleshall and Stoke on Trent both contained twenty-five town-ships in 1604 (Peel 1911). The county had a few examples of three-field systems. Most of it had irregular fields lying among pastures and wastes. The use of the terms 'field' and 'furlong' was variable.

Plot (1686, 339–55) provides much information about Staffordshire agriculture. The fields of the county were discussed by Gray (1915), Roberts (1973), and Birrell (1979). The Staffordshire VCH volumes contain many references to open fields in their eco-nomic sections. An Elford map of 1719 shows the township to be about 60 per cent open with four named fields and a holding dispersed in all four (Thomas 1931, 89 and map). Some open-field is marked on six tithe maps dated 1839–50 (Kain and Oliver 1995, 471–9).

Much of the county has irregular fields lying among extensive commons and assarts. These fields were most likely run as a one-field system with convertible husbandry, ei-ther within them or by occasional use of the waste. They were early enclosed. Elsewhere there were three-field systems, as well as demesnes run on a three-year tilth but not necessarily having blocks of furlongs arranged in great fields.

Gray (1915, 497–8) listed seventeen places with evidence of three fields, but several have only small areas of open land and eight are extents with a third of the land 'waste and in common', which may not be three-field systems. Only four of them are satisfac-tory examples of three fields. Several other townships have three-tilth fourteenth-century demesnes (Anon., SHC 1913: 61–135; Madaley, Bentley, Essington, Ashley, Walsall, Drayton Bassett, Wiggington, and Handsacre) and Thomas (1931, 71–2) lists others of various dates. Keele in 1729 had a three-field system, but tenants had large areas of enclosed land (E. M. Yates 1975, *NSJFS* 15: 29). The Table identifies only seven ex-amples of three-field (three of them in townships with much enclosure) and no two-field systems. Demesnes of both dispersed and block form are known (text). Yardlands were of variable size (pp. 99, 103, 114).

Table App.16. Staffordshire field systems.

Place	Demesne	Field No.	Date	Sources
Cannock		3	1419	Plus piece in a 'riding', Birrell 1979, 14, from Staffordshire CRO, D(W) 1734/2/1 rot 7
Clayton		3	1360	F. Parker 1890, SHC 11: 328
Himley	dispersed	3	fifteenth century	Birrell 1979, 13; equal number of selions in each field
Perton		3	1663	Thomas 1931; about 60 per cent enclosed; small fields
Seisdon		3	1780	Thomas 1931; about 80 per cent enclosed; small fields
Sugge		3	fourteenth century	M. E. Cornford and E. B. Miller, SHC 1921, 15; terrier of nine acres
Trescot		3	1663	Thomas 1931; about 50 per cent enclosed; small fields

Irregular fields are more commonly recorded. Gray noted that away from the Trent Valley in the south-west fields were irregular. He gave details of surveys of Rolleston and two other places (Gray 1915, 87, 445, 514–15). Wootton under Weaver, in 1545, had four areas called 'fields' and another arable area called 'legh'. Eight of the nine tenants who had land in legh had unequal amounts in all the other four fields. The six tenants who had no land in legh had unequal amounts in sometimes two, three, or four of the other fields. The names were therefore probably locational only. Rocester, in *c.*1540, had eight tenants with irregular holdings in five named fields. Five had lands in varying combinations of three fields, two in two fields, and one in one field. It is likely that the field names were locational without cropping significance.

Furlongs were often called 'fields' and the mixed terminology has been noted for Whitgreave (p. 79). A rental of 1416 named the (furlong) location of each holding. The term 'field' was used twelve times in Bromley and eight times in the hamlet of Hurst. Some fields were large and contained many small pieces belonging to several tenants, from one rood to several acres. Other fields were small and belonged to one tenant, for example Lucyfield was 4½ acres in extent and Sharpsbyfield 4 acres 2½ roods. Other named pieces were possibly part of the open fields or were enclosed. Uttoxeter had a similar survey made in 1414 (Birrell 1979, 4).

Complex land-use operated at Huntington in 1598 (H. L. E. Garbett, SHC 1928, 142–3). A farm consisted of six crofts, three doles of meadow, four closes of 'land or pasture' (most probably convertible), and twenty-eight selions dispersed in nine pieces lying in four places, with three pieces (ten acres) enclosed from a field called *Pytchefeld*. A Wetton survey by Senior is described in the text.

Work-service

Customary tenants of the Prior of Lappeley (1338) performed one day's ploughing and harrowing at the winter and Lenten sowings, and cut hay and carried for one day (Anon., SHC 1913, 69). Eccleshall work-service including ploughing and reaping, and the tenants also paid towards the expenses of eight men working in the park of Blore (Anon., SHC 1913, 77). A Burton Abbey survey of 1307 listed work-service for the usual round of agricultural production, with the abbey providing food and cider. The villeins also had to find a man to hunt for three days; this was commuted in 1307 for a payment, but care was required for hound pups until they were a year old (Birrell 1979, 27).

Field orders

A series of by-laws were agreed at Alrewas in 1342 (Birrell 1979). Court orders of the fourteenth and fifteenth centuries show that there were fences around the arable fields of Orgreave in 1351, St Michael's Lichfield in 1421, and Haywood in 1413; all were to be made sound on Lady Day, and anyone leaving gaps was fined (Birrell 1979, 16). Trysull's three fields each had a surrounding hedge in 1669, maintained by a rate of twelvepence on each yardland. Fines collected for not repairing parts of the hedge on a tenant's own land went towards the cost of making gates (Thomas 1931, 81). The hedges were presumably to control animals on the fallow field and commons.

Cropping and farming management

Oxen were used for ploughing at Burton in *c*.1115, there being sixteen oxen for two ploughs. Similar details are given for other Burton manors of Branston, Stretton, and Wetmore with some details of work-service (C. G. O. Bridgeman, SHC 1916, 209–19). Many areas had large extents of pasture, some from assarts. In 1155 the bishop of Coventry and Lichfield had a grant of 1,500 acres assarted from Cannock. During the late thirteenth century, 485 acres were assarted by various villages on the south of Cannock forest (Birrell 1979, 6–7). The prior of Tutbury in *c*.1235 granted a pasture at Calwich near the priory's wood that lay next to assarts of other men in return for some work-service of ploughing and reaping. Common of pasture was retained in the priory wood, assarts, and furlongs (Saltman 1962, 200–1).

There were large numbers of assarts during the thirteenth century made for conversion to arable, but not all were successful. In the early fourteenth century grants of 756 and 450 acres made at Cannock for cultivation had not been taken up because deer and horses still used them (Birrell 1999, 182). The extent of pasture and woodland in Cannock Chase has been reconstructed for the year 1554 (C. Harrison 1999, SHC, 4th ser., 19: 104).

Other activities providing manorial income were mining: iron and sea-coal were obtained at Sedgley in 1291 (Anon. 1911, SHC, NS, 14: 202) and Longden in the early fourteenth century (Birrell 1979, 12). Ironstone mines were recorded at Rugeley in 1311 (Birrell 1979, 12) and Newcastle under Lyme in 1423 (J. C. Wedgwood, SHC 1912, 219). There were marl pits at Bushbury in 1435 (G. P. Mander, SHC 1928, 83).

Convertible husbandry in the county has been described (pp. 93–4).

Ridge and furrow

Ridge and furrow occurred over most of the county except for the southern former forested area and the northern moorlands (HER). Much survived (2004) in the Dove Valley. A fine vertical photograph of Stramshall (1948) shows ridge and furrow lying in small fields with curved hedges (VCH, *Staffordshire*, vi (1979): 129). Stanton had similar survival in 1966 (SK 127 464) (CUCAP, AOA48). Most of the hedges mask walls through which they are growing. Other examples viewed in a site visit of 2001 were at Okeover (SK 15 48) and Ilam (SK 135 517). A survey of a section of broad ridge was made at Burton on Trent (SK 236 256) in 2000 (J. Thomas, University of Leicester Archaeological Services Report 2000/4 (copy at Staffordshire County Council HER)). A charter of Wombourne refers to several 'ridges' in 1316 (G. P. Mander, SHC 1928, 25).

Strip lynchets are recorded in the Dove Valley and Peak District (D. J. Robinson et al. 1983, *NSJFS* 3: 93–103), as at Hilderstone (SJ 94 34) (HER). CUCAP have photographs of strip lynchets at Wetton (SK 102 545) (AOA 38 (1966)) and Mayfield (SK 153 464) (AHE 54 (1963)).The enclosed fields at Groundslow Farm near Tittensor had curved furlong-type hedges before 1830 (C. R. J. Currie, VCH, *Staffordshire*, vi (1979): 100 (plan)). Curved hedges survive west of Denstone, near Alton (SK 08 41).

BIBLIOGRAPHY

NSJFS *North Staffordshire Journal of Field Studies*
SHC   Staffordshire Historical Collections, published by the William Salt Archaeological Society and Staffordshire Record Society

Birrell, J. (1999), *The Forests of Cannock and Kinver 1235–1372*, SHC, 4th ser., 18.

Peel, A. (1911), 'A Puritan survey of the Church in Staffordshire', *English Historical Review*, 26: 338–52.

Saltman, A. (1962), *The Cartulary of Tutbury Priory*, SHC, 4th ser., 4 (Kendal: Staffordshire Record Society).

Thomas, H. R. (1931), 'The enclosure of open field and commons in Staffordshire', in SHC 1931, 59–99.

Yates, E. M. (1974), 'Enclosure and the rise of grassland farming in Staffs.', *NSJFS* 14: 46–60.

ACKNOWLEDGEMENTS

Christopher Welch, County Archaeologist, for HER data, and John Darlington, Stafford Borough Council.

32. SUFFOLK

Suffolk has a central area with clay soils and sand at the west and east—the Brecklands and Sandlings. The county includes fenland at Lakenheath and Mildenhall. There are is much dispersed medieval settlement and many greens. Few of the many churches mentioned in 1086 are associated with large greens, which seem to be later. An archaeological survey of the South Elmham parishes revealed that the large Gresham Green at St Cross has much medieval pottery around it, and more settlement lies along the Beck Valley. St James and the other villages were similarly dispersed with concentrations around greens (M. J. Hardy and E. A. Martin 1986, 147–50, 232–5, 315–17, plans). In 1305 'new rents' were paid for forty cottages without land at Hadleigh that may represent this type of green settlement (Pigot 1863, 246–8). Borough English was practised on many Suffolk manors (G. R. Corner 1859, *PSIA* 2: 236–41), and at some places partible inheritance operated (see p. 120). Suffolk field systems fall into three main types: irregular common fields with partially regulated cropping and dispersed holdings, on which sheepfolding occurred; irregular systems with interspersed crofts of arable and large pieces of demesne that had partially regulated cropping; and non-common subdivided fields, where holdings consisted of both open and enclosed pieces and had a very limited dispersion. No common rights existed over the closes and rights were limited elsewhere.

There are several surveys and field books from the sixteenth century onwards. Ickworth (Hervey 1893) also has a 1665 survey, and there are surveys for Norton and Thrandeston, and an atlas of the Stanhope estates, dating from 1600–2 (D. P. Dymond 1973–6, 195). Tithe maps of 1838 and 1840 show some open-field at Baytham and Moulton.

Suffolk fields have been described by Postgate (1973), Bailey (2007), and Martin and Satchell (2008). Gray (1915, 331, 334) found that the cropping of forty acres at Bedingfield had only two acres of fallow in *c.*1396 (implying intensive agriculture) and at Bawdsey in 1431 two holdings lay irregularly in five named areas called 'fields' as well as in enclosures. More details of Suffolk fields have been given.

Convertible husbandry

Postgate (1962) studied the field systems of the Breckland, the sandy region of south-west Norfolk and north-west Suffolk, as described (pp. 88–9). In many places the fold-course system equated to a convertible-husbandry system.

It is difficult to distinguish the system of convertible husbandry from infield and out-field cultivation. The terms were used at Croxton in 1327, and in 1560 Stanford had six outfield shifts cropped in rotation with not more than two successive crops and four infield shifts, of which one was left fallow each year. The term 'infield' at Icklingham referred to how the land was cultivated, not its location next to the settlement (Postgate 1962, 90–6). Sheepfolding has been described (p. 72); it is very much involved with infield and outfield, and caused increased field regulation in the early modern period. Demesnes in both dispersed and block forms are known.

Tenementa *(yardlands) and partible inheritance*

The limited dispersion of holdings at Theberton and Dennington has been described (p. 67). The medieval farm was often called a *tenementum* and there is evidence that some were originally undispersed pieces, as shown by the example of Hadleigh. In 1305 the freeholds were all named pieces, mainly having personal names. Most were held by several people, some of whom were related to each other, suggesting that they had been split by partible inheritance. One had been customary land. The tenants' lands were even more divided (Pigot 1863, 230–49). There had been a stated 22½ holdings called 'lands', but sometime before 1305 they had been split up into thirty parts, each bearing a personal name. These thirty consisted of several whole lands with fractions of half and a quarter of a land, in all adding up to twenty. In 1305 these thirty fractions were further divided between very many people, occasionally two, but up to twenty tenants in a whole land. As with the freeholds, some of the holders had the same surname as the land-name. Other examples are Badwell Ash in 1299, where thirty-five villein holdings were held jointly by between two and seven people (Powell 1910, 76–7), and Walsham-le-Willows in 1365 and 1578 (Dymond 1976, 201).

The holding at Darsham identified as belonging to a freeman of 1086 has been described (p. 67). Douglas (1927, 27, 37–43) discussed the sizes and terminology of East Anglian customary holdings.

Work-service

Hardwick work-services were described in the early thirteenth century. They included mowing each week (in season) one acre of wheat, barley, rye, or oats, or half an acre of peas, for which there was an allowance of two sheaves. What was mowed was to be tied up and set in shocks, but not carried. After threshing the villeins could have as much straw as could be collected with one rake (called a 'helm'). Two days of work were required for mowing the meadows and hoeing, and five fold-hurdles were to be made out of timbers from Hardwick wood. Dung had to be loaded on carts and spread over half an acre (Tymms 1853, *PSIA* 1: 184–5).

The customary tenants of Hadleigh, in 1305, for each holding called a 'land' owed ploughing service with their team for six acres, and the same ground was to be harrowed and sown with seed brought from the grange. The holders were to perform twelve other unspecified works during the year. They also had to turn and carry hay, reap four acres at harvest time, carry the lord's corn, send two men to collect straw, and carry manure. They were to send two men to help with carriage of wheat and oats (Pigot 1863, 238).

Farming

The medieval economy and land-use are fully described by Bailey (2007, 67–115). At Hadleigh in 1305, the land quality was suitable for sowing on each acre 2½ bushels of corn, 2½ bushels of rye, 2 bushels of peas and beans, 4 bushels of oats, and 4 bushels of barley. Each plough was yoked with four oxen and four draught animals, and normally ploughed one acre daily (Pigot 1863, 230).

Ridge and furrow

Little ridge and furrow is known in Suffolk, since the sketch plough was used. Of the fourteen records in the HER (2000), four are likely to be post-medieval cultivation ridges, two of them in woodland. There is reference to ridges at Horningsheath (1638; V. B. Redstone 1903, *PSIA* 11: 288). Flat ploughed strips were often separated by a grass balk, as at Westhorpe (Martin and Satchell 2008, 29). Furlong boundaries surviving as broad low banks have been observed at Stanton (TL 968 731) (D. P. Dymond 1973–6, 204).

Long curved hedge lines of 'coaxial' fields occur in various parts of the county, mainly on the flat claylands of High Suffolk, notably at the South Elmhams (for a complete distribution see Martin and Satchell 2008, 196, Type 1.3). A reconstructed furlong plan, mapped by fieldwork, was prepared for South Elmham St Michael in 2010 (Figure 5.3). The boundaries underlie the long hedges. The hedges are therefore the result of later enclosure, presumably following furrows of an early large-scale layout of medieval fields. The hedges acknowledge the existence of the furlong boundaries by slight kinks where they cross over them.

BIBLIOGRAPHY

PSIA *Proceedings of the Suffolk Institute of Archaeology*

Dodd, K. M. (1974), *The Field Book of Walsham-le-Willows, 1577*, Suffolk Record Society, 17.
Hervey, J. (1893), *Ickworth Survey Boocke anno 1665* (Ipswich: Alfred Martin Sparks).

ACKNOWLEDGEMENT

I am grateful to Edward Martin for HER information and discussion.

33. SURREY

Surrey has the Thames lowlands at the north, the North Downs and Greensand in the centre, and a wide belt of Wealden Clay at the south. There are extensive sandy soils with heaths and nearly 25 per cent of the modern administrative county is covered by woodland. Many parishes are long, traversing varying terrain and soils, for example Great Bookham, five miles by half a mile to one mile wide. Component townships are a common feature; Godalming parish, for instance, contained Tuesley, Hurtmore, Farncombe, Shackelford, as well as Godalming, each with its own fields (H. E. Malden 1912, VCH, *Surrey*, iv: 409). Surrey field systems were irregular with limited dispersion that sometimes involved the use of closes, and had no communal cropping regulation. There were many undispersed holdings and no evidence for open fields on the Weald.

The county has several maps with significant areas of open field. Two plans have been published showing Great Bookham, one of 1614 and another of 1798 (with a field book) (J. H. Harvey 1968, *Proceedings of the Leatherhead and District Local History Society*, 3: 79–83). Tithe maps of 1837–48 show some open field in seventeen places (Kain and Oliver 1995). Ewell and Cuddington reconstructed fields of 1408 show commons at the north and downs at the south with arable furlongs around the villages (C. F. Titford 1973, *SAC* 69: 27–35). Waddon township in Croydon, shown on a map of 1692, has the same arrangement: central open field strips and an enclosed arable block, with common at the north and south (Blair 1991, 68; map).

Irregular fields

The fields of Surrey have been discussed by Gray (1915), Bailey and Galbraith (1973), and Blair (1991). Gray (1915, 355–69, 549) examined many terriers of the thirteenth and fourteenth centuries and found only irregular field systems involving arable closes, and lands dispersed irregularly in many fields or in just a single field and sometimes in one piece—an undispersed yardland. Blair (1991, 66–7 with plan) investigated the evidence for open fields and found no references to dispersed fields on the Weald. Most open fields were on good quality soils in the centre of the county. Division into, for example, north and south fields was merely topographical. The Table gives other examples of irregular fields ranging from one to six in number. Putney is described in the main text (p. 81).

Yardlands

Yardlands have been discussed (pp. 114–15). The complexity and variability of Surrey yardlands are illustrated by the example of *Yngefelde* in Egham. A detailed survey of yardlands made in 1378 described one that was about 20 acres dispersed: 14 roods and a piece lying in Ermeshe Field, 46 roods and a piece in Southcroft, and 14½ roods, a piece, and 4 moors (*c.*½–1 acre) lying elsewhere. There were also three enclosed crofts containing three acres. Another yardland consisted of 12 acres and half a yardland was 8.75 acres plus a 4½-acre purpresture croft of enclosed heath (H. Jenkinson1958, *SRS* 12: i, 34–40).

Farming

Demesnes of both block and dispersed forms are known (pp. 67, 103 and Table App. 17). Those of Cheam, Charlwood, and Merstham were run on a two-course tilth in

Table App.17. Surrey field systems.

Place	Demesne type	Field no.	Date	Sources
Banstead	dispersed		1368 and 1680	Lambert 1912, 34–6, 61–3, 124–5, 316, 194–5
Epsom		2	1496 and 1679	Bailey and Galbraith 1973, 76; holdings irregularly distributed
Ewell	block	1	1350 and 1476	Meekings and Shearman 1968, xxvi, 50–61, two fields, one demesne
Great Bookham		1	1614	C. K. Curie 2000, *SAC* 87: 56–9
Maldon	block	3	1627	H. Lambert 1933, *SAC* 41: 35; small remnant fields
Mortlake		2	1617	Northamptonshire CRO, Spencer 7j5; locational fields, irregularly dispersed holdings
Putney		4	1617	Northamptonshire CRO, Spencer 7j5; locational fields, irregularly dispersed holdings
Reigate		several	1623	W. Hooper 1945, 39–42, 199
Roehampton		1	1497	Northamptonshire CRO, Spencer SOX 136; field book used for enclosure in 1590
Wimbledon		6	1498	Bailey and Galbraith 1973, 77–9; holdings irregularly distributed

1211 (Sewill and Lane 1951, 8–9). More common were three-season tilths, such as Paddington (Abinger), Dorking in 1349–50 (H. E. Maldon 1927, *SAC* 37: 251–4), and Oxted in 1370. The names of the 'fields' and their crops are given in account rolls for 1361–3. The yields were very low, 1.96 for wheat and 1.09 for oats (W. F. Mumford 1966, *SAC* 63: 75, 78–1). Demesne farming in the fourteenth century has been studied for Farleigh and Tilling Down in Tandridge (M. Saaler 1996, *SAC* 83: 57–71; 1991, *SAC* 81: 19–40).

Work-service of the manors belonging to Canterbury (mainly Cheam, Croyden, and Mortlake) in the thirteenth century is described by Du Boulay (1966, 164–74). Some work-service in 1325 for Banstead included carrying wood from the park (Lambert 1912, 324).

Field orders

Ewell farmers had commoning rights after haytime and corn harvest in 1255 (Meekings and Shearman 1968, xxv). Carshalton tenants had, in 1369, 200 sheep trespassing on the lord's pasture; horses, cows, three young oxen, and hogs in the meadow; hogs in the lord's vetch and oats; and they carried thorns from the lord's land. In 1361 beasts were in the lord's wood and in 1446 tenants fished in the lord's water, made purprestures, and broke the pound (Guiseppi 1916, 59). Banstead tenants in 1403–4 were fined for poaching in the lord's park and warren with bows, arrows, and dogs, and putting cattle, pigs, and horses on his pasture (Lambert 1929, *SAC* 37: 168). In 1575, Ashtead cattle were not to be put in the cornfields until the crop was carried, except on the proprietors' own land provided there was no disadvantage to neighbours. Tenants were fined for ploughing and sowing twelve acres of the common down in 1643 (Bailey and Galbraith 1973, 75–7).

Ridge and furrow

No wide ridge is known in the county. Reverse-S hedge patterns occur on 1870s Ordnance Survey 1:10,560 maps, mainly in the north (HLC, i, 19). Very few survive. An example is a small area at West Clandon (approximately TL 005 052), south of Woking. There are 'ladder' field patterns at Horley (TV 29 45) (both mapped in HLC i, figs. 22–3).

BIBLIOGRAPHY

HLC N. R. Bannister and P. M. Wills (2001), 'The Surrey Historic Landscape Characterisation' (Surrey County Council Report, at HER)
SAC *Surrey Archaeological Collections*
SRS Surrey Record Society

Bailey, K. A., and Galbraith, I. G. (1973), 'Field systems in Surrey: an introductory survey', *SAC* 69: 73–87.
Guiseppi, M. S. (ed.) (1916), *The Court Rolls of the Manor of Carshalton*, SRS 2.
Hooper, W. (1945), *Reigate: Its Story through the Ages* (Guildford: Surrey Archaeological Society).
Sewill, R., and Lane, E. (1951), *The Free Men of Charlwood* (Charlwood: Sewill and Lane).

ACKNOWLEDGEMENTS

David G. Bird, Principal Archaeologist, and Emily Brants, HER Officer, Surrey County Council, for information.

34. SUSSEX

Sussex has two bands of high ground running approximately east to west; at the north are situated the South Downs and the High Weald. The Weald has dispersed settlement, partly caused by many parishes having detached pieces in the Weald from Saxon times. To Tarring (Worthing) belonged Tarring Marlpost (in Horsham), and Broadwater had Sedgwick alias Little Broadwater near Horsham (E. H. W. Dunkin 1892; *SxAC* 38: 150–1). Saxon charters describe many of these as swine pastures, Stanmere, near Brighton, having amongst others, Lindfield north of Wivelsfield, nine miles distant in AD 775 (S 50; E. Barker 1947–8, *SxAC* 86: 85–90, 87, 113–14). Assarting occurred in the Weald (P. F. Brandon 1969, *Transactions of the Insitute of British Geographers*, 48: 135–53). Many townships had low percentages of arable open fields: for instance, Iford and Swanborough (2,173 acres) had approximately 500 acres of arable. Most Sussex fields were irregular, with intermixed usage of 'field' and 'furlong'. A few townships have 'great fields', most of which are unlikely to have had a regular cropping arrangement. Sheepfolding was practised on the Downs.

An Atherington map of 1606 indicates small areas of open field (J. M. Johnston 1901, *SxAC* 44: 147). The Sutton map of 1608 by Ralph Treswell shows the extent of open fields with many commons and downs amounting to 51 per cent of the total (Leconfield 1956, 2 and end). Brighton has a 1792 field book and map based on an earlier field book of 1739. Small land units were called 'pauls' (i.e. poles) that were an eighth of an acre, but in practice most parcels were 'paul pieces' consisting of two or four pauls, i.e. a rood or half an acre. The total pasture was 861 acres out of 1,562 acres (Farrant 1978, 67–81; map p. 69). Kingston by Lewes has a field book and a fine map of 1799 (East Sussex CRO, ADA/MSS 51, fols. 14–55) reproduced by Thorburn (2001, 8). Open-field is marked on eighteen tithe maps dated 1838–46 (Kain and Oliver 1995; see also Steer 1962).

Some coastal districts were still open in *c.*1600, with arable divided into 'laines', pasture on downs, and meadows called 'brooks' or 'brookland'. The word 'laine' was sometimes used for a furlong and sometimes for a group of furlongs that lie together in one location. The locational use could be applied in a Midland sense for a great field (Godfrey 1928, 107). Chichester glebe surveys of 1724 refer to 'plain land' (meaning open field) and meadow called 'brooks' (Ford 1992, 57–61). Open field holdings were also called 'tenantry land' (G. R. Corner 1853, *SxAC* 6: 240).

Gray (1915, 498) records two two-field systems, neither adequate since Amberley refers to demesne in 1378 and Broadwater, in 1802, had much enclosure. He lists five three-field arrangements in addition to those in the Table. The field systems noted in the Table record one two-field, six three-field, three four-field, one five-field, and one seven-field. It is unlikely in most cases that these relate to a cropping arrangement. Demesnes occur both in block and dispersed forms (five and six examples, respectively); some have been described more fully (pp. 99, 103). Yardlands and wists have also been described (pp. 110, 115).

Irregular fields are likely to have been widespread. Tithe disputes on the South Downs refer to many furlong and piece names, but never to 'great fields': Washington and Sullington, 1246 (Magd. Washington charter 5), Buddington, *c.*1250 (Magd. charter Buddington 2), and Findon in 1256 and 1477 (Salzman 1923, 21–5, 91–3). The term 'field' was used for 'furlong' at Buddington *c.*1250 (ibid.) and at West Grinstead in 1270 (Magd. charter Binelands and Grinstead 11). The long descriptions of the arable fields of Findon made in 1256 and 1477 have no reference to great fields (Salzman 1923, 21–5, 91–2). The Findon lands of 1256 refer to the lands of fifty-seven people, but no place called a furlong. Of 10 persons with holdings of 4¾—19 acres, 3 lay in one place, 2 in two, 4 (1 described as a yardland) in three, and 1 holding lay in four places. There was thus very little dispersion. In 1477 the term 'furlong' was used eleven times and 'field' seven, both referring to limited areas. Many areas (furlongs) were called 'fields' at Willingdon in 1292 and at Dadesham in 1321 (Wilson 1961, 23, 56–9).

Work-service

A very detailed description of work-service for a ferling at Wiston in 1358 lists the usual grain-production tasks for corn, rye, and peas; the tenants also had to carry two loads of wood, cart the lord's manure, gather apples and make cider, wash and shear sheep, dig ditches, gather broom, heather, or bracken, carry and process flax, and gather nuts. The lord required two 'herdwikes', where two flocks of 200 sheep each were folded on ten acres of stubble to be milked for making butter and cheese. If the lord was unable to provide the sheep, those of the tenants were to be used. Oxen were used for ploughing (W. Hudson 1910, *SxAC* 53: 143–82).

Many other examples of work-service providing information about farming have been published, such as the FitzAlan surveys of 1301 (Clough 1969), those for Laughton, Willington, and Goring (Wilson 1961), and those for the Sussex manors of Canterbury. For example Wadhurst had a holding called *Drofland*, ten acres for which the work-service was to drive animals as required between manor and the pastures, as well as clean out houses at the manorial court. Each hide contained four virgates, and each virgate, as well as the usual arable services, had to provide one hurdle seven feet long with eight 'pieces' (i.e. bars), carry wood to the court at Malling, and hunt with a man without bow and arrow for twelve days (Redwood and Wilson 1958, 35–9). Battle Abbey custumals for Alciston, Appledram, Marley, and elsewhere describe the usual agricultural round (Searle 1974), and those for Eastbourne in 1253 are very detailed (*Calendar of Inquisitions Miscellaneous 1219–1307*, i (1916), no. 188).

Farming

Wiston demesne lay in a block. There were seventeen pieces in all, some divided into parts making twenty-one cropping units. Details of cropping arrangements and yields for each piece are given in account rolls from 1376–7 to 1386–7 (Godman 1911, 133 and his table 1). There is no trace of a regular cycle of cropping. Six pieces were cropped for nine or more of the eleven years (only one in all of them). Five pieces were cropped consecutively for three years and left for at least five years without a crop (eight years in one case). This seems to be a system of near continuous cropping mixed with a cycle of convertible husbandry.

Sheep and their pastures were an important part of the Sussex open-field economy. Some information is given in the text and much more can be found in the manorial custumals noted above. The custumal of a yardland at Ratton, Eastbourne (1248), included washing and shearing ten sheep (Salzman 1923, 51).

Convertible husbandry was practised on the South Downs. At Clapham in 1150, tithes were due from newly cultivated ground (Magd. charter Clapham 1). Similarly tithes were due from West Grinstead in 1241, when the pasture called *Fursfeld* was ploughed up and cultivated (Magd. charter Grinstead and Stanford 1). Ashurst also had a common called *Fyresfeld* in 1460 (Magd. charter Ashurst and Lancing 6). At Findon, in 1256, tithe of demesne pastures previously ploughed were to go to Sele Priory, and tithes of lands in pastures that might be tilled after 1256 were to go to the church. Another Findon tithe agreement of 1477 noted that the 'newe broken' land of the demesne ('curtland') had not been sown for five years (Salzman 1923, 24, 92), in other words, it had reverted back to pasture.

Table App.18. Sussex field systems.

Place	Demesne	Holding (acres)	Sheep commons	Field No.	Date	Sources
Alfriston				3	1432	Gray 1915, 33, 443
Atherington in Climpimg				3	1606	P. M. Johnston 1901, 44, 147; two-thirds enclosed
Brighton	block	7		4	1792	Farrant 1978, 67–81; field book and map; unequal laines but lands fully dispersed Yardland seven acres
East Blatchington				3	1432	Gray 1915, 33
Ertham				4	fourteenth century	Peckham 1946, 186–8
Gate				7	1229	Salzman 1903, SxRS 2: 9; unequal field sizes
Iford		10 [yardlands]		3	c.1565	Northamptonshire CRO, Westmorland (Apethorpe) 5.ix
Kingston by Lewes				3	1799	Thorburn 2001, 8 ex East Sussex CRO, ADA/MSS 51, fols. 14–55
Ovingdean				3	1444	Gray 1915, 499
Rottingdene	block	10 [yardlands]		5	c.1566	Northamptonshire CRO, Westmorland (Apethorpe) 5.ix; yardlands in three seasons; also a courtlane, forty-nine acres
Steyning		26	250	4	1550	J. E. Ray 1930, SxRS 36: 79; unequal field sizes
Stokes				2	1257–8	Salzman 1907, SxRS 7: 9
Angemeryng	dispersed				1400	M. Clough 1969, SxRS 67: 122
Burpham		12½	100		1724	Ford 1992, 61.
Clymping	block				1604	P. M. Johnson 1901, SxAC 44: 147
Crokeburst [in Durrington]	block	229			1254	Salzman 1923, 5–6

Dadesham	dispersed			1321	Wilson 1961, SxRS 60: 56–9
Dichelinge	dispersed	94	400	1588?	Godfrey 1928, SxRS 34: 107–10; sheep on a several down
Durrington	dispersed			1257	Magd. Charter Durrington 5
Edbourghton	dispersed			1343	Salzman 1934, SxRS 40: 106–9; measured in 'palls'
Laughton	block	213½		1292	Wilson 1961, SxRS 60: map 2
Lewes		8	60	1150	Salzman 1932, SxRS 38: 62; common for four beasts
Radmell Beverington	dispersed			1588	Godfrey 1928, SxRS 34: 107–10; in crofts, laynes, and furlongs
Shermanbury, *Iwhurst* in	dispersed			1352	Magd. Charter Grinstead and Stanford 6
Southwick		8	100	1550	J. E. Ray 1930, SxRS 36: 88; eight oxen, four cows, one horse;
Sutton	block	137		1608	demesne in three pieces: Leconfield 1956, 15

Ridge and furrow

No ridge and furrow for Sussex is recorded at Cambridge and none was found during a search of aerial photographs by Nash (A. Nash 1983, *SxAC* 121: 109).

Iford's fields have been reconstructed by a fieldwork survey of furlong boundaries and correspond exactly to the tithe map of 1840 (Figure App.4). The north could not be satisfactorily mapped because of long crop growth. At the far south of Kingston by Lewes a few small lynchets lie on the slopes of the down, recorded as open-field strips on a 1799 survey.

BIBLIOGRAPHY

SxAC   *Sussex Archaeological Collections*
SxRS   Sussex Record Society

Clough, M. (1969), *Two Estate Surveys of the Fitzalen Earls of Arundel*, SxRS 67.
Farrant, J. and S. (1978), *Aspects of Brighton, 1650–1800*, University of Sussex Occasional Paper, no. 8.
Ford, W. K. (1992), *Chichester Diocesan Surveys, 1686 and 1724*, SxRS 78.
Gardiner, M. (1996), 'The geography and peasant rural economy of the eastern Sussex high weald 1300–1420', *SxAC* 134: 125–39.
Godman, P. S. (1911), 'On a series of rolls of the Manor of Wiston', *SxAC* 54: 131–82.
Leconfield, H. (1956), *Sutton and Duncton Manors* (London: Oxford University Press).
Steer, F. W. (1962), *A Catalogue of Sussex Estate and Tithe Award Maps*, SxRS 61.
Searle, E. (1974), *Lordship and Community: Battle Abbey and its Banlieu, 1066–1538* (Toronto: Pontifical Institute of Mediaeval Studies, 1974).

Iford

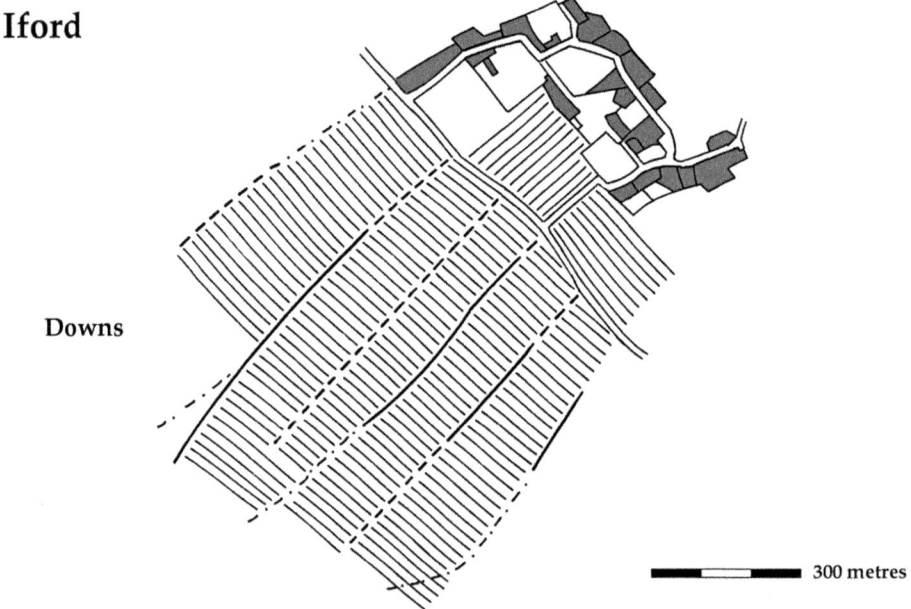

Downs

300 metres

Fig. App.4. Iford furlongs, Sussex. Mapped in 2004; *opposite:* an extract of the Tithe Map, 1840 (courtesy of East Sussex Record Office, TD/E61/1).

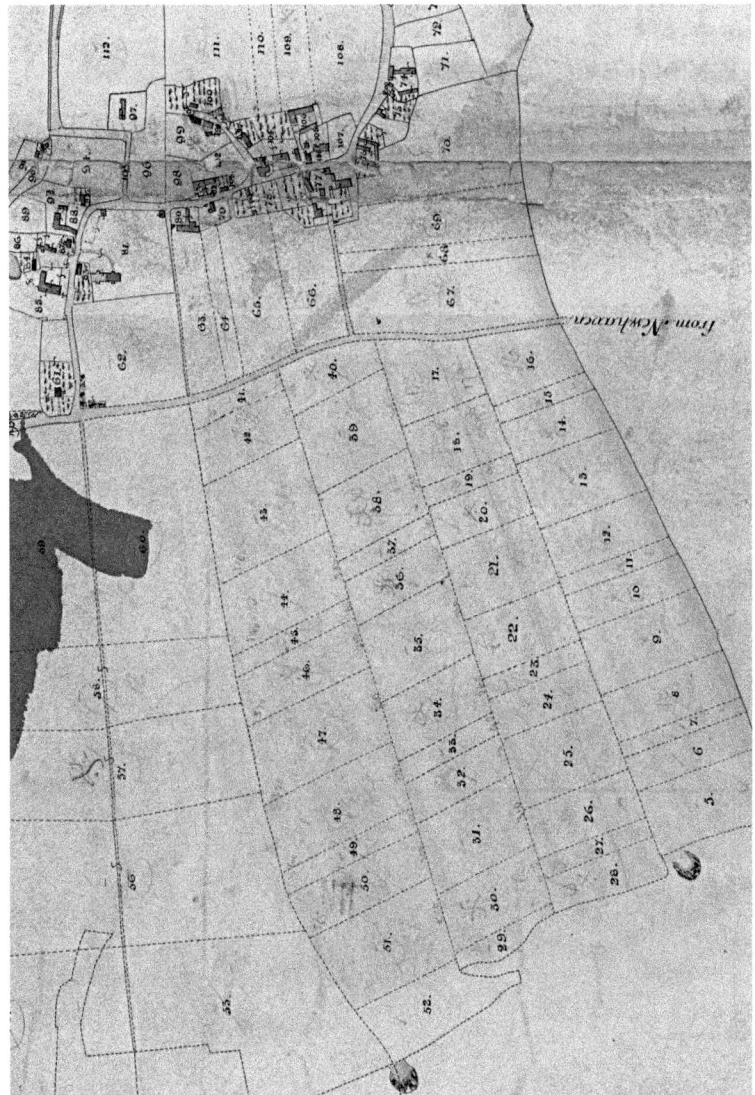

Fig. App.4. Iford Tithe Map 1840.

Thorburn, M. (2001), *An Account of the Manor of Hyde, Kingston near Lewes, Sussex* (Kingston: Thorburn). [East Sussex CRO, LIB0502946]

ACKNOWLEDGEMENTS

The County Archaeologists of East Sussex, Andrew Woodcock, and of West Sussex, Mark Taylor, for information from the HER. Margaret Thorburn, Lewes, for hospitality and fieldwork arrangements at Iford and Kingston. J. H. Robinson, Esq., for permission to survey Iford, and Elizabeth Hughes, East Sussex County Archivist, for permission to publish the Iford Tithe Map.

35. WARWICKSHIRE

Warwickshire falls into two main topographical parts—the northern Arden, which had dispersed settlement and was early enclosed, and the southern Feldon, with extensive medieval arable. Many authors, from Leyland in the sixteenth century on, have commented and written about this division. The dispersed settlement pattern of the north-western half of the county, reconstructed for *c*.1300, is illustrated by Dyer (1996, *TBWAS* 100, 120). Detailed studies of Yardley (Skipp 1970), and Wishaw and Middleton in Arden, describe many ends and greens (M. A. Hodder 1991–2, *TBWAS* 97: 104). Hooke (1996, *TBWAS* 100: 99–111) discussed the formation of the county, using charter and place-name evidence. Place-names with *ton* are mainly found in the south Feldon region and names containing the *leigh* element (wood or wood-pasture) in the northern wooded region. Gelling (1974, *TBWAS* 86: 59–68) showed the presence of some Welsh names in the Birmingham area. Warwickshire field systems follow the topography—regular extensive fields in the south with irregular arrangements in the north, often involving crofts and greens.

The fields have been discussed by Gray (1915), Barratt (1955), and Roberts (1973). Thurlaston has a field book and open-field map of 1717 (VCH, *Warwickshire*, vi (1951): opp. p. 78). The vill and its closes occupied only 2.4 per cent of the township, the rest being open (Northamptonshire CRO, Buccleuch maps 4550/1 and 2; 13). An Ilmington map of 1778 was partly illustrated by Beresford and St Joseph (1979, 26–9). A 1754 map of Kinwarton fields is reproduced by Hooke (1998, 118). Tithe maps of 1841–50 show that there was some open field left in five townships (Kain and Oliver 1995, 534–9).

Two- and three-field systems

Gray (1915, 70, 499–500) provided details of twelve townships that had two fields in the Middle Ages, mostly lying in the south (mapped by Roberts (1973, 222), drawing upon the work of J. B. Harley). Lapworth and Hampton Lucy demesnes both lay in two fields in 1299 (Hollings 1934–50, 189, 263). The Table lists seven two-field, one three-field, and two four-field systems. Yardlands have been discussed (p. 115). Gray (1915, 500) gave details of three townships with three fields in the Middle Ages. Roberts (1973, 202–9) mapped the distribution of three-field systems; they occurred mainly in the east of the county in the early modern period. In the north, Warton on Polesworth in the Tame Valley had three fields until 1772 covering 80 per cent of the township.

Four fields

The fallow of Hampton Lucy was sown in 1476 and Grandborough had an area called *The Hiche* that was continuously cultivated (Dyer 1981, 28). Such cultivation probably led to the formation of four-field systems. Many examples of four-fields, generally called 'quarters' are given in glebe terriers (Barratt 1955). Roberts (1973, 195–205, with county plans) showed that there were many four-field systems in the south of the county, the reconstructed plan of Crimscote by D. J. Pannett being illustrated.

Multiple and irregular fields

Details of surveys showing irregular fields at Hampton in Arden and Langdon in the early seventeenth century were given by Gray (1915, 86–7, 512–13). At Hampton there were seven named areas, five called 'fields', that had holdings distributed in a variable number of them. Langdon also had some enclosed arable pieces.

Sutton Coldfield (13,030 acres), lying next to Cannock Chase, had large areas of 'waste' as well as the hamlets of Hill, Little Sutton, and Walmley. There was so much waste that anyone willing to build a house could enclose sixty acres around it (1528). Large 700-acre blocks of the 4,000 acres of waste were divided by lot amongst the inhabitants and kept in tillage for five years, after which it reverted to pasture and another 700 acres were taken in. Sutton had eighteen widely scattered open areas, ten called 'fields' in the eighteenth century, some of them assigned to the use of the three settlement groups (M. W. Beresford 1946, *TBWAS* 64: 101–8).

Henley had twenty-nine pieces of land, eleven called 'fields', with holdings lying in a few fields only. Similarly Erdington in 1463 had eighty-seven pieces, of which eight were called 'field'; tenancies were irregular. The Westminster manor at Knowle had more than seventy pieces named in 1408, twenty-two called 'fields', which were roughly equivalent to East Midland furlongs. There was no arrangement into regular groupings of large fields (Hilton 1950, 23). Stoneleigh, lying in the Arden, had many woods and irregular fields in its 9,908 acres. In the twelfth century it contained six settlements and there were nine in 1305, as well as several granges. Lands lay in complex and often large pieces bearing no relation to the uniform dispersion found in the south (Hilton 1960, xlv–xlvi). A twenty-acre half-yardland was distributed mainly in acre pieces very unequally in six areas called 'fields' (Hilton 1960, 139,

Table App.19. Warwickshire field systems.

Place	Yardland size	Field no.	Date	Sources
Barton in Blakewell	20	2	1196	Wellstood 1932, 3
Lighthorne		2	1586–1714	Barratt 1955, 142–56
Cherington		2	1616	Barratt 1955, 76–83
Oxhill		2	1585	Barratt 1971, 25–36; four fields in 1714
Whichford		2	1585	Barratt 1971, 131–67
Newbold Pacey	*c.*16	2	1608	Barratt 1971, 14; thirty sheep and four beasts per yardland
Brailes		2	seventeenth century	Gray 1915, 29, 437
Aston Cantlow		3	1585	Barratt 1955, 17
Admington		4	*c.*1550	Gray 1915, 88–9
Thurlaston		4	1717–19	Northamptonshire CRO, Buccleuch Map 4550/1

135–7). A detailed terrier of the prior's land refers to six fields and three other named areas with very irregular amounts of land. There was some organization of cropping because a croft next to *Culvecroft* was sown with the *Wodefeld* (Wood Field) in the fourteenth century (ibid. 116–20). Two yardlands at Finbury lay as 1½ yardlands in two closes, and the other half, 22 acres, in the common fields in three pieces (Hilton 1966, 121).

The area around Yardley and Solihull was complicated with many small fields, crofts, and greens (Skipp 1970 and 1981). Tanworth-in-Arden had some strip fields by 1271 that were referred to as a 'common field' in *c*.1500 but without evidence of any rotational practice. Assarts, enclosed and held in severalty, lay adjacent to the fields and others were made in the waste. A reconstructed plan shows the landscape in *c*.1350 being largely assarts with some wood, a park, heath, and many settlements and farmsteads (Roberts 1968).

Work-service and field orders

Sutton Coldfield customary tenants, in 1329, provided two days' service for a yardland, driving game when the lord came to hunt. All of them had pasturage all year round in the outwoods of Cannock Chase and could take dead wood for fuel; forest fences were to be maintained, traps set, and trackways had to be cut (M. W. Beresford 1946, *TBWAS* 64: 101–3). A Coventry court leet book describes extensive commons in 1423. There were regulations not to surcharge them, but no communal field orders. The parish contained the several townships of Radford, Coundon, Stivichall, Keresley, Exhall, and Foleshill (Harris 1907–13, 7–19, 45, 100). At Henley-in-Arden, Beaudesert court rolls of 1442 refer to drains needing attention, stray animals, and pasture overburdened (W. Cooper 1931, 6–7), but has no orders referring to communal agriculture. Orders for 1592 and later are similar (W. Cooper 1946, 25–9). Sheldon in the woody land of the north has seventeenth-century field orders that refer to the making of fences around the cornfields before they were sown (Roberts 1973, 215–16). In the south, there were detailed orders controlling grazing and other open-field matters at Lower Shuckburgh in the fourteenth century and Long Marston yardlands in 1543 (Dyer 1981, 31–2).

Farming

In the seventeenth and eighteenth centuries many field systems in the south of the county had grass furrows a few feet wide between every land, for example at Brailes (1672) and Little Wolford (1750) (Roberts 1973, 198). At Kinwarton, in 1722, tithe was taken from grass growing on the heads and furrows in every field (Tate 1949, *TBWAS* 65: 104). The green furrows of Crimscote are illustrated by an excellent aerial photograph published by the Orwins (1938, pl. 11). Crimscote also illustrates an area of cow pasture formed by permanent conversion to pasture of lands lying far-distant from the settlement. At Brailes in 1672, the manorial court ordered that grass 'hades' (heads) at the ends of lands should be left to form routeways around the furlongs. Court orders for Kinwarton describe commons usage and animal stinting in 1722 (Tate 1949, *TBWAS* 65: 103–4).

Ridge and furrow

Broad, curved ridges occur in the eastern part of the county and some in the north. Cherington glebe lay in ridges in 1585 and 1616 (Barratt 1955, 76). A map of ridge and furrow for the county in the late 1940s has been published (Harrison et al. 1965), and can be compared with that surviving in the eastern part of the county in 1990 (Hall 2001, 36). Isham and Palmer (1990 and HER) extended the study to all the county on GIS format. Significant areas of ridge and furrow (20–70 per cent) survived at Arlscote in Warmington, Ladbroke, Little Lawford, Napton on the Hill, Radway, Tysoe, and both Upper and Lower Shuckburgh.

Detailed plans of Crimscote (Roberts 1973, 196) and Admington (C. Dyer 1996, *TBWAS* 100: 125) have been published. A photograph of Crimscote is well known (Orwin and Orwin 1938, pl. 11). Cropmark ridge and furrow overlies Roman and earlier sites at Barford (A. Oswald 1966–7, *TBWAS* 83: 5 and pl. 1). The Thurlaston map of 1717 shows many furlong alignments, up to five and six, with lands orientated in the same direction.

BIBLIOGRAPHY

TBWAS *Transactions of the Birmingham and Warwickshire Archaeological Society*

Cooper, W. (1931), *Records of Beaudesert* (Leeds: J. Whitehead & Son).
Cooper, W. (1946; facs. 1992), *Henley in Arden* (Birmingham: Cornish Bros. Ltd).
Harris, M. D. (1907–13), *Coventry Leet Book, 1420–1535* (London: Early English Text Society).
Harrison, M. J., Mead, W. R., and Pannett, D. J. (1965), 'A Midland ridge and furrow map', *Geographical Journal*, 131: 366.
Hilton, R. H. (1950), *The Social Structure of Rural Warwickshire in the Middle Ages*, Dugdale Society, Occasional Paper, 9 (Oxford: Dugdale Society).
Isham, A., and Palmer, R. (1990), *Medieval Settlements Research Group Annual Report*, 5: 14–17.
Wellstood, F. C. (1932), *Warwickshire Feet of Fines*, Dugdale Society, Record Series Publications, 11 (London: Dugdale Society).

ACKNOWLEDGEMENTS

I am grateful to Helen McLagen, Andy Isham, Emma Jones, and Nicholas Palmer for access to information in the HER.

36. WESTMORLAND

Westmorland is largely mountainous with the major valleys of the Eden in the north and the Kent in the south, where lie the principal settlements of Penrith, Appleby, and Kendal. High ground in the fells throughout the county was used for cattle and sheep rearing, often by transhumance. Herders moved cattle for the summer and lived in cottages called 'shielings', returning in the autumn. By the seventeenth century many shielings developed into permanently settled farms (Ramm et al. 1970, 1–12, 38–9). Parishes are often large and comprise many townships. Kendal was about 60,000 acres and contained twenty-seven townships that took three years to survey for a tithe dispute with Trinity College, Cambridge. Only 13 per cent of it was arable in *c.*1835 (E. J. Evans 1974, *TCWAAS*, NS, 74: 176–9). Elliot (1973, 82–4) provided some information on Westmorland. The field systems were of limited extent, mostly irregular, and there were undispersed assarts.

Sill Field in Preston Patrick, open in 1771, has been described (Atkin 1993, 146). A 1764 estate map of Murton shows three large open fields divided into 231 riggs (Elliot 1973, 82). At Kentmere intermixed strips marked on a map of 1834 were dispersed along the dale (M. A. Atkin 1991, *TCWAAS*, NS., 92: 70–1). Eight places had some open-field marked on tithe maps of 1838–42 (Kain and Oliver 1995, 544–8). Rydal had strips surviving as late as 1916 (Rawnsley 1916, 58).

Murton and Hilton fields, lying below the fells east of Appleby, were studied by B. Tyson (1992, *TCWAAS*, NS, 92: 161–82). Murton had three unequal fields; the Great Field on the north had dispersed ownership and extended to about 175 acres divided into seven furlongs. There were two other fields lying south of the village called Little and Moor Fields, much smaller and divided into four and two furlongs respectively. Large areas of pasture with fell on the east surrounded the fields; the names are likely to be locational rather than implying a three-course rotation. Neighbouring Hilton field, mapped in 1764, extended to about 450 acres. It was divided into twenty furlongs and a holding was fairly uniformly dispersed. The furlong pattern is substantially 'rectangular' and looks planned.

Examples of undispersed holdings, possibly assarts occur at New Biggin in Hutton Roof township in Kirkby Lonsdale, where seven acres of land lay in one ridding with its bounds described in *c.*1225. Another at Lupton, in the same parish, had ten acres in one ridding and ten acres in another, both with their bounds described (Farrer 1905, 984, 991–2). Oxgang sizes have been given (see p. 115).

Field orders

Court proceedings at Troutbeck, in 1560, referred to disturbing neighbours' cattle, fishing in the lord's water, and having unringed pigs. Peat production at Levens was controlled by an agreement with the lord in 1676. Each tenant was to maintain the principal drainage channel and make a causeway over it to his moss room and to take part in a moss jury (Curwen 1923, 50, 139). Pastoral farming in Troutbeck during 1550–1750 showed that Woundale outpasture was used for summer grazing. Manorial courts restricted the pasture of the fells for cattle or sheep (M. A. Parsons 1993, *TCWAAS*, NS, 93: 115–30).

Ridge and furrow

Two examples of broad ridge are recorded at CUCAP: Dalton (SD 543 758) (BAW 42–6 (1970)) and Hartley (NY 786 093) (BMD 95 (1973)). Broad ridges five to eight yards wide were observed in 2004 at Stephen Kirkby (NY 770 082) and at Nateby (NY 0773 069), with hedges following their course.

Broad ridge occurs in back paddocks at Maulds Maeburn (NE 62 17) (B. K. Roberts, 1996, *TCWAAS*, NS, 96: 46). Similar strips were recorded in paddocks at Melkonthorpe (NY 55 25), Baber (NY 79 11), and Rookby (NY 80 11) (B. K. Roberts 1993, *TCWAAS*, NS, 93: 131–44). Long lands up to 400 yards have been mapped at Waitby and Smardale townships in Kirkby Stephen parish (B. K. Roberts 1993, *Archaeological Journal*, 150: 433–55, plan p. 439). Newby has curved hedges more than 800 yards long (B. K. Roberts 1988, *Geografiska Annaler*, 70B: 169).

BIBLIOGRAPHY

TCWAAS   *Transactions of the Cumberland and Westmorland Antiquarian and Archaeological Society*, in three series

Farrer, W. (1905), *The Chartulary of Cockersand Abbey*, iii.1, Chetham Society, 57.
Rawnsley, W. F. (ed.) (1916), *Rydal* (Kendal: T. Wilson).

ACKNOWLEDGEMENT

Bette Hopkins for providing HER information.

37. WILTSHIRE

The topography and geology of Wiltshire are dominated by two outcrops of chalk. Between these and the Cotswolds, at the north-west, a belt of clay runs from Swindon to Warminster. Many parishes had component townships. Downton contained six, each with its own yardlands, in 1376–7 (Lennard 1916, 5) and Bishopstone's 1677 glebe terrier referred to five hamlets (Hobbs 2003, 38). The greens of Semley have been described. Wiltshire is predominantly a county with two- and three-field systems and the extensive downland commons allowed a high number of animal stints for each yardland. Downland sheepfolding was practised.

The open fields of Wiltshire were discussed by Gray (1915, 24–33, 421–30) and by R. Scott and E. Kerridge (1959, VCH, *Wiltshire*, iv: 7–64). Much information is available in the two surveys of the Earl of Pembroke's lands made in 1562 and 1631 (Straton 1909; Kerridge 1953), and in the glebe terriers edited by Steven Hobbs (2003). Tithe maps of the 1840s show open-field land at twenty-eight places.

Field numbers

Gray (1915, 501–2) found twenty-five Glastonbury Abbey manors that were two-field in 1517. Fox (in Rowley 1981, 80) listed five townships with two-season demesnes in the early fourteenth century. Gray (1915, 501–2) recorded twenty-four townships with three-field systems and nine that had demesne on a three-season system (1915, 501–2). Recent work (Hall unpublished) has identified two-fields for eleven places and twenty-six three-field, with ten having irregular three to six fields. The glebe terriers, where the evidence is clear, record seventeen two-field, eighteen three-field, and six with multiple fields (Hobbs 2003).

Yardlands and field structure

Of the fifteen demesnes listed, all but two are dispersed. The demesne of Corsham, in the late fourteenth century, also seems to have been mainly in block form (Gray 1915, 74–5). Most yardlands varied in size from seventeen to thirty acres and have stints generally for thirty great beasts (kine), two to four horses, with twenty-four to sixty sheep, and a few of seventy to a hundred sheep. Most places have downland and the sheep commons were specified as being in the fields and on the downs. The yardlands of Wylye and Broad Chalke had a 'cote' (presumably a sheepcote). Larger yardlands of around forty-one acres were recorded at Lacock and Natton in 1305 (WRS 34: 50).

Tenurial sequence

Three neighbours of a terrier of Amesbury held thirty-four of the forty-three named positions (79 per cent), indicating a regular tenurial order in 1428 (Pugh 1947, WRS 3: 18–19). Glebe terriers of Chisledon (1608), Codford St Mary (1609), and Sopworth (1704) had very few neighbours, also suggesting a regular tenurial order (Hobbs 2003, 95–6, 103–4, 388–9).

Field regulations

Offences are recorded in court rolls between 1275 and 1278 at Sevenhampton and Stratton. Cows and sheep trespassed in the crops, calves went into the hay meadows, and geese damaged pasture and corn. Sheep ordered to be put in the fold failed to arrive. Boundaries were ploughed up and there were encroachments on the lord's land (Pugh 1970).

Detailed field regulations were made for Winterbourne Stoke in 1574 (Goddard 1906, *WAM* 34: 208–15). They give the customs of commoning for the principal farm animals; additional orders stated that no geese, goats, or unringed pigs were to enter the commons. Each yardland owner was to provide four hurdles and struts to make a large common fold on the fallow. Lands were said to lie in ridges. There were grass grounds called 'linches' (probably the risers of lynchets, or possibly leys, which were also called 'lanes' in some places) which were not to be used to feed horses until the corn was carried. On High Down a conygarth was made within a ditched earthwork (presumably prehistoric) that enclosed several barrows.

Farming

Open-field farming accounts giving details of crops sown and yields in 1633–9 survive for Shrewton and Netton (Kerridge 1952, *WAM* 54: 416–28). Crops planted and harvested are detailed at Sevenhampton and Stratton in 1271–7 (M. W. Farr 1959, WRS 14: 1–31). At Durrington, account rolls list seeds planted and crop yields; peas and vetch were about 15 per cent of the total during 1324–41 (J. N. Hare 1974, *WAM* 74: 139). A marl pit at Heddington 120 by 100 feet was to be used to marl specified lands in the thirteenth century (WRS 34: 54, no. 247).

Brixton Deverill has a very detailed description of the work-service due for a yardland in *c*.1225 (J. R. Pierrepont 1984, *WAM* 78: 55–61). Laycock Cartulary gives similar information for Bishopstrow, Heddington, Hatherop, and Lacock (Clark-Maxwell

1902, *WAM* 32: 311–46). The manors of Bec Abbey have detailed customs and work-service recorded in the thirteenth century (Chibnall 1951). At Ogbourne St Andrew, in addition to the agricultural round, they included washing and shearing sheep, and making hurdles and wattles for the fold, which was to be carried from place to place. Sheepfolding was an important method of farming in Wiltshire and is discussed in the main text (pp. 54, 55). Sheep were washed and the lord's fold moved at Bromham in the fourteenth century (Searle 1974).

The Glastonbury Abbey survey of 1189 records some work-service for the usual farming round, including washing and shearing sheep at Winterbourne Monkton. The smith had to repair the lord's three ploughs and shoe affers, as well as collect fleeces from sheep when sheared (Jackson 1882, 123–4). Work-service for Erlestoke is described in an extent of 1309 (Watson-Taylor 1906, *WAM* 34: 86). Detailed work-service at South Newton was recorded in 1315. At Stoford, in the same parish, one of the services was to find one man to act as a 'dencher', twenty-four days yearly, to work mornings. In the early seventeenth century, 'denchering' was the technique of preparing rough grassy ground by skimming with a breast plough and burning the sods; seeds were sown after ploughing (Straton 1909, 343). Since one of the open fields of Stoford was called *Burn-bake Field* (Straton 1909, pl. opp. p. 542), it seems very likely that the term 'dencher' meant the same in 1315. Remnant work-service on the Pembroke manors in the 1560s included at Netherhampton washing and shearing the lord's sheep (Straton 1909, 16).

Physical remains

Ridge and furrow lies almost entirely in the clay belt with very little on the chalklands, which were probably only slightly ridged. A good example of such slightly ridged chalkland strips can be seen on the edge of Overton Down (Fowler and Blackwell 1998, colour pl. 15). At Little Woodbury 'parallel grooves' in the chalk corresponded to a map of 1600 (*Proceedings of the Prehistoric Society*, 6 (1940): 31–4, pls. I and III). Furrows at Ashton Keynes, gouged into the subsoil as they approached a headland, had been realigned towards the left, so increasing the reverse-S form (Figure 5.2). 'Ridge and furrow' on Thornham Down, partly overlying Celtic Fields, is of irregular form and is likely to be of early modern date (McOmish et al. 2002, 111–12).

Lynchets of various types are a prominent feature of the chalk downlands and have been discussed by many authors. Most strip lynchets are likely to be medieval in origin. At Hatchbury, in the fourteenth century, two acres of arable lying in *Kenescumbe* were stated to lie between two unploughed 'lynches' (J. L. Kirby 1993, WRS 49: 130). The glebe of Winterbourne Stoke refers to three acres of arable with 'linchyardes' on each side in 1609 (Hobbs 2003, 475). An excavation report on lynchets at Bishopstone and in the Vale of Pewsey discusses their nature (P. Wood and G. Whittingham 1957–60, *WAM* 57: 335–8), and Crawford and Keiller (1928, pls. 28 and 29, pp. 166–8) had earlier showed that the lynchets of Calstone could be exactly related to the strips marked on an early eighteenth-century map. The tithe maps of Coombe Bisset, Calne, and East Knoyle indicate lynchets.

BIBLIOGRAPHY

WAM   *Wiltshire Archaeological and Natural History Magazine*
WRS   Wiltshire Record Society

Fowler, P. J., and Blackwell, I. (1998), *An English Countryside Explored* (Stroud: Tempus).

Pugh, R. B. (1970), *Court Rolls of the Wiltshire Manors of Adam de Stratton*, WRS 24 (Devizes).

Straton, C. R. (1909), *Survey of the Lands of William, 1st Earl of Pembroke* (Oxford: Printed for the Roxburghe Club).

ACKNOWLEGEMENT

Roy Canham for providing information from the HER, including the remarkable photograph used as Figure 5.2.

38. WORCESTERSHIRE

Worcestershire is split by the valley of the River Severn, with the Avon in the south-east, and high ground at the west and north. Many parishes contain more than one settlement, such as Bredon with seven hamlets, five of which were 'in open field' in 1813 (Pitt 1813, 55), and Ombersley, where a survey of 1605 refers to fifteen hamlets in its 6,926 acres (Yelling 1968, 165–6). Hanbury has been described (Dyer 1991). Worcestershire fields fall into two types. In the south-east and the Severn Valley, regular Midland-type arrangements occur, whereas on the higher ground of the west and north there were irregular fields.

Gray described the field systems of the county (1915, 61, 503–4) and they were further discussed by Roberts, drawing on the work of Yelling and others (Roberts 1973, 188–231). Open-field maps are held in the County Record Office for several places. Among them are Cleve Prior (1772), which had an arrangement consisting of networks of rectangular furlongs. The glebe was fairly uniformly distributed, being 45½ acres in 123 parcels, of which most were single lands (copy in HER; Yelling 1977, 68–9, 124). Some open-field land is shown on the tithe maps (1838–47) of nineteen places (Kain and Oliver 1995).

Gray (1915, 503–4) listed four early two-field systems and a single three-field one at Shurnock (1237). He noted that in 1365 the bishop of Worcester's demesnes for twelve manors had one-third waste. Many of the bishop's demesne lands had been cultivated on a two-tilth system in 1299 (Northwick, Bredon, Fladbury, Ripple, Hanbury), but this may not mean there were two regular fields (Hollings 1934–50).

The Table lists six two-field, three three-field, and three four-field systems. Two-field systems readily developed into four fields, as explained in the main text (p. 43). Much of the eastern part of the county appears to have converted to four fields by the middle of the sixteenth century (Yelling 1969, 31–3).

Multiple and irregular fields

Parts of north and south-west Worcestershire did not have regular cropping systems and 'furlongs' were often referred to as 'fields'. At Ankerdine, nine selions lay in a field called *Haselfurlung*…and the field called *Bernefurlung* in 1286 (Worc. Cath. Mun.,

Table App.20. Worcestershire field systems.

Place	Yardland size	Demesne type	Field no.	Date	Sources
Bredicot			2	1308–23	Worcs. Cath. Mun., B 120 and B 124; North and South Fields
Harvington			2	1240	W. H. Hale 1865, 62
Kempsey			2	1240	Hilton 1966, 122
Netherton			2	thirteenth century	Worcs. Cath. Mun., B 557; Lower and Upper Fields
Overbury			2	1280	Worcs. Cath. Mun., B 120 and B 606; Northern and Southern Fields
Tredington		dispersed	2	1299	Hollings 1934–50, 279; Tredington was four-field in 1801 (K. G. Davies and G. E. Fussell 1951, *TWAS* 28: 53–8)
Berrow			3	1801	K. G. Davies and G. E. Fussell 1951, *TWAS* 28: 53–8
Eckington			3	1325	Worcs. Cath. Mun., B 235b; *Wodefeld*, *Middulfeld*, and *Nethurfeld*
Hardwick in Bedwardine			3?	1275	Worcs. Cath. Mun., B 477a; *Blakefeld*, *Middlefeld*, and *Longenacre*
Cleve Prior		block	4	1772	Yelling 1977, 68–9; 124; demesne enclosed before 1650
Flyford Favell	40		4	1595	J. Yelling 1969, 31–3
Shipston on Stour			4	1801	K. G. Davies and G. E. Fussell 1951, *TWAS* 28: 53–8

B34; also main text, pp. 79, 84). Hanley Castle had six main fields associated with three settlements in the Middle Ages, but the charters refer to many other 'field' names, with some arable lying in pieces rather than individual strips (Toomey 2001, map p. xvi). Castlemorton's twenty-four-acre yardland was dispersed in five areas, none called a field (L. Fullbrook-Leggatt 1946–8, 286).

Glebe terriers refer to multiple fields at Himbledon, Salwarpe, and Rushock in the seventeenth century; Himbleton glebe (forty-seven parcels) was irregularly distributed in six fields (Yelling 1969, 33–4). The Clifton on Teme manor of Hamme had land that had been split into seven holdings by the fifteenth century which were identifiable with later enclosed farms, implying there were then consolidated holdings (R. G. Griffiths 1931, *TWAS* 8: 58–62; R. G. Griffiths 1932 *TWAS* 9: 77–9). A 1649 survey showed that often there was a multiplicity of names. Broadwas had ten named fields, one of them, *Berry Field*, likely part of the demesne. Cropthorne had four fields; Hillhampton five named fields (Cave and Wilson 1924, 23–5, 40, 79–80).

Yardland sizes have been listed (p. 115). The stints for sheep during 1443–1539 for the bishop of Worcester's estates were mainly 50–60 per yardland, but Stoke was 80 and Whitstones and Wick was 120 plus 30 beasts and 2 horses (Dyer 1980, 325). In 1585 Flyford Flavell had forty sheep and eight beasts per forty acres; at Grafton Flyford forty-five-acre farms supported fifty sheep and three beasts; and Abberton twenty-five acres could pasture fifty sheep, six beasts, and two horses (Yelling 1977, 154).

Work-service and farming

The agricultural routine was recorded in *c.*1240 for the manors of the Priory of St Mary (Hale 1865). Work-service for villein yards is fully listed in the 1299 surveys of the bishop of Worcester's manors of Northwick and Kempsey (Hollings 1934–50, 13, 64). Similar items are recorded for Shelsley Beauchamp (1286) and Chaddesley Corbett (1290), the last including carriage to Tewksbury bridge to get cloth (Bund 1909, 26, 31). Hanley tenants made hurdles in the early fourteenth century (Toomey 2001, 117, 144–6).

Court orders for Elmley were concerned with forbidding untethered animals to go in the harvest fields (1373 and later). In 1412 at least twenty selions were to be cleared before cattle were put in the stubble, and gleaning was not allowed until all sheaves had been carted away (Field 2004, 20, 79). By-laws for the episcopal estates in the fifteenth century are mostly concerned with the regulation of animals, for example ensuring that pigs were ringed and that fences were maintained between cultivated land and meadow (Dyer 1980, 330–1). Bromsgrove courts of 1495–1503 only recorded matters relating to drains and overstocking commons (Barber 1963, 105–34).

Rye was grown on poor land at Pensham, 1794–1807, on a system of crop and fallow 'by reason of the badness of the lands', but the other land there was cropped on a four-tilth cycle. Some places had limited areas of 'every year's land' in the eighteenth century, for example Pebworth and Church Lench (Yelling 1977, 160–1). Harvington also had every year's land in 1714. It was heavily manured or planted with (corn) crops containing clover as part of the cycle (Roberts 1973, 201).

Ridge and furrow

Broad ridge and furrow occurs over most of the county but is sparse in the west and north-west and not known in the Teme or Wye valleys (HER distribution map, Feb. 2004), but survives in the south-east of the county near the Severn and Avon Valleys, for example Inkberrow (SP 01 58), Cookhill (SP 05 57), and White Ladies Aston (SO 92 53). Extensive broad-ridge systems are recorded on oblique aerial photographs for Cotheridge (SO 785 545) and Beckford (SP 158 708) (CUCAP, AQJ 49 and AUA 82 (1966 and 1969)).

Early modern narrow ridges are known at Malvern. They generally occur on what was common land and lie mostly in the north and west of the county. West of Worcester in the region of old enclosure are various types of early modern ridging. Wide, low-profile, straight ridges lie near Tinkers Cross (SO 794 561), and at Blackfield Farm (SO 780 565) wide ridges are preserved. Near the Little Cobb House (SO 772 578), ridges of different widths occur. On the south-west side narrow, slightly curved, low-profile ridges have 'heads' at the lower end against a small stream. These are likely to be of seventeenth- or early eighteenth-century date. Wide lynchets lie in Suckley Wood (SO 735 527), one of them being further subdivided into ridges aligned along the terrace.

Many enclosure and tithe maps can be accessed and their data studied on the County website at <http://www.worcestershiremaps.org.uk>. Irregular hedged fields of likely medieval date, as mapped in the nineteenth century, occurred at Caines, near Worcester (Rippon et al. 2012, 61 (fig. 5)).

BIBLIOGRAPHY

TBGAS	*Transactions of the Bristol and Gloucestershire Archaeological Society*
TWAS	*Transactions of the Worcestershire Archaeological Society*
WHS	Worcester History Society
Worc. Cath. Mun.	Worcester Cathedral Muniments (calendar online)

Barber, A. F. C. (1963), *Court Rolls of Bromsgrove and Kings Norton, 1494–1504*, WHS 44.

Bund, J. W. Willis (1909), *Inquisitiones Post Mortem for the County of Worcester*, WHS 26.

Dyer, C. (1980), *Lords and Peasants in a Changing Society: The Estates of the Bishopric of Worcester, 680–1540* (Cambridge: Cambridge University Press).

Hale, W. H. (1865), *Registrum et Consuetudinarius Prioratus Beatae Mariae Wigorniensis*, Camden Society, 91 (London).

Pitt, W. (1813), *A General View of the Agriculture of the County of Worcester* (repr. New York: A. M. Kelley, 1969).

Toomey, J. P. (2001), *Records of Hanley Castle, Worcestershire, c.1147–1547*, WHS 59.

ACKNOWLEDGEMENTS

Malcolm Atkin, County Archaeologist; Jeremy Brotherton for HER data; and Peter Walker for a field visit.

39. YORKSHIRE

Yorkshire (3,879,979 acres) has a wide range of topography. The extensive lowlands of the Vale of York extend from the Tees Valley to the Humber and link with the Vale of Pickering. At the north-east lie the North York Moors, with the Wolds in the east, and the lowlands of Holderness in the south-east. The Dales and the Pennines lie on the west with dispersed settlement in upland townships, such as Northowram in the manor of Wakefield, studied by Faull and Moorhouse (1981, 603–6; Map 26). In its 3,400 acres there were thirty-five settlements plus another eleven with the location lost or only approximately known. Seven of these have records of their own field systems. Birstall parish, lying south-east of Bradford with 13,974 acres, consisted of 8 townships, 6 of them recorded in 1086 (Cradock 1933, 1). Yorkshire field systems fall into three main types. The Wolds and Holderness in the east had extensive fields with long strips controlled by communal regulations; the Derwent and Ouse valleys had fields similar to Midland types; and the west and north had small irregular fields involving convertible husbandry.

The field systems of Eastern Yorkshire have been recently described (Hall 2012, in Wrathmell 2012) and much Yorkshire evidence has been used earlier (e.g. p. 45); a brief summary is given here. Previous accounts were published in 1973 by June Sheppard, who used Beresford's analyses of the evidence from glebe terriers (Beresford 1948–51) and regional studies such as those for the East Riding published by Harris (1959 and 1961). Modern West Yorkshire was described by Faull and Moorhouse (1981, 659–67). Much open-field history is described in the volumes of the VCH *Yorkshire*, primarily for the East Riding (vols. ii–v, vii, and viii). For the Dales, there is information about the physical aspects of lynchets and ridge and furrow in the *Yorkshire Dales Mapping Project* Report (Horne and Macleod 1995).

Two- and three-field systems

Two- and three-field systems occur mainly on the lowland and on the Wolds. Gray (1915, 504–9) found ten townships with satisfactory evidence of two fields before 1510, six of them being demesne. There were twenty-five examples of three fields before 1610, of which twelve were samples of demesne lands. Beresford (1948–51, 348–9) studied glebe terriers, mainly of the seventeenth century. The East Riding had most surviving open field, and as would be expected at that date, there were few open fields in the upland parishes of the north and west of the county. Most townships were then two- and three-field. An early example of a two-field system occurs at Marton (near Bridlington (ER)) in *c*.1188, where there were twenty acres 'on each side' (Farrer 1916, 462). Pontefract provides an early example of a three-field arrangement in *c*.1180, where three acres lay in three named fields (Farrer 1916, 260).

Irregular fields and convertible husbandry

The North York Moors and the Pennines fall outside the region with 'Midland-type' fields. Townships lying on or near high ground often had small and irregular field systems that were enclosed at an early date. It is commonly found in such field systems that the terms 'field' and 'furlong' are interchangeable names and do not relate to cropping arrangements. Examples are Rastrick near Halifax, 1399 (Clay 1924), and Gilling in Richmondshire, 1540 (Brown 1922, 42–3).

Convertible husbandry and infield–outfield was used in townships with abundant pasture. Under such an arrangement the long period of rest allowed the land to recover, and it was manured by grazing animals. In the southern part of the West Riding, part of the common at Hitchells in Bessacarr could be brought into tillage in 1187 (Farrer 1915b, 163). The field system of Wigtwizzle (SK 250 957), in Bradfield, on the edge of the Pennines near Sheffield, has been described (Innocent 1924, 276–8). The Wigtwizzle system was convertible husbandry practised alongside an 'infield' arrangement. The Dearne Valley in the West Riding had small irregular fields and much pasture in *c*.1700 (Harvey 1974).

Oxgangs and field structure

The oxgang lands of the customary tenants were dispersed in the fields and lordly demesnes lay either in a block or were dispersed in small pieces among the tenants' lands. The oxgang size varied between about seven acres at Hutton Magna in 1254 (Brown 1892, 40–1) and nineteen acres at Ormesby (Brown 1889, 278). The field systems of the Wolds and Holderness were laid out in a regular, planned manner. Two bovates, part of ten, were the outermost two on the west side at Kirkby Grindalythe in *c*.1200 (Farrer 1915b, 385). Beresford found clear evidence of a regular order at Langtoft, Wetwang, and Great Givendale in the sixteenth and seventeenth centuries. At Wetwang two oxangs lay 'throughout all the fields lying next to one oxgang of land of William Moores' (Beresford 1948–51, 337). Regular order was further studied by S. Göransson (1961, 85–98). Harvey (1980 and 1981, 192–7) found evidence for a regular ordering of oxgangs through the fields of some Holderness townships. An approximate relationship was shown between township areas and the carucate areas of 1086, and implied that there was a connection between taxation and the area of the fields in some cases (Harvey 1978). The fields were regulated with by-laws. Barley (1943, 35–60) studied East Yorkshire for the years 1594–1856. At Bridlington, seventeenth-century courts recorded orders and offences, including animals trespassing in the cornfield, tethering of horses on the balks, and breaches of the pound (Purvis 1926, 110–11, 242–3).

Field patterns

Ridge and furrow occurs in many parts of the county. In the lowlands a chequerboard furlong pattern is found, contrasting with many field systems of the East Riding that have a remarkably simple form. Some townships have open-field plans. The Wharram Percy area has been mapped by archaeological survey. The hedges of early enclosed fields sometimes reflected curved strip patterns as at Snapethorpe Hall estate, Wakefield, mapped in 1601 (W. B. Crump 1939, *YAJ* 34: 359). Crayke has been described.

BIBLIOGRAPHY

YAJ   *Yorkshire Archaeological Journal*
YAS   Yorkshire Archaeological Society

Barley, M. W. (1943), 'East Yorkshire manorial by-laws', *YAJ* 35: 35–60.
Brown, W. (1922), *Deeds*, iii, YAS, Record Series, 63.
Clay, C. T. (1924), *Deeds*, iv, YAS, Record Series, 65.
Cradock, H. C. (1933), *A History of the Ancient Parish of Birstall, Yorkshire* (London: Society for Promoting Christian Knowledge).
Faull, M., and Moorhouse, S. A. (eds.) (1981), *West Yorkshire: An Archaeological Survey to* AD *1500* (Wakefield: West Yorkshire Metropolitan County Council).
Harvey, J. C. (1974), 'Common field and inclosure in the Lower Dearne Valley', *YAJ* 46: 110–27.
Horne, P. D., and Macleod, D. (1995), *The Yorkshire Dales Mapping Project*, RCHME.
Purvis, J. S. (1926), *Bridlington Charters, Court Rolls and Papers* (London: A. Brown).

ACKNOWLEDGEMENTS

Linda Smith and Robert White for HER information, and John Hurst and Maurice Beresford for introducing me to the Wolds. Lord Middleton of Birdsall for access to Wharram and Sir Richard and Lady Storey for access to Settrington. The owners and farmers of Butterwick, North Grimston, and Raisthorpe.

Glossary

abuttal: land or property bordering a given holding to assist in identifying its location.

acre (divided into 4 roods): 0.4047 hectares, 4,840 square yards.

aftermath: the right to graze on stubble ground or meadow after the hay was removed; the second or later mowing of meadow or pasture;

balk: a narrow strip of grass between arable lands.

breck: a temporary intake of common, pasture, or waste for arable cultivation.

burn-bake: a method of breaking up rough ground involving burning turf; see the Gazetteer for Cornwall, Hampshire, and Wiltshire.

champion: (medieval open) field areas.

cotsetland: property belonging to the holding of a cottar or cottager (labourer).

demesne: the 'home farm', land belonging to a manor-house.

field: a group of furlongs in the Central Region; elsewhere often an alternative name for a furlong.

field book: a survey of every land, furlong, and field in a township.

furlong: a group of lands lying together. Other regional names were bydales, falls, flatts, furshotts, rivings, sheths, shots, and wongs.

furlong boundary: the edge of a furlong; a headland or joint. Boundaries survive as low linear banks of soil in modern arable fields.

glebe: land belonging to a rectory or vicarage.

hain: to keep land (pasture or arable) separated from cattle.

headland: the first land in a furlong lying at right angles to its neighbour (see Plate 2); a boundary between two furlongs; the name sometimes used for the type of boundary later called a joint.

hide: a unit of taxation on arable land (before AD 1200); it had nominal area of 120 acres.

hundred: group of villages united for taxation and administration purposes.

inhoc, inhok, inhoke: land temporarily enclosed from fallow and put under cultivation.

joint: boundary between two furlongs with lands lying in the same orientation (see Plate 2).

land: the smallest unit of arable cultivation, usually one or two roods in area; *c.*220 yards in length and 5½ or 11 yards wide. Other regional names were lants, lawns, loons, byerdoles, quillets, ranes, reins, ridges, riggs. Also an arable strip ridged up by ploughing.

ley: a 'land' set down to pasture.

lot, lotting: meadow or other land assigned to a given owner by lottery, often annually.

mear, *meer*, *mere*: a boundary or landmark.
mile: 1.6 kilometres, 1,760 yards.
moss: a peat bog.

neat: bovine animals.

oxgang: the term used in Northern England for a villein or customary holding. It was often about half the size of a yardland, eight to twelve acres being common values.

penfold, *pinfold*: compound or pen to confine stray animals.
ploughland: an assessment of arable land used at the Domesday Survey of 1086.

quality: a valuation and survey of open-field holdings made prior to parliamentary enclosure.

reine: a balk.
rig(g): land ploughed into a ridge; ridge and furrow.
ringyard: a fence, hedge, or wall surrounding arable land to protect it from grazing animals.
rood: one-quarter of an acre.

several, *severalty*: (land) in separate or private ownership, usually free of common rights.
slade: a wet low place; a small valley.
soke: an area of jurisdiction.
solskifte: the system of identifying parcels that always lay either towards the east or south of a particular piece, or on the west and north.
stint: allowance of sheep, cattle, and other stock (for an oxgang or a yardland).

tenurial-cycle: regular arrangement of lands among tenants in the open fields.
terrier: a detailed description of a scattered holding giving the area, location, and often naming the neighbouring holders. Furlong and great field names are usually given.
tilth: an area used for arable cultivation in a particular year.
township: the smallest unit of land possessing a complete and independent field system.

vill: a medieval settlement.
villein: a medieval farmer holding an oxgang or yardland.
virgate: one-quarter of a hide in 1086; later, the Latin name for a yardland.

wrest plough: a plough that could be reversed so that the surface remained flat and was not ridged up; used primarily in Kent and East Anglia.

yard: 0.91 metres.
yardland: the villein or customary holding in Southern England. In the Central Region a series of scattered lands, commonly about sixty, totalling around twenty-five acres, but variable. Other regional names were eriung, ferling, husbandland, *tenementum*, wholeland, wist, and yokeland (as well as oxgang).

Bibliography

Record society publications have been treated as journals for bibliographical purposes.

Aalbersberg, G., and Brown, T. (2011), 'The environment and context of the Glastonbury lake village: a re-assessment', *Journal of Wetland Archaeology*, 10: 136–51.

Abrams, J., and Ingham, D. (2008), *Farming on the Edge*, East Anglian Archaeology, 123 (Bedford: Albion Archaeology).

Adams, K. A. (1990), 'Monastery and village at Crayke North Yorkshire', *Yorkshire Archaeological Journal*, 62: 29–50.

Addy, S. O. (1922), 'Taxation by the oxgang', *Derbyshire Archaeological Journal*, 44: 58–91.

Addyman, P. V. (1964), 'A dark-age settlement at Maxey, Northants.', *Medieval Archaeology*, 8: 20–73.

Alcock, N. N. (1970), 'An East Devon manor in the later Middle Ages', *Report and Transactions of the Devonshire Association*, 102: 151–7.

Allison, K. J. (1957), 'The sheep-corn husbandry of Norfolk in the sixteenth and seventeenth centuries', *Agricultural History Review*, 5: 12–30.

Andrews, J. H. B. (1962), 'Chittlehamholt', *Report and Transactions of the Devonshire Association*, 94: 233–331.

Arnold, C. J., and Wardle, P. (1981), 'Early medieval settlement patterns in England', *Medieval Archaeology*, 25: 145–9.

Ashikaga, K. and Henstock, A. (1996), 'The East Bridgford estate maps of 1612–14', *Transactions of the Thoroton Society of Nottinghamshire*, 100: 77–93.

Ashworth, H., and Turner, C. (1999), 'Land off Dando Close, Wollaston, Northamptonshire: evaluation report', Heritage Network 224; NS, Report No. 65; copy at Northamptonshire HER.

Aston, M. (ed.) (1988), *Aspects of the Medieval Landscape of Somerset* (Taunton: Somerset County Council).

Aston, M. (1994), 'The medieval settlement studies in Somerset', in M. Aston and C. Lewis (eds.), *The Medieval Landscape of Wessex*, Oxbow Monograph, 46 (Oxford: Oxbow), 219–37.

Aston, M., Austin, D., and Dyer, C. (eds.) (1989), *Rural Settlements in Medieval England* (Oxford: Blackwell).

Atkin, M. A. (1985), 'Some Settlement Patterns in Lancashire', in Hooke 1985b, 170–85.

Atkin, M. A. (1993), 'Sillfield, Preston Patrick: a double-oval type of field pattern,' *Transactions of the Cumberland and Westmorland Antiquarian and Archaeological Society*, NS, 93: 145–53.

Atkinson, J. C. (ed.) (1878–81), *Cartularium Abbathiae de Whiteby, ordinis S. Benedicti, fundatae anno MLXXVIII*, 2 vols., Surtees Society, 69 and 72.

Atkinson, J. C. (ed.) (1886–7), *The Coucher Book of Furness Abbey*, vol. 1, pts. 1–3, Chetham Society, NS, 9, 11, and 14.

Attenborough, F. L. (ed. and trans.) (1922), *The Laws of the Earliest English Kings* (Cambridge: Cambridge University Press).

Attwood, G. (1963), 'A study of Wiltshire water meadows', *Wiltshire Archaeological and Natural History Magazine*, 58: 403–13.

Ault, W. O. (1972), *Open-Field Farming in Medieval England* (London: Allen and Unwin).

Bailey, F. A. (1937), *A Selection from the Prescot Court Leet and Other Records, 1447–1600*, Lancashire and Cheshire Record Society, 89.

Bailey, M. (1990), 'Sand into gold', *Agricultural History Review*, 38: 40–57.

Bailey, M. (2007), *Medieval Suffolk* (Woodbridge: Boydell).

Baker, A. H. R. (1963), 'The field system of an East Kent parish (Deal)', *Archaeologia Cantiana*, 78: 96–117.

Baker, A. H. R. (1964), 'Open fields and partible inheritance on a Kent manor', *Economic History Review*, 17: 1–23.

Baker, A. H. R. (1965a), 'Some fields and farms in medieval Kent', *Archaeologia Cantiana*, 80: 152–74.

Baker, A. H. R. (1965b), 'Field patterns in seventeenth-century Kent', *Geography*, 50: 18–30.

Baker, A. H. R. (1966), 'Field systems in the Vale of Holmesdale', *Agricultural History Review*, 14: 1–24.

Baker, A. H. R. (1970), 'Contracting arable lands in 1341', *Bedfordshire Historical Record Society*, 49: 7–18.

Baker, A. H. R. (1973), 'A relatively neglected field form: the headland ridge', *Agricultural History Review*, 21: 47–52

Baker, A. R. H., and Butlin, R. A. (eds.) (1973), *Field Systems in the British Isles* (Cambridge: Cambridge University Press).

Bannister, A. T. (1929), *A Transcript of "The Red Book": A Detailed Account of the Hereford Bishopric Estates in the Thirteenth Century*, Camden Miscellany, 15, 1–33 (London: Royal Historical Society).

Barber, A., and Watts, M. (2008), 'Excavations at Saxon's Lode Farm, Ripple, 2001–2', *Transactions of the Worcestershire Archaeological Society*, 3rd ser., 21: 1–90.

Barnes, P. (1997), 'The adaptation of open-field farming in an East Nottinghamshire parish: Orston 1641–1793,' *Transactions of the Thoroton Society of Nottinghamshire*, 101: 125–32.

Barratt, D. M. (1955–71), *The Ecclesiastical Terriers of Warwickshire Parishes*, i: *Parishes A to L*; ii: *Parishes Lo to W*, Dugdale Society, 22 and 27.

Bates, E. H. (1899), 'The five hide unit in the Somerset Domesday', *Proceedings of the Somerset Archaeological and Natural History Society*, 45: 51–107.

Bates, E. H. (1912), [Review], *Proceedings of the Somerset Archaeological and Natural History Society*, 57: 130; see also ibid. (2003), 145: 6–7.

Beckwith, I. (1967), 'The remodelling of a common-field system', *Agricultural History Review*, 15: 108–12.

Bedingfeld, A. L. (1966), *Cartulary of Creake Abbey*, Norfolk Record Society, 35 (Norwich).

Beecham, H. A. (1956), 'A review of balks as strip boundaries in the open fields', *Agricultural History Review*, 4: 22–44.

Bell, M., and S. Limbrey (eds.) (1982), *Archaeological Aspects of Woodland Ecology*, BAR International Series, 146 (Oxford: British Archaeological Reports).

Beresford, M. W. (1947–52 and 1953–60), 'Glebe terriers and open-field Buckinghamshire', *Records of Buckinghamshire*, 15: 283–98 and 16: 5–28.

Beresford, M. W. (1948), 'Glebe terriers and open field Leicestershire', *Transactions of the Leicestershire Archaeological and Historical Society*, 24: 77–126.

Beresford, M. W. (1948–51), 'Glebe terriers and the open field', *Yorkshire Archaeological Journal*, 37: 325–68.

Beresford, M. W. (1964), 'Dispersed and grouped settlement in medieval Cornwall', *Agricultural History Review*, 12: 13–27.

Beresford, M., and Hurst, J. G. (1971), *Deserted Medieval Villages* (London: Lutterworth Press).

Beresford, M. W., and St Joseph, J. K. S. (1979), *Medieval England*, 2nd edn (Cambridge: Cambridge University Press).

Bettey, J. (ed.) (2007), *Archives and Local History in Bristol and Gloucestershire: Essays in Honour of David Smith* (Bristol: Bristol and Gloucestershire Archaeological Society).

Bettey, J. H. (1982), 'Sheep farming in Dorset during the seventeenth century', *Proceedings of the Dorset Natural History and Archaeological Society*, 102: 1–5.

Bigmore, P. (1979), *The Bedfordshire and Huntingdonshire Landscape* (London: Hodder & Stoughton).

Birch, W. de Gray, (1885–93), *Cartularium Saxonicum,* 3 vols. (London: Whiting and Co.).

Birrell, J. (1979), 'Medieval agriculture', in VCH, *Staffordshire*, vi: 1–48.

Birrell, J. (1999), *The Forests of Cannock and Kinver: Select Documents 1235–1372*, Staffordshire Record Society, Collections for the History of Staffordshire, 4th ser., 18.

Bishop, T. M. (1935–6), 'Assarting and the growth of open fields', *Economic History Review*, 6: 13–29.

Bishop, T. M. (1938–9), 'The rotation of crops at Westerham, 1297–1350', *Economic History Review*, 8: 38–44.

Blair, W. J. (1991), *Early Medieval Surrey* (Stroud: Alan Sutton).

Blair, W. J. (1994), *Anglo-Saxon Oxfordshire* (Stroud: Alan Sutton).

Blake, E. O. (1962), *Liber Eliensis*, Camden 3rd Series, 92 (London: Royal Historical Society).

Bolton, D. K. (1980), 'Finchley', VCH, *Middlesex*, vi: 38–100.

Booth, P. H. W., and Dodd, J. P. (1979), 'The manor and fields of Frodsham', *Transactions of the Historic Society of Lancashire and Cheshire*, 128: 27–57.

Bouch, C. M. L., and Jones, G. P. (1961), *A Short Economic History of the Lake Counties, 1500–1830* (Manchester: Manchester University Press).

Bowen, H. C. (1961), *Ancient Fields* (London: British Association for the Advancement of Science).

Bowen, H. C., and Fowler, P. J. (eds.) (1978), *Early Land Allotment in the British Isles*, British Archaeological Reports, British Series, 48 (Oxford: British Archaeological Reports).

Bowman, P., and Liddle, P. (2004), *Leicestershire Landscape* (Leicester: Leicestershire County Council).

Bradford, J. (1957), *Ancient Landscapes* (London: Bell).

Brandon, P. F. (1971), 'Agriculture and the effects of floods and weather at Barnhorne', *Sussex Archaeological Collections*, 109: 69–93.

Bridgeman, E. R. O. (1899), *Weston under Lizard*, Collections for a History of Staffordshire, NS, 2 (Stafford: William Salt Archaeological Society).

Bridgeman, G. T. O. (1881), *Collections for a History of Staffordshire*, 2: 118–20.

Bridges, J., (1791), *The History and Antiquities of Northamptonshire*, ed. P. Whalley (Oxford: D. Prince and J. Cooke).

Briston, M., and Halliday, T. M. (2008), *The Pilsgate Manor of the Sacrist of Peterborough*, Northamptonshire Record Society, 43.

Britnell, R. H. (1983), 'Agriculture in a region of old enclosure, 1185–1500,' *Nottingham Medieval Studies*, 27: 38–55.

Britnell, R. H. (1988), 'The fields and pastures of Colchester, 1280–1350', *Essex Archaeology and History*, 19: 159–65.

Britnell, R. H. (2004), 'Fields, farms and sun-division in a moorland region', *Agricultural History Review*, 52: 26–31.

Brown, A. G. (2006), 'The environment of the Raunds area', in Parry 2006, 19–31.

Brown, W. (1889), *Cartularium Prioratus de Gyseburne*, Surtees Society, 86.

Brown, W. (1892), *Yorkshire Inquisitions of the Reigns of Henry II and Edward I*, Yorkshire Archaeological Society Record Series, 12.

Buckberry, J. L., and Hadley, D. M. (2001), 'Fieldwork at Chapel Road, Fillingham', *Lincolnshire History and Archaeology*, 36: 11–18.

Buckley, D. G. (ed.) (1980), *Archaeology in Essex to AD 1500*, CBA Research Report, 34 (London: Council for British Archaeology).

Bull, E. J. (1995), 'The bi-axial landscape of prehistoric Buckinghamshire', *Records of Buckinghamshire*, 35: 11–19.

Bush, R. J. E. (1978), 'Martock Hundred', in VCH, *Somerset*, iv: 78–109.

Buteux, S., and Chapman, H. (2009), *Where Rivers Meet: The Archaeology of Catholme and the Trent–Tame Confluence*, CBA Research Report, 161 (York: Council for British Archaeology).

Butlin, R. A. (1973), 'Field systems of Northumberland and Durham', in Baker and Butlin 1973, 93–144.

Campbell, B. M. S. (1980), 'Population change and the genesis of the commonfields on a Norfolk manor', *Economic History Review*, 33: 174–92.

Campbell, B. M. S. (1981a), 'Commonfield origins—the regional dimension', in Rowley 1981: 112–29.

Campbell, B. M. S. (1981b), 'The regional uniqueness of English field systems? Some evidence from eastern Norfolk', *Agricultural History Review*, 29: 16–28.

Campbell, B. M. S. (1981–3), 'The extent and layout of common fields in eastern Norfolk', *Norfolk Archaeology*, 38: 5–32.

Campbell, B. M. S. (2000), *English Seigniorial Agriculture, 1250–1450* (Cambridge: Cambridge University Press).

Casselden, P. (1987), 'Chartridge and Pednor hedgerows: a landscape survey', *Records of Buckinghamshire*, 29: 133–59.

Cave, T., and Wilson, R. A. (eds.) (1924), *The Parliamentary Survey of the Lands and Possessions of the Dean and Chapter of Worcester: Made in or about the Year 1649 in Pursuance of an Ordinance of Parliament for the Abolishing of Deans and Chapters*, Worcestershire Historical Society, 31.

Chapman, D., and Gardiner, M. (2005), 'Rethinking the early medieval settlement of woodlands', *Landscape History*, 27: 33–49.

Chapman, J., and Seeliger, S. (1997a), *A Guide to Enclosure in Hampshire 1700–1900*, Hampshire Record Series, 15 (Winchester: Hampshire County Council).

Chapman, J., and Seeliger, S. (1997b), *Formal and Informal Enclosures in Hampshire 1700–1900*, Hampshire Papers, 12 (Winchester: Hampshire County Council).

Chapman, V. (1953), 'Open fields in west Cheshire', *Transactions of the Historic Society of Lancashire and Cheshire*, 104: 35–59.

Chibnall, A. C. (1965), *Sherington: Fiefs and Fields of a Buckinghamshire Village* (Cambridge: Cambridge University Press).

Chibnall, A. C. (1979), *Beyond Sherington* (Chichester: Phillimore).

Chibnall, M. (1951), *Select Documents of the English Lands of the Abbey of Bec*, Camden 3rd Series, 73 (London: Royal Historical Society).

Christie, N., and Stamper, P. (eds.) (2011), *Rural Medieval Settlement: Britain and Ireland, AD 800–1600* (Oxford: Windgather Press).

Churchill, I. J., Griffin, R., and Hardman, F. W. (1956), *Calendar of Kent Feet of Fines, To the End* of *Henry III's Reign*, Kent Records, 15 (Ashford: Printed for the Records Branch by Headley Bros.).

Clark, G. N. (1925), 'Open fields and inclosure at Marston, Oxfordshire', in F. W. Weaver and G. N. Clark (eds.), *Churchwardens' Accounts of Marston, Spelsbury, Pyrton*, Oxfordshire Record Society, 6 (Supplement), 1–24.

Cobbett, W. (1822 [1930]), *Rural Rides*, ed. G. D. H. Cole and M. Cole, i (London: Peter Davies).

Cooper, G. M. (1853), 'Berwick parochial records', *Sussex Archaeological Collections*, 6: 232 and 240–1.

Corbett, W. J. (1897), 'Elizabethan village surveys', *Transactions of the Royal Historical Society*, NS, 11: 66–87.

Corbett, W. J., and Methold, T. T. (1900), 'The rise and devolution of the Manor of Hepworth, Suffolk,' *Proceedings of the Suffolk Institute of Archaeology*, 10: 19–48.

Costen, M. (1992), *The Origins of Somerset* (Manchester: Manchester University Press).

Cox, A., Fox, J., and Thomas, G. (1999), 'Sedgeford historical and archaeological research project, 1997 interim report', *Norfolk Archaeology*, 43: 172–7.

Cox, B. (1994), *The Place-Names of Rutland*, 3 vols. in 1, English Place-Name Society, 67–9 (Nottingham: English Place-Name Society).

Craster, H. H. E. (1909), *History of Northumberland*, ix (London: Northumberland County History Committee).

Craster, H. H. E. (1914), *History of* Northumberland, x (London: Northumberland County History Committee).

Crawford, O. G. S. (1937), *The Strip Map of Litlington* (London: HMSO).

Crawford, O. G. S., and Keiller, A. (1928), *Wessex from the Air* (Oxford: Clarendon Press).

Cromarty, D. (1966), *The Fields of Saffron Walden in 1400*, Essex Record Office Publications, 43 (Chelmsford: Essex County Council).

Crook, B. (1947), 'Newnham Priory: a rental of Biddenham manor, 1505–6', *Publications of the Bedfordshire Historical Record Society*, 25: 82–94.

Crowson, A., Lane, T., Penn, K., and Trimble, D. (2005), *Anglo-Saxon Settlement of Eastern England*, Lincolnshire Archaeological and Heritage Reports Series, 7 (Heckington: Heritage Trust for Lincolnshire).

Crummy, P. (1979), 'Crop marks at Gosbecks, Colchester', *Aerial Archaeology*, 4: 77–82.

Cunliffe, B. (1972), 'Saxon and medieval settlement-pattern in the region of Chalton, Hampshire', *Medieval Archaeology*, 16: 1–12.

Cunningham, W. (1910a), *The Growth of English Industry and Commerce*, 5th edn (Cambridge: Cambridge University Press).

Cunningham, W. (1910b), 'Common rights at Cottenham and Stretham in Cambridgeshire', Camden Society, Miscellanea, XII, item no. 4: 169–287.

Curwen, E. E. (1927), 'Prehistoric agriculture in Britain', *Antiquity*, 1: 261–89.

Curwen, J. F. (ed.) (1923), *Records Relating to the Barony of Kendale*, i, Cumberland and Westmorland Antiquarian and Archaeological Society, Record Series, 4.

Dahlman, C. J. (1981), *The Open Field System and Beyond* (Cambridge: Cambridge University Press).

Darby, H. C. (1977), *Domesday England* (Cambridge: Cambridge University Press).

Darby, H. C., and Terrett, I. B. (1971), *The Domesday Geography of Midland England* (Cambridge: Cambridge University Press).

Darlington, R. R. (1945), *Cartulary of Darley Abbey* (Kendal: T. Wilson & Son).

Davison, A. (2003), 'The archaeology of West Acre', *Norfolk Archaeology*, 44: 202–21.

Davison, A., and Cushion, B. (1998–2001), 'The archaeology of the Hargham estate', *Norfolk Archaeology*, 43: 257–74.

Deegan, A., and Foard, G. (2007), *Mapping Ancient Landscapes in Northamptonshire* (Swindon: English Heritage).

Dendy, F. W. (1894), 'The ancient farms of Northumberland', *Archaeologia Aeliana*, NS, 16: 121–56.

Denman, D. R., Roberts, R. A., and Smith, H. J. F. (1967), *Commons and Village Greens* (London: L. Hill).

DeWindt, E. B. (1976), *Liber Gersumarum of Ramsey Abbey: A Calendar and Index of B. L. Harley MS 445* (Toronto: Pontifical Institute of Mediaeval Studies).

Dils, J. (1998), *Historical Atlas of Berkshire* (Reading: Berkshire Record Society).

Dodds, M. H. (1926), *History of Northumberland*, xii (London: History of Northumberland Committee).

Dodgshon, R. A. (1980), *The Origin of British Field Systems* (London: Academic Press).

Dottie, R. G. (1986), 'Seventeenth-century Chidwall', *Transactions of the Historic Society of Lancashire and Cheshire*, 135: 18–31.

Douglas, D. C. (1927), *The Social Structure of Medieval East Anglia* (Oxford: Clarendon Press).

Draper, S. (2009), 'Continuity or congruity?', in A. Smith (ed.), *The Last of the Britons* (Taunton: Council for British Archaeology South West and Somerset Archaeological and Natural History Society), 28–36.

Drury, P., and Rodwell, W. (1978), 'Investigations at Asheldham, Essex', *Antiquaries Journal*, 58: 133–51.

Drury, P., and Rodwell, W. (1980), 'Settlement in the Later Iron Age and Roman periods', in Buckley 1980, 59–75.

Du Boulay, F. R. H. (1959), 'Late-continued demesne farming at Otford', *Archaeologia Cantiana*, 73: 116–24.

Du Boulay, F. R. H. (1961a), *Medieval Bexley* ([Bexleyheath]: Bexley Corporation Public Libraries).

Du Boulay, F. R. H. (1961b), 'Denns, droving and danger', *Archaeologia Cantiana*, 76: 75–87.

Du Boulay, F. R. H. (1966), *The Lordship of Canterbury* (London: Nelson).

Dyer, C. (1981), *Warwickshire Farming, 1349–c.1520*, Dugdale Society Occasional Paper, 27.

Dyer, C. (1987), 'The rise and fall of a medieval village: Little Aston', *Transactions of the Bristol and Gloucester Archaeological Society*, 105: 165–81.

Dyer, C. (1991), *Hanbury: Settlement and Society in a Woodland Landscape*, Leicester University Department of English Local History Occasional Papers, 4th ser., 4 (Leicester: Leicester University Press).

Dyer, C. (1995), 'Sheepcotes: evidence for medieval sheepfarming', *Medieval Archaeology*, 39: 136–64.

Dyer, C. (2002), 'Villages and non-villages in the medieval Cotswolds', *Transactions of the Bristol and Gloucestershire Archaeological Society*, 120: 7–16.

Dyer, C. (2007), 'Landscape and Society in Bibury, Gloucestershire', in Bettey 2007, 62–77.

Dymond, D. P. (1973–6), 'The parish of Walsham-le-Willows and their medieval background: two Elizabethan surveys', *Proceedings of the Suffolk Institute of Archaeology*, 33: 195–210.

Dymond, D. P. (1995), *The Register of Thetford Priory, 1518–1540*, 2 vols., Norfolk Record Society, 59–60.

Dymond, D., and Martin, E. (1999), *An Historical Atlas of Suffolk*, 3rd edn (Ipswich: Suffolk County Council and Suffolk Institute of Archaeology and History).

Ekwall E. (1974), *Concise Oxford Dictionary of English Place-Names*, 4th edn (Oxford: Clarendon Press).

Eland, G. (ed.) (1946), 'Wootton Underwood in 1657', *Records of Buckinghamshire*, 14: 133–48.

D'Elboux, R. H. (ed.) (1947), *Survey of the Manor of Robertsbridge*, Sussex Record Society, 47.

Ellaby, R. (2000), 'The Horley demesne of Reigate Priory', *Surrey Archaeological Collections,* 87, 145–55, plan.

Elliot, G. C. (1959), 'The system of cultivation and evidence for enclosure in Cumberland open fields in the sixteenth century', *Transactions of the Cumberland and Westmorland Antiquarian and Archaeological Society*, NS, 59: 85–104.

Elliot, G. C. (1960), 'The enclosure of Aspatria', *Transactions of the Cumberland and Westmorland Antiquarian and Archaeological Society*, NS, 60: 97–108.

Elliot, G. C. (1973), 'The field systems of Northwest England', in Baker and Butlin 1973, 41–92.

Elrington, C. R. (2003), *Abstracts of Feet of Fines for Gloucestershire, 1199–1299*, Gloucestershire Record Series, 16 ([Gloucester]: Bristol and Gloucestershire Archaeological Society).

Elvey, E. M. (1953–60), 'Buckinghamshire in 1086', *Records of Buckinghamshire*, 16: 342–62.

Elvey, E. M. (1961–5), 'The abbot of Missenden's estates in Chalfont St. Peter', *Records of Buckinghamshire*, 17: 20–40.

Evans, C., and Standring, R. (2012), 'A landscape corridor: A14 improvements investi-gations', *Proceedings of the Cambridge Antiquarian Society*, 101: 81–104.

Everson, P. (1977), 'Excavations in the vicarage garden at Brixworth, 1972', *Journal of the British Archaeological Association*, 130: 56–122.

Faith, R. (1996), 'The topography and social structure of a small soke in the Middle Ages: The Sokens, Essex', *Essex Archaeology and History*, 27: 206–12.

Faith, R. (1997), *The English Peasantry and the Growth of Lordship* (London: Leicester University Press).

Faith, R. (2006), 'Worthys and enclosures', *Medieval Research Group Annual Report*, 21: 9–14.

Faith, R. (2007), 'Worthy farms on the edge of Dartmoor', *Medieval Research Group Annual Report*, 22: 57.

Farnham, G. F. (1930–3), *Leicestershire Medieval Village Notes*, 6 vols. (Leicester: Privately printed).

Farrer, W. (1897), *The Court Rolls of the Honor of Clitheroe*, i (Manchester: Emmott & Co.).

Farrer, W. (1899), *Final Concords of the County of Lancaster,* Lancashire Record Society, 39 (London).

Farrer, W. (1902), *Lancashire Pipe Rolls* (Liverpool: H. Young & Sons).

Farrer, W. (1907), *Lancashire Inquests, Extents and Feudal Aids*, Part 2: *AD 1310–AD 1333*, Lancashire Record Society, 54.

Farrer, W. (1914), *Early Yorkshire Charters*, i (Edinburgh: Printed for the editor by Bal-lantine, Hanson & Co.).

Farrer, W. (1915a), *Lancashire Inquests, Extents, and Feudal Aids, AD 1313–1355*, Part 3: *AD 1313–AD 1355*, Lancashire Record Society, 70.

Farrer, W. (1915b), *Early Yorkshire Charters*, ii (Edinburgh: Printed for the editor by Ballantine, Hanson & Co.).

Farrer, W. (1916), *Early Yorkshire Charters*, iii (Edinburgh: Printed for the editor by Ballantine, Hanson & Co.).

Ferguson, R. S. (1877), *Miscellany Accounts of the Diocese of Carlisle with the Terriers* (London: G. Bell).

Field, R. K. (2004), *Court Rolls of Elmley Castle, Worcestershire, 1347–1564*, Worces-tershire Record Society, NS, 20.

Fieldhouse, R. T. (1980), 'Agriculture in Wensleydale from 1600', *Northern History*, 16: 169–95.

Figg, W. (1850), 'Manorial customs of Southese-with-Heighton, near Lewes', *Sussex Archaeological Collections*, 3: 249–52.

Figg, W. (1851), 'Tenantry customs in Sussex, the Drinker Acres', *Sussex Archaeological Collections*, 4: 305–8.

Finberg, H. P. R. (1951), *Tavistock Abbey: A Study in the Social and Economic History of Devon* (Cambridge: Cambridge University Press).

Finberg, H. P. R. (1952), in Hoskins and Finberg 1952, 265–88.

Finberg, H. P. R. (1961), *The Early Charters of the West Midlands* (Leicester: Leicester University Press).

Finberg, H. P. R. (1964), *The Early Charters of Wessex* (Leicester: Leicester University Press).

Finberg, H. P. R. (1971), 'Ayshford and Boehill', *Report and Transactions of the Devon-shire Association*, 103: 19–24.

Finberg, H. P. R. (1972), *The Agricultural History of England and Wales, AD 43–1042*, i.2 (Cambridge: Cambridge University Press).

Finberg, H. P. R. (1975), *The Gloucestershire Landscape* (London: Hodder and Stoughton).

Finberg, H. P. R., and Hoskins, W. G. (1952), *Devonshire Studies* (London: Cape).

Finch, M. E. (1956), *The Wealth of Five Northamptonshire Families*, Northamptonshire Record Society, 19.

Fisher, H. A. L. (1911), *The Collected Papers of Frederick William Maitland*, ii (Cam-bridge: Cambridge University Press), 344–5.

Fisher, H. A. L. (1928), *The Collected Papers of Paul Vinogradoff*, i (Oxford: Clarendon Press), 286–96.

Fisher, J. L. (1951), 'The Leger Book of St. John's Abbey, Colchester', *Transactions of the Essex Archaeological Society*, 24: 77–127.

Fleming, L. (1960), *Cartulary of Boxgrove Priory*, Sussex Record Society, 59.

Foard, G. (1978), 'Systematic fieldwalking and the investigation of Saxon settlement in Northamptonshire', *World Archaeology*, 9: 357–74.

Foard, G., Hall, D., and Partida, T. (2009), *Rockingham: An Atlas of the Medieval and Early Modern Landscape*, Northamptonshire Record Society, 44.

Ford, S. (1995), 'The excavation of a Saxon settlement and a Mesolithic flint scatter at Northampton Road, Brixworth, Northamptonshire', *Northamptonshire Archaeology*, 26: 79–108.

Ford, S., and Hazel, A. (1990), 'Trial trenching of a Saxon pottery scatter at North Stoke, South Oxfordshire, 1988', *Oxoniensia*, 55: 169.

Foster, C. W. (1920), *Final Concords of the County of Lincoln*, Lincolnshire Record Society, 17.

Fowkes, D. V. (1977), 'The Breck system of Sherwood Forest', *Transactions of the Thoroton Society of Nottinghamshire*, 81: 55–61.

Fowkes, D. V., and Potter, G. R. (1988), *William Senior's Survey of the Estates of the First and Second Earls of Devonshire c.1600–28*, Derbyshire Record Society, 13.

Fowler, G. H. (1919), *A Calendar of the Feet of Fines for Bedfordshire*, Bedfordshire Historical Record Society, 6.

Fowler, G. H. (1926), *Cartulary of the Priory of Dunstable*, Bedfordshire Historical Record Society, 10.

Fowler, G. H. (1930), *Cartulary of the Abbey of Old Warden*, Bedfordshire Historical Record Society, 13.

Fowler, G. H. (1935), *Records of Harrold Priory*, Bedfordshire Historical Record Society, 17.

Fowler, G. H. (1936), *Four Pre-Enclosure Village Maps*, Bedfordshire Historical Record Society, Quarto Memoirs, 2.

Fowler, J. T. (1891–3), *The Coucher Book of Selby*, 2 vols., Yorkshire Archaeological Society Record Series, 10 and 13.

Fowler, P. J. (1966), 'Two finds of Saxon domestic pottery in Wiltshire', *Wiltshire Arch-aeological and Natural History Magazine*, 61: 31–7.

Fowler, P. J. (2002), *Farming in the First Millennium AD* (Cambridge: Cambridge University Press).

Fox, H. S. A. (1972), 'Field systems of east and south Devon, Part I: East Devon', *Report and Transactions of the Devonshire Association*, 104: 81–135.

Fox, H. S. A. (1973), in M. Havinden (ed.), *Husbandry and Marketing in the South-West, 1500–1800*, Exeter Papers in Economic History, 8 (Exeter: Department of Economic History, University of Exeter), 19–38.

Fox, H. S. A. (1975), 'The chronology of enclosure and economic development in medieval Devon,' *Economic History Review*, 2nd ser., 28: 181–201.

Fox, H. S. A. (1981), 'Approaches to the adoption of the Midland System', in Rowley 1981, 64–111.

Fox, H. S. A. (1986), 'The alleged transformation from two-field to three-field systems in medieval England', *Economic History Review*, 2nd ser., 39: 526–48.

Fox, H. S. A., and Padel, O. (2000), *The Cornish Lands of the Arundells of Lanherne, 14th to 16th Centuries*, Devon and Cornwall Record Society, NS, 41.

Freeman, J. (1987), 'Semley', in VCH, *Wiltshire*, xiii: 66–79.

Fullbrook-Leggatt, L. E. W. O. (1946–8), 'Medieval Gloucester: II', *Transactions of the Bristol and Gloucestershire Archaeological Society*, 67: 217–306.

Fyfe, R. M., Brown, A. G., and Rippon, S. J. (2003), 'Mid- to late-Holocene vegetation history of Greater Exmoor, UK: estimating the spatial extent of human-induced vegetation change', *Vegetation History and Archaeobotany*, 12: 215–32.

Fyfe, R. M., and Rippon, D. (2004), in R. and G. Collins (eds.), *Debating Late Antiquity in Britain*, British Archaeological Reports, British Series, 356 (Oxford: British Archaeological Reports).

Gardiner, M. (1984), 'Saxon settlement and land division in the western Weald', *Sussex Archaeological Collections*, 122, 75–83.

Gardiner, M. (2011), 'South-East England: Forms and Diversity in Medieval Rural Settlement', in Christie and Stamper 2011, 100–17.

Gawne, E. (1970), 'Field patterns in Widdecombe Parish and the Forest of Dartmoor', *Report and Transactions of the Devonshire Association*, 102: 49–69.

Gaydon, A. T. (1959), *The Taxation of 1297: A Translation of the Local Rolls of Assessment for Barford, Biggleswade and Flitt Hundreds, and for Bedford, Dunstable, Leighton Buzzard and Luton*, Bedfordshire Historical Record Society, 39.

Gelling, M. (1968), 'The Charter Bounds of Aescebyrig and Ashbury,' *Berkshire Archaeological Journal*, 54: 13–38.

Gelling, M. (1976), *The Place-Names of Berkshire*, Part 3, English Place-Name Society, 51 (Cambridge: Cambridge University Press).

Gerard, C., and Aston, M. (2007), *Shapwick Project, Somerset: A Rural Landscape Explored*, Society for Medieval Archaeology, Monograph 25 (Leeds).

Godber, J. (1968), 'The Travel Journal of Philip Yorke 1744–63', *Bedfordshire Historical Record Society*, 47: 125–63.

Goddard, C. V. (1906), 'Customs of the Manor of Winterbourn Stoke', *Wiltshire Archaeological and Natural History Magazine*, 34: 208–15.

Godfrey, W. H. (1928), *The Book of John Rowe 1597–1622*, Sussex Record Society, 34.

Godman, P. S. (1911), 'Rolls from the Manor of Wiston', *Sussex Archaeological Collections*, 54: 130–82.

Gomme, G. L. (1890), *The Village Community* (London: Walter Scott).

Göransson, S. (1961), 'Regular open-field pattern in England, and the Scandinavian Solskifte', *Geografiska Annaler*, 43: 80–104.

Gover, J. E. B., Mawer, A., and Stenton, F. M. (1931–2), *The Place-Names of Devon*, 2 vols., English Place-Name Society, 8–9 (Cambridge: Cambridge University Press).

Gover, J. E. B., Mawer, A., and Stenton, F. M. (1939), *The Place-Names of Wiltshire*, English Place-Name Society, 16 (Cambridge: Cambridge University Press).

Graham, T. H. B. (1908), 'The common fields of Hayton', *Transactions of the Cumberland and Westmorland Antiquarian and Archaeological Society*, NS, 8: 340–51.

Graham, T. H. B. (1910), 'The townfields of Cumberland', *Transactions of the Cumberland and Westmorland Antiquarian and Archaeological Society*, NS, 10: 118–34.

Graham, T. H. B. (1913), 'The townfields of Cumberland, Part II', *Transactions of the Cumberland and Westmorland Antiquarian and Archaeological Society*, NS, 13: 1–30.

Graham, T. H. B. (1934), *The Barony of Gilsland, Lord William Howard's Survey Taken in 1603*, Cumberland and Westmorland Antiquarian and Archaeological Society, Extra Series (Kendal: Printed by Titus Wilson & Son).

Grainger, F., and Collingwood, W. G. (1929), *The Register and Records of Holm Cultram* (Kendal: T. Wilson & Son).

Gray, H. L. (1915), *English Field Systems* (Cambridge, MA: Harvard University Press).

Green, C. (1972), 'Castor', *Medieval Archaeology*, 16: 158.

Greenwell, W. (1852), *The Boldon Buke*, Surtees Society, 25.

Greenwell, W. (1872), *Feodarium Prioratus Dunelmensis*, Surtees Society, 58.

Gretton, R. H. (1910), 'Lot-meadow customs at Yarnton, Oxon.', *Economics Journal*, 20: 38–45.

Gretton, R. H. (1912), 'Historical notes on the lot-meadow customs at Yarnton, Oxon.', *Economics Journal*, 22: 53–62.

Griffiths, M. (1980), 'Kirtlington manor court 1500–1650', *Oxoniensia*, 45: 260.

Grundy, F. G. (1934), 'The Saxon charters of Dorset', *Proceedings of the Dorset Natural History and Archaeological Society*, 56: 127–8.

Gurney, F. G. (1946), 'An agricultural agreement of the year 1345 at Mursley and Dunton—with a note upon Walter of "Henley"', *Records of Buckinghamshire*, 14: 245–64.

Hailstone, E. (1873), *History of the Parish of Bottisham* (Cambridge: Cambridge Antiquarian Society).

Hale, W. H. (1858), *The Domesday of St. Paul's of the Year M.CC.XXII*, Camden Society, 69 (London).

Hall, D. (1972), 'Modern surveys of medieval field systems', *Bedfordshire Archaeological Journal*, 7: 53–66.

Hall, D. (1973), '"Newton-in-the-Willows", in "Medieval Britain in 1972"', *Medieval Archaeology*, 17: 147 [note].

Hall, D. (1974), 'Medieval pottery from the Higham Ferrers Hundred', *Journal of the Northampton Museum*, 10: 38–58.

Hall, D. (1975), 'Hartwell, Northamptonshire, a parish survey', *CBA Group 9, Newsletter*, 5: 7–9.

Hall, D. (1977), *Wollaston: Portrait of a Village* (Wollaston: Wollaston Society).

Hall, D. (1979), 'New evidence of modifications of open-field systems', *Antiquity*, 53: 222–4.

Hall, D. (1981a), 'Fieldwork and documentary evidence', in K. Biddick (ed.) *Archaeological Approaches to Medieval Europe* (Kalamazoo, MI: Western Michigan University), 43–68.

Hall, D. (1981b), 'The changing landscape of the Cambridgeshire silt fens', *Landscape History*, 3: 37–49 (plan, pp. 44–5).

Hall, D. (1982), *Medieval Fields* (Aylesbury, Shire Publications, repr. Oxford, 2010).

Hall, D. (1983), 'Fieldwork and fieldbooks', in Roberts and Glasscock 1983, 115–31.

Hall, D. (1985), 'Late Saxon topography and early medieval estates', in Hooke 1985b, 61–9.

Hall, D. (1988), 'The Late Saxon countryside: villages and their fields', in Hooke 1988, 99–122.

Hall, D. (1989), 'Field systems and township structure', in Aston et al. 1989, 191–205.

Hall, D. (1995), *The Open Fields of Northamptonshire*, Northamptonshire Record Society, 38. [Available on the Society's Website <http://www.northamptonshirerecordsociety.org.uk>.]

Hall, D. (1996), *Cambridgeshire Survey, the Isle of Ely and Wisbech* (Fenland Project 10), East Anglian Archaeology, 79 (Cambridge: Cambridgeshire Archaeological Committtee).

Hall, D. (2001), *Turning the Plough* (Northampton: English Heritage and Northamptonshire County Council).

Hall, D. (2006), 'Aynho fields, open and enclosed', *Northamptonshire Past and Present*, 56: 7–22.

Hall, D. (2009), 'Open fields' and 'Medieval atlas', in Foard et al. 2009, 29–36 and 73–101.

Hall, D. (2012), 'Field Systems and Landholdings', in Wrathmell 2012, 278–89.

Hall, D. (2013), 'Forest and woodland' and 'Open fields', in Partida et al. 2013, 17–31 and 32–50; maps 1M–86M.

Hall, D., and Coles, J. (1994), *Fenland Survey: An Essay in Landscape and Persistence* (London: English Heritage).

Hall, D., and Hutchings, J. (1972), 'The distribution of archaeological sites between the Nene and Ouse valleys', *Bedfordshire Archaeological Journal*, 7: 1–16.

Hall, D., and Martin, P. W. (1979), 'Brixworth, Northamptonshire: an intensive archaeological survey,' *Journal of the British Archaeological Association*, 122: 1–6.

Hall, D., and Martin, P. W. (1980), 'Fieldwork survey of the Soke of Peterborough', *Durobrivae*, 8: 13–14.

Hall, D., and Palmer, R. (1996), in Hall 1996.

Hall, H. (ed.) (1903), *The Pipe Roll of the Bishopric of Winchester for the Fourth Year of the Pontificate of Peter des Roches, 1208–1209* (London: P. S. King & Son, for the London School of Economics).

Hallam, H. E. (1963), 'The fen bylaws of Spalding and Pinchbeck', *Reports and Papers of the Lincolnshire Architectural and Archaeological Society*, 10: 40–56.

Hallam, H. E. (1965), *Settlement and Society: A Study of the Early Agrarian History of South Lincolnshire* (Cambridge: Cambridge University Press).

Hamerow, H. F. (1991), 'Settlement mobility and the Middle Saxon shift', *Anglo-Saxon England*, 20: 1–19.

Hamerow, H. F. (1993), *Excavations at Mucking*, ii: *The Anglo-Saxon Settlement*, English Heritage Archaeological Report, 21 (London: English Heritage).

Hamerow, H. F. (2002), *Early Medieval Settlements* (Oxford: Oxford University Press).

Hamerow, H. F. (2012), *Rural Settlements and Society in Anglo-Saxon England* (Oxford: Oxford University Press).

Hamerow, H. F., Hinton, D. A., and Crawford, S. (2010), *The Oxford Handbook of Anglo-Saxon Archaeology* (Oxford: Oxford University Press).

Hanna, K. A. (1988), *The Cartulary of Southwick Priory*, Hampshire Record Society, 9.

Hardy, A., Mair, C. B., and Williams, R. J. (2007), *Death and Taxes: The Archaeology of a Middle Saxon Estate Centre at Higham Ferrers* (Oxford: Oxford Archaeology).

Hardy, M. J., and Martin, E. A. (1986), 'Field surveys' *Proceedings of the Suffolk Institute of Archaeology*, 36: 147–50, 232–5, 315–17, with plans.

Harper-Bill, C. (1981), *Blythburgh Priory Cartulary*, Part ii, Suffolk Record Society, 2.

Harris, A. (1955), 'The 'land' and oxgang in the East Riding of Yorkshire', *Yorkshire Archaeological Journal*, 38: 529–35.

Harris, A. (1959), *The Open Fields of East Yorkshire*, East Yorkshire Local History Society Publication No. 9 (Hull: East Yorkshire Local History Society).

Harris, A. (1961), *The Rural Landscape of the East Riding of Yorkshire, 1700–1850* (East Ardsley: S. R. Publishers).

Hart, C. R. (1966), *The Early Charters of Eastern England* (Leicester: Leicester University Press).

Hart, C. R. (1970), *The Hidation of Northamptonshire*, Department of English Local History, Occasional Papers, 2nd ser., 3 (Leicester: Leicester University Press).

Hart, C. R. (1974), *The Hidation of Cambridgeshire*, Department of English Local History, Occasional Papers, 2nd ser., 6 (Leicester: Leicester University Press).

Hart, C. R. (1975), *The Early Charters of Northern England and the North Midlands* (Leicester: Leicester University Press).

Hart, W. H. (1863–7), *Historia et Cartularium Monasterii Sancti Petri Gloucestrii*, 3 vols., Rolls Series, 3 (London: HMSO; Wiesbaden: Kraus facsimile, 1971).

Hart, W. H., and Lyons, P. A. (1884–93), *Cartulary of Ramsey Abbey*, 3 vols., (London: Longman; [Wiesbaden]: Kraus Reprint, 1965).

Harvey, M. (1978), *The Morphological and Tenurial Structure of a Yorkshire Townships: Preston in Holderness, 1066–1750*, Queen Mary College Occasional Paper in Geography, 13 (London: Queen Mary College).

Harvey, M. (1980), 'Regular field and tenurial arrangements in Holderness, Yorkshire', *Journal of Historical Geography*, 6: 3–16.

Harvey, M. (1981), 'The origin of planned field systems in Holderness, Yorkshire', in Rowley 1981, 184–201.

Harvey, M. (1982), 'Regular open-field systems on the Yorkshire Wolds', *Landscape History*, 4: 29–39.

Harvey, P. D. A. (1965), *A Medieval Oxfordshire Village: Cuxham 1240–1400* (Oxford: Oxford University Press).

Haverfield, F. (1918), 'Centuriation in Roman Britain', *English Historic Review*, 33: 289–96.

Hawkes, S., and Gray, M. (1969), 'The early Anglo-Saxon settlement at New Wintles Farm, Eynsham', *Oxoniensia*, 34: 1–4.

Hayes, P. P., and Lane, T. W. (1992), *Lincolnshire Survey, the South-West Fens*, East Anglian Archaeology, 55 (Sleaford: Heritage Trust for Lincolnshire).

Heaton, M. (2009), *A Farming History of Spratton, 1766–1914* (Spratton: Spratton Local History Society).

Henman, W. N. (1947), 'Newnham Priory: a Bedford rental, 1506–7', *Bedfordshire Historical Record Society*, 25: 63.

Hesse, M. (1998), 'Medieval field systems and land tenure in South Creake, Norfolk', *Norfolk Archaeology*, 43: 79–97.

Hewlett, G. (1973), 'Reconstructing a historical landscape from field and documentary evidence: Otford in Kent', *Agricultural History Review*, 21: 94–110.

Hey, D. (1974), *An English Rural Community: Myddle under the Tudors and Stuarts* (Leicester: Leicester University Press).

Hey, G. (2004), *Yarnton: Saxon and Medieval Settlement and Landscape from 1989 to 1998* (Oxford: Oxford Archaeology).

Higham, N. J., and Ryan, M. J. (eds.) (2010), *The Landscape Archaeology of Anglo-Saxon England* (Woodbridge: Boydell).

Higham, N. J., and Ryan, M. J. (eds.) (2011), *Place-Names, Language and the Anglo-Saxon Landscape* (Woodbridge: Boydell).

Hill, J. H. (1867), *The History of the Parish of Langton* (Leicester: Printed for the subscribers by Ward and Sons).

Hilton, R. H. (1954), 'Medieval agricultural history', in VCH, *Leicestershire*, ii: 145–98.

Hilton, R. H. (1959), 'Old enclosure in the West Midlands: a hypothesis about their late medieval development', *Géographie et histoire agraires: Annales de l'Est*, 21: 272–83.

Hilton, R. H. (1960), *Stoneleigh Leger Book*, Dugdale Society, 24 (Oxford: Printed for the Dugdale Society at the University Press).

Hilton, R. H. (1966), *A Medieval Society: The West Midlands at the End of the Thirteenth Century* (London: Weidenfeld and Nicolson).

Hinton, D. (1997), 'The "Scole-Dickleburgh field system" examined', *Landscape History*, 19: 5–12.

Historical Manuscripts Commission (1911), *Manuscripts of Lord Middleton, Wollaton Hall, Nottinghamshire* (London: HMSO).

Hoare, C. M. (1918), *The History of an East Anglian Soke* (Bedford: Bedfordshire Times).

Hobbs, S. (ed.) (1998), *The Cartulary of Forde Abbey*, Somerset Record Society, 85.

Hobbs, S. (ed.) (2003), *Wiltshire Glebe Terriers, 1588–1827*, Wiltshire Record Society, 56.

Hogan, M. P. (1988), 'Clays, culturae and their cultivation at Wistow', *Agricultural History Review*, 36: 117–31.

Holden, P., Herring, P., and Padel, O. J. (2010), *The Lanhydrock Atlas* (Fowey: Cornwall Editions).

Hollings, M. (1934–50), *The Red Book of Worcester*, 2 vols. (London: Printed for the Worcestershire Historical Society by Mitchell, Hughes and Clarke).

Homans, G. C. (1941), *English Villagers of the Thirteenth Century* (Cambridge, MA: Harvard University Press).

Hooke, D. (1981), 'Open-field agriculture—the evidence from the pre-Conquest charters of the West Midlands', in Rowley 1981, 39–63.

Hooke, D. (1983), *The Landscape of Anglo-Saxon Staffordshire: The Charter Evidence* (Keele: Department of Adult Education, University of Keele).

Hooke, D. (1985a), *The Anglo-Saxon Landscape: The Kingdom of the Hwicce* (Manchester: Manchester University Press).

Hooke, D. (1985b), 'Village development in the West Midlands', in D. Hooke (ed.) *Medieval Villages*, Oxford University Committee for Archaeology, Monograph No. 5 (Oxford: Oxford University Committee for Archaeology), 125–54.

Hooke, D. (1987), 'Anglo-Saxon estates in the Vale of the White Horse,' *Oxoniensia*, 52: 133–43.

Hooke, D. (1988), 'Regional variation in southern and central England in the Anglo-Saxon period', in D. Hooke (ed.), *Anglo-Saxon Settlements* (Oxford: Basil Blackwell), 123–49.

Hooke, D. (1994), *The Pre-Conquest Charter Bounds of Devon and Cornwall* (Woodbridge: Boydell Press).

Hooke, D. (1998), *The Landscape of Anglo-Saxon England* (London: Leicester University Press).

Hooke, D. (1999), 'Saxon Conquest and Settlement', in Kain and Ravenhill 1999, 95–104.

Hoskins, W. G. (1937), 'The fields of Wigston Magna', *Transactions of the Leicestershire Archaeological and Historical Society*, 19: 145–8.

Hoskins, W. G. (1950), *Essays in Leicestershire History* (Liverpool: Liverpool University Press).

Hoskins, W. G. (1955), *The Making of the English Landscape* (London: Hodder & Stoughton).

Hoskins, W. G. (1958–9), 'Review', *Economic History Review,* 2nd ser., 11: 160.

Hoskins, W. G., and Finberg, H. P. R. (1952), *Devonshire Studies* (London: Cape).

Hudson, J. R. (1932–3), *Register of the Abbey of St Benedict of Holme*, 2 vols., Norfolk Record Society, 2–3.

Hudson, W. (1898–1900), 'Three manorial extents of the thirteenth century', *Norfolk Archaeology*, 14: 1–56.

Hudson, W. (1910), 'On a series of rolls of the manor of Wiston', *Sussex Archaeological Collections*, 53: 143–82.

Hudson, W. (1917–19), 'The Anglo-Danish village community of Martham, Norfolk', *Norfolk Archaeology*, 20: 273–316.

Hughes, C. J. (1967–8), 'Hides, carucates and yardland in Leicestershire', *Transactions of the Leicestershire Archaeological and Historical Society*, 43: 19–23.

Hughes, M. W. (1940), *Calendar of the Feet of Fines*, Buckinghamshire Archaeological Society, 4.

Hull, P. L. (1971), *The Caption of Seisin of the Duchy of Cornwall, 1337*, Devon and Cornwall Record Society, 17.

Hulton, W. A. (1853), *Documents relating to the Priory of Penwortham*, Chetham Society, 30.

Hunt, W. (1893), *Two Chartularies of the Priory of St Peter at Bath*, Somerset Record Society, 7.

Hunter, J. (1999), *The Essex Landscape: A Study of its Form and History* (Chelmsford: Essex Record Office Publications).

Hunter, J., and Ralston, I. (1999), *The Archaeology of Britain* (London: Routledge).

Hurst, J. G. (1967–8), 'Saxon and medieval pottery from Kirby Bellars', *Transactions of the Leicestershire Archaeological and Historical Society*, 43: 10–18.

Hylton, L. (1907), 'A rental of the Manor of Merstham in 1522', *Surrey Archaeological Collections*, 20: 90–114.

Innocent, C. F. (1924), 'The Field-system of Wigtwizzle', *Transactions of the Hunter Archaeological Society*, 2: 276–8.

Ivens, R. J., Busby, P., Shepherd, N., Hurman, B., and Mills, J. (1995), *Tattenhoe and Westbury: Two Deserted Medieval Settlements in Milton Keynes*, Buckinghamshire Archaeological Society Monograph, 8 (Aylesbury: Buckinghamshire Archaeological Society).

Jackson, J. E. (1882), *Liber Henrici de Soliaco, Abbatis Glaston.* (London: Nichols and Sons).

James, W., and Malcolm, J. (1794), *A General View of the Agriculture of the County of Buckingham* (London: Board of Agriculture).

Jenkins, J. G. (1935), *History of the Parish of Penn* (London: Saint Catherine Press).

Jones, L., Woodward, A., and Buteux, S. (2006), *Iron Age, Roman and Saxon Occupation at Grange Park*, British Archaeological Reports, British Series, 425 (Oxford: British Archaeological Reports).

Jones, M. U. (1974), 'Excavations at Mucking, Essex: an interim report', *Antiquaries Journal*, 54: 183–9.

Jones, R., and Hooke, D. (2011), 'Methodological approaches to medieval rural settlements and landscapes', in Christie and Stamper 2011, 31–42.

Jones, R., and Page, M. (2006), *Medieval Villages in an English Landscape: Beginnings and Ends* (Macclesfield: Windgather Press).

Jones, R. A. (1985), 'The map of 1614', *Proceedings of the Hampshire Field Club*, 41: 195–7.

Jones, R. H. (1870), 'An Anglo-Saxon charter relating to the parish of Stockton in Wiltshire', *Wiltshire Archaeological and Natural History Magazine*, 12: 216–20.

Kain, R., and Ravenhill, W. (eds.) (1999), *Historical Atlas of South-West England* (Exeter: University of Exeter Press).

Kain, R. J. P., and Oliver, R. R. (1995), *The Tithe Maps of England and Wales* (Cambridge: Cambridge University Press).

Kemble, J. M. (1839–48), *Codex Diplomaticus Aevi Saxonici*, 6 vols. (London: Sumptibus Societatis [English Historical Society]).

Kerr, B. (1968), 'Dorset fields and their names', *Proceedings of the Dorset Natural History and Archaeological Society*, 89: 251–6.

Kerr, W. J. B. (1925), *Higham Ferrers and its Ducal and Royal Castle and Park* (Northampton: R. Harris and Son).

Kerridge, E. (1951), 'Ridge and furrow and agrarian history', *Economic History Review*, 2nd ser., 4: 14–36.

Kerridge, E. (1953), *Surveys of the Manors of Philip First Earl of Pembroke and Montgomery, 1631–3*, Wiltshire Record Society, 9.

Kerridge, E. (1955), 'A reconsideration of some former husbandry practices', *Agricultual History Review*, 3: 32–40.

Kerridge, E. (1959), 'Agriculture', in VCH, *Wiltshire*, iv: 43–64.

Kerridge, E. (1967), *The Agricultural Revolution* (London: Allen and Unwin).

Kerridge, E. (1992), *The Common Fields of England* (Manchester: Manchester University Press).

Kettle, A. J. (1979), 'Agriculture 1500 to 1793', in VCH, *Staffordshire*, vi: 49–90.

King, E. (1983), 'Estate Records of the Hotot Family', in *A Northamptonshire Miscellany*, Northamptonshire Record Society, 32: 1–58.

Kirby, T., and Oosthuizen, S. (2000), *An Atlas of Cambridgeshire and Huntingdonshire History* (Cambridge: Centre for Regional Studies, Anglia Polytechnic University).

Kirk, R. E. G. (ed.) (1899–1910), *Feet of Fines for Essex*, 6 vols. (Colchester: Essex Archaeological Society).

Lambert, H. C. M. (1912), *History of Banstead in Surrey* (London: Oxford University Press).

Larking, L. (1860), 'Pedes Fines', *Archaeologia Cantiana*, 3: 209–40.

Leadam, I. S. (1897), *The Domesday of Inclosures, 1517–1518* (London: Longmans, Green).

Leaver, R. A. (1988), 'Five hides in ten counties; a contribution to the regressional data', *Economic History Review*, 2nd ser., 41: 525–42.

Leeds, E. T. (1923, 1927, 1947), 'A Saxon village near Sutton Courtenay, Berkshire', *Archaeologia*, 73: 147–92, 76: 12–80, and 92: 73–94.

Lees, B. A. (1935), *Records of Templars in England in the Twelfth Century* (London: Oxford University Press).

Lennard, R. (1916), *The Black Death*, Oxford Studies in Economic History, 5 (Oxford: Clarendon Press).

Lennard, R. (1932–4), 'English agriculture under Charles II', *Economic History Review*, 4: 23–45.

Levett, A. E. (1938), *Studies in Manorial History* (Oxford: Clarendon Press).

Lewis, C. (2011), 'Test pit excavation within currently occupied rural settlement—2010', *Medieval Settlement Research*, 26: 48–59; see also <http://www.access.arch.cam.ac.uk>.

Lewis, C., Mitchell-Fox, P., and Dyer, C. (1997), *Village, Hamlet and Field* (Manchester: Manchester University Press).

Leys, A. M. (1938–41), *The Sandford Cartulary*, 2 vols., Oxfordshire Record Society, Oxfordshire Record Series, 19 and 22.

Lindley, E. S. (1952), 'A history of Wortley', *Transactions of the Bristol and Gloucestershire Archaeological Society*, 69: 91–195.

Lipson, E. (1937), *The Economic History of England*, 7th edn (London: A. and C. Black).

Lobel, M. D. (1935), *The History of Dean and Chalford*, Oxfordshire Record Society, Oxfordshire Record Series, 17.

Losco-Bradley, S., and Kinsley, G. (eds.) (2002), *Catholme: An Anglo-Saxon Settlement on the Trent Gravels in Staffordshire* (Nottingham: University of Nottingham).

Loud, G. A. (1989), *Introduction to Somerset Domesday* (London: Alecto Historical Editions).

Loveluck, C., and Atkinson, D. (2007), *The Early Medieval Settlement Remains from Flixborough, Lincolnshire: The Occupation Sequence, c. AD 600–1000* (Oxford: Oxbow).

Lucy, S., Tipper, J., and Dickens, A. (2009), *The Anglo-Saxon Settlement and Cemetery at Bloodmoor Hill, Carlton Colville, Suffolk*, East Anglian Archaeology, 131 (Cambridge: Cambridge Archaeological Unit).

Lumby, J. H. (1936), *Calendar of the Deeds and Papers in the Possession of Sir James de Hoghton*, Lancashire Record Society, 88.

MacDermot, E. T. (1911), *The History of the Forest of Exmoor* (Taunton: Barnicott & Pearce).

Mackreth, D. F. (1996), *Orton Hall Farm: A Roman and Early Anglo-Saxon Farmstead*, East Anglian Archaeology, 76 (Manchester: Nene Valley Archaeological Trust).

Maclean, J. (1873–6), *History of the Deanery of Trigg Minor*, i–ii (London: Nichols and Sons).

Macnair, A., and Williamson, T. (2010), *William Faden and Norfolk's 18th-Century Landscape* (Oxford: Oxbow).

McOmish, D., Field, D., and Brown, G. (2002), *The Field Archaeology of Salisbury Plain Training Area* (Swindon: English Heritage).

Maitland, F. W. (1889), 'The surnames of English villages', *Archaeological Review*, 2; repr. in Fisher 1911, ii, 337–63.

Maitland, F. W. (1897 [1969]), *Domesday Book and Beyond* (Cambridge: Cambridge University Press; repr. London: Fontana).

Maitland, F. W. (1898 [1997]), *Township and Borough* (Cambridge: Cambridge University Press; repr. London: Routledge/Thoemmes Press).

Major, K. (1950), *Registrum Antiquissimum of the Cathedral Church of Lincoln*, Lincoln Record Society, 41.

Malim, T. (1997), 'New evidence on the Cambridgeshire Dykes and Worstead Street Roman road', *Proceedings of the Cambridge Antiquarian Society*, 85: 27–122.

Malim, T. (2000), 'The Anglo-Saxon dykes', in Kirby and Oosthuizen 2000, no. 27.

March, H. C. (1903), 'The problem of lynchets', *Proceedings of the Dorset Natural History and Antiquarian Field Club*, 24: 66–92.

Margery, I. D. (1940), 'Roman centuriation at Ripe', *Sussex Archaeological Collections*, 81: 31–41.

Martin, E., and Satchell, M. (2008), *East Anglian Fields*, East Anglian Archaeology, 124 (Ipswich: Suffolk County Council).

Martin, S. H. (1952–4), 'The Ballingham charters', *Transactions of the Woolhope Naturalists' Field Club*, 34: 70–5.

Mawer, A., and Stenton, F. M. (1925), *The Place-Names of Buckinghamshire*, English Place-Name Society, 2 (Cambridge: Cambridge University Press).

Mawer, A., and Stenton, F. M. (1927), *The Place-Names of Worcestershire*, English Place-Name Society, 4 (Cambridge: Cambridge University Press).

Mawer, A., and Stenton, F. M. (1934), *The Place-Names of Surrey*, English Place-Name Society, 11 (Cambridge: Cambridge University Press).

Maztat, W. (1988), 'Long strip field layouts and their later subdivisions', *Geografiska Annaler*, 70B: 133–47.

Mead, W. R. (1954), 'Ridge and furrow in Buckinghamshire', *Geographical Journal*, 120: 34–42.

Medlycott, N., and Germany, M. (1994), 'Archaeological fieldwalking in Essex 1985–1993: interim results', *Transactions of the Essex Society for Archaeology and History*, 25: 14–27.

Meekings, C. A. F., and Shearman, P. (1968), *Fitznells Cartulary*, Surrey Record Society, 26.

Mellows, W. T. (1949), *The Chronicle of Hugh Candidus* (London: Oxford University Press).

Millett, M., and James, S. (1983), 'Excavations at Cowdery's Down, Basingstoke, 1978–1981', *Archaeological Journal*, 140: 151–279.

Mills, J., and Palmer, R. (2007), *Populating Clay Landscapes* (Stroud: Tempus Publishing).

Milne, J. G. (1940), 'Muniments of Holy Trinity Priory Wallingford,' *Oxoniensia*, 5: 50–77.

Milne, J. G. (1942), 'The Berkshire muniments of Corpus Christi College, Oxford', *Berkshire Archaeological Journal*, 46: 32–44, 78–87.

Moore, J. S. (1964), 'The Domesday teamland: a reconsideration', *Transactions of the Royal Historical Society*, 5th ser., 14: 109–30.

Moore, N. (1966), 'A possible case of Saxon monastic estate planning', *Norfolk Research Committee Bulletin*, 16: 8–9.

Moreton, C. E., and Rutledge, P. (1997), *John Skayman's Book, 1516–18*, Norfolk Record Society, 61.

Morris, B. (2005), 'The Roman to medieval transition in the Essex landscape', *Medieval Settlement Research Group, Annual Report*, 20: 37–44.

Mortimer, R. (1979), *Leiston Abbey Cartulary and Butley Priory Charters*, Suffolk Record Society, 1.

Mortimer, R. (2000), 'Village development and ceramic sequence: the Middle to Late Saxon village at Lordship Lane Cottenham, Cambridgeshire', *Proceedings of the Cambridge Antiquarian Society*, 89: 5–33.

Morton, J. (1712), *The Natural History of Northamptonshire* (London: R. Knaplock and R. Wilkin).

Murray, J., and MacDonald, T. (2005), 'Excavations at Station Road, Gamlingay, Cambridgeshire', *Anglo-Saxon Studies in Archaeology and History*, 13: 173–330.

Needham, S., and Macklin, M. G. (eds.) (1992), *Alluvial Archaeology in Britain* (Oxford: Oxbow).

Neilson, N. (1898), *Economic Conditions on the Manors of Ramsey Abbey* (Philadelphia: Press of Sherman & Co.).

Neilson, N. (1920), *A Terrier of Fleet* (London: Oxford University Press).

Newton, K. C. (1970), *The Manor of Writtle: The Development of a Royal Manor in Essex, c.1086–c.1500* (London: Phillimore).

Nichols, N. (1987), *Local Maps of Nottinghamshire to 1800: An Inventory* (Nottingham: Nottingham County Council Leisure Services).

Nitz, H.-J. (1988a), 'Settlement structure and settlement systems of the Frankish central state in Carolingian and Ottonian times', in Hooke 1988, 249–73 and fig. 12.2.

Nitz, H.-J. (1988b), 'The international spread of common-field systems', *Geografiska Annaler*, 70B: 148–59.

Northamptonshire Historic Database. A GIS data-set held by Archaeological Data Services: <http://archaeologydataservice.ac.uk>; search on 'Northamptonshire Historic Database': see Foard et al. 2009: 9–10, and Partida et al. 2013: 5–16.

Oakden, J. P. (1984), *The Place-Names of Staffordshire*, Part 1, English Place-Name Society, 55 (Nottingham: English Place-Name Society).

Oosthuizen, S. (1993), 'Saxon commons in south Cambridgeshire', *Proceedings of the Cambridge Antiquarian Society*, 82: 93–100.

Oosthuizen, S. (1997), 'Prehistoric fields into medieval furlongs', *Proceedings of the Cambridge Antiquarian* Society, 86: 145–52.

Oosthuizen, S. (1998), 'The origins of Cambridgeshire', *Antiquaries Journal*, 78: 85–109.

Oosthuizen, S. (2003), 'The roots of the common fields: linking prehistoric and medieval field systems in west Cambridgeshire', *Landscapes*, 4: 40–64.

Oosthuizen, S. (2010), 'Anglo-Saxon fields', in Hamerow et al. 2010, 377–401.

Oosthuizen, S. (2013), *Tradition and Transformation in Anglo-Saxon England* (London: Bloomsbury).

Orwin, C. S., and Orwin, C. S. (1938), *The Open Fields* (Oxford: Clarendon Press).

Oxford Archaeology (1991), 'Land opposite Windmill Banks, Higham Ferrers', *Northamptonshire Archaeology*, 31: 177.

Padel, O. J. (1985), *Cornish Place-Name Elements*, English Place-Name Society, 56/7 (Nottingham: English Place-Name Society).

Padel, O. J. (1999), 'Place-names', in Kain and Ravenhill 1999, 88–94 and Maps 13.1, 13.3.

Page, W. (1893), *The Chartulary of Brinkburn Priory*, Surtees Society, 90.

Palmer, R. (1995), 'Air photo interpretation and the Lincolnshire fenland', *Landscape History*, 18: 5–16.

Palmer, R. C. (1984), *The Whilton Dispute, 1264–1380* (Princeton: Princeton University Press).

Parker, F. (1887), 'Chartulary of St Thomas' Priory, near Stafford', *Collections for a History of Staffordshire*, 8: 125–201.

Parker, F. (1890), 'Chartulary of the "Austin" Priory of Trentham', *Collections for a History of Staffordshire*, 11: 213–336.

Parry, S. (2006), *Raunds Area Survey: An Archaeological Study of the Landscape of Raunds, Northamptonshire, 1985–94* (Oxford: Oxbow).

Partida, T., Hall D., and Foard, G. (2013), *Northamptonshire: An Atlas of the Medieval and Early-Modern Landscape* (Oxford: Oxbow).

Patten, R. (2012), 'An Iron Age and Roman settlement at Summersfield, Papworth Everard', *Proceedings of the Cambridge Antiquarian Society*, 101: 115–42.

Pawson, E. (1979), *The Early Industrial Revolution* (London: Batsford Academic).

Peckham, W. D. (1925), *Sussex Custumals of the Bishop of Chichester*, Sussex Record Society, 31.

Peckham, W. D. (1946), *Chartulary of the High Church of Chichester*, Sussex Record Society, 46.

Pelham, R. A. (1931), 'Studies in the historical geography of medieval Sussex', *Sussex Archaeological Collections*, 72: 156–84.

Pelham, R. A. (1934), 'The distribution of sheep in Sussex in the early fourteenth century', *Sussex Archaeological Collections*, 75: 128–33.

Peterson, W. (1914 [1970]), *Germania*, in *Cornelius Tacitus: Opera Minora*, rev. E. H. Warmington (Cambridge, MA: Harvard University Press), 129–214.

Pettit, P. A. J. (1968), *The Royal Forests of Northamptonshire*, Northamptonshire Record Society, 22.

Pigot, H. (1863), 'Extent of Hadleigh Manor 1305', *Proceedings of the Suffolk Institute of Archaeology*, 3: 229–49.

Plot, R. (1686), *The Natural History of Staffordshire* (Oxford: Printed at the Theater).

Pocock, E. A. (1968), 'The first fields in an Oxfordshire parish', *Agricultural History Review*, 16: 85–100.

Postgate, M. R. (1962), 'The field systems of Breckland', *Agricultural History Review*, 10: 80–101.

Postgate, R. W. (1973), 'Field Systems of East Anglia', in Baker and Butlin 1973, 281–324.

Pounds, N. J. G. (1945), 'The Lanhydrock atlas', *Antiquity*, 19: 20–6.

Powell, E. (1910), *A Suffolk Hundred in the Year 1283* (Cambridge: Cambridge University Press).

Powlesland, D. (2003), *Twenty-Five Years of Archaeological Research on the Sands and Gravels of Heslerton* (Colchester: Landscape Research Centre).

Pratt, C. W. M. (2000–2), 'Interpretation of a Tudor manorial survey of Acton Beauchamp', *Transactions of the Woolhope Naturalists' Field Club*, 50: 184–223.

Prescott, J. E. (1897), *Register of the Priory of Whetheral* (Kendal: T. Wilson).

Preston-Jones, A., and Rose, P. (1986), 'Medieval Cornwall', *Cornish Archaeology*, 25: 35–185.

Pryor, F. (1985), *Archaeology and Environment in the Lower Welland Valley*, East Anglian Archaeology, 27 (Cambridge: Cambridgeshire Archaeological Committee in conjunction with the Fenland Project Committee and the Scole Archaeological Committee).

Raban, S. (ed.) (2001), *The White Book of Peterborough*, Northamptonshire Record Society, 41.

Raban, S. (ed.) (2011), *The Accounts of Godfrey, Abbot of Peterborough, 1299–1321*. Northamptonshire Record Society, 45.

Rackham, H. (ed. and trans.) (1961), *Pliny: Natural History*, v: *Libri XVII–XIX* (London: Heinemann).

Rackham, J. (ed.) (1994), *Environment and Economy in Anglo-Saxon England*, CBA Research Report, 89 (York: Council for British Archaeology).

Rackham, J., Brown, G., and Leary, J. (2004), *Tatberht's Lundenwic: Archaeological Excavations in Middle Saxon London*, Pre-Construct Archaeology Monograph 2 (London: Pre-Construct Archaeology).

Rackham, O. (1986), *The Woods of South-Eastern Essex* (Rochford: Rochford District Council).

Raine, J. (1864–5), *The Priory of Hexham*, 2 vols., Surtees Society, 44 and 46.

Ramm, H. G., McDowall, R. W., and Mercer, E. (1970), *Shielings and Bastles*, RCHME (London: HMSO).

Ratcliff, S. C. (1946), *Elton Manorial Records, 1279–1351* (Cambridge: Privately Printed for presentation to members of the Roxburghe Club).

RCHM (1960), *A Matter of Time* (London: HMSO).

Reaney, P. H. (1935), *The Place-Names of Essex*, English Place-Name Society, 12 (Cambridge: Cambridge University Press).

Reaney, P. H. (1943), *The Place-Names of Cambridge and the Isle of Ely*, English Place-Name Society, 19 (Cambridge: Cambridge University Press).

Redwood, B. C., and Wilson, A. E. (1958), *Custumals of the Sussex Manors of the Archbishop of Canterbury*, Sussex Record Society, 57.

Reece, R. (1983), 'Continuity on the Cotswolds: some problems of ownership, settlements and hedge survey between Roman Britain and the Middle Ages', *Landscape History*, 5: 11–19.

Rees, U. (1985), *The Cartulary of Haughmond Abbey* (Cardiff: Shropshire Archaeological Society and University of Wales Press).

Reichel, O. J. (1909), 'A batch of old deeds relating to Buckland Filleigh', *Report and Transactions of the Devonshire Association*, 41: 241–55.

Reid, M. (1997), *Buckinghamshire Glebe Terriers, 1578–1640*, Buckinghamshire Record Society, 30.

Renes, H. (2010), 'Grainlands: the landscape of open fields in a European perspective', *Landscape History*, 31: 37–71.

Rippon, S. (1991), 'Early planned landscapes in south-east Essex', *Essex Archaeology and History*, 22: 46–60.

Rippon, S. (1996), 'Essex c.700–1066', in O. Bedwyn (ed.), *The Archaeology of Essex: Proceedings of the Writtle Conference* (Colchester: Essex County Council), 117–28.

Rippon, S. (2004), *Historic Landscape Analysis: Deciphering the Countryside* (York: Council for British Archaeology).

Rippon, S. (2008), *Beyond the Medieval Village* (Oxford: Oxford University Press).

Rippon, S., Smart, C., and Pears, B. (2012), 'Inherited landscapes: "The Fields of Britannia Project"', *Medieval Settlement Research*, 27: 57–64.

Rippon, S., Smart, C., and Pears, B. (2013), 'The Fields of Britannia', *Landscapes*, 14: 33–53.

Roberts, B. K. (1968), 'A study of the medieval cultivation in the Forest of Arden, Warwickshire', *Agricultural History Review*, 16: 101–13.

Roberts, B. K. (1973), 'Field systems of the West Midlands', in Baker and Butlin 1973, 188–230.

Roberts, B. K. (1988), 'Norman village plantations and long strip fields in northern England', *Geografiska Annaler*, 70B: 169–77.

Roberts, B. K. (1996), 'The great plough', *Landscape History*, 18: 17–30.

Roberts, B. K., and Glasscock, R. E. (1983), *Villages, Fields and Frontiers*, British Archaeological Reports, International Series, 185 (Oxford: British Archaeological Reports).

Roberts, B. K., and Wrathmell, S. (2000), *An Atlas of Rural Settlement in England* (London: English Heritage).

Roberts, B. K., and Wrathmell, S. (2002), *Region and Place: A Study of English Rural Settlement* (London: English Heritage).

Robertson, A. J. (1956), *Anglo-Saxon Charters*, 2nd edn (Cambridge: Cambridge University Press).

Roden, D. (1973), 'Field systems of the Chilterns', in Baker and Butlin 1973, 325–76.

Roderick, A. J. (1949), 'Open-field agriculture in Herefordshire in the later Middle Ages,' *Transactions of the Woolhope Naturalists' Field Club*, 33: 55–67.

Roffe, D. (1989), *Introduction to Huntingdonshire Domesday* (London: Alecto Historical Editions).

Roffe, D. (2007), *Decoding Domesday* (Woodbridge: Boydell Press).

Rogerson, A. (1995), 'Fransham: an archaeological and historical survey of a parish on Norfolk boulder clay', Unpub. PhD thesis, University of East Anglia. [Full report forthcoming in the East Anglian Archaeology series.]

Rogerson, A. (1997), *Barton Bendish and Caldecote*, East Anglian Archaeology, 80 (Dereham: Norfolk Museums Service).

Rose-Troup, F. (1934), 'Medieval customs and tenure in the Manor of Ottery St Mary', *Report and Transactions of the Devonshire Association*, 66: 211–33.

Ross, C. D. (1964), *The Cartulary of Cirencester Abbey, Gloucestershire*, i–ii [iii by M. Devine, 1977] (London: Oxford University Press).

Ross, M. S. (1995), 'The water meadows of the River Stirchel, Dorset', *Proceedings of the Dorset Natural History and Archaeological Society*, 116: 27–32.

Round, J. H. (1895), *Feudal England* (London: Swan Sonnenschein).

Round, J. H. (1900a), *Feet of Fines for the Tenth Year of Richard I, AD 1198–AD 1199*, Pipe Roll Society, 24 (London: Printed by Love & Wyman).

Round, J. H. (1900b), 'The hidation of Northamptonshire', *English Historical Review*, 15: 78–86.

Round, J. H. (1902), 'The Northamptonshire Survey', in VCH, *Northamptonshire*, i: 365–92.

Rowley, T. (ed.) (1981), *The Origins of Open Field Agriculture* (London: Croom Helm).

Royal Commission (1807), *Nonarum Inquisitiones in Curia Scaccarii*, Record Commission Publications, 3 ([London]: Printed by G. Eyre and A. Strahan).

Rutledge, P. (1995), 'Colkirk: a north Norfolk settlement pattern', *Norfolk Archaeology*, 41: 15–34.

Ryan, M. J. (2011), 'That "dreary old question": the hide in early Anglo-Saxon England', in Higham and Ryan 2010, 203–23.

Ryder, I. E. (2006), *Common Right and Private Interest: Rutland's Common Fields and their Enclosure*, Rutland Local History and Record Society, Occasional Publication, 8.

Salter, H. E. (1921), *Newington Longeville Charters*, Oxfordshire Record Society, 1, item no. 3.

Salter, H. E. (1930), *The Boarstall Cartulary,* Oxford Historical Society, 88.

Salter, H. E. (1934), *Cartulary of Osney Abbey*, iv, Oxford Historical Society, 97.

Salter, H. E. (1935), *Cartulary of Osney Abbey*, v, Oxford Historical Society, 98.

Salter, H. E. (1947), *Thame Cartulary*, Oxfordshire Record Society, 25.

Saltmarsh, J., and Darby, H. C. (1935), 'The infield–outfield system on a Norfolk manor', *Economic History*, 3: 30–44.

Salzman, L. F. (1923), *The Chartulary of the Priory of St Peter at Sele* (Cambridge: Heffer).

Salzman, L. F. (1934), *The Chartulary of the Priory of St. Pancras of Lewes*, Part 2, Sussex Record Society, 40.

Sawyer, P. (1968), *Anglo-Saxon Charters: An Annotated List and Bibliography* (London: Royal Historical Society); now available online in rev. edn, <http://www.esawyer.org.uk>.

Sawyer, P. H. (1979), *Charters of Burton Abbey* (Oxford: Oxford University Press for the British Academy).

Scargill-Bird, S. R. (1887), *Custumals of Battle Abbey*, Camden Society, NS, 41 (London: Camden Society).

Scott, R. (1959), 'Medieval agriculture', in VCH, *Wiltshire*, iv: 7–42.

Searle, C. E. (1993), 'Customary tenants and enclosure of Cumbrian commons', *Northern History*, 29: 126–53.

Seebohm, F. (1883), *The English Village Community* (London: Longmans, Green).

Sharples, N. M. (1991), *Maiden Castle: Excavation and Field Survey 1985–6* (London: English Heritage).

Shaw, M. (1993), 'Warmington', *Medieval Settlement Research Group Report*, 8: 41–7.

Shaw, M. (1993–4), 'The discovery of Saxon sites below fieldwalking scatters', *Northamptonshire Archaeology*, 25: 77–92.

Sheppard, J. A. (1973), 'Field systems of Yorkshire', in Baker and Butlin 1973, 145–87.

Sheppard, J. A. (1974), 'Metrological analysis of regular village plans in Yorkshire', *Agricultural History Review*, 22: 118–35.

Sheppard, J. A. (1979), *The Origins and Evolution of Field and Settlement Patterns in the Herefordshire Manor of Marden*, Occasional Paper, 15 (London: Department of Geography, Queen Mary College, London).

Silvester, R. J. (1988), *Marshland and the Nar Valley, Norfolk*, Fenland Project, 3; East Anglian Archaeology, 45 (Dereham: Norfolk Museums Service).

Skipp, V. (1970), *Medieval Yardley* (Chichester: Phillimore).

Skipp, V. (1981), 'The evolution of settlement and open-field topography in North Arden down to 1300', in Rowley 1981, 162–83.

Slater, G. (1907), *English Peasantry and the Enclosure of Common Fields* (London: Constable).

Slee, H. (1952), 'The open fields of Braunton', *Report and Transactions of the Devonshire Association*, 84: 142–9.

Smith, A. H., and Baker, G. M. (1990), *The Papers of Nathan Bacon of Stiffkey*, iii: *1586–95*, Norfolk Record Society, 53.

Spratt, D. A., and Harrison, B. J. D. (1989), *The North York Moors Landscape Heritage* (Newton Abbot: David and Charles).

Spufford, M. (1965), *A Cambridgeshire Community: Chippenham from Settlement to Enclosure*, University of Leicester, Department of Local History, Occasional Paper, 20 (Leicester: Leicester University Press).

Spufford, M. (1974), *Contrasting Communities: English Villages in the Sixteenth and Seventeenth Centuries* (Cambridge: Cambridge University Press).

Spufford, M. (2000), 'General view of the rural economy of the County of Cambridge', *Proceedings of the Cambridge Antiquarian Society*, 89: 69–85.

Stamper, P. (1989), 'Agriculture', in VCH, *Shropshire*, iv: 20–118.

Stamper, P. (1999), 'Landscapes in the Middle Ages', in Hunter and Ralston 1999, 247–63.

Stanes, R. (1994), 'Braunton Great Field management study.' Unpublished report for North Devon District Council (Devon, HER).

Stenton, F. M. (1920), *Documents Illustrative of the Social and Economic History of the Danelaw* (London: Oxford University Press).

Stenton, F. M. (1922), *Transcripts of Charters relating to the Gilbertine Houses*, Lincolnshire Record Society, 18.

Stevenson, W. H., and Salter, H. E. (1939), *The Early History of St. John's College, Oxford*, Oxford Historical Society, 102.

Stewart-Brown, R. (1917), 'The townfield of Liverpool, 1207–1807', *Transactions of the Historic Society of Lancashire and Cheshire*, 68: 35–67.

Stitt, F. B. (1952), 'A Kempston Estate in 1342', *Bedfordshire Historical Record Society*, 32: 71–91.

Stocks, G. A., and Tait, J. (1921), *Dunkenhalgh Deeds, 1200–1600*, Chetham Society, NS, 80.

Stoertz, C. (1997), *The Ancient Landscapes of the Yorkshire Wolds* (Swindon: RCHME).

Sylvester, D. (1957), 'The open fields of Cheshire,' *Transactions of the Historic Society of Lancashire and Cheshire*, 108: 1–33.

Sylvester, D. (1959), 'A note on medieval three-course arable systems in Cheshire', *Transactions of the Historic Society of Lancashire and Cheshire*, 110: 183–6.

Sylvester, D. (1969), *The Formation of the Rural Landscape of the Welsh Borderland* (London: Macmillan).

Tait, J. (1920–3), *The Chartulary or Register of the Abbey of St Werburgh, Chester*, 2 vols., Chetham Society, NS, 82 and 85.

Tate, W. E., and Turner, M. E. (1978), *A Domesday of English Enclosure Acts and Awards* (Reading: University of Reading).

Taylor, C. (1983), *Village and Farmstead: A History of Rural Settlement in England* (London: George Philip).

Taylor, C. S. (1893–4), 'The pre-Domesday hide in Gloucestershire', *Transactions of the Bristol and Gloucestershire Archaeological Society*, 18: 288–319.

Taylor, E. G. (1888), 'Domesday survivals', in P. E. Dove (ed.) *Domesday Studies*, i (London: Longmans, Green).

Taylor, G. (2003), 'An early to middle Saxon settlement at Quarrington, Lincolnshire', *Antiquaries Journal*, 83: 231–80.

Thirsk, J. (1964), 'The common fields', *Past and Present*, no. 29: 3–29.

Thirsk, J. (1973), 'Field systems of the East Midlands', in Baker and Butlin 1973, 232–80.

Thomas-Stanford, C. (1921), *An Abstract of the Court Rolls of the Manor of Preston (Preston Episcopi)*, Sussex Record Society Record Series, 27.

Thompson, A. H. (1933), *A Calendar of Charters and Other Documents belonging to the Hospital of Wyggeston at Leicester* (Leicester: Published for the Corporation of the City of Leicester by E. Backus).

Timson, R. T. (1973), *The Cartulary of Blyth Priory*, 2 vols., Thoroton Society Records Series, 27 and 28.

Titow, J. Z. (1965), 'Medieval England and the open-field system', *Past and Present*, no. 32: 86–102.

Torns, E. (ed.) (1954), *Chertsey Abbey Court Rolls Abstract*, Surrey Record Society, 21: 5–18.

Travers, A. (1989), *A Calendar of the Feet of Fines for Buckinghamshire, 1259–1307*, Buckinghamshire Record Series, 25.

Tupling, G. H. (1927), *The Economic History of Rossendale*, Chetham Society, 86.

Tyack, G. C. (1976), 'Hendon', in VCH, *Middlesex*, v: 1–49.

Tyler, S. A. (2011), 'Early to Middle Saxon Settlement in the Chelmer–Blackwater River Valley, Essex', in *Studies in Early Anglo-Saxon Art and Archaeology: Papers in Honour of M. G. Welsh*, British Archaeological Report, British Series, 527 (Oxford: British Archaeological Reports), 121–30.

Tymms, T. (1853), 'Customs of Hardwick', *Proceedings of the Bury and West Suffolk Archaeological Institute*, 1: 177–86.

Upex, S. G. (2002), 'Landscape continuity and the fossilization of Roman fields', *Archaeological Journal*, 159: 77–108.

Upex, S. G. (2003), 'A migration period site at Polebrook', *South Midlands Archaeology*, 33: 41–51.

Upex, S. G. (2008), *The Early Origins of Polebrook and Oundle* (Oundle: The Rockingham Forest Trust).

Venn, J. A. (1923), *Foundations of Agricultural Economics* (Cambridge: Cambridge University Press; 2nd edn (1933)).

Vinogradoff, P. V. (1892), *Villeinage in England* (Oxford: Clarendon Press).

Vinogradoff, P. V. (1908), *English Society in the Eleventh Century* (Oxford: Clarendon Press).

Wade, K. (1983), 'The Saxon Period', in A. J. Lawson (ed.), *The Archaeology of Witton*, East Anglian Archaeology, 18 (Dereham: Norfolk Museums Service), 50–77.

Wade-Martins, P. (1980a), *Excavations in North Elmham Park, 1967–72*, 2 vols., East Anglian Archaeology, 9 (Gressenham, Dereham: Norfolk Archaeological Unit).

Wade-Martins, P. (1980b), *Village Sites in Launditch Hundred*, East Anglian Archaeology, 10 (Gressenham, Dereham: Norfolk Archaeological Unit).

Wade-Martins, P. (1994), *An Historical Atlas of Norfolk*, 2nd edn (Norwich: Norfolk Museums Service).

Walker, D. (1998), *The Cartulary of St Augustine's Abbey, Bristol*, Bristol and Gloucestershire Record Series, 10.

Walker, V. W., and Gray, D. (1940), *Newstead Priory Cartulary, 1344*, Thoroton Society Record Series, 8.

Ward, G. (1930), 'A note on the yokes at Otford', *Archaeologia Cantiana*, 42: 147–56.

Ward, G. (1932–4), 'A Roman colony near Brancaster', *Norfolk Archaeology*, 25: 373–85.

Warner, P. (1987), *Greens, Commons and Clayland Colonization*, University of Leicester, Department of English Local History Occasional Paper, 4th ser., 2 (Leicester: Leicester University Press).

Watson, C. E. (1939), 'The Spillman Cartulary', *Transactions of the Bristol and Gloucestershire Archaeological Society*, 61: 74–87.

Weaver, F. W. (1909), *A Cartulary of Buckland Priory in the County of Somerset*, Somerset Record Society, 25.

West, S. (1990), *West Stow, Suffolk: The Prehistoric and Romano-British Occupations*, East Anglian Archaeology, 48 (Bury St Edmunds: Suffolk County Planning Department).

Whitehead, B. J. (1968), 'The management and land-use of water meadows in the Frome Valley, Dorset', *Proceedings of the Dorset Natural History and Archaeological Society*, 89: 257–81.

Whitelock, D. (1930), *Anglo-Saxon Wills* (Cambridge: Cambridge University Press).

Whitfield, M. (1981), 'The medieval fields of south-east Somerset', *Proceedings of the Somerset Archaeological and Natural History Society*, 125: 17–29.

Wilkinson, T. J. (1988), *Archaeology and Environment in South Essex*, East Anglian Archaeology, 42 (Chelmsford: Essex County Council).

Williams, A., and Martin, G. H. (2003), *Domesday Book: A Complete Translation* (London: Penguin).

Williams, J. H. (ed.) (2007), *The Archaeology of Kent to AD 800* (Woodbridge: Boydell and Kent County Council).

Williams, P. (1970–2), 'Land tenure in the Bishop's manor of Whitbourne', *Transactions of the Woolhope Naturalists' Field Club*, 40: 333–55.

Williams, R. J. (1993), *Pennyland and Hartigans: Two Iron Age Sites in Milton Keynes*, Buckinghamshire Archaeological Society Monograph Series, 4.

Williamson, T. (1982), 'The development of settlement in north-west Essex', *Essex Archaeology and History*, 17: 120–33.

Williamson, T. (1987), 'Early co-axial field systems of the East Anglian boulder clays', *Proceedings of the Prehistoric Society*, 53: 419–31.

Williamson, T. (1993), *The Origins of Norfolk* (Manchester: Manchester University Press).

Williamson, T. (1998), 'The "Scole–Dickleburgh field system" revisited', *Landscape History*, 20: 19–28.

Williamson, T. (2003), *Shaping Medieval Landscapes* (Macclesfield: Windgather Press).

Williamson, T., Liddiard, R., and Partida, T. (2013), *Champion: The Making and Unmaking of the English Midland Landscape* (Manchester: Manchester University Press).

Willis, D. S. (ed.) (1916), *The Estate Book of Henry de Bray, co. Northants (c.1289–1340)*, Camden Society, 3rd ser., 27.

Wilson, A. E. (1961), *Custumals of Laughton, Willington and Goring*, Sussex Record Society, 60.

Winchester, A. (1987), *Landscape and Society in Medieval Cumbria* (Edinburgh: Donald).

Wise, C. (1899), *The Compotus of the Manor of Kettering for AD 1292* (Kettering: W. E. and J. Goss).

Woodward, P. (1991), 'The documented and monumental sequence', in Sharples 1991, 17–21.

Wrathmell, S. (ed.) (2012), *Wharram: A Study of Settlement on the Yorkshire Wolds*, xiii, York University Archaeological Publications, 15 (York: York University).

Wright, G. T. (1906), *Longstone Records, Derbyshire* (Bakewell: Benjamin Gratton).

Wrottesley, G. (1906), 'The Chartulary of Dieulacres Priory', *Stafford Historical Collections*, NS, 9: 293–365.

Yelling, J. A. (1968), 'Common land and enclosure in East Worcestershire, 1540–1870', *Transactions of the Institute of British Geographers*, 45: 157–68.

Yelling, J. A. (1969), 'The combination and rotation of crops in East Worcestershire 1540–1660', *Agricultural History Review*, 17: 24–43.

Yelling, J. A. (1977), *Common Field and Enclosure in England, 1450–1850* (London: Macmillan).

Youd, G. (1962), 'The common fields of Lancashire', *Transactions of the Historic Society of Lancashire and Cheshire*, 113: 1–40.

Index

References to figures are **bold**.

Place names in the Gazetteer are not indexed unless they are referred to in the text, or are connected with a particular topic. Neither are all place-names in the county summary lists indexed (pages 38–40; 59; 90–4; 96–9; 101–3; 110–15; 124–6; 142–7; 167–8: 171–2). No tables are indexed. Readers interested in a particular county therefore need to consult these sections.

Each Gazetteer entry has information additional to the text on demesne types, yardland or oxgang sizes, regular tenurial structure, work-service, farming, assarts, and items unique to a particular county. The entries conclude with a description of the physical remains of ridge and furrow and strip lynchets.